Neurohumoral and Metabolic Aspects of Injury

ADVANCES IN EXPERIMENTAL MEDICINE AND BIOLOGY

Editorial Board:

Nathan Back	*Chairman, Department of Biochemical Pharmacology, School of Pharmacy, State University of New York, Buffalo, New York*
N. R. Di Luzio	*Chairman, Department of Physiology, Tulane University School of Medicine, New Orleans, Louisiana*
Alfred Gellhorn	*University of Pennsylvania Medical School, Philadelphia, Pennsylvania*
Bernard Halpern	*Collège de France, Director of the Institute of Immuno-Biology, Paris, France*
Ephraim Katchalski	*Department of Biophysics, The Weizmann Institute of Science, Rehovoth, Israel*
David Kritchevsky	*Wistar Institute, Philadelphia, Pennsylvania*
Abel Lajtha	*New York State Research Institute for Neurochemistry and Drug Addiction, Ward's Island, New York*
Rodolfo Paoletti	*Institute of Pharmacology and Pharmacognosy, University of Milan, Milan, Italy*

Volume 1
THE RETICULOENDOTHELIAL SYSTEM AND ATHEROSCLEROSIS
Edited by N. R. Di Luzio and R. Paoletti • 1967

Volume 2
PHARMACOLOGY OF HORMONAL POLYPEPTIDES AND PROTEINS
Edited by N. Back, L. Martini, and R. Paoletti • 1968

Volume 3
GERM-FREE BIOLOGY: Experimental and Clinical Aspects
Edited by E. A. Mirand and N. Back • 1969

Volume 4
DRUGS AFFECTING LIPID METABOLISM
Edited by W. L. Holmes, L. A. Carlson, and R. Paoletti • 1969

Volume 5
LYMPHATIC TISSUE AND GERMINAL CENTERS IN IMMUNE RESPONSE
Edited by L. Fiore-Donati and M. G. Hanna, Jr. • 1969

Volume 6
RED CELL METABOLISM AND FUNCTION
Edited by George J. Brewer • 1970

Volume 7
SURFACE CHEMISTRY OF BIOLOGICAL SYSTEMS
Edited by Martin Blank • 1970

Volume 8
BRADYKININ AND RELATED KININS: Cardiovascular, Biochemical, and Neural Actions
Edited by F. Sicuteri, M. Rocha e Silva, and N. Back • 1970

Volume 9
SHOCK: Biochemical, Pharmacological, and Clinical Aspects
Edited by A. Bertelli and N. Back • 1970

Volume 10
THE HUMAN TESTIS
Edited by E. Rosemberg and C. A. Paulsen • 1970

Volume 11
MUSCLE METABOLISM DURING EXERCISE
Edited by B. Pernow and B. Saltin • 1971

Volume 12
MORPHOLOGICAL AND FUNCTIONAL ASPECTS OF IMMUNITY
Edited by K. Lindahl-Kiessling, G. Alm, and M. G. Hanna, Jr. • 1971

Volume 13
CHEMISTRY AND BRAIN DEVELOPMENT
Edited by R. Paoletti and A. N. Davison • 1971

Volume 14
MEMBRANE-BOUND ENZYMES
Edited by G. Porcellati and F. di Jeso • 1971

Volume 15
THE RETICULOENDOTHELIAL SYSTEM AND IMMUNE PHENOMENA
Edited by N. R. Di Luzio and K. Flemming • 1971

Volume 16A
THE ARTERY AND THE PROCESS OF ARTERIOSCLEROSIS: Pathogenesis
Edited by Stewart Wolf • 1971

Volume 16B
THE ARTERY AND THE PROCESS OF ARTERIOSCLEROSIS: Measurement and Modification
Edited by Stewart Wolf • 1971

Volume 17
CONTROL OF RENIN SECRETION
Edited by Tatiana A. Assaykeen • 1972

Volume 18
THE DYNAMICS OF MERISTEM CELL POPULATIONS
Edited by Morton W. Miller and Charles C. Kuehnert • 1972

Volume 19
SPHINGOLIPIDS, SPHINGOLIPIDOSES AND ALLIED DISORDERS
Edited by Bruno W. Volk and Stanley M. Aronson • 1972

Volume 20
DRUG ABUSE: Nonmedical Use of Dependence-Producing Drugs
Edited by Simon Btesh • 1972

Volume 21
VASOPEPTIDES: Chemistry, Pharmacology, and Pathophysiology
Edited by N. Back and F. Sicuteri • 1972

Volume 22
COMPARATIVE PATHOPHYSIOLOGY OF CIRCULATORY DISTURBANCES
Edited by Colin M. Bloor • 1972

Volume 23
THE FUNDAMENTAL MECHANISMS OF SHOCK
Edited by Lerner B. Hinshaw and Barbara G. Cox • 1972

Volume 24
THE VISUAL SYSTEM: Neurophysiology, Biophysics, and Their Clinical Applications
   Edited by G. B. Arden • 1972

Volume 25
GLYCOLIPIDS, GLYCOPROTEINS, AND MUCOPOLYSACCHARIDES
OF THE NERVOUS SYSTEM
   Edited by Vittorio Zambotti, Guido Tettamanti, and Mariagrazia Arrigoni • 1972

Volume 26
PHARMACOLOGICAL CONTROL OF LIPID METABOLISM
   Edited by William L. Holmes, Rodolfo Paoletti,
   and David Kritchevsky • 1972

Volume 27
DRUGS AND FETAL DEVELOPMENT
   Edited by M. A. Klingberg, A. Abramovici, and J. Chemke • 1973

Volume 28
HEMOGLOBIN AND RED CELL STRUCTURE AND FUNCTION
   Edited by George J. Brewer • 1972

Volume 29
MICROENVIRONMENTAL ASPECTS OF IMMUNITY
   Edited by Branislav D. Janković and Katarina Isaković • 1972

Volume 30
HUMAN DEVELOPMENT AND THE THYROID GLAND: Relation to Endemic Cretinism
   Edited by J. B. Stanbury and R. L. Kroc • 1972

Volume 31
IMMUNITY IN VIRAL AND RICKETTSIAL DISEASES
   Edited by A. Kohn and M. A. Klingberg • 1973

Volume 32
FUNCTIONAL AND STRUCTURAL PROTEINS OF THE NERVOUS SYSTEM
   Edited by A. N. Davison, P. Mandel, and I. G. Morgan • 1972

Volume 33
NEUROHUMORAL AND METABOLIC ASPECTS OF INJURY
   Edited by A. G. B. Kovach, H. B. Stoner, and J. J. Spitzer • 1972

Volume 34
PLATELET FUNCTION AND THROMBOSIS: A Review of Methods
   Edited by P. M. Mannucci and S. Gorini • 1972

Neurohumoral and Metabolic Aspects of Injury

Proceedings of the IUPS Satellite Symposium held
August 3-7, 1971, in Budapest, Hungary

Edited by
Arisztid G. B. Kovách
Experimental Research Department
Semmelweis Medical University
Budapest, Hungary

H. B. Stoner
MRC Toxicology Unit
Medical Research Council Laboratories
Carshalton, Surrey, United Kingdom

and
John J. Spitzer
Department of Physiology and Biophysics
Hahnemann Medical College and Hospital
Philadelphia, Pennsylvania

PLENUM PRESS • NEW YORK-LONDON • 1973

Library of Congress Catalog Card Number 72-93823
ISBN 0-306-39033-7

© 1973 Plenum Press, New York
A Division of Plenum Publishing Corporation
227 West 17th Street, New York, N.Y. 10011

United Kingdom edition published by Plenum Press, London
A Division of Plenum Publishing Company, Ltd.
Davis House (4th Floor), 8 Scrubs Lane, Harlesden, London NW10 6SE, England

All rights reserved

No part of this publication may be reproduced in any form without written permission from the publisher

Printed in the United States of America

PREFACE

Under the auspices of the International Union of Physiological Sciences, the Federation of Hungarian Medical Societies and the Hungarian Physiological Society a Symposium was organized on the "Neurohumoral and Metabolic Responses to Injury" in Budapest August 3-7, 1971 as a satellite symposium to the XXV Congress of Physiology.

The aim of this multidisciplinary Symposium was to concentrate on the neurohumoral and hormonal aspects of the biological response to trauma including resistance and adaptation to injury. It is becoming increasingly apparent that the changes which occur after injury reflect the disturbance of regulatory processes and control mechanisms concerned with the preservation of homeostatic equilibrium. One of the main aims of the Symposium was to consider the effect of injury on control mechanisms.

It was thought advisable to include speakers more closely associated with the study of control mechanisms under normal conditions in order to ensure cross-fertilisation of ideas between the physiological and pathophysiological fields.

Recent progress in biochemistry, physiology biophysics and pharmacology has tended to clarify the chain of events that are produced after different stress conditions such as haemorrhage, trauma, burns, bacterial infections, etc. The Symposium gathered together eminent scientists from the various disciplines all focusing on the questions of the bodily reaction to injury and the development of the irreversible state. The discussion of the problem of shock in a symposium needs multidisciplinary attendance, but the clarification of the problem needs more basic research on the signficant cellular metabolic reactions to trauma and hypoxia.

The Symposium provided an excellent setting for discussions both inside and outside the meeting. These could not be included in the report of the proceedings but it is hoped that they will lead to renewed efforts in research into these often neglected, but clinically important, problems.

The organising Committee of the Symposium wish to thank the following who supported it with their donations:

International Union of Physiological Sciences
The Wellcome Trust, U. K.
The Medical Commission on Accident Prevention, U. K.
Chemical Works of Gedeon Richter Ltd., Hungary
Chinoin Pharmaceutical Works, Hungary
Coca Cola Export Corporation, Italy
EG YT Pharmacochemical Works, Hungary
Gebr. Giulini G.M.B.H., German Federal Republic
Imperial Chemical Industries Ltd., U. K.
Lufthansa, German Federal Republic
Medicor, Hungary
Pharmacia, Uppsala, Sweden

The method of publication used requires that the typescript, figures, etc. are put in a form which can be sent directly to the printers. This has proved not to be the easiest form of publication, particularly for a multinational Symposium. The Editors have tried their best to produce some order out of a chaos of typewriters. They were, however, unable to reduce the papers to complete uniformity and the different English usages to be found in the text reflect the variations in those of the Editors and the authors. The Editors excuse is that it gives the reader a taste of the original. An important duty of the Editors is to convey their very sincere gratitude to Mrs. D. Edwards, Mrs. N. P. Brewster, and Mrs. K. Bodolay, who bore the brunt of the secretarial work of this publication and to join with the authors in thanking all those other secretaries who have been concerned with the papers at different stages and without whom the Symposium could not have taken place.

                                                Arisztid G. B. Kovách
                                                H. B. Stoner
                                                John J. Spitzer

CONTENTS

List of Contributors xvii

CONTROL OF BLOOD FLOW

Control of Organ Blood Flow Following
       Hemorrhage 1
  A. G. B. Kovách, A. Mitsányi, E. Monos,
  I. Nyáry, and A. Sulyok

Cerebral Blood Flow and Metabolism in
       Hemorrhagic Shock in Baboons 19
  M. Reivich, A. G. B. Kovách, J. J. Spitzer,
  and P. Sandor

Adrenocortical Blood Supply During Hemorrhage
       in Dogs 27
  E. Monos, A. G. B. Kovách, Z. Áprili,
  Zs. Suba, and L. Cserényi

Microcirculation and Hypoxia 33
  D. W. Lübbers

The Microcirculation of Normal and Injured
       Tissue 45
  Paul C. Johnson

Further Experiments on the Role of Phosphate
       in the Local Regulation of Blood Flow 53
  S. M. Hilton

Uterine Circulation in Haemorrhage 57
  L. Takács and L. A. Debreczeni

Blood Microrheology and the Development of
    "Stasis" in the Microvasculature
    After Injury   65
  Holger Schmid-Schönbein, E. Volger,
  H. J. Klose, and J. Weiss

Blood Rheology after Hemorrhage and Endotoxin   75
  S. Chien, S. Usami, J. Dellenback, and
  V. Magazinovic

Rheology of Red Cell Suspensions in
    Experimental Dehydration   95
  F. Torres and L. -E. Gelin

FLUID TRANSFER

Regulation of Interstitial Fluid Volume and
    Pressure   111
  Arthur C. Guyton, Aubrey E. Taylor,
  Harris J. Granger, and W. Harry Gibson

Influence of Histamine on Transport of Fluid
    and Plasma Proteins into Lymph   119
  Eugene M. Renkin and Reginald D. Carter

Effect of Vasoconstrictors on Lymph Oxygen
    Tension in Normo- and Hypovolemia   133
  T. Barankay, S. Nagy, G. Horpácsy,
  and A. Petri

Transvascular Fluid Fluxes During Severe,
    Prolonged Canine Hemorrhagic or
    Endotoxin Hypotension   143
  George J. Grega and Francis J. Haddy

Capillary Filtration Rate in Cat Hind Limb
    Muscle during Hemorrhage   149
  Susanna Biró and A. G. B. Kovách

HUMORAL INFLUENCES

The Adenohypophysial Response to Stress and
    Its Hypothalamic Control   157
  S. M. McCann, L. Krulich, P. Illner,
  M. Quijada, K. Ajika, and C. P. Fawcett

Thyroid Function after Injury   165
  Roger Kirby and Ivan D. A. Johnston

Insulin Secretion during Hemorrhagic Shock 179
    M. Vigas, R. E. Haist, F. Bauer, and
    W. R. Drucker

The Effect of Persisting Hypovolemic Shock
        on Pancreatic Output of Insulin 187
    W. R. Drucker, B. L. Gallie, T. S. Lau,
    G. Farago, R. A. Levene, and R. E. Haist

The Decline in Pancreatic Insulin Release
        during Hemorragic Shock in the
        Baboon 189
    G. S. Moss, G. Cerchio, D. C. Siegel,
    P. C. Reed, A. Cochin, and V. Fresquez

Glucose Homeostasis in the Postoperative State 209
    J. Tiefenbrun, S. Finkelstein, and
    W. C. Shoemaker

Some In Vitro and In Vivo Effects of a New
        Prostaglandin Derivative 213
    B. David Polis, Anne Marie Grandizio,
    and Edith Polis

METABOLIC INTERRELATIONS

Non-Esterified Fatty Acid (FFA) Metabolism
        Following Severe Hemorrhage in
        the Conscious Dog 221
    John J. Spitzer, Roslyn Wiener, and
    Eugene H. Wolf

Alterations of Mitochondrial Structure and
        Energy-Linked Functions in Hemorrhagic
        Shock and Endotoxemia 231
    Leena M. Mela, Leonard D. Miller,
    Leonardo V. Bacalzo Jr., Kenneth Olofsson,
    and Raleigh R. White IV

Effect of Hemorrhagic Shcok on Gluconeogensis,
        Oxygen Consumption and Redox State of
        Perfused Rat Liver 243
    A. G. B. Kovách and P. Sándor

Potential Relationships of Changes in Cell
        Transport and Metabolism in Shock 253
    A. E. Baue, M. M. Sayeed, and M. A. Wurth

Dexamethasone (DXM) Effect on the Histamine-
        Releasing Activity of Endotoxin 263
   William Schumer, R. P. Obernolte, and
   P. R. Erve

Liver Metabolism after Injury 271
   D. F. Heath

Aerobic and Anaerobic Glycolysis in the
        Liver after Hemorrhage 277
   M. Kessler, H. Lang, H. Starlinger,
   J. Höper, and M. Thermann

Sequential Circulatory and Metabolic Changes
        in the Liver and Whole Body During
        Hemorrhagic Shock 293
   W. C. Shoemaker, L. J. Stahr, S. I. Kim,
   and D. H. Elwyn

Effect of Porto-Caval Shunt on Splanchnic
        Blood Flow in Hemorrhage 313
   J. Hamar, M. Gergely, I. Nyáry, and
   A. G. B. Kovách

Adipose Tissue and Hemorrhagic Shock 323
   S. Rosell, P. Sándor, and A. G. B. Kovách

Influence of Endotoxin on Adipose Tissue
        Metabolism 337
   J. A. Spitzer, A. G. B. Kovách, S. Rosell,
   P. Sandor, J. J. Spitzer, and R. Storck

Characterization of Rat Brain Mitochondria,
        the Effect of Injury 345
   J. Somogyi, Jill E. Cremer, and
   Kornélia Ikrényi

Brain and Liver Intracellular Compartmental
        Redox States in Hypoxia, Hypocapnia,
        and Hypercapnia 353
   A. T. Miller Jr. and Francis M. H. Lai

Peripheral and Cardiac Factors in Experimental
        Septic Shock 363
   Lerner B. Hinshaw

The Lysosomal Protease - Myocardial Depressant
        Factor System in Circulatory Shock 367
   Allan M. Lefer and Thomas M. Glenn

CONTENTS

Myocardial Metabolism during Endotoxic Shock 375
   John C. Scott, Jen Tsoh Weng, and
   John J. Spitzer

The Activity of Lipoprotein Lipase in Rat
      Heart after Tourniquet Stress 387
   J. Borbola, A. Gecse, and S. Karady

Time Course of Metabolite and Enzyme Changes
      in Hypoxic Conditions 395
   G. Horpácsy, T. Barankay, K. Tárnoky,
   S. Nagy, and G. Petri

Energy and Tissue Fuel in Human Injury and
      Sepsis 401
   John M. Kinney, Frank E. Gump, and
   Calvin L. Long

Environmental Temperature and Metabolic
      Response to Injury Protein,
      Mineral and Energy Metabolism 409
   D. P. Cuthbertson, G. S. Fell, A. G. Rahimi,
   and W. J. Tilstone

Environmental Temperature and the Metabolic
      Response to Injury - Plasma Proteins 417
   A. Fleck, F. C. Ballantyne, W. J. Tilstone,
   and D. P. Cuthbertson

Metabolic, Hormonal and Enzymatic Functions of
      Rats during Recovery from Injury 423
   S. Németh, M. Vigas, and A. Straková

Decreased 2,3-Diphosphoglycerate and Reduced
      Oxygen Delivery Following Massive
      Transfusion and Septicemic Shock 429
   Harvey J. Sugerman, Leonard D. Miller,
   Maria Delivoria-Papadopoulos, and
   Frank A. Oski

Complete Oxyhemoglobin Dissociation Curves in
      Thermal Trauma 441
   G. Arturson

NEURAL REGULATION

Baro- and Chemoreceptor Mechanisms in
      Haemorrhage 455
   Joan C. Mott

The Influence of Blood Pressure and Blood Flow
 on the Local Tissue Po$_2$ of the Carotid
 Body 463
D. W. Lübbers, H. Acker, H. P. Keller,
and E. Seidl

Sympathetic Nervous Activity after Hemorrhage 473
L. Fedina, M. Kollai, and A. G. B. Kovách

Hypothalamic and Cortical Evoked Potentials
 in Hemorrhagic Shock 481
E. Dóra, A. G. B. Kovách, and I. Nyáry

Central Nervous Mechanisms in Resistance and
 Adrenocortical Reactions to Trauma 489
K. Murgas, S. Németh, and M. Vigas

Thermoregulation after Injury 495
H. B. Stoner

Effect of Stress on Various Characteristics of
 Norepinephrine Metabolism in Central
 Noradrenergic Neurons 501
Anne-Marie Thierry

Hypothalamic Catecholamines and Adrenergic
 Innervation of Hypothalamic Blood
 Vessels. Potential Role in Shock 509
E. T. Angelakos, M. P. King, and L. Carballo

Activity of Adrenal Catecholamine-Producing
 Enzymes and Their Regulation after
 Stress 517
Richard Kvetnanský and Irwin J. Kopin

The Role of Lysosomal Hydrolases in the
 Mechanism of Shock 535
James J. Smith, Daniel J. Loegering,
Daniel J. McDermott, and Michael J. Bonin

Reticuloendothelial Function in Experimental
 Injury and Tolerance to Shock 545
Burton M. Altura and S. G. Hershey

RESISTANCE AND ADAPTATION

Mechanisms of Hypothermic Protection against
 Anoxia 571
James A. Miller Jr. and Faith S. Miller

CONTENTS

Resistance of the New-Born to Injury 587
   R. A. Little

Comparative Physiology of Hemorrhage. A Study
      of Adrenergic Receptor Stimulation
      and Blockade in Pigeon and Mammal 593
   G. Rubányi, L. Huszár, and A. G. B. Kovách

Biosynthesis of Adrenal Catecholamines during
      Adaptation of Rats to Immobilization
      Stress 603
   Richard Kvetnanský

Phylogenetic Aspects of Adrenocortical Activity
      during the Process of Adaptation 619
   M. Juráni, L. Mikulaj, and K. Murgas

Endocrine Functions during Adaptation to Stress 631
   L. Mikulaj and A. Mitro

Metabolic and Endocrine Functions of Rats
      Conditioned to Noble-Collip Drum
      Trauma 639
   S. Németh and M. Vigas

Carbohydrate Metabolism in Shock Resistance 649
   A. Bardoczi, J. Arokszallasi, and S. Karady

Central Nervous Influence upon the Adrenocortical
      Reaction during Stress Situations 655
   K. Murgas and V. Jonec

INDEX 661

PARTICIPANTS

Altura, B.M.	U.S.A.
Arokszállási, I.	Hungary
Angelakos, E.	U.S.A.
Arturson, G.	Sweden
Barcroft, H.	United Kingdom
Bardóczi, A.	Hungary
Baue, A.E.	U.S.A.
Beaumariage, M.L.	Belgium
Bermudez, E.	Cuba
Biró, Zs.	Hungary
Brooks, S.,Ch.	U.S.A.
Caldwell, F.T.	U.S.A.
Chien, Shu	U.S.A.
Cuthbertson, D.P.	United Kingdom
Dietzman, R.H.	U.S.A.
Donhoffer, Sz.	Hungary
Dóra, E.	Hungary
Drucker, W.	Canada
Dupuis, S.	France
Faragó, G.	Canada
Fedina, L.	Hungary
Fleck, A.	United Kingdom
Fonyó, A.	Hungary
Gaary, E.	U.S.A.
Gelin, L.E.	Sweden
Gergely, M.	Hungary
Grega, G.J.	U.S.A.
Guyton, A.	U.S.A.
Halmágyi, M.	German Fed.Rep.
Hamar, J.	Hungary
Heath, D.F.	United Kingdom
Hilton, S.M.	United Kingdom
Hinshaw, L.B.	U.S.A.
Honig, A.	German Dem.Rep.
Horpácsi, G.	Hungary
Huszár, L.	Hungary
Ikrényi, K.	Hungary
Irving, M.	United Kingdom
Johnson, P.	U.S.A.
Jordanov, M.	Bulgaria
Kabal, J.	U.S.A.
Karády, I.	Hungary
Kessler, M.	German Fed.Rep.
Kinney, J.M.	U.S.A.
Kirby, R.	United Kingdom
Koizumi, K.	U.S.A.

Kollai, M.	Hungary
Kovach, A.	Hungary
Kvetnansky, R.	Czechoslovakia
Lefer, A.M.	U.S.A.
Linderholm, H.	Sweden
Linke, P.G.	German Dem.Rep.
Little, R.A.	United Kingdom
Lübbers, D.W.	German Fed.Rep.
Magazinovits, V.	Yugoslavia
Mc Cann, S.M.	U.S.A.
Mela, L.	U.S.A.
Mendau-Wolf, M.	German Dem.Rep.
Mikulaj, L.	Czechoslovakia
Miller, A.T. Jr.	U.S.A.
Miller, J.A. Jr.	U.S.A.
Monos, E.	Hungary
Moss, G.	U.S.A.
Mott, J.C.	United Kingdom
Murgas, K.	Czechoslovakia
Nagy, S.	Hungary
Németh, S.	Czechoslovakia
Nyáry, I.	Hungary
Polis, B.D.	U.S.A.
Reivich, M.	U.S.A.
Renkin, E.M.	U.S.A.
Roheim, P.	U.S.A.
Rosell, S.	Sweden
Rubányi, G.	Hungary
Rubinstein, E.H.	U.S.A.
Sayeed, M.	U.S.A.
Sándor, P.	Hungary
Schmid-Schönbein, H.	German Fed.Rep.
Shoemaker, W.C.	U.S.A.
Simon, L.	Hungary
Smith, J.J.	U.S.A.
Somogyi, J.	Hungary
Sonnenschein, R.	U.S.A.
Spitzer, J.A.	U.S.A.
Spitzer, J.J.	U.S.A.
Stoner, H.B.	United Kingdom
Straub, F.B.	Hungary
Sugerman, H.J.	U.S.A.
Sulyok, A.	Hungary
Szántó, Gy.	Hungary
Szentágothai, J.	Hungary
Takács, L.	Hungary
Thierry, A.M.	France
Tiefenbrun, J.	U.S.A.
Tomka, N.	Hungary

PARTICIPANTS

Véghelyi, P.	Hungary
Vigas, M.	Czechoslovakia
Volger, E.	German Fed.Rep.
Wiener, R.	U.S.A.

CONTROL OF ORGAN BLOOD FLOW FOLLOWING HEMORRHAGE[1]

A.G.B. Kovách[2], A. Mitsányi, E. Monos, I. Nyáry, A. Sulyok

Experimental Research Dept. Semmelweis Medical University

Budapest, Hungary, Dept. of Physiol. Hahnemann Med. College

and Hosp., and Johnson Research Foundation, Med. School,

Univ. of Pennsylvania, Philadelphia, Pa. 19104

The circulatory responses to hemorrhage involve a pattern of selective vasoconstriction and vasodilatation which produces a redistribution of the diminished cardiac output (Blalock and Levy 1937, Frank et al. 1956, Corday and Williams 1960, Sapirstein et al. 1960, Gregg 1962, Takács et al. 1962, Smith et al. 1965, Fell 1966, Neutze et al. 1968, Kovách et al. 1968, Kovách 1970). Hinshaw et al. (1961) observed in the dog that loss of 10 per cent of the blood volume produced a 21 per cent fall in cardiac output, but a decline of only 7 per cent in the arterial pressure. A 20 per cent hemorrhage caused a 45 per cent fall in cardiac output, but only 15 per cent drop in arterial blood pressure. The physiologic compensatory mechanisms activated by acute volume deficit favor the maintenance of pressure at the expense of flow.

The control of blood flow is exerted by the smooth muscle cells of the peripheral blood vessels, i.e. the arterioles, precapillary sphincters and venules. The smooth muscle elements are subject to physiological regulation both by the vegetative nervous system and by local chemical and physical factors.

Review of the literature leaves no doubt as to the importance

1. These studies were supported by the Hungarian Medical Research Council and partly by the John A. Hartford Foundation.
2. Senior Foreign Scientist of the National Science Foundation, Washington, D.C. Permanent Address: Budapest, VIII. Üllői ut 78/a.

of vasoconstriction and of reduced effective regional blood flow in the development of shock. The importance of the adrenergic vasoconstriction in the development of shock is emphasized by the results of infusion of sympathomimetics (Remington et al. 1950, Erlanger and Gasser, 1919, Freedman et al.1941, Yard and Nickerson, 1956, Lillehei and MacLean 1959, Kovách et al.1966) or of activation of sympathetic nervous system that can produce shock (Overman and Wang 1947, Freeman 1933, Smiddy et al. 1958, Fedina et al. 1965).

The measurement of total blood flow in individual vascular areas do not give a measure of the degree of local capillary tissue perfusion. Not all blood measured flows through the exchange vessels, but part of it can pass through anatomical or physiological A-V anastomoses.

Because most probable the inadequate exchange of metabolites between blood and tissue cells in the ultimate impairment in shock the local tissue blood flow measurements could give a better understanding of the importance of the areas of the vascular bed having the greatest sensitivity to damage.

In our present studies we were interested in the following questions in respect of hemodynamic failure after standardized hemorrhage.

1. The distribution of blood flow and the local flow in organs representing large tissue masses (muscle, adipose tissue). The flow restriction of these areas can produce generalized irreversible damage through metabolites released from the hypoxic or ischaemic tissues.
2. Tissue blood flow in vital organs and the adrenergic control of the vessels after hemorrhage.
3. Cerebral regional blood flow changes after hemorrhage, local flow reduction that can produce impairment of neural control mechanisms and affect the homeostatic capacity of the whole body.
4. Blood flow changes in endocrine organs as by this means hormone production and release can be affected during hemorrhage and reinfusion.

Methods and procedure. The experiments were performed on dogs under chloralose (Merck) anesthesia. For inducing hemorrhagic shock a modified Lamson technique was used. As a standard procedure the blood pressure of the dog was kept on the stabilized mean arterial blood pressure levels of 55 and 35 mmHg respectively (BI, BII) by bleeding into a reservoir (Engelking and Willig,1958). Each period was maintained for 90 minutes. After these hypovolemic periods all the blood remaining in the bottle was reinfused. Blood coagulation was prevented with 5 mg/kg heparin (Richter Gedeon,

Budapest). The arterial, the right atrial pressure, the ECG in standard lead II, respiration and the electrocorticographic activity were recorded simultaneously. The cardiac output, the head-forelimb, the splanchnic, and the pelvic-hindlimb blood flow were measured by the thermodilution method (Fronek and Ganz, 1960) as modified in our Laboratory (Kovách and Mitsányi, 1964).

Blood flow in the tissue was measured by adapting three methods: one method was that described by Hensel and Ruef (1954) based on the measurement of the thermal conductivity of the tissues. Continuous recording was carried out with a direct writing two channel device (Fluvograph, Hartmann and Braun, Frankfurt). The measuring probes were made according to Betz et al.(1961). Blood flow records were evaluated in terms of the difference in relative (percentage) thermal conductivity ($\Delta\lambda$) at the beginning of the experiment and after death i.e. the difference between those is referred to as 100 %.

The H_2 washout method described by Aukland, Bower and Berliner (1964) was also adapted for the measurement of local blood flow. Two or three platinum electrodes 0.1-0.5 mm in diameter were introduced into the tissues. The electrodes were insulated with Insl-x (Insl-x Product Corp.Yonkers, N.Y.) leaving 0.8 mm bare at the top. The indifferent electrode of the circuit was a calomel electrode introduced under the skin via a polyethylene tubing KCl-agar bridge. A positive polarizing potential of 250 mV was supplied to the measuring electrodes. Hydrogen saturation of the tissue was attained by inhaling the gas and by local intraarterial injection of a hydrogen containing bolus. The exponential desaturation curves were recorded on a Kipp BD-3 micrograph (Delft) and replotted to semilogarithmic paper; the slope of the line thus obtained is proportional to blood flow and can be calculated in ml/min/100 g according to Kety (1960).

In our experiments on the adipose tissue, muscle and adrenal gland the blood flow was measured by the venous outflow technique and recorded with a coordinate writer (Jacquet, Bale) on a kymograph.

Phenoxybenzamine (Dibenzyline Smith Kline and French, Philadelphia) was injected in 4-5 mg/kg body wt. i.v. or 0.5 mg/kg into the common carotid artery, 2 hours before the bleeding started. Dichlorisoproterenol was administered in 1 mg/kg, atropine sulphate in 3 mg/kg i.v. The statistical method employed was Student's t-test.

Results. In Figure 1 the cardiac output and its distribution to the head-forelimb, splanchnic and pelvic-hindlimb regions are indicated before and after treatment with alfa and beta adrenergic receptor blockers and atropine. Before drug treatment 40-45 % of the cardiac output (2.93 1/m^2 body surface) is flowing through the head-forelimb fraction, 35-40 % through the splanchnic organs,

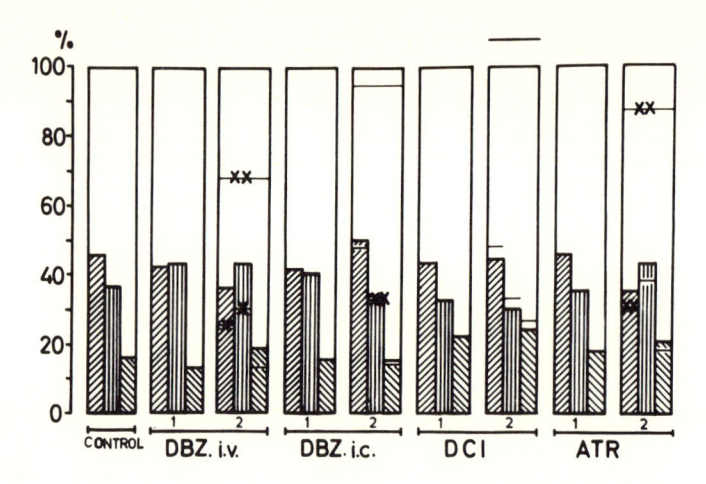

Fig.1.:Change in cardiac output and its regional distribution after the administration of Phenoxybenzamine (PBZ),Dichlorizoproterenol (DCI) and Atropinesulphat (ATR). The changes are expressed in two ways a/ in % of the control values (horizontal lines)and b/in % of the cardiac output 30 minutes after drug treatment (columns). Cardiac output □ ;head forelimb ▨ ;splanchnic area ▥ ;pelvic-hindlimbs ▧ ;self control (1); after drug treatment (2); ✘ :p<0.05; ✘✘:p<0.01. DBZ has been administered either in 5 mg/kg intravenously or 0.5 mg/kg into the common carotid artery;DCI in 1 mg/kg; ATR in 3 mg/kg intravenously.

and 18-25 % through the pelvic-hindlimb region. Phenoxybenzamine (i.v.5 mg/kg) reduced the cardiac output and its head-forelimb and splanchnic fractions significantly. Injecting 0.5 mg/kg Phenoxybenzamine into the common carotid artery affected only the splanchnic blood flow significantly. Dichlorisoproterenol injection exerted no significant effect on the hemodynamics.In an other series of experiments we have found that propranolol in 1 mg/kg doses did not significantly reduce the cardiac output. The splanchnic and hindlimb blood flow fractions were significantly reduced in propranolol treated dogs. Atropine (3 mg/kg) decreased the cardiac output and its head-forelimb fraction significantly.

In Fig.2 the local blood flow changes in the hypothalamus and in the liver are demonstrated. The blood flow decreased to 70 % in the hypothalamus during BI and to 40 % during BII. There was no difference between the untreated and PBZ treated group during bleeding. The blood flow in the hypothalamus increased after reinfusion but did not achieve the original flow values. The recovery in flow was more complete in the PBZ treated group after reinfusion.

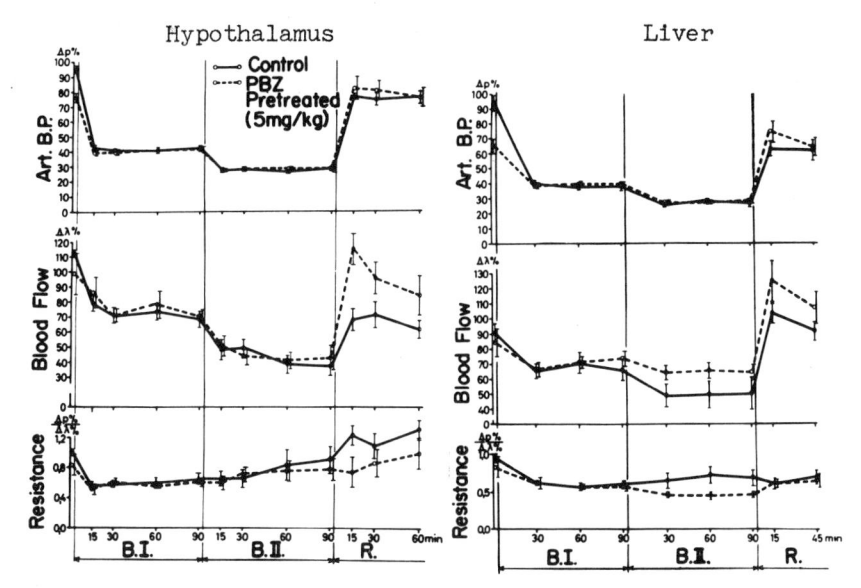

Fig.2.: Changes in hypothalamic (left) and liver (right) blood flow mean arterial blood pressure and peripheral resistance before, during hemorrhage and after reinfusion, in control and Phenoxybenzamine (PBZ) treated dogs. Blood flow measured with the heated thermocouple method. BI=first bleeding period, BII=second bleeding period, R= reinfusion. Blood flow records were evaluated in terms of relative (percentage) thermal conductivity ($\Delta\lambda$) at the beginning of the experiment and after death is referred to as 100 %.

Liver local flow was 70 % of the control level in BI and 50 % in BII. After reinfusion the flow values returned to the initial level. PBZ had no effect in BI but during BII, there was no further reduction in flow (see Fig.2).

In the gracilis muscle flow decreased to 70 % during BI and to 60 % during BII. Reinfusion returned the flow to the initial level. In the PBZ treated group the drug reduced the arterial blood pressure and also the flow. During bleeding the fall in blood flow was less pronounced than in control. It was 80 % in BI and 75 % in BII the difference in flow between control and PBZ treated group was not significant during bleeding (see Fig.3). The peripheral vascular resistance did not increase during bleeding.

The myocardial flow was reduced (see Fig.3 right side) in the control group during bleeding to 60-70 % of the starting values. Temporary recovery was observed after reinfusion. PBZ had a very pronounced effect on myocardial blood flow; it was elevated by 40 %. The blood flow remained above the original value during the

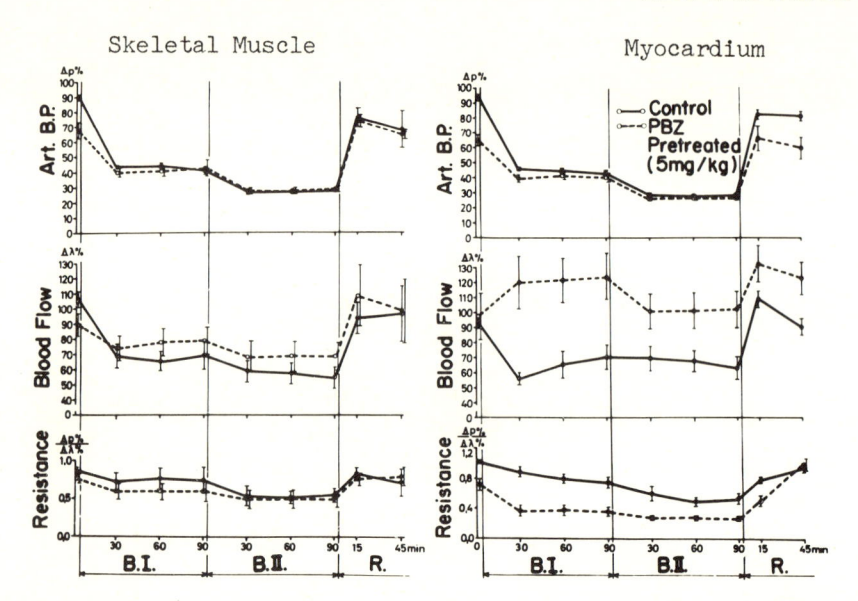

Fig.3.: Changes in the skeletal muscle (m.gracilis left) and myocardial (right) blood flow, mean arterial blood pressure, peripheral vascular resistance, in control and Phenoxybenzamine (PBZ) pretreated dogs before, during hemorrhage and after reinfusion. Blood flow measured with the heated thermocouple method. BI-first bleeding period, BII=second bleeding period. R= reinfusion. Evaluation as in Fig.2.

whole hypotensive period. Vascular resistance was significantly lower in the PBZ treated group during bleeding.

The renal cortical blood flow (see Fig.4) was 40 % of the control level in BI and 25 % in BII. After reinfusion the original blood flow levels were obtained. PBZ treatment diminished the blood flow reduction following bleeding both in BI and BII significantly ($P < 0.001$). The significant elevation in vascular resistance after bleeding found in the control group was abolished through the PBZ treatment (Mitsányi et al. 1972).

After bleeding to a mean arterial blood pressure of 55 mmHg, blood flow in subcutaneous adipose tissue decreased from 6+0.9 to 0.6+0.21 ml/min/100 g ($P < 0.001$), and remained at that low level during bleeding to 35 mmHg (Fig.4). In 5 out of 9 experiments the blood flow ceased completely. After reinfusion there was a temporary regeneration but 60 minutes later the blood flow was again significantly lower than in the control, in 2 cases there was no blood

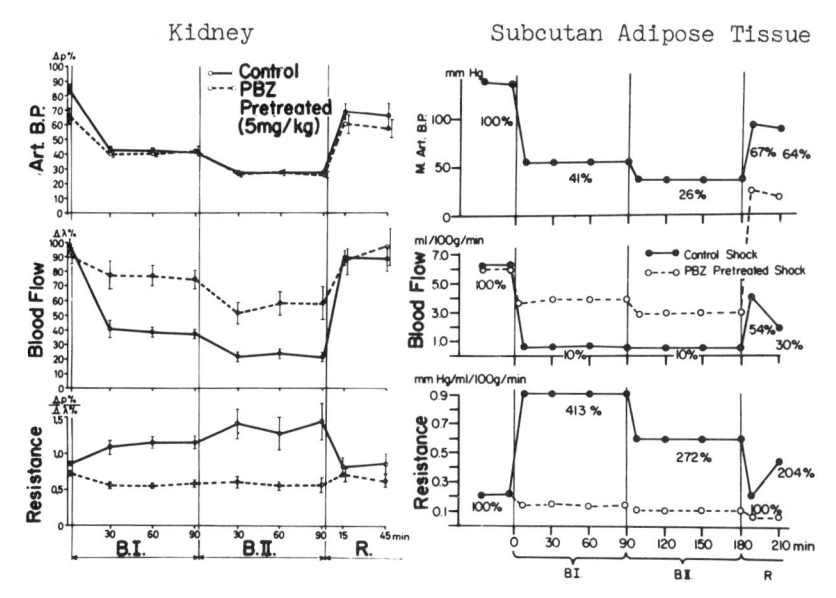

Fig.4.: Changes in the kidney cortical (left) and subcutaneous adipose tissue (right) blood flow, mean arterial blood pressure, peripheral vascular resistance, in control and Phenoxybenzamine (PBZ) pretreated dogs, before, during hemorrhage and after reinfusion. Renal blood flow was measured by the heated thermocouple (evaluation as in Fig.2) the suprapubic subcutaneous adipose tissue blood flow by venous outflow method. Ordinate: absolute values in mmHg, ml/100 g/min, and mmHg/ml.100 g.min. Numbers in the figure percentual values. BI=first bleeding, BII=second bleeding, R=reinfusion.

flow at all after reinfusion. In the PBZ (5 mg/kg i.v.) treated group the decrease in blood flow was significantly smaller than in control. In BII the flow was still 2.6 ml/min/100 g. The vascular resistance elevated from the control 100 % to 413 % during bleeding. This was absent in the PBZ treated group (Kovách, Rosell et al. 1970).

The intestinal (ileum) blood flow measured by the venous outflow method decreased to 29 % during BI and to 14.6 % during BII (see Fig.5). The recovery after reinfusion was not complete (87.8%) and only temporary. PBZ pretreatment elevated the blood flow from 40 to 52 ml/min/100 g. The blood flow was both in BI (46%) and BII (30%) above the values of the untreated group. Recovery was complete in the PBZ treated group. PBZ treatment completely abolished the elevation in vascular resistance during bleeding, which was significant in the untreated bled group ($P < 0.001$).

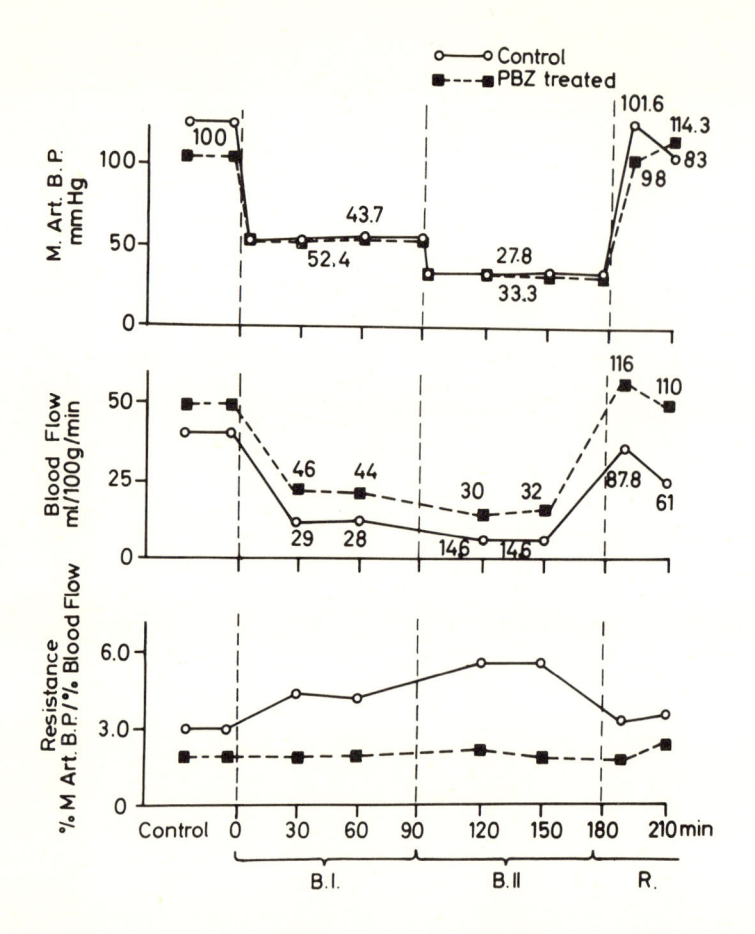

Fig.5.: Changes in the intestinal (ileum) blood flow mean arterial blood pressure and peripheral vascular resistance before, during hemorrhage and after reinfusion in control and Phenoxybenzamine (PBZ) 5 mg/kg i.v. pretreated dogs. Blood flow was measured with the venous outflow method. Ordinate in absolute values, percentage changes are indicated in the figure with numbers.

The hypophyseal blood flow measured by the Hydrogen washout method was 49.1 ml/min/100 g (see Fig.6). Bleeding to 55 mmHg had only minor effect on blood flow and it decreased significantly only 60 minutes after the bleeding started ($P < 0.05$). During BII the blood flow was 55 % of the control value. Interestingly the flow remained at this level after reinfusion with restored arterial blood pressure. The vascular resistance decreased during bleeding, but increased significantly ($P < 0.01$) after reinfusion. PBZ treatment increased the prebleeding blood flow in the hypophysis to 71.0

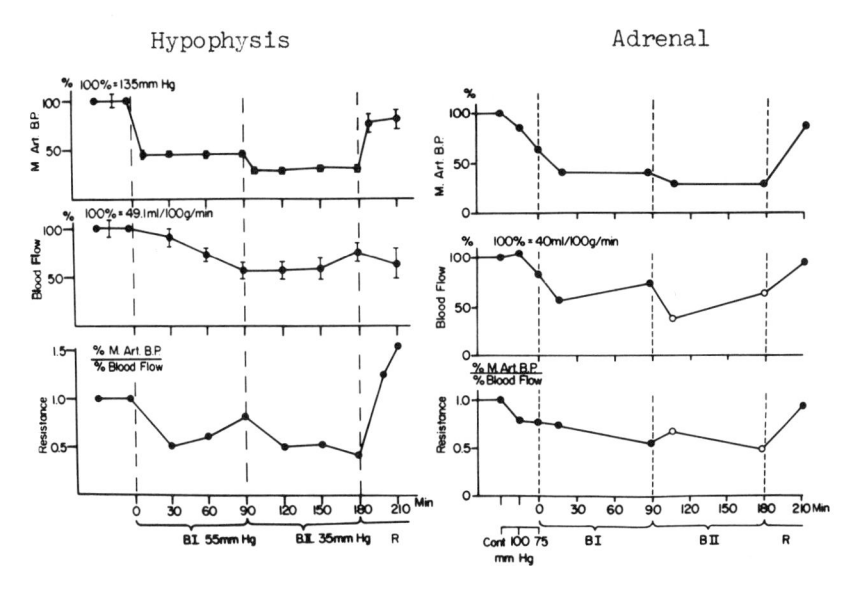

CONTROL OF ORGAN BLOOD FLOW FOLLOWING HEMORRHAGE

Fig.6.: Changes in the hypophyseal (left) and adrenal (right) blood flow, mean arterial blood pressure and peripheral vascular resistance before, during hemorrhage and after reinfusion in dogs. Hypophyseal blood flow was measured by the Hydrogen washout method, in the adrenal gland by venous outflow method. Ordinate in %, the absolute values are indicated with numbers on the figure. BI= first bleeding, BII= second bleeding, R= reinfusion.

ml/100 g/min, and the flow remained above the initial control level throughout the whole hypovolemic state. In this group the vascular resistance did not increase in oposit to the untreated animals.

Fig.6 shows the effect of blood loss on the adrenal gland blood flow measured using the venous outflow method. The initial adrenal gland blood flow was 40 ml/min/100 g before bleeding. Hemorrhage to 100 mmHg arterial pressure did not lower flow and the latter started to fall at 75 mmHg blood pressure. During BI the flow decreased significantly ($p < 0.02$) to 56 % of the starting level, and was slightly increasing reaching 75 % at the end of BI. After bleeding to 35 mmHg the flow fell again to 41 % of the control and was continuously increasing to 62 % at the end of BII. Reinfusion accomplished a complete restoration in flow. During the whole oligemic period the peripheral resistance was significantly reduced in the adrenal vascular bed.

Thyroid blood flow was 88.4 ml/100 g/min in the prebleeding

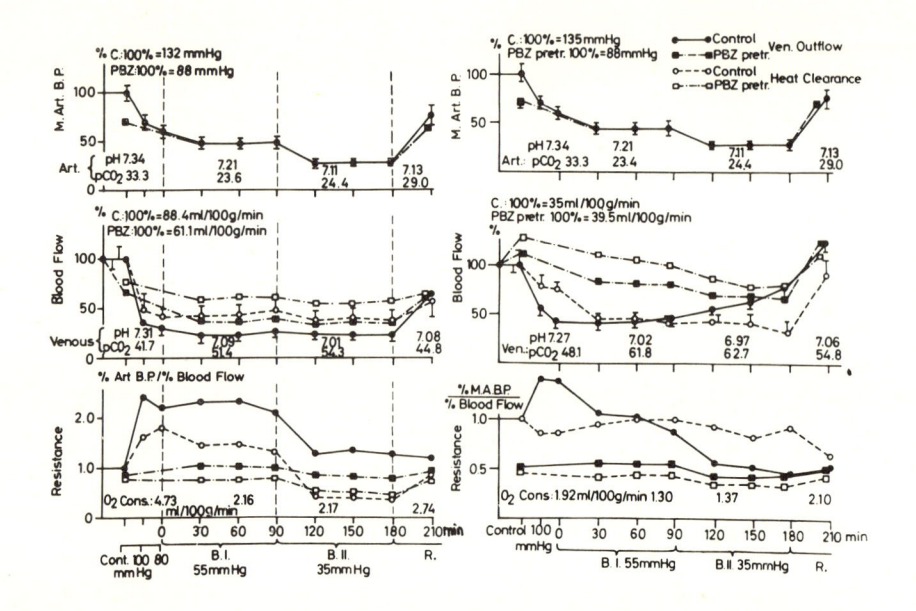

Fig.7.: Changes in the thyroid (left) and submandibular salivary gland (right) blood flow, arterial blood pressure and peripheral vascular resistance in control and Phenoxybenzamine (PBZ) pretreated dogs, before, during hemorrhage and after reinfusion. Blood flow was measured simultaneously by the heated thermocouple (evaluation as in Fig.2) and the venous outflow method in both organs. Ordinate in %, absolute values indicated with numbers in the figure. Arterial and venous blood pH, pCO_2 values and the O_2 consumption (ml/100 g/min) indicated in the figure. BI= first bleeding, BII=second bleeding, R=reinfusion.

state (Fig.7). On bleeding the dogs to 100 mmHg arterial blood pressure the blood flow decreased to 38 % of the control value, measured by the venous outflow and to 49 % by the heated thermocouple method. Bleeding to 80 mmHg decreased further venous outflow to 25 % and tissue flow to 45 %. The same flow was maintained during 55 and 35 mmHg blood pressure. After reinfusion the blood flow increased but the original flow values were not reestablished. PBZ pretreatment reduced the arterial pressure from 132 to 88 mmHg and the blood flow of the thyroid gland was 61.1 ml/100 g/min. PBZ decreased the vascular resistance significantly ($P < 0.02$). Bleeding the PBZ treated dogs to 55 mmHg the venous outflow dropped to 43 % and remained throughout the whole oligemic period between 38-45 %. The fall in nutritive blood flow was less affected during bleeding in the PBZ treated group and was 60 % during BI and 55 % during BII significantly (P< 0.01) higher compared with the untreated group. The prebleeding values were obtained in the PBZ group after reinfusion. The peripheral vascular resistance increased significantly (P< 0.01) after blood loss: maximal values were reached already at 100 mmHg mean arterial blood pressure. PBZ treatment abolished the elevation in vascular resistance during hypovolemia.

Salivary gland (submandibular) blood flow changes during hemorrhage are summarized on Fig.7. The control, 100 % flow was 35 ml/100 g/min. The venous outflow fell to 58 % following bleeding to 100 mmHg, and to 45 at 55 mmHg blood pressure. The venous outflow was continuously increasing during BII. The increase was significant (P< 0.02) at 60 minutes of BII. The tissue flow was showing continuous decrease. The original flow was obtained with both methods after reinfusion. After PBZ treatment the arterial blood pressure fell from 135 to 88 mmHg, and the salivary gland blood flow increased.The peripheral resistance decreased to 50 % of its control level. In PBZ treated group the blood flow measured by both methods remained at the starting level during BI. The flow decreased to 75 % of control value by the venous outflow and to 82 % by the tissue flow measurement in BII. Only in the venous outflow measurements was a short lasting elevation in vascular resistance at the first 20 minutes of hemorrhage. Later the resistance fell below the control value. The resistance calculated from the nutritive flow values, and in both PBZ group remained unchanged throughout the whole bleeding procedure.

Discussion and conclusions. Our data were obtained on dogs using the same standardized hemorrhagic shock model with identical blood flow methods. It is apparent from the results obtained that both the total and the nutritive blood supply responds in a specific way from organ to organ after hemorrhage. An extreme example is the heart muscle on the one side with maximal compensation possibility, on the other side is the subcutaneous adipose tissue with very low flow rate and irreversible vascular damage after hemorrhage.

Since the adipose tissue constitutes approximately 15 % of the body weight and has a resting flow of 6 to 9 ml/min/100 g which is diminished greatly (to 10 %) following bleeding, the tissue may function as a blood reserve by shifting blood during short term emergency periods to other organs. On the other hand, if the vasoconstriction in adipose tissue persists for a long time, this may have local metabolic and vascular consequences (Kovách et al.1970).

In previous studies of cerebral blood flow in hemorrhagic and other forms of shock either measurement of total blood flow or the window technique measuring the changes in the calibre of the cerebrocortical vessels was used. Jourdan et al.(1950) did not find alterations after hemorrhage in the cortical vessel calibre. Kovách et al.(1958) found that the total blood flow through the head decreased 20-25 % in tourniquet and hemorrhagic shock but these changes alone could not be responsible for the demonstrated neural impairment (electrocorticogram, Kovách et al.1958). Similar flow changes were published by Green and Rapela (1968). It has been established that cerebral blood vessels respond with marked autoregulation of flow to acute changes in cerebral perfusion pressure (Lassen 1959; Rapela and Green 1964; Harper 1965). These results do not guarantee that the blood flow distribution to all areas within the brain would be even and sufficient. To compare the distribution of blood in the capillary circulation in different parts of the brain, we studied the cerebral circulation of cats in ischemic and hemorrhagic shock by the benzidine method of Slonimski and Cunge, (1937). It was found that in both types of shock, substantial regional differences occurred the most pronounced changes were ischaemia of the cerebral cortex and breaking up the normal capillary pattern; empty capillaries and intravascular "sludge" formation were seen (Kovách 1961). Using ^{32}P labelled red blood cells it was demonstrated that part of the blood in the brain did not take part in the active circulation (Erdélyi et al. 1958).

Our studies on the local blood flow measurements (H_2 washout, and thermal conductivity methods) demonstrated a pronounced fall in flow in the hypothalamus (Kovách 1970). These results suggest that the hypothalamic blood flow autoregulation differs from that in other parts of the brain. A similar blood flow decrease was observed also in the frontal cortex. The hypothalamic and cerebrocortical impairment may lead to deficiency in central nervous regulatory functions: circulatory control, temperature regulation, neurohormonal secretion. Impaired hypothalamic cardiovascular control in shock had been previously demonstrated (Kovách et al. 1962).

The comparison of blood flow restoration in different organs after reinfusion has shown that in the heart muscle, liver, skeletal muscle, intestine, renal cortex, adrenal and salivary gland the general conditions deteriorate although the initial flow values were obtained temporarily. This could suggest the absence of vascular damage or occlusion, in contrast to the events in hypothalamus, hypophysis, thyroid gland and adipose tissue. The pronounced fall in nutritive blood flow and oxygen delivery to the cells during hypovolemia and the lack of capability to achieve the initial blood flow in these last four tissues after reinfusion suggests irreversible vascular impairment. The local damage may influence the condition of the whole animal and its ability to restore normal functions and adequately operating neural and hormonal control mechanisms.

PBZ treatment diminished the peripheral vascular resistance in all ten tissues studied but the extent of the fall in resistance was not equal. The most pronounced and significant decrease in resistance was in the myocardioum, hypophysis, adipose tissue and salivary gland. The blood flow increased significantly after PBZ administration in the myocardium, hypophysis, salivary gland 30-40 % and intestine with 10 %. The subcutaneous adipose tissue blood flow remained unaltered by PBZ application. The renal cortical flow fell with 8 % in the hypothalamus and liver with 10 %, in the skeletal muscle with 18 % and in the thyroid gland with 26 %.

The effect of alfa-adrenergic blockade (administration of PBZ) on blood flow pattern of different organs is also not uniform after hemorrhage. In some organs (skeletal muscle, hypothalamus, occipital cortex, liver), the circulatory conditions were identical in the untreated and PBZ treated dogs. In other organs (kidney, adipose tissue, thyroid, intestine) the intensive vasoconstriction developing after bleeding was abolished by PBZ treatment. The third type of tissue was the myocardium, hypophysis and salivary gland, in which the blood flow remained above the original level - in the PBZ treated group - during the whole hypovolemic period.

The differences cannot be explained alone by the alfa-adrenergic receptor blockade and the consequent inhibition of vasoconstriction. It is more probable that the known secondary vasodilatation after adrenergic stimulation (Berne 1958) is the explanation of the phenomenon specially in the myocardium. Phenoxybenzamine does not block the inotropic and chronotropic effects (Moran and Perkins 1961; Nickerson and Chan 1961) in contrary it has a positive effect (Chang 1968). The existence of beta-receptors in the coronary bed has been repeatedly confirmed (Douthell et al. 1964, Gaal et al. 1966, Parratt 1965, 1967). Phenoxybenzamine is elevating the blood catecholamine level during hemorrhage (Millar et al.1959). According to the above mentioned data, PBZ on one hand inhibits the primary vasoconstrictive effect of the sympathetic hyperactivity, on the other hand it does not influence the vascular dilatation produced either by beta-receptor stimulation or by tissue metabolites.

PBZ pretreatment did not influence hypothalamic blood flow changes in hemorrhage, but prevented the electrophysiological deteriorations. This finding suggested another way of protection, through the metabolic effect of the drug (Kovách et al.1971).

PBZ treatment influenced beneficially the restoration of flow in the hypothalamus and adipose tissue, but did not change it in the thyroid gland, where the flow was reduced already after PBZ administration, and in the hypophysis where the flow was above the control level during the whole hypovolemic period. These results are suggesting that in part of the tissues the vascular bed is only affected by the adrenergic hyperactivity (renal cortex, adipose tissue) compared to other tissues where sympathetic hyperactivity is not able to explain the changes (heart muscle, hypophyses, salivary gland).

Regarding the Phenoxybenzamine effect the literature agrees that the drug abolishes the vasoconstriction of the resistance vessels and this way it increases the blood flow through the organs (Chien 1967). It was proved by Gourzis and Nickerson (1965) that PBZ increases in the first place the nutritive blood flow. This finding was confirmed by microcirculatory studies (Altura 1967). Our results in addition are suggesting that the effect of PBZ in hemorrhagic shock is not homogenous in all vascular beds. The results on the metabolic effect of PBZ (Kovách et al. 1971) are suggesting further metabolic considerations to explain the exact mechanism of the vascular behaviour in shock and the mechanism of Phenoxybenzamine action.

REFERENCES

Altura, B.M. (1967) Amer. J. Physiol. 212:1447.

Aukland, K., Bower, B. F., Berliner, R.W. (1964) Circ. Res.14:164.

Berne, R.M. (1958) Circ. Res. 6:644.

Betz, E., Braasch, D., Hensel, H. (1961) Arzneimittel-Forsch. 11:333.

Blalock, A., Levy, S.E. (1937) Amer. J. Physiol. 118:734.

Chang, P. (1968) Europ. J. Pharmacol. 4:240.

Chien, S. (1967) Physiol. Rev. 47:214.

Corday, E., Williams, J.H., Jr. (1960) Amer. J. Med. 29:228.

Doutheil, U., Ten Bruggencate, H.G., Kramer, K. (1964) Pflügers Arch. 281:181.

Engelkind, R., Willig, F. (1958) Pflügers Arch. 267:306.

Erdélyi, A., Kovách, A.G.B., Menyhárt, J., Petőcz, L., Nagy, I., Csuzi, S. (1958) Acta physiol. Acad. Sci. hung, Suppl.12:5.

Erlanger, J., Gasser, H. S. (1919) Amer. J. Physiol. 49:345.

Fedina, L., Kovách, A.G.B., Vanik, M. (1965) Acta physiol. Acad. Sci. hung. 26 Suppl.35.

Fell, C. (1966) Amer. J. Physiol. 210:863.

Frank, E.D., Frank, H.A., Jacob, S., Weizel, H.A.E., Korman, H., Fine, J. (1956) Amer. J. Physiol. 186:74.

Freeman, N.E. (1933) Amer. J. Physiol. 103:185.

Freeman, N.E., Freedman, H., Miller, C.C. (1941) Amer. J. Physiol. 131:545.

Fronek, A., Ganz, W. (1960) Circ. Res. 8:175.

Gaal, P.G., Kattust, A.A., Kolin, A., Ross, G.(1966) Brit. J. Pharmacol. 26:713.

Gourzis, J.T., Nickerson, M. (1965) In: L.C.Mills, J.H.Moyer (eds) Shock and Hypotension. Grune and Stratton, New York p. 289.

Green, H.D., Rapela, C.E.(1968) In: Microcirculation as Related to Shock. Ed. D. Shepro, G.P. Fulton. New York-London, Academic Press p.93.

Gregg, E.D.(1962) In: Schock Pathogenese und Therapie (Hrsg.K.D. Bock) Berlin, Springer. p.57.

Harper, A.M. (1965) Intern.Symp. Univ. Lund. Copenhagen, Munksgaard. p. 94.

Hensel, H.,Ruef, J. (1954) Pflügers Arch. 259:267.

Hinshaw, D.B., Peterson, M., Huse, W.M., Stafford, C.E., Joergenson, E.J. (1961) Amer. J. Surg. 102:224.

Jourdan, F., Collet, A., Masbernard, A.(1950) C.R.Soc.Biol.(Paris) 144:1507.

Kety, S.S.(1960) Meth.med.Res.8:228.

Kovách, A.G.B.(1970) J.clin.Path.23. Suppl.4.202.

Kovách, A.G.B.(1968) Proc.IUPS Congress, Washington, D.C.25.Vol.6.35.

Kovách, A.G.B.(1961) Fed. Proc.20. Part 3.122.

Kovách, A.G.B., Fedina, L., Mitsányi, A., Naszlady, A., Biró, Zs. (1962) In: Proc.of 22nd IUPS Congress, Leiden. Excerpta Med.Int. Congr. Ser.48:678.

Kovách, A.G.B., Koltay, E., Fonyó, A., Kovách, E.(1971) Biochem. Med.5:384.

Kovách, A.G.B., Lányi, M., Bach, I.(1957) Cited A.G.B.Kovách,1966. Orvosképzés. 41:324.

Kovách, A.G.B., Mérey, F., Grastyán, E.(1958) Cited A.G.B.Kovách and A. Fonyó. In:The Biochemical Response to Injury. Oxford,Blackwell. 1960. p.133.

Kovách, A.G.B., Mitsányi, A.(1964) Wiener Med.Wochenschr.114:401.

Kovách, A.G.B., Róheim, P.S., Irányi, M., Kiss, S., Antal, J.(1958) Acta physiol.Acad.Sci.hung. 14:231.

Kovách, A.G.B., Rosell, S., Sándor, P., Koltay, E., Kovách, E., Tomka, N.(1970) Circ. Res.26:733.

Lassen, A. (1959) Physiol. Rev.39:183.

Lillehei, R.C., MacLean, L.D. (1959) Fed. Proc.18:415.

Millar, R.A., Keener, E.B., Benfey, B.G.(1959) Brit.J.Pharmacol.14:9.

Mitsányi, A., Kovách, A.G.B., Hámori, M.(1972) Acta physiol.Acad. Sci.hung. In press.

Moran, N.C., Perkins, M.E.(1961)J.Pharmacol.133:192.

Neutze, J.M., Wyler, F., Rudolph, A.M.(1968)Amer.J.Physiol.215:857.

Nickerson,M., Chan, G.C.M.(1961)J.Pharmacol.133:186.

Overman, R.R., Wang, S.C.(1947) Am.J.Physiol.148:289.

Parratt, J.R.(1965) Brit. J.Pharmacol. 24:601.

Parratt, J.R.(1967) Amer. Heart J.73:137.

Rapela,C.E., Green, H.D.(1964) Circ.Res. 15.Suppl. 205.

Remington, J.W.,Hamilton, W.F., Boyd, G.H.Jr.,Hamilton, W.F.Jr., Caddell, H.M. (1950) Amer.J.Physiol. 161:116.

Sapirstein, L.A.,Sapirstein, E.H., Bredemeyer,A.(1960)Circ.Res.8:135.

Slonimski, P., Cunge, M.(1936-1937) Fol.Morph. 7:126.

Smiddy, F.G.,Segel,D.,Fine, J.(1958)Proc.Soc.Exp.Biol.Med.97:584.

Smith,L.L.,Reeves, C.D., Hinshaw, D.B.(1965)In:L.C.Mills,J.H.Moyer (eds.)Shock and Hypotension.Grune and Stratton,New York.p.373.

Takács, L., Kállay, K., Skolnik, J.H. (1962) Circ.Res.$\underline{10}$:753.

Yard, A.C., Nickerson, M. (1956) Fed. Proc. $\underline{15}$:502.

CEREBRAL BLOOD FLOW AND METABOLISM IN HEMORRHAGIC SHOCK IN BABOONS

M. Reivich[1], A.G.B. Kovach, J.J. Spitzer & P. Sandor

Cerebrovascular Research Lab. Department of Neurology,
University of Pennsylvania, Philadelphia, Pennsylvania

The effects of hemorrhagic shock on cerebral hemodynamics and metabolism have been studied in baboons in an attempt to understand better the pathophysiology of shock in primates.

METHODS

Eight animals have been studied, three of which were control animals and five experimental animals. They were anesthetized with Sernylan, 1 mg/kg, with additional doses as needed, and paralyzed with gallamine. Respiration was controlled and end-tidal PCO_2 continuously monitored with an infra-red CO_2 analyzer.

Catheters were placed in the femoral artery and vein for infusion of drugs, blood sampling and monitoring of arterial blood pressure. A catheter in the superior saggital sinus was used to monitor cerebral venous pressure and to obtain blood samples. Arterial and cerebral venous blood was analyzed for pH, PCO_2, PO_2, O_2 content, O_2 saturation, glucose, lactate, pyruvate, hemoglobin and hematocrit.

After the preparation was completed the animal was given 5 mg/kg. of heparin.

Regional cerebral blood flow (rCBF) was measured by 2 techniques. Repeated measurements were made by monitoring the

[1] Recipient of U.S.P.H.S. Career Research Development Award 5 K03 HE-11, 896-06.

cerebral clearance of Xe^{133} with 8 collimated NaI crystals 5 mm diam. The Xe^{133} was injected as a bolus into the internal carotid artery via a catheter in a branch of the external carotid artery. The position of the eight probes is shown in Fig. 1.

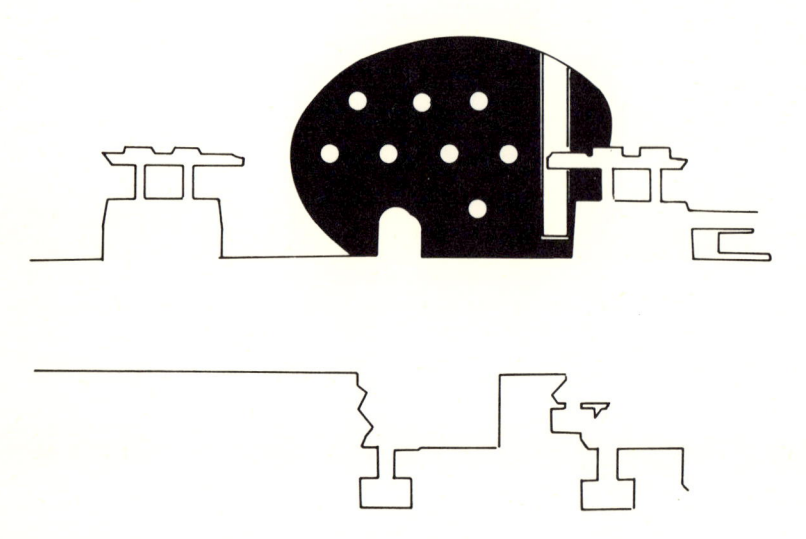

Fig. 1. Diagram of head in stereotatic holder with the position of the 8 regional flow probes indicated.

With this technique the mean flow in a core of brain tissue under each probe can be determined. This flow is the weighted mean of the flows in the white and grey matter of that region.

At the end of each experiment rCBF was also determined by an autoradiographic technique utilizing C^{14}-antipyrine [1].

Cardiac output was determined by the cardiogreen indicator dilution technique. In the control animals repeated measurements were made of rCBF and cerebral metabolism over a period of 5-6 hr. In the experimental animals after 2 or 3 control determinations of rCBF and metabolism were obtained, the animal's mean arterial blood pressure was reduced to 55 mm Hg by bleeding and was maintained at that level for 90 min. while repeated measurements of rCBF and metabolism were performed. The mean arterial pressure was then lowered further to 35 mm Hg and maintained at that level for 90 min and measurements of rCBF and metabolism were obtained.

RESULTS

The hemodynamic changes in these experiments are shown in

Table 1. Mean arterial pressure was controlled at 56 ± 1 and 34 ± 1 mm Hg respectively during the two phases of bleeding.

Heart rate increased significantly from a control value of 122 ± 3 beats/min to 245 ± 18 beats/min during hemorrhage.

Hematocrit fell progressively from 39 ± 1% during the control period to 28 ± 4% at the end of the second period of bleeding. This difference was significant.

TABLE 1

Hemodynamic Changes During Hemorrhagic Shock

	Control Period	First Bleeding period	Second Bleeding Period
Mean Arterial Pressure mm Hg	121 ± 3	56 ± 1	34 ± 1
Heart Rate beats/min.	122 ± 3	227 ± 10	245 ± 18
Hematocrit %	39 ± 1	36 ± 2	28 ± 4
Cardiac Output ml/min/kg	137 ± 5	46 ± 11	34 ± 4
Total Peripheral Resistance mm Hg(ml/min)	0.070 ± .006	0.091 ± .016	0.072 ± .010
Cerebrovascular Resistance mm Hg/(ml/min/100 g)	3.09 ± 0.22	1.36 ± 0.20	1.09 ± 0.13
Mean Cerebral Blood Flow ml/min/100 g	38 ± 3	32 ± 5	28 ± 4

Values are means ± S.E.

Cardiac output was 137 ± 5 ml/min/kg during the control period and fell significantly to 34 ± 4 ml/min/kg during the second bleeding period.

Because of the fall in cardiac output concommitant with the

decrease in mean arterial blood pressure, total peripheral resistance did not change significantly from the control value of 0.070 ± 0.006 mm Hg/(ml/min).

A bleeding volume of approximately 22 ml/kg was necessary to reduce the animal's mean arterial blood pressure to 56 mm Hg. There was a tendency for some of this blood to be taken up again by the animal during the second bleeding phase but not a significant amount.

In the control animals mean CBF did not change significantly during the 5-6 hr of observation. A progressive reduction in mean CBF was observed during bleeding. In the second bleeding period CBF was reduced to 74% of control values.

Cerebral vascular resistance was significantly reduced during the first bleeding period from a control value of 3.09 ± 0.22 to 1.36 ± 0.20 mm Hg/(ml/min/100g). A further reduction to 1.09 ± 0.13 mm Hg/(ml/min/100g) occurred during the second bleeding period.

The metabolic changes during these experiments are shown in Table 2.

TABLE 2
Metabolic Changes During Hemorrhagic Shock

	Control Period	First Bleeding Period	Second Bleeding Period
Arterial PCO_2 mm Hg	33 ± 2	32 ± 3	33 ± 3
Arterial pH	7.42 ± .02	7.26 ± .05	7.06 ± .06
Arterial HCO_3 mEq/L	22.9 ± 0.6	15.6 ± 2.6	9.1 ± 0.8
Arterial lactate mM	1.29 ± 0.18	4.69 ± 1.60	12.84 ± 2.74
Arterial pyruvate mM	0.05 ± 0.02	0.16 ± 0.04	0.28 ± 0.03
Arterial glucose mg%	107 ± 13	259 ± 64	193 ± 49
CMR_{gl} mg/100gm/min	4.44 ± 0.30	4.47 ± 0.74	4.84 ± 0.95
CMR_{lac} μMoles/100gm/min	1.06 ± 3.10	3.78 ± 6.56	10.93 ± 6.98
CMR_{pyr} μMoles/100gm/min	1.58 ± 0.34	1.16 ± 0.80	0.34 ± 0.56
CMR_{o2} ml/100gm/min	3.93 ± 0.35	2.97 ± 0.23	2.64 ± 0.38

Values are means ± S.E.

Mean arterial PCO_2 was maintained between 32 ± 3 and 33 ± 3 mm Hg during the control and experimental observations by changing the respiratory rate. A metabolic acidosis developed during bleeding with arterial pH falling significantly from a control value of $7.42 \pm .02$ to $7.26 \pm .05$ at the end of the first bleeding period and to $7.06 \pm .06$ at the end of the second bleeding period.

Arterial bicarbonate concommitantly decreased significantly from a control value of 22.9 ± 0.6 mEq/L to 15.6 ± 2.6 and 9.1 ± 0.8 mEq/L at the end of the first and second bleeding periods respectively.

A significant lactic acidosis developed during hemorrhage with arterial lactate concentration rising from a control value of 1.29 ± 0.18 mM to 12.84 ± 2.74 mM at the end of the second bleeding period.

There was also a significant increase in arterial pyruvate concentration from $0.05 \pm .02$ to $0.28 \pm .03$ mM by the end of the bleeding periods.

In addition arterial glucose concentration increased during bleeding.

There was no significant change in cerebral glucose uptake (CMR_{gl}) which remained between 4.4 and 4.8 mg/100 g/min during the study.

Cerebral lactate output (CMR_{lac}) increased from 1.06 ± 3.10 to 10.93 ± 6.98 µMoles/100 g/min during hemorrhage but the increase did not reach the level of significance.

Cerebral pyruvate output (CMR_{pyr}) decreased from 1.58 ± 0.34 to 0.34 ± 0.56 µMoles/100 g/min during bleeding but this change was not significant.

Cerebral O_2 consumption fell significantly during the second bleeding period from a control value of 3.93 ± 0.35 to 2.64 ± 0.38 ml/100 g/min. If this reduction in $CMRO_2$ was due to ischemia secondary to the reduced CBF, one might expect to see a widening of the (A-V) O_2 across the brain. However this did not occur. The (A-V) O_2 remained at about 10 vol. % during the entire study.

This was due to the marked hemodilution which occurred during hemorrhage and the consequent fall in O_2 content of the arterial blood from 18.6 ± 0.8 to 12.2 ± 1.5 vol. %. Cerebral venous O_2 content fell to 2.6 ± 0.3 vol. % by the end of the second bleeding period so that the brain was extracting about as much of the available O_2 as it could.

In spite of this, cerebral venous PO_2 fell to only 21 ± 4 mm Hg at the end of the 2nd bleeding period, due to a shift of the O_2 disassociation curve to the right secondary to the metabolic acidosis present.

DISCUSSION

Judging from the hemodynamic changes observed in these studies a relatively great stress was produced by the hemorrhage in spite of the fact that the bleeding volume represented only about 25% of the animal's blood volume. A marked hemodilution was also present which undoubtedly tended to maintain the animal's circulatory blood volume if no significant number of RBC's were sequestered from the circulation.

When considering the metabolic results it is found that the data in the literature concerning the effects of hemorrhagic shock on cerebral O_2 consumption are contradictory. Beecher and Craig (2) found no changes in the O_2 uptake of the cerebral cortex in cats. Wilhelmi et al.(3) also noted no consistent changes in cerebral O_2 consumption in rats. Kovach et al.(4) obtained evidence from O_2 tension measurements suggesting that O_2 utilization was increased in the hypothalamus of dogs after bleeding.

The present studies reveal a significant reduction in overall $CMRO_2$ during hemorrhagic shock. Several factors may account for these differences among which must be considered species differences, the anesthetic agent used and the fact that arterial PCO_2 was controlled in the present studies. In the control studies, Sernylan did not reduce $CMRO_2$ significantly from values obtained in awake man. The fact that respiration was controlled prevented the hyperventilation that usually occurs with hemorrhagic shock. This therefore may have allowed a more severe cerebral acidosis to be present than would have occurred if respiration were not controlled. This point will be considered further below.

A further explanation for the low $CMRO_2$ observed could be an artifact due to an underestimation of the true mean flow calculated from the regional flow data. This could occur if the probes preferentially saw regions with low flow. The preliminary data from the autoradiographic rCBF method do show patchy regions of reduced cortical flow. However, there is no reason to suspect that these regions happened to be preferentially seen by the flow probes.

The level of cerebral venous PO_2 present in these studies is not usually associated with a reduction in $CMRO_2$ when produced by hyperventilation or by lowered arterial PO_2. Manifest tissue hypoxia has been considered to occur only when cerebral venous PO_2 falls below 20 mm Hg. The present results might be explained on the basis of inhomogeneous tissue perfusion as shown by the autoradiographic data. Local regions of critically reduced tissue PO_2 may be present producing a reduction in $CMRO_2$ in those areas. It might also be postulated that the metabolic acidosis produced during bleeding may have a detrimental effect on the energy metabolism of the brain as a whole so as to produce a decrease in cerebral metabolic rate at a relatively high tissue PO_2.

Although there was a significant reduction in CVR during bleeding, maximum cerebral vasodilation was not produced (5). This raises the question of whether factors tending to oppose vasodilation, such as increased sympathetic tone, were operative. There is evidence in the literature of hyperactivity of the sympathoadrenal system in shock (6). This coupled with the growing evidence that the sympathetic innervation of the cerebral vessels may play some role in the regulation of CBF gives this possibility some weight (7,8).

In summary we have obtained evidence that significant changes occur in cerebral hemodynamics and metabolism during hemorrhagic shock in the primate. The exact mechanisms underlying these changes must await further experimentation.

REFERENCES

1. Reivich et al. (1969). J. Appl. Physiol. 27: 296

2. Beecher, H.K. & Craig, F.N. (1943). J. Biol Chem. 148:383

3. Wilhelmi, A.E., Russell, J.A., Long, C.N.H. & Engel, M.G. (1945). Am. J. Physiol. 144:683.

4. Kovach, A.G.B., Mitsanyi, A. & Stekiel, W. (1965). Acta Physiol. Acad. Sci. Hung. 26: (Suppl.) 35.

5. Reivich, M. (1964). Am. J. Physiol. 206:25.

6. Fedina, L., Kovach, A.G.B. & Vanick, M. (1965). Acta Physiol. Acad. Sci. Hung. 26: (Suppl.) 35.

7. James, I.M., Millar, R.A. & Purves, M.J. (1969). Circu. Res. 25:77.

8. Meyer, J.S., Yoshida, K. & Sakamoto, K. (1967). Neurol. 17:638

Acknowledgement

These studies were performed with the excellent technical assistance of Ms. Elaine Shipko and Ms. Sue Pierce.

This work was supported by the John A. Hartford Foundation.

ADRENOCORTICAL BLOOD SUPPLY DURING HEMORRHAGE IN DOGS

E. Monos, A.G.B. Kovách, Z. Áprili, Zs. Suba, and

L. Cserényi

Experimental Research Department, Semmelweis Medical

University, Budapest, Hungary

In recent years we have found that venous outflow from the adrenal gland did not reflect reliably the effective blood supply of cortical and medullary tissues in dogs (Monos et al. 1969,1970; Sulyok et al.1970). The venous outflow was several times higher than the effective flow, and their proportion could be shifted by such interferences as venous stagnation, or administration of pituitary hormones. Figure 1 is meant to demonstrate quantitatively some of these results. As it is indicated, at control state the mean adrenal venous outflow was more than 4 ml/min/g, but the cortical and medullary blood supply determined by H_2-desaturation technique was below 1 ml/min/g. After artificial obstruction of the vein, while total outflow decreased to 25 % of the initial value, the decrease in cortical and medullary flow amounted 40-50 % only.

To understand the hormone producing capacity of the adrenal gland at various degrees of hemorrhage, it is important to have more information on its blood supply. While there are several studies on total flow of the gland in similar pathological state (Sapirstein et al. 1960; Monos et al. 1962, 1965; Stark et al. 1965), changes in cortical blood supply have not been examined so far. Therefore it seemed rewarding to investigate this problem.

Methods. The experiments were performed on 15 dogs under Nembutal(30 mg/kg b.w.) anesthesia. For estimation of the effective blood supply of the adrenal cortex, local H_2-desaturation technique as described by Aukland et al. (1964) was applied. Simultaneously the venous outflow reflecting the total blood flow of the gland was measured by a drop-counter device via canulation of the

Fig. 1.

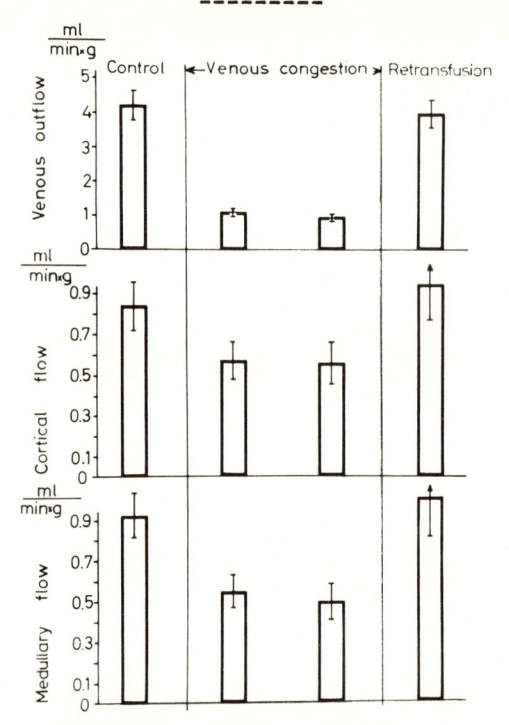

Effect of venous congestion on total, cortical and medullary blood supply to the adrenal gland (data derived from 11 experiments on dogs). Duration of congestion was 1 hour, during which two measurements were performed. Means and S.E.M.

lumboadrenal vein. Placing of the platinum electrode and of the venous canula is shown in Fig. 2. The arterial blood pressure was measured continuously throughout the experiments. After various degrees of bleeding, the arterial pressure was stabilized at 100, 75 and 50 mmHg levels by a pressure-bottle set, while each hypotensive period lasted for about half an hour.

Results. In accordance with our earlier experiments, there was a marked difference between total and cortical blood flow in the normotensive (about 120 mmHg) state (Fig.3), the total flow per unit tissue wt being about four times the cortical flow. This observation reflects the presence of functioning shunts of considerable size in the adrenal gland. Relevant data (Sapirstein et al. 1960; Monos et al. 1962; 1965; Stark et al. 1965) indicate, that after a mild hemorrhage total blood flow of the adrenals does not decrease, but in some cases it even increases.

Fig. 2.

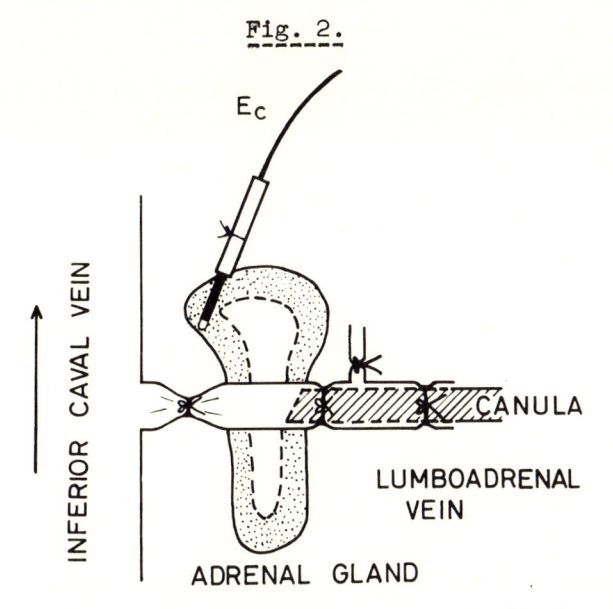

Position of adrenocortical platinum electrode (E_c) and of the canula in the lumbo-adrenal vein.

Fig. 3.

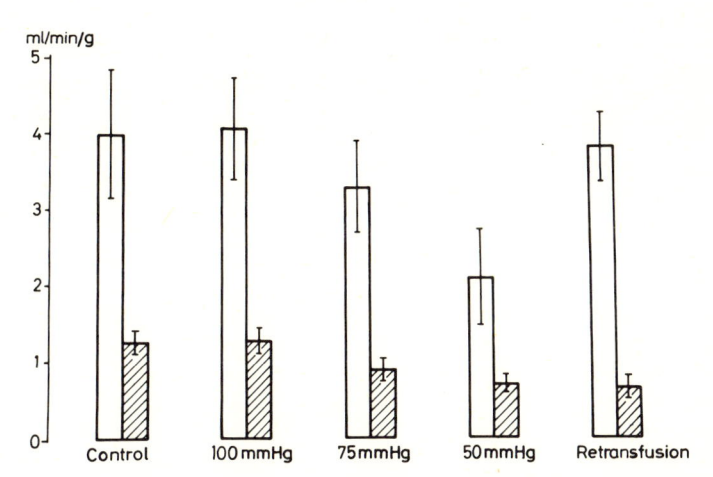

Effect of graded hemorrhage on total (white columns) and cortical (shaded columns) blood flow of the adrenal gland. Means and S.E.M.

These present experiments with bleeding to a 100 mmHg pressure level yielded similar results. The venous outflow at 120, and 100 mmHg arterial pressure levels were 3,98 (S.E.M.:+o,86), and 4,06 (S.E.M.:+0,71) ml/min/g adrenal weight respectively. Under these conditions also the adrenocortical blood supply remained at the same level as in the control state (Fig.3). The related mean adrenocortical local flow values were 1,22 (S.E.M.:+0,15), and 1,27 (S.E.M.:+0,17) ml/min/g tissue weight. After bleeding to 75, or 50 mmHg, both blood flow parameters decreased nearly parallel with the arterial blood pressure. There was no further autoregulation. Reinfusion of the shed blood (about 100 mmHg arterial blood pressure level) enhanced only the venous outflow, but not the local cortical blood supply. Consequently the extracortical fraction of the adrenal blood flow, i.e. the difference between the white and shaded columns on Figure 3, increased significantly.

The calculated vascular resistance values of the adrenal gland before and after hemorrhage are shown on Fig.4. Mild hemorrhage to 100 mmHg arterial blood pressure level elicited a maximal decrease in both total and cortical resistance, indicating a statistically significant change in both variables. After reinfusion the resistance to flow in the cortical vessels reached a very high level, while the total resistance of the gland did not change significantly.

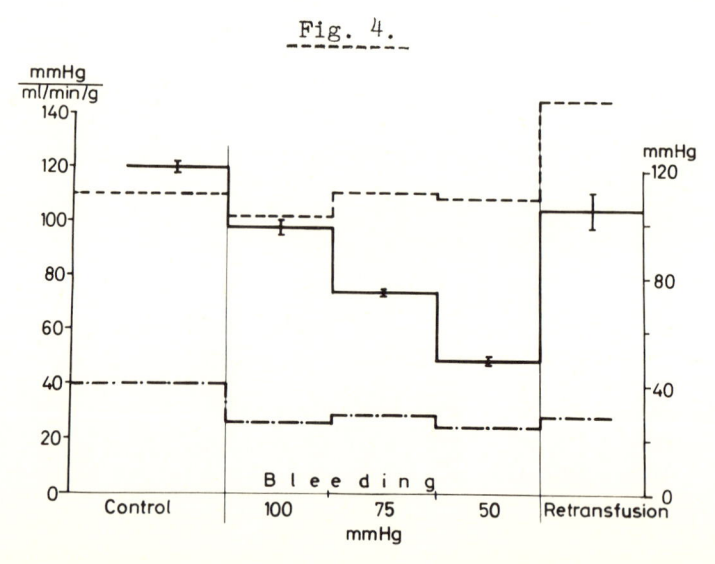

Fig. 4.

Changes in cortical (broken line) and total adrenal resistance to flow (dotted line) during graded hemorrhagic hypotension. Solid line: arterial pressure. Means and S.E.M.

Fig. 5.

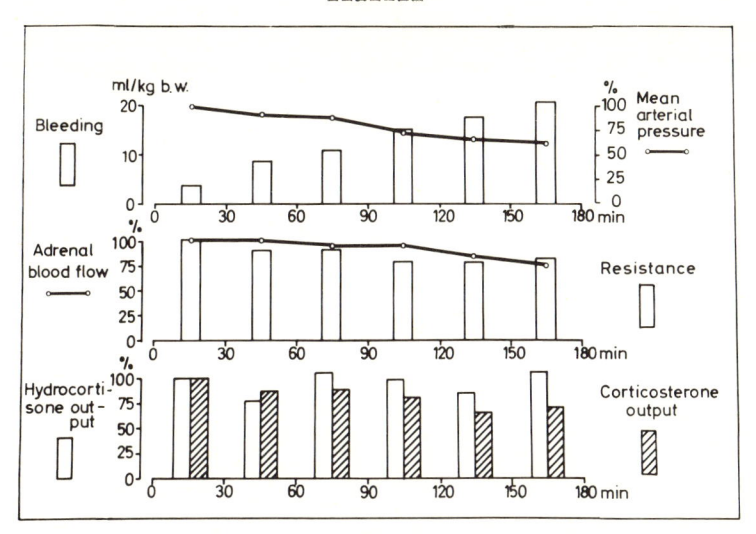

Effect of continuous blood loss on arterial pressure, adrenal venous flow and corticoid output (mean values derived from 16 experiments on dogs).

Conclusions. It is indicated, that a mild hemorrhage to 100 mmHg arterial blood pressure reduced neither total nor cortical blood flow, consequently the blood supply of the hormone producing tissues could be well maintained in this state. This finding is supported by results obtained from earlier experiments (Monos et al. 1962). As it can be seen in Fig.5., during continuous blood loss to a 20 ml/kg body weight level, with arterial blood pressure decreasing by 35 per cent, total glycocorticoid output did not change significantly.

Reinfusion after a deeper and long-lasting hypotension elicited a large increase in cortical vascular resistance which indicates a substantial deterioration of cortical blood supply. This latter phenomenon was not reflected on the venous outflow. Presumedly, hemodynamic changes before or during reinfusion can led to plugging the small vessels in the adrenal cortex.

It is concluded, that the comparative evaluation of metabolic and blood flow relations in the adrenals under various conditions should not be attempted without measuring the effective blood supply. Cortical and total adrenal blood flow can react differently from each other after changes in arterial blood pressure.

REFERENCES

Aukland, K., Bower, B.F. & Berliner, R.W. (1964) Circ. Res. 14:164.

Monos, E., Biró, Zs. & Kovách, A.G.B. (1970) Acta physiol. Acad. Sci. hung. 36:379.

Monos, E., Biró, Zs., Sulyok, A. & Kovách, A.G.B. (1969) Acta physiol. Acad. Sci. hung. 35:321.

Monos, E., Kovách, A.G.B. & Koltay, E. (1962) Proc. Union physiol. sci. 22. International Congress, Leyden. Excerpta med. Vol. 2 No. 222.

Monos, E., Szatlóczky, E., Koltay, E., Varga, E. & Kovách, A.G.B. (1965) In: A shock élettana és klinikuma. Symposium. Budapest.

Sapirstein, L.A., Sapirstein, E.H. & Bredemeyer, A. (1960) Circ. Res. 8:135.

Stark, E., Varga, B., Ács, Zs. & Papp, M. (1965) Pflügers Arch. 285:296.

Sulyok, A., Monos, E., Biró, Zs., Sándor, P. & Kovách, A.G.B. (1970) In: The 34th Annual Conference of the Hungarian Physiological Society. Ed. K. Lissák. Akadémiai Kiadó, Budapest. p. 114.

MICROCIRCULATION AND HYPOXIA

D. W. Lübbers

Max-Planck-Institut für Arbeitsphysiologie

Dortmund, West Germany

Microcirculation brings about the convective transport of O_2 through the capillary network of the tissue. From the capillaries into the tissue O_2 is transported mainly by diffusion. Within the tissue O_2 reacts with cytochrome oxidase which is part of the respiratory chain. The respiratory chain is situated in the mitochondria. Consuming O_2 and substrate H_2, it produces ATP-energy. Aerobically 1 mol glucose will produce 2 + 36 = 38mol ATP, but anaerobically only 2 mol ATP. The energy production in warm-blooded animals depends largely on aerobic energy production. Therefore, in most organs a lack of O_2 produces quickly an energy deficiency.

In the following we will discuss in which way the convective transport, the diffusive transport and the chemical reaction with cytochrome oxidase contribute to the development of a lack of O_2 in the tissue.

1) Influence of Chemical Reactions of O_2 on Tissue Hypoxia.

We begin with the reaction between O_2 and cytochrome oxidase. This reaction has been studied by measuring the dependence of the O_2 consumption of isolated mitochondria on the oxygen pressure of the incubation medium. All authors (1,2,3,6,7,10,11,13,14,17,29) found 2 phases: in the first phase the O_2 consumption is independent of the O_2 pressure (zero order reaction). In the second phase the decreasing Po_2 slows down respiration, i.e. the availability of O_2 limits the respiration. The point of transition from the first to the second phase is called critical mitochondrial O_2 pressure. In the literature different critical Po_2 values are reported. Mostly these values lie between 0.5 and 12 Torr (2,3, 24). Only Chance et al., found extremely low values of about

0.01 Torr Po$_2$ and they used a luminescence method for measuring such small amounts of Po$_2$.

Our measurements with the platinum electrode reveal a major difficulty: the respiration of isolated mitochondria is not sufficiently constant to use the decrease of respiration as a criterion of the critical O$_2$ pressure. It depends on the substrate, but with the same substrate respiration might first increase and later decrease or vice versa. We could not predict this behaviour (27). By means of an automatic data recording system we measured together with Piroth & Starlinger the Po$_2$ decrease and calculated the momentary respiration by a computer. In 80 per cent of all curves there was a statistically normal distribution of the respiration values; in the rest there was sometimes a distribution with two peaks and sometimes statistical analysis was not possible (22). The instability of the respiration makes it somehow arbitrary to define a critical O$_2$ pressure only by measuring respiration.

If the O$_2$ pressure is below 1 Torr, then the platinum electrode works at the border of its sensitivity (see 15). Especially good multi-wire Pt electrodes are able to detect Po$_2$ changes of 0.1 to 0.05 Torr (27), but the electrodes very often drift about as much, thereby disturbing exact kinetic measurements. Furthermore, the time responses of these electrodes are slow compared with the change of oxygen tension during transition from phase one to phase two. From all these experiments we have to conclude that respiration measurements, and especially those with the platinum electrode, are not the right ones to measure the critical O$_2$ tension in isolated mitochondria. To analyze the critical O$_2$ pressure of isolated mitochondria the redox state of the respiratory chain, for example of cytochrome oxidase and cytochrome c, has to be measured. In recent preliminary experiments with Starlinger, we found the same low critical Po$_2$ values as Chance et al. have reported.

The critical O$_2$ tension of the tissue has also been determined by measuring the decrease of O$_2$ tension in the tissue itself. If the flow of a hemoglobin-free perfused heart is suddenly stopped a linear decrease of the O$_2$ pressure is first observed, at least in 50% of the experiments. The bending of the curves starts between 30 and 40 Torr, indicating that myoglobin releases its chemically bound O$_2$. Let us assume that tissue respiration is constant enough at least until a Po$_2$ of 10 Torr, then, knowing the dissociation curve of myoglobin and its O$_2$ binding capacity, it is possible to calculate the myoglobin concentration in the heart (4,5). Knowledge of the myoglobin concentration allows us to calculate the Po$_2$ decrease without myoglobin. One observes that between 3 and 5 Torr the respiration slows down which is indicated as an apparent increase in myoglobin concentration. Corrected for the time response of the electrode the values are 1 to 2 Torr. By means of a rapid spectrophotometer we measured simultaneously the de-oxygenation of myoglobin and the reduction of cytochrome oxidase. When myoglobin was reduced by 50% (corresponding to a Po_2 of 3 Torr) the cytochrome oxidase was reduced to about 20% (19).

All these tissue measurements gave a critical O_2 pressure of 1 to 5 Torr. The limit of these measurements in tissue is that they integrate over a certain tissue area, for our Po_2 measurements over an area of 50 μm^2, for the photometric measurements over an area of 4 mm^2. The actual value, therefore, must be lower than the measured one. I think that these values agree sufficiently well with the value of 0.01 Torr found in isolated mitochondria, so it is reasonable to assume that the mitochondria in situ have the same critical Po_2 as those in vitro.

2) The Influence of the Diffusive Transport of O_2 on the Development of Tissue Hypoxia.

In tissue O_2 is transported mainly by diffusion. It is known that for consciousness in man a cerebral venous Po_2 of 18 to 19 Torr is necessary (2,20). This is independent of the amount of O_2 which is available in blood, for example at different hemoglobin concentrations. The Po_2 of 18 to 19 Torr is the critical O_2 transport pressure of the brain which is necessary for the O_2 transport by diffusion. Taking into account only the diffusion perpendicular to the capillary wall, the decrease in O_2 pressure depends on respiration rate, capillary distance and the diffusion properties of the tissue as described by the Krogh diffusion equation (18, 26). Around the capillaries an oxygen tension field exists in which all Po_2 values from the arterial to the lowest tissue values are found. Such an O_2 tension field can be measured with Pt microelectrodes (12, 28). Fig. 1 shows the O_2 tension changes measured with a Pt microelectrode of diameter 1 - 2 μm introduced perpendicularly into the brain cortex. The values agree with predictions of the diffusion theory. Our experiments showed that at present a Po_2 histogram describes best the O_2 supply to tissue.

Fig 2 shows the Po_2 histogram of a normally supplied brain. Lowering of the O_2 supply increases the proportion of low Po_2 values, so that the maximum moves to the left. If the respiration increases or the arterial Po_2 decreases, O_2 lack may develop. Regarding the Po_2 transport by diffusion, it is characteristic that hypoxia of different degrees begins always in small spots and not in the whole tissue.

Although the Po_2 gradients flatten towards the border of the supply area, the border-line between normoxia and hypoxia is rather sharp. Using a parallel cylinder model for the capillary

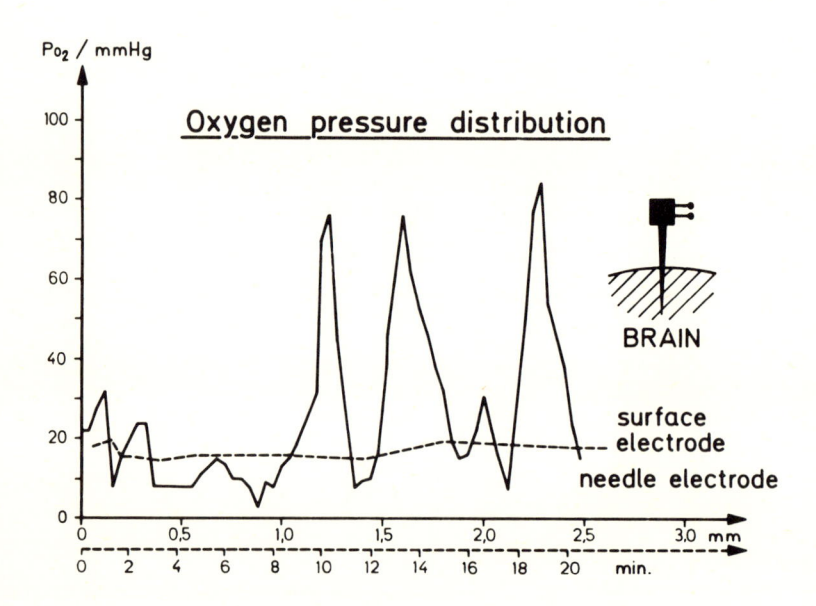

Fig. 1. Local Tissue Po$_2$ in the Brain Cortex (Cat)

Abscissa: solid line - depth of puncture; broken line - time in minutes; ordinate - local O$_2$ tension in Torr;
solid line - needle electrode. The perpendicularly puncturing electrode measures the local Po$_2$ in the brain. The local Po$_2$ changes depend on the distance from the capillaries. One observes some very low Po$_2$ values between the capillaries and high Po$_2$ values close to the arterial Po$_2$. Broken line - Po$_2$ control measurements by means of a Po$_2$ surface electrode. The Po$_2$ remains rather constant during the measurement. Insert - the puncturing electrode penetrating perpendiculary the brain cortex.

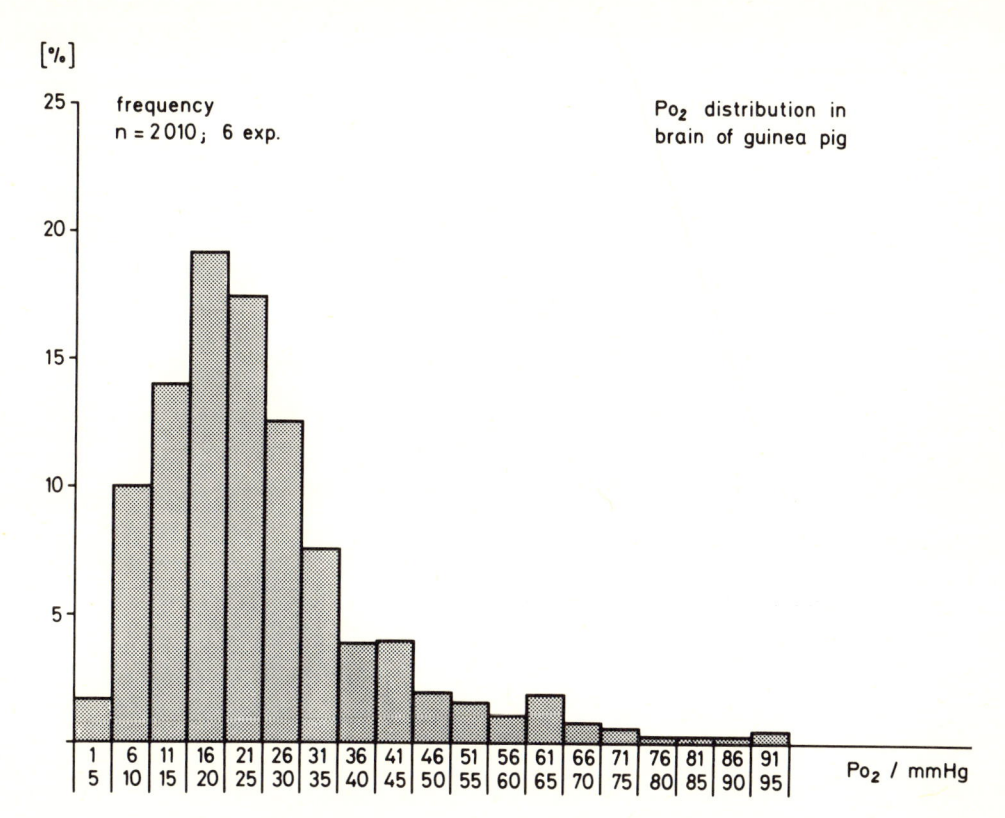

Fig. 2. Frequency Distribution of the Local Tissue Po_2 (Guinea-pig).

Abscissa: Po_2 classes in Torr; ordinate: frequency percentage; total number of measurements: 2010, 6 different experiments. The frequency distribution corresponds to a brain cortex normally supplied with O_2. In hypoxia the proportion of low Po_2 values would increase.

network where $pr_{1,2} = P_{O_2}$ at distances r_1 or r_2 from the centre, R_z = radius of the cylinder; $k = O_2$ conductivity; A = respiration; $a = r_1/R_z$, the P_{O_2} difference between 2 points r_1 and r_2 from the centre is

$$pr_1 - pr_2 = \frac{A}{2k} R_z^2 \ln \frac{r_2}{r_1} + \frac{r_1^2 - r_2^2}{2}$$

If $r_2 = R_z$, then

$$pr_1 - pR_z = \frac{A}{2k} \left[R_z^2 \ln \frac{R_z}{r_1} + \frac{r_1^2 - R_z^2}{2} \right]$$

Choosing the r_1 as a fraction of R_z we obtain with

$$r_1 = aR_z$$

$$pr_1 - pR_z = \frac{A}{2k} R_z^2 \left[-\ln a - \frac{(a^2 - 1)}{2} \right]$$

Using the approximation $\ln a = (a - 1) - \frac{1}{2}(a - 1)^2$ (for $0 < a \leq 2$) we can estimate the P_{O_2} gradient at the border of the cylinder by means of the following simple formula:

$$p_1 - pR_z = \frac{A}{k} (a - 1) R_z^2$$

For example, taking $A = 0.1$ cm^3 O_2/cm^3/min; $k = 2.7 \times 10^{-5}$ cm^3 O_2/cm/min/atm and $R_z = 3 \times 10^{-3}$ cm – which is about correct for the brain – we find for an $a = 0.9$ about $pr_1 - pR_z = 2.5$ Torr. This means that a P_{O_2} decrease of 2.5 Torr takes place in the last 3 μm (= 10% of the total radius). When $a = 0.99$, i.e. along the last 1/100 of the radius, the decrease is 0.25 Torr.

If in our example the P_{O_2} at the border of the cylinder is zero (= anoxia), then normoxia begins at a distance of only about 0.03 μm towards the center. The change of respiration, A, and the factor, a, affect ΔP_{O_2} linearly and that of the size of the supply area (R_z) according to a square-law.

In organs with myoglobin facilitated diffusion (23,25,30) may increase the supply area, since at low tissue P_{O_2} O_2 can diffuse directly not only as dissolved O_2, but also as O_2 reversibly bound to myoglobin. Longmuir thinks that cytochrome P450 could possibly have a similar function in the liver. The physiological importance of facilitated diffusion for the development of hypoxia has not been experimentally clarified.

3. The Influence of the Convective Transport of O_2 on the Development of Tissue Hypoxia.

The capillary network is more complicated than the earlier diffusion calculations assumed. Krogh supposed that in muscle all capillaries run parallel and are parallelly perfused (cylinder model) (12). Diemer assumed the existence of countercurrent systems (cone model) (17). In reality mixed systems are mostly found. The calculation of such a mixed system has been carried out by Grunewald (8,9). Together with Grunewald we were able to show that there exist asymmetric structures which allow an optimum O_2 supply. Under the same conditions in the countercurrent system an area of hypoxia exists, but not in the asymmetric system (16) (Fig. 3)

Therefore, we have to conclude that the three-dimensional pattern of the capillary network influences tissue O_2 transport

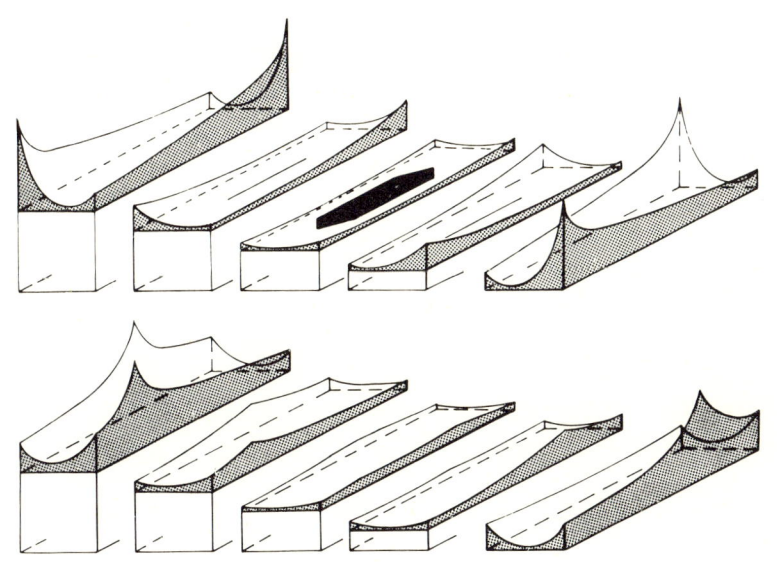

Fig. 3. Calculated Po_2 in the Tissue for Symmetric and Asymmetric Capillary Pattern.

The tissue block is cut into 5 slices. Over each section its O_2 profile is drawn. The capillaries run along the four edges of the tissue block. In the upper part the capillaries run parallel, but the flow direction is antiparallel. Arterial and venous ends are situated opposite (countercurrent system). In the lower part the arterial inflows (visible as peak in the Po_2 profile) are arranged asymmetrically. Po_2 values below 1 Torr are marked in black. Under the same supply conditions the lower structure is more effective with regard to O_2 supply, since black areas exist only in the pure countercurrent system.

very much. In addition to the geometrical pattern, the direction of flow is of utmost importance. Hypoxia can develop, either when the geometrical pattern is disturbed, for example by closing a capillary, or simply when the flow direction changes. Closing a capillary is especially efficient, since the Po_2 decrease is proportional to the square of the distance.

The pure countercurrent system can have an important disadvantage. If, with regard to O_2 consumption in tissue, the Po_2 gradient between the arterial and the venous part of the capillary is large, then O_2 can reach the vein directly, bypassing the tissue (O_2 shunt diffusion). Large gradients can be avoided if the necessary O_2 transport capacity of the perfusion medium is reached at relatively low Po_2 values. Here hemoglobin plays an important role. According to Henry's Law the amount of O_2 in a solution is given by:

$$u_{O_2} = \alpha \cdot Po_2.$$

The slope of the straight line,

$$\alpha (= u_{O_2}/Po_2)$$

is the solubility coefficient. The O_2 binding curve of hemoglobin relates the amount of O_2 which is chemically bound and physically dissolved to the O_2 pressure of the solution or the blood. Its slope expresses the momentary O_2 capacity of the solution which can be defined as the apparent solubility coefficient, $\alpha^+(21)$, i.e.

$$\alpha^+ = 7.6 du_{O_2}/dPo_2 \quad (\text{units ml } O_2/\text{ml/atm})$$

where u_{O_2} is in vol%.

For example human blood has an O_2 transport capacity which is dependent on the Po_2 range; it is about 30 to 150 times larger than that for serum which is only 0.023 ml O_2/ml/atm at 37°. In the Po_2 range of 0 to 25 Torr α^+ for blood is 1.7-3.5. It rises between 30 and 65 Torr Po_2 to 3.7 and decreases at higher Po_2 values (between 65 and 80 Torr Po_2 α^+ is 3.2-2.0, between 80 and 90 Torr Po_2 is 1.5-1.0, between 90 and 95 Torr Po_2 is 0.63). Only this high O_2 transport capacity of hemoglobin allows counter-current systems of capillaries without too much O_2 shunt diffusion.

Hemoglobin contributes to the O_2 supply in other forms also.
1) The large O_2 capacity of blood allows a relatively small blood flow, and so economizes the O_2 supply.

2) The large O_2 capacity of hemoglobin is usable in a Po_2 range which is perfectly adapted to the atmospheric Po_2: The loading of the blood with O_2 in the lung uses the atmospheric pressure as energy source.

3) The O_2 binding curve of hemoglobin has such a form that it provides the necessary transport Po_2 at a relatively high O_2 capacity, so that the gradient necessary for Po_2 transport into tissues is maintained, even though O_2 is continuously delivered to the tissue.

4) The O_2 binding curve of myoglobin has the form of a hyperbola. It is half saturated at a Po_2 of about 3 Torr, at which the hemoglobin has lost more than 95% of its O_2. So the oxygenated myoglobin is a perfect sink for the O_2 which is delivered by the hemoglobin. It maintains the gradient during the resaturation phase and therefore the O_2 flow into the tissue.

5) At the low critical mitochondrial Po_2 hemoglobin is almost completely desaturated. Therefore, the total O_2 tension is usable for the intercapillary O_2 transport.

The analysis has shown the interdependence of the different mechanisms which contribute to the O_2 supply to tissue. The chemical reaction of O_2 with cytochrome oxidase proceeds at an extremely low O_2 pressure, so the total O_2 pressure in the capillary can be used for diffusive O_2 transport in the tissue. The O_2 transport capacity of the blood due to hemoglobin guarantees maintenance of the O_2 gradient, even though O_2 is continuously delivered to the tissue. This mechanism can be improved by myoglobin. The pattern of the capillary network influences strongly the Po_2 distribution in the capillaries and consequently in the tissue. There are capillary patterns which specially are suitable for O_2 transport, but at the moment we have no systematical knowledge about the occurrence or formation of such structures.

REFERENCES

1. Bänder, A. & Kiese, M. (1955) Naunyn Schmiedebergs Arch. 224, 312.
2. Chance, B. & Pring, M. (1968) In: Biochemie des Sauerstoffs, ed. by B. Hess and Hj. Staudinger, Springer, Berlin--Heidelberg, New York, p. 102
3. Chance, B., Schoener, B. & Schindler, F. (1964) In: Oxygen in the Animal Organism, ed. by F. Dickens & E. Neil, Pergamon Press, London, p. 367.
4. Fabel, H., Lübbers, D.W. & Rybak, B. (1964) Pflugers Arch. 279, R 32.

5. Follert, E. & Lübbers, D.W. (1972) In: Oxygen Supply. Theoretical and Practical Aspects of Oxygen Supply and Microcirculation of Tissue, ed. by M. Kessler, D.F. Bruley, L.C. Clark Jr., D.W. Lübbers, I.A. Silver & J. Strauss, Urban & Schwarzenberg, Munich and Berlin, University Park Press, Baltimore (in print).
6. Frimmer, M., Hegner, D. & Winkelmann, W. (1963) Klin. Wschr. 41, 715.
7. Greven, K., Gaebell, H. & Schuster, J. Pflügers Arch. 279, 214.
8. Grunewald, W. (1968) In: Oxygen Transport in Blood and Tissue ed. by D.W. Lübbers, Thieme, Stuttgart, p.100.
9. Grunewald, W. (1971) Bedeutung der Kapillarstrukturen fur die O_2-Bersorgung der Organe und ihre Analyse anhand digital simulierter Modelle Habil. Schrift Bochum.
10. Hegner, D. & Glossmann, H.Z. (1965) Z. Naturforsch 20 B, 234.
11. Kessler, M. & Lübbers, D.W. (1964) Pflügers Arch. 281, 50.
12. Krogh, A. (1919) J. Physiol. (London) 52, 391.
13. Longmuir, J.S. (1954) Biochem.J. 57, 81.
14. Longmuir, J.S. (1964) In: Oxygen in the Animal Organism, ed. by F. Dickens & E. Neil, Pergamon Press, London, p. 219.
15. Lübbers, D.W. (1966) In: A Symposium on Oxygen Measurements in Blood and Tissue and their Significance, ed. by J.P. Payne & D.W. Hill, Churchill, London, p. 103.
16. Lübbers, D.W. (1967) In: Marburger Jahrbuch 1966/67, ed by C.G. Wendt Elwert, Marburg, p.305.
17. Lübbers, D.W. (1968) In: Biochemie des Sauerstoffs, ed. by B. Hess & Hj.Staudinger, Springer, Berlin, p.67.
18. Lübbers, D.W. (1969) In: Oxygen Pressure Recording in Gases, Fluids and Tissues, ed. by F. Kreuzer & H. Herzog, Karger, Basel, p.112.
19. Lübbers, D.W. & Fabel, H. (1964) Verh.Dtsch.Ges.Kreislaufforsch, 30, 198.
20. Opitz, E. & Schneider, M. (1950) Ergebn. Physiol. 46, 126.
21. Opitz, E.&Thews, G. (1952) Arch.Kreislaufforsch.18,137.
22. Piroth, D. (1970) Probleme und Moglichkeiten der digitalen Datenerfassung bei der Analyse biologischer Reaktionskinetiken am Beispiel der Bestimmung des kritischen Sauerstoffdruckes. Dissertation, Marburg/Lahn.
23. Roughton, F.J.W. (1932) Proc.Roy.Soc. (London) B,111, 1.
24. Schindler, F.J. (1964) Oxygen Kinetics in the Cytochrome Oxidase - Oxygen Reaction Dissertation Philadelphia.
25. Scholander, P.F. (1960) Science 131, 585.
26. Silver, I.A. (1966) In: A Symposium on Oxygen Measurements in Blood and Tissues and their Significance, ed. by J.P. Payne & D.W. Hill, Churchill, London, p. 135.
27. Starlinger, H. & Lübbers, D.W. - Unpublished results.
28. Thews, G. (1968) In: Oxygen Transport in Blood and Tissue ed. by D.W. Lübbers, Thieme, Stuttgart, p.1.

29. Weiss, Chr. (1968) In: Oxygen Transport in Blood and Tissue, ed. by D.W. Lübbers, U.C. Luft, G. Thews & E. Witzleb. Tissue, Thieme, Stuttgart, p. 227.
30. Wittenberg, J.B. (1966) J. Biol. Chem. 241, 104.

THE MICROCIRCULATION OF NORMAL AND INJURED TISSUE

Paul C. Johnson

Department of Physiology, University of Arizona College of Medicine, Tucson, Arizona 85724

The response of the microcirculation after tissue injury is recognized as a critical link in the chain of events which leads to tissue repair. However, tissue injury is but one of a number of perturbations with which the microcirculation must cope if the organism is to survive. General or localized changes in arterial and venous pressure for example constitute another potentially life-threatening situation. Appropriate homeostatic mechanisms exist to maintain a suitably low capillary pressure in the liver and intestine where the capillary filtration coefficient is high (i.e. 20-40 times greater than skeletal muscle) during venous pressure elevation in that region. Adequate flow must be maintained to vital organs during localized or systemic hypotension. A variety of mechanisms exist in the microcirculation to maintain constancy of capillary flow and pressure under such circumstances. Local regulation of blood flow must also be linked to the metabolic requirement of the tissue.

In the past 20 years there has been an explosion of knowledge regarding the hemodynamics of the peripheral circulation, made possible by improved techniques for pressure and flow measurement in the large vessels. This information has led to a better understanding of peripheral circulatory function. It has also led to a heightened curiosity regarding the behaviour of the individual microcirculatory vessels. Stimulated by this interest, a number of laboratories are now heavily engaged in quantitative studies of the microcirculation. Ultimately these studies should provide the basis for an improved understanding of normal and inflammatory reactions of the microcirculation.

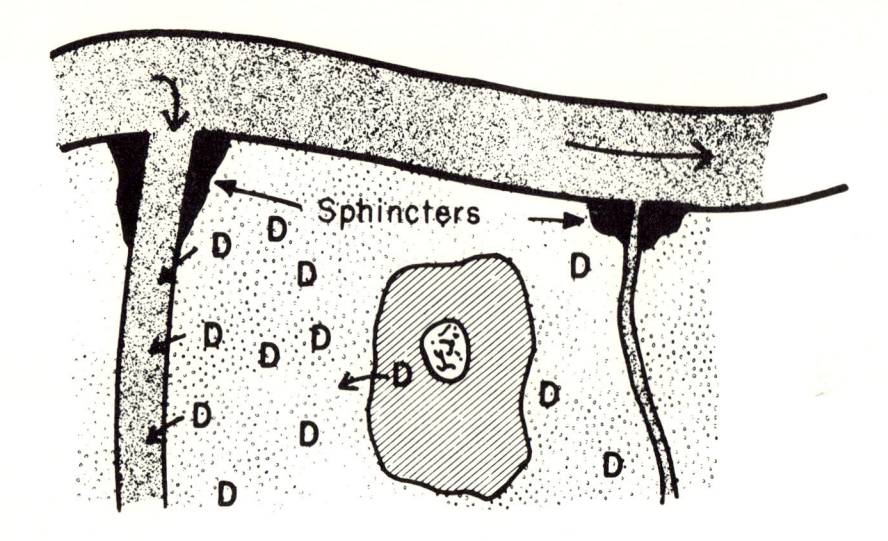

Fig 1 Local regulation of blood flow at the capillary level.
Periodic flow is associated with sequential accumulation and
washout of vasodilator metabolites. (From Winsor, T. &
Hyman, C: A Primer of Peripheral Vascular Diseases.
Philadelphia: Lea & Febiger, p. 67, Fig 6-2. Reproduced by
permission).

    The classic description of local regulation of blood flow
by August Krogh (1) envisioned the individual capillaries as
supplying a restricted cylinder of tissue around the capillary.
Further, he postulated that the individual capillaries would open
and close in accordance with local metabolic requirements. Under
resting conditions anaerobic metabolites were believed to
accumulate in the vicinity of a closed capillary causing the
vessel to open. The resulting flow would wash out the vasodilator
metabolites causing the vessel to close. In the present day
adaptation of this model (Fig 1) the metabolites are believed to
act upon the precapillary sphincter, since Krogh's suggestion of
capillary contractility is no longer accepted. During periods
of increased metabolism all capillaries are believed to be open
and flowing continuously, providing a greater surface area for

exchange, a shorter average tissue pathway for diffusion, and an increased rate of O_2 and nutrient delivery by the blood stream.

If the Krogh hypothesis is correct we can also expect increased numbers of open capillaries in reactive hyperemia as well, since all capillaries should be engulfed by a rising tide of anaerobic metabolites during the period of ischemia. Alternatively if the Krogh hypothesis is not correct and capillaries are flowing continuously, then we would expect a large augmentation of flow in individual capillaries during reactive hyperemia.

We have performed experiments aimed at testing the Krogh hypothesis for reactive hyperemia. Specifically, we studied the red cell velocity in individual capillaries of skeletal muscle during reactive hyperemia. The results (2) do not support the Krogh hypothesis.

Methods

Red cell velocity was measured in capillaries of the perfused cat sartorius muscle. The cats were anesthetized with α-chloralose (75 mg/Kg). The muscle was surgically isolated and removed with care taken to preserve all branches of the femoral artery and vein to the muscle and to avoid tissue trauma. The muscle was continuously superfused with warm Ringer's solution during surgery. The artery was cannulated and connected to the contralateral femoral artery by polyethylene tubing. The vein was also cannulated and the venous drainage from the muscle returned to the jugular vein through a photoelectric drop counter.

The muscle was mounted on the heated stage of a Leitz Panphot microscope. The tissue was moistened with Ringer's solution and covered with Saran Wrap. The muscle was stretched to its approximate *in vivo* dimensions on a rigid frame with adjustable clamps. The red cell velocity was measured by the dual slit photometric method of Wayland and Johnson (3). This method uses twin phototubes which measure light intensity at two positions along the capillary. As red cells pass through the vessel, signals are produced in the phototubes which are similar but displaced in time. The time difference depends upon the distance between the sensors and the velocity of the cells. The distance being known, velocity can be calculated continuously as the reciprocal of the time difference. In our studies red cell velocity, gross flow and arterial and venous pressure were recorded continuously.

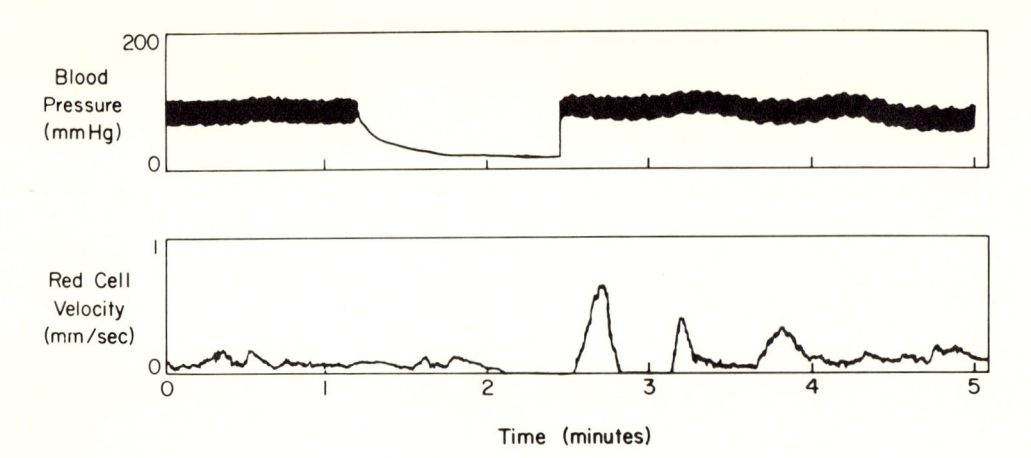

Fig 2 Capillary flow associated with short period of arterial inflow occlusion. The under-damped response seen after release in this capillary was observed in 37% of the sartorius muscle capillaries studied.

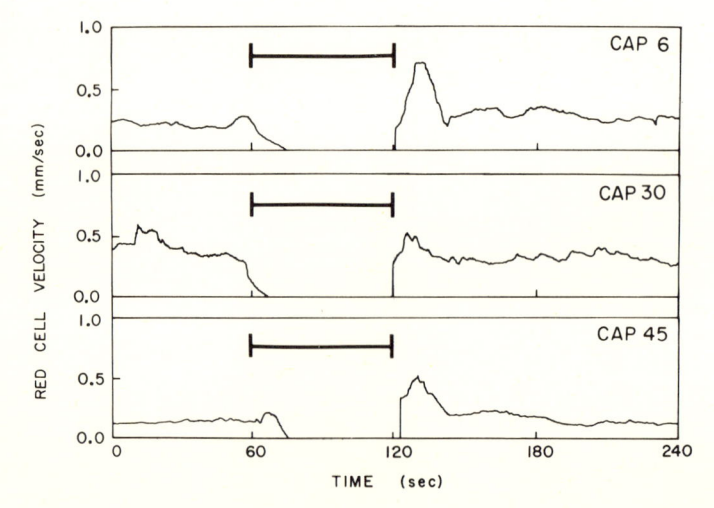

Fig 3 Red cell velocity in three capillaries associated with a one minute occlusion. This type of response appears to be critically damped and most nearly resembles the gross flow pattern. Arterial inflow occlusion indicated by the heavy horizontal bar. (From Burton, K.S. & Johnson, P.C. Am. J. Physiol. in press. Reproduced by permission).

RESULTS

We were able to identify 4 separate patterns of flow behaviour in individual capillaries. The pattern most frequently observed (37% of the total) consisted of a rapid rise to peak flow, a precipitous drop to zero and a slower return to control (Fig 2). Sometimes the on-off pattern was repeated 1 to 2 times before steady flow was restored. The peak velocity was typically about 4 times greater than control in this group. Flow stopped on the average for about 7 sec. This stoppage occurred during the period of reactive hyperemia in the gross flow pattern. A somewhat different pattern of response was seen in 21% of the capillaries (Fig 3); in these vessels flow rose to a high level rapidly but fell back slowly, much like gross flow reactive hyperemia without a period of flow stoppage.

About one-fourth of the capillaries showed no reactive hyperemia and in a small number (14%) hyperemia developed slowly and persisted for about 60 sec. This is about twice the duration of reactive hyperemia recorded in the gross flow. The responses varied from one capillary to the next but seemed reasonably consistent for any given capillary.

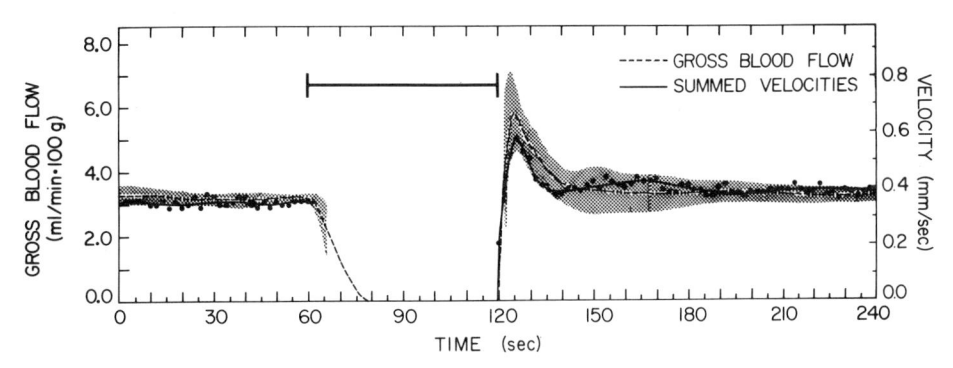

Fig 4 Average gross flow and summated capillary flow in cat cartorius muscle. Control flow and velocity are normalized to the same initial level to compare the magnitude of increase in each during reactive hyperemia. (From Burton, K.S. & Johnson, P.C. Am. J. Physiol. in press. Reproduced by permission).

To test the Krogh hypothesis we summed the individual capillary flow patterns and compared the resultant curve with the average gross flow pattern. This comparison is shown in Fig 4.

This is the summated behaviour of capillaries which were flowing more or less continuously during the control period. The peak velocity is about 1.5 times above the control value. By comparison peak flow is 1.8 times greater than control in the gross flow record. These data suggest that perhaps 70% of the gross flow increase can be attributed to hyperemia in capillaries already flowing. Since the variability in gross flow is considerable, the difference in peak flow is not significant. Therefore we cannot conclude that static capillaries do in fact begin to flow during reactive hyperemia.

DISCUSSION

The capillary patterns of reactive hyperemia suggest that the microcirculatory mechanisms involved in this response do not behave in a well-modulated fashion. Earlier workers have commented upon the seemingly exponential decay of reactive hyperemia; a process which appears to fit well with the idea of washout of metabolites accumulated during ischemia. These observations were of course based upon the gross flow which represents summated behaviour of all of the individual capillaries. It should also be noted that when we summated individual capillaries, the resultant curve has an exponential character to it. It would appear from our studies that the exponential decay in gross flow does not in itself represent typical behaviour in individual capillaries. The typical behaviour in fact resembles an under-damped mechanical system. Possibly the sudden rise of pressure triggers a strong myogenic response in the arterioles, leading to flow stoppage momentarily. Alternatively, rapid washout of metabolites during peak flow might also lead to a strong vasoconstriction.

It has been noted previously that intravascular pressure plays a role in active hyperemia. If venous pressure is elevated during arterial occlusion the subsequent hyperemia is much reduced (4). In other studies on skeletal muscle we have noted that graded reduction of arterial pressure sometimes leads to an increase in flow in certain capillaries. We also noted this phenomenon several years ago in similar experiments on cat mesentery (5). For these reasons we are inclined to look upon the highly reactive behaviour seen in many of these capillaries after occlusion as indicative of a myogenic response.

Under-damped flow behaviour has also been described in studies of gross flow in kidney and intestine (6,7). Both organs are believed to possess a strong myogenic type of response. Possibly in such organs the type of behaviour which we noted in about 40% of skeletal muscle capillaries occurs in the great majority of intestinal and renal capillaries.

The concepts which emerge from these studies are quite different from the classical Krogh model. We believe that skeletal muscle capillaries are open and flowing continuously, rather than being intermittently active. Flow increases principally by augmentation rather than by recruitment. Secondly, the flow through the capillary is determined by the local pressure stimulus as well as by local metabolic factors. The exact site of action of these stimuli and their relative potency are appropriate subjects for future investigation. Current studies in our laboratory with micro-occlusion methods suggest that flow is not regulated at the level of individual capillaries. Occlusion of an arteriole leads to reactive hyperemia while occlusion of a single capillary does not.

REFERENCES

1. Krogh, A. (1950) Isis, 41:14.

2. Burton, K.S. & Johnson, P.C. Amer. J. Physiol. (In press).

3. Wayland, H. & Johnson, P.C. (1967) J. Appl. Physiol. 22:333.

4. Patterson, G.C. (1956) Clin. Sci. 15:17.

5. Johnson, P.C. & Wayland, H. (1967) Amer. J. Physiol. 212:1405.

6. Waugh, W.H. (1964) Circ. Res. 14 (Suppl 1) 156.

7. Johnson, P.C. (1967) Gastroenterology, 52:435.

Supported by NIH Grant AM 12065 and a Grant-in-Aid from the American Heart Association.

FURTHER EXPERIMENTS ON THE ROLE OF PHOSPHATE IN THE LOCAL REGULATION OF BLOOD FLOW

S. M. Hilton

Department of Physiology, The Medical School

Birmingham, B15 2TJ, England

Adenosine and its compounds, in particular ADP and ATP, have long been known to exert strong vasodepressor effects. These compounds have often been suggested to play a role in fundamental circulatory responses: their possible association with exercise hyperaemia in skeletal muscle has been mooted ever since the first tentative suggestions of this kind were made by Zipf (1931) and Rigler (1932); and their possible contribution to traumatic shock since the work of Bennet & Drury (1931) and Zipf (1931). Zipf (1931) also posed the question of their release following simple circulatory arrest. The local dilator effect of ATP on resistance vessels was clearly demonstrated by Stoner & Green (1945) who described erythema and a rise in temperature of forearm skin following close arterial injections of this compound. This vasodilator effect has since been described and measured many times in experiments on animals as well as man.

Whenever these compounds appear in body fluids, inorganic phosphate will almost certainly be present as well. It may therefore be pertinent to the major problems being discussed at this symposium to present results which go to show that inorganic phosphate may be the initiator of the strong, local vasodilatation which develops rapidly in a working muscle after the onset of contraction. This idea concerning inorganic phosphate stemmed from a strange finding, that the soleus muscle of the cat, which is a slow muscle, is quite unlike fast muscles in having no significant functional vasodilatation (Hilton, 1968; Hilton & Vrbova, 1968; Hilton, Jeffries & Vrbova, 1970). For if a substance could be found which is released in large amounts by fast muscles when they contract, but not in significant amounts by slow muscles, this could be the agent which has been sought for so long. We then showed that contracting fast muscles release large amounts of inorganic phosphate into the venous effluent, while slow muscles do not, and that the amount released by fast muscles is related to the frequency of contraction (Hilton & Vrbova, 1970). Since then a very close correlation has been found between the amounts released by fast muscles, and the extent of the hyperaemia that develops within them (Hilton & Hudlicka, 1972). Moreover, in those soleus muscles in which a small hyperaemia does develop with contraction, the relationship with phosphate output is significant and consistent. No significant relationship is found between the hyperaemia and the increase in osmolarity of the venous blood from working muscles, fast or slow; and the relationship with potassium release is also not regular.

It was found in other experiments that the effects of series of twitch contractions can be matched by infusions of sodium dihydrogen phosphate (Hilton & Vrbova, 1970); and, using the technique of retrograde arterial injection, it was shown that small amounts of phosphate can mimic exactly the vasodilator effect of a brief tetanus, a property not shared by either potassium or hyperosmolar solutions.

In suggesting that inorganic phosphate, which is inevitably released from fast muscles, is the initiator of the vasodilatation that occurs within them, we must also account for the fact that the amounts which have to be infused to mimic the effects of contraction are some 50 times as great as the amounts which are released into the blood during contractions. In a recent study in which it was shown that phosphate was not demonstrated, as large enough amounts were not used for fear of exceeding the solubility product of calcium phosphate (Barcroft, Foley & McSwiney, 1971). In our most recent experiments, the rate of passage of phosphate across the capillaries of skeletal muscle, in the direction of blood to interstitial fluid, has been measured. It has been found that the movement of phosphate is sufficiently slow for the figure given above to be the actual concentration difference between the two compartments when equilibrium has been reached (Hilton, Hudlicka & Jackson, unpublished observations). In other words, the phosphate concentration in the interstitial fluid of a contracting muscle can be over 50 times as high as that in the venous (or capillary) blood. If equilibrium is not rapidly reached in a working muscle under normal conditions, then the concentration difference could be even greater. There is evidence suggesting that equilibrium is reached relatively slowly; for a muscle twitching at a constant frequency, over a period of upto 10 min, shows a progressively increasing phosphate concentration in the venous blood (Hilton & Vrbova, unpublished observations).

All the evidence obtained so far therefore points to the conclusion that inorganic phosphate could be not only the initiator but also the main cause of the functional vasodilatation in skeletal muscle. To complete the picture, it should be added that, if contractions are prolonged, there may be sufficient release of potassium to help maintain the vasodilatation, and if the need for oxygen greatly exceeds the supply, a further boost to the vasodilatation may be provided by the accumulation of adenosine (Berne, Rubio, Dobson & Curnish, 1971). But the conclusion reached in relation to inorganic phosphate raises the question of its possible involvement in generating some of the vascular components of at least some shock states.

REFERENCES

Barcroft, H., Foley, T. H. & McSwiney, R. R. (1971). J. Physiol. 213, 411-420.

Bennet, D. W. & Drury, A. N. (1931). J. Physiol. 72, 288-320.

Berne, R. M., Rubio, R., Dobson, J. G. & Curnish, R. R. (1971). Circ. Res. 28, Suppl. 1, 115-119.

Hilton, S. M. (1968). In Circulation in Skeletal Muscle, Proceedings of an International Symposium (1966), edited by O. Hudlicka, pp 137-144, London, Pergamon Press.

Hilton, S. M. & Hudlicka, O. (1971). J. Physiol. 219, 25-26P.

Hilton, S. M., Jeffries, M. G. & Vrbova, G. (1970). J. Physiol. 206, 543-562.

Hilton, S. M. & Vrbova, G. (1968). J. Physiol. 194, 86-87P.

Hilton, S. M. & Vrbova, G. (1970). J. Physiol. 206, 29-30P.

Rigler, R. (1932). Arch. Exp. Path. Pharmak. 167, 54-56.

Stoner, H. B. & Green, H. N. (1945). Clin. Sci. 5, 159-175.

Zipf, K. (1931b). Arch. Exp. Path. Pharmak. 160, 579-598.

UTERINE CIRCULATION IN HAEMORRHAGE

L. Takács and L.A. Debreczeni

Second Department of Medicine Semmelweis University
Medical School, Budapest, Hungary

In the present experiments the circulatory effects of pregnancy were studied in rats under control conditions as well as in haemorrhagic hypotension.

Methods

The experiments were performed on inbred albino rats. The following groups were used: control non-pregnant (dioestrus) rats, non-pregnant rats bled to 40 mmHg or 70 mmHg, control pregnant rats, pregnant rats bled to 40 mmHg or 70 mmHg. Na pentobarbital anaesthesia (50 mg/kg I.P.) was used.

Pregnancy was checked empirically by palpation and was later verified by dissection. The pregnant rats were randomly selected and the weight of the uterus varied between 10 and 60 g. In the following "uterus" denotes the organ with its contents.

Blood pressure was measured in a carotid artery with a mercury manometer. Cardiac output was determined by Evans blue dilution. The organ fractions of the cardiac output were estimated by the ^{86}Rb fractionation technique (Sapirstein 1956, 1958; Takács et al. 1964). According to Deutsch and Dreichlinger (1964) the extraction rate of ^{86}Rb by the placenta and several other tissues (heart, kidney, etc.) remained the same during pregnancy (3-20 days). The ^{86}Rb uptake of the uterine complex of the rats was considered to be the sum of the nutritive flows of the myometrium, placenta and foetuses. No attempt was made to measure the intra-uterine blood flow distribution.

Hypotension was induced in 2 groups by bleeding to 70 mmHg or to 40 mmHg blood pressure in 10 min and was kept constant for 20 min. The maximum bleeding volume was measured.

The results were evaluated by Student's "t" test, and by correlation analysis. Regression equations were calculated only when significant correlations had been found and are indicated by "Y" in the Figures. All the regression coefficients differed significantly from zero. When no difference was found between the regression coefficients, the parallel line assay was performed; the results are shown at the bottom of the Figures.

Results and Discussion

The circulatory parameters as functions of the increasing uterine weight are presented in the Figures.

(a) Untreated, control pregnant rats

Cardiac output (21·3 ml/min/100 g body weight), blood pressure (114·0 mmHg) and total peripheral resistance (442 x 10^3 CGS units) in pregnant animals did not differ significantly from the same parameters of the nonpregnant animals. In the control pregnant rats an increase in the uterine weight was not accompanied by a proportionate increase in cardiac output.

The uterine and carcass fractions of the cardiac output showed a positive, whereas the renal and splanchnic fractions revealed a negative correlation with the increasing uterine weight in the last third of pregnancy (Fig 1). At the end of the pregnancy the increase in the uterine and carcass fractions and the decrease in the renal and splanchnic fractions amounted to 10-11% of the cardiac output. These changes in the cardiac output fractions without any substantial alterations in organ blood flow rates are probably due to the haemodynamic effect of the pregnant uterus. It should be emphasized that in spite of a significantly decreased renal fraction (12·3 %) in pregnant rats the renal blood flow (341·0 ml/min/100 g organ weight) did not differ from that of control non-pregnant animals (16·5 % and 371·9 ml/min/100 g organ weight).

(b) Haemorrhagic hypotension

The blood pressure of pregnant rats was lowered to about 40 mmHg or 70 mmHg by a smaller blood loss (16·1 and 21·8 ml/kg) than in the case of non-pregnant animals (20·7 and 26·5 ml/kg).

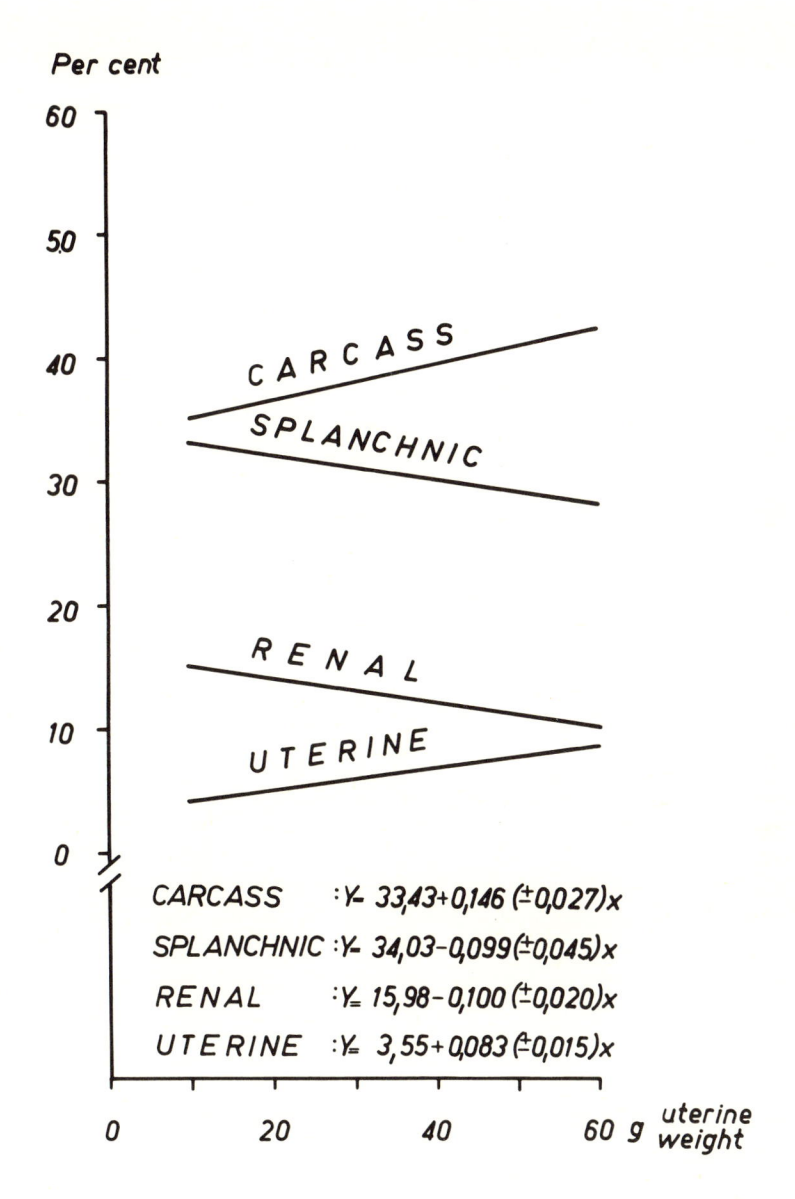

Fig 1. The relationship between the organ fractions of the cardiac output and increasing uterine weight in the last third of pregnancy.

CARCASS : $Y = 33.43 + 0.146 (\pm 0.027)x$
SPLANCHNIC : $Y = 34.03 - 0.099 (\pm 0.045)x$
RENAL : $Y = 15.98 - 0.100 (\pm 0.020)x$
UTERINE : $Y = 3.55 + 0.083 (\pm 0.015)x$

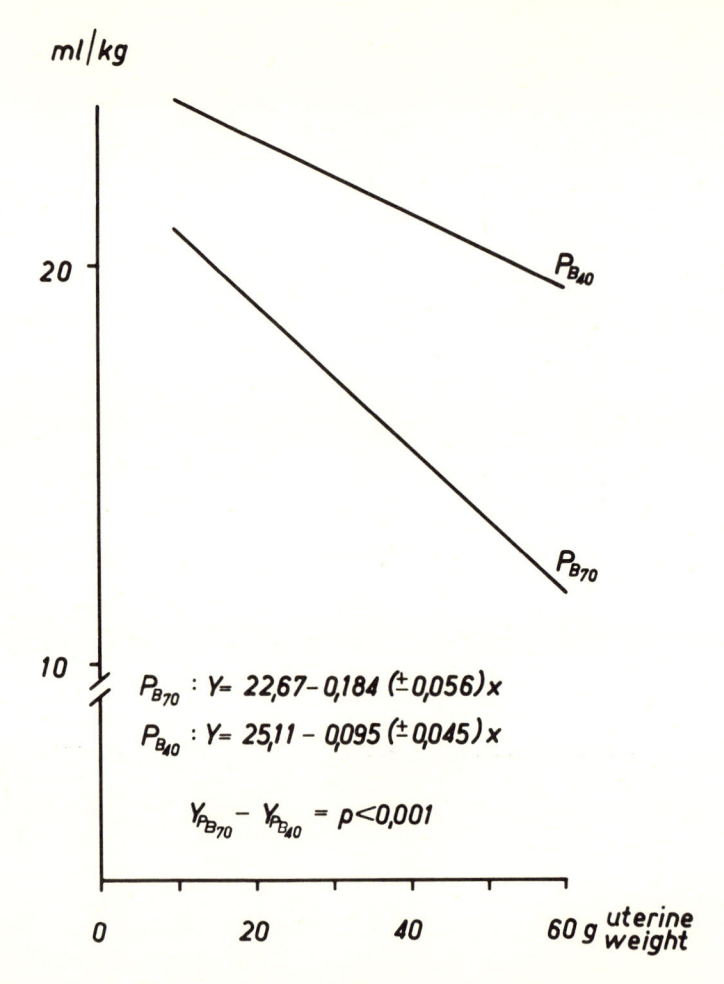

Fig 2. The relationship between the maximal bleeding volume and increasing uterine weight. P_{B70} refers to the pregnant group bled to 70 mmHg, P_{B40} refers to the pregnant group bled to 40 mmHg. The result of parallel line assay analysis is indicated at the bottom of the figures (Y - Y).

$P_{B70}: Y = 22.67 - 0.184 (\pm 0.056)x$
$P_{B40}: Y = 25.11 - 0.095 (\pm 0.045)x$
$Y_{P_{B70}} - Y_{P_{B40}} = p < 0.001$

The maximal bleeding volume of pregnant rats was correlated with the increase in uterine weight in both groups of animals.

Although regression coefficients were similar, the respective equations were different (Fig 2).

After haemorrhage the cardiac output decreased markedly. The circulatory pattern of posthaemorrhagic changes (the decrease in the renal and skin fractions of the cardiac output and the increase in the coronary and carcass fractions) in pregnant and non-pregnant rats was similar to that observed in male ones (Sapirstein et al. 1960, Takács et al. 1962).

The differences between non-pregnant and pregnant groups, both bled to 70 mmHg, are characterized by the smaller total peripheral resistance. higher blood flow and lower resistance in all the organs examined, and by the smaller carcass and greater uterine fractions in the pregnant animals.

In the 40 mmHg group, the total peripheral resistance was lower, the cardiac output and the blood flow to various organs were higher and organ resistances were smaller in the pregnant than in the non-pregnant animals. The carcass fraction was smaller, whereas the kidney, uterine and skin fractions were higher in the pregnant rats.

(c) Uterine circulation

The uterine blood flow calculated per 100 g weight showed a significant negative correlation with increasing uterine weight. This relationship could not be demonstrated after bleeding. In each group of pregnant animals the increase in uterine weight ran parallel with the changes of the total uterine blood flow (ml/min; not calculated per 100 g organ weight; Fig 3). Uterine blood flow was 0·19 ml/min in dioestrual rats with a 0·67 g uterine weight. In the control pregnant group the blood flow was 3·4 ml/min with a mean uterine weight of 36·8 g. While the weight of the pregnant uterus was about 55 times that of the control uterus, the blood flow in the former was only 18-fold the mean dioestrual value.

The uterine blood flow in pregnant bled rats amounted to only 20-50 per cent of that in the control pregnant rats. The regression coefficients for control pregnant rats and pregnant rats bled to 40 mmHg differed significantly. The parallel line assay indicated differences between the controls and those bled to 70 mmHg and between those bled to 70 mmHg and those bled to 40 mmHg.

The relationships between the uterine fractions and weights were similar in the control and bled groups. The corresponding regression coefficients and the results of the parallel line assay showed no significant difference (Fig 4).

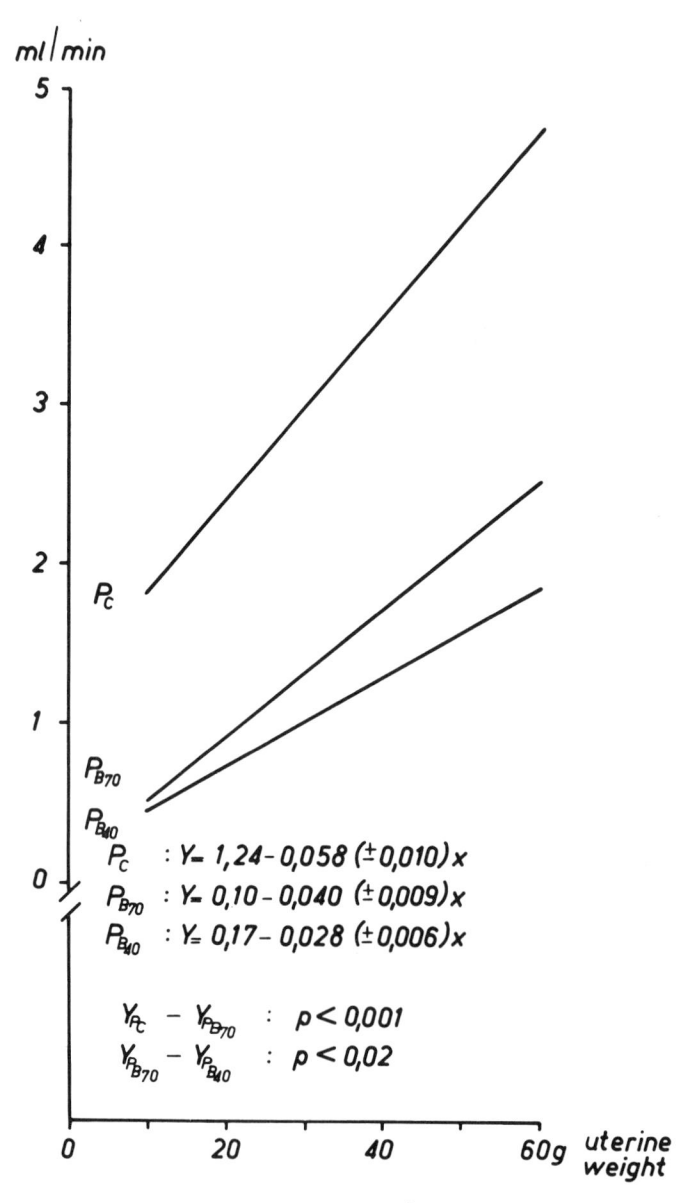

Fig 3 The total uterine blood flows (not calculated for 100 g organ weight) as functions of increasing uterine weights. P_C refers to the control pregnant group. The signs and symbols are as in Fig 2.

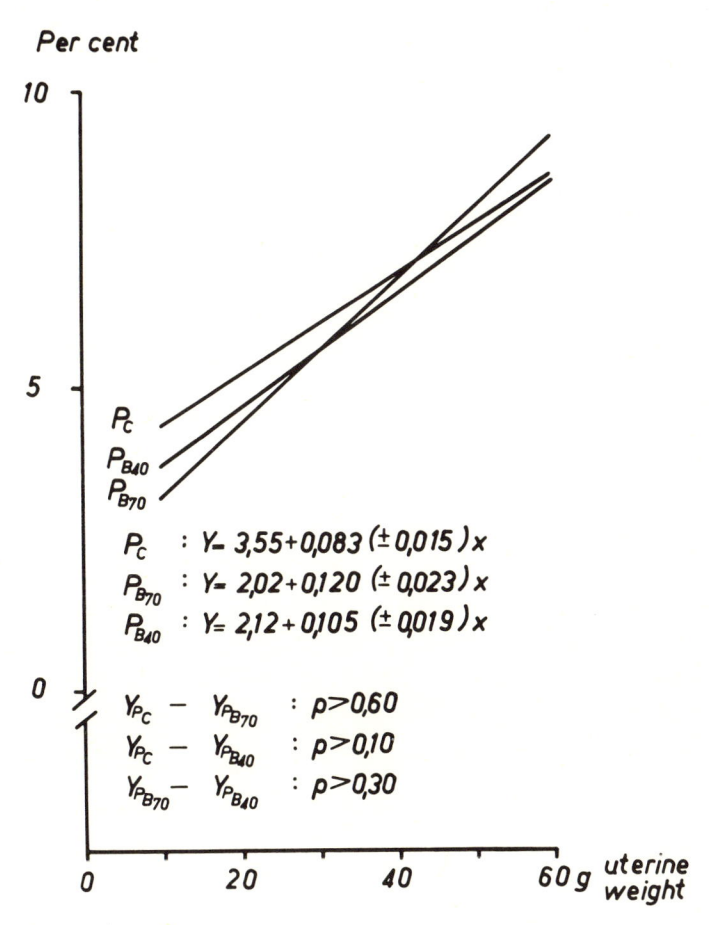

Fig 4 The relationship between uterine fractions of the cardiac output and increasing uterine weight. The signs and symbols are as in Figs 2 & 3.

P_C : $Y = 3.55 + 0.083 (\pm 0.015) x$
P_{B70} : $Y = 2.02 + 0.120 (\pm 0.023) x$
P_{B40} : $Y = 2.12 + 0.105 (\pm 0.019) x$

$Y_{P_C} - Y_{P_{B70}}$: $p > 0.60$
$Y_{P_C} - Y_{P_{B40}}$: $p > 0.10$
$Y_{P_{B70}} - Y_{P_{B40}}$: $p > 0.30$

The blood flow changes were proportional to those in the cardiac output. Accordingly, the uterine fractions of cardiac output was left essentially unchanged by haemorrhage. The results clearly indicate that the pregnant uterus is not involved in the posthaemorrhagic redistribution of the cardiac output.

Summary

In the last third of pregnancy of rats the cardiac output did not differ from that of nonpregnant animals. The renal and carcass fractions of the cardiac output were lower, the uterine and splanchnic fractions were higher than those of nonpregnant rats and these fractions changed proportionally with uterine weight.

The blood pressure of pregnant rats was lowered to about 70 or 40 mmHg by a smaller blood loss than the pressure in nonpregnant animals. A significant negative correlation between increasing uterine weight and maximal bleeding volume has been found.

The posthaemorrhagic changes in the circulatory pattern in pregnant and nonpregnant rats were similar to those observed in male ones. In pregnant rats, however, the diminution of the renal fraction of cardiac output and the elevation of carcass fraction were smaller than in nonpregnant rats. The circulation of the pregnant uterus did not take part in posthaemorrhagic homeostasis.

REFERENCES

Deutsch, G. & Dreichlinger, O. (1964) Revue roumaine d'embryologie et de cytologie. Serie d'embryologie 1:139.

Sapirstein, L.A. (1956) Circ. Res. 4:689.

Sapirstein, L.A. (1958) Amer. J. Physiol. 193:161.

Sapirstein, L.A., Sapirstein, E.H. & Bredemeyer, A. (1960) Circ. Res. 8:135.

Takács, L., Kállay, K. & Skolnik, J.H. (1962) Circ. Res. 10:753.

Takács, L., Kállay, K. & Karai, A. (1964) Acta Physiol. hung. 25:389.

BLOOD MICRORHEOLOGY AND THE DEVELOPMENT OF "STASIS" IN THE

MICROVASCULATURE AFTER INJURY

      Holger Schmid-Schönbein, E. Volger, H.J. Klose and
      J. Weiss

      Department of Physiology, University of Munich, Fed.
      Rep. Germ.

    The development of intravascular blood stasis in the venules after injury has been observed in vivo without exception since the original description by Cohnheim (3). Despite great progress in knowledge of the biochemical processes accompanying cell injury, this biophysical phenomenon is still being placed in the center of the tissue response. Letterer (9) thinks that vascular stasis is the principle cause of necrosis due to prolonged anoxia, in spite of a normal macrocirculation.

    In an attempt to explain changes in the flow behaviour of blood, Braasch (1) has pointed to a loss of red cell flexibility after injury, while Knisely (8) and Gelin (6) have emphasized the significance of pathological red cell aggregation. In one of his last papers, the Swedish pathologist and pioneer rheologist R. Fahraeus (4) attempted to relate tissue injury to a loss of red cell filtrability.

    Intravascular flow phenomena and their quantitative interpretation have been the subject of extensive studies in the field of hemorheology. The quantitative study of the microscopic flow properties of red cells and red cell aggregates under defined conditions of flow (Microrheology) has supplemented the traditional study of the so-called "anomalous viscosity of blood" (macrorheology). Such experiments are not only much more relevant to microvascular flow, but have recently yielded a more general understanding of the factors facilitating or restricting the flow of blood in all kinds of vessels (2,3).

The key factor responsible for the unusual fluidity of blood is continuous red cell deformation in flow. The physical properties of the mammalian red cell - a membrane shell filled incompletely with a viscous fluid (5) - give rise to an unparalleled deformability or "fluidity" (12). This property of the cell allows unique adaptation to the flow regime and the participation of the cell contents in flow in contrast to the strong disturbances of flow caused by non-deformable particles. All factors that interfere with this adaptation greatly increase the viscous resistance of blood, especially in narrow vessels and when flow forces are reduced. Immobilization of the cells in aggregates occurs as a physiological phenomenon in most mammalian species when the flow forces are reduced. The reversible aggregation of red cells into rouleaux and of rouleaux into three dimensional structures not only substantially increases apparent blood viscosity in prestatic flow but allows uncoagulated blood in stasis to withstand finite forces without flowing or yielding. Blood thereby assumes the properties of a solid. The higher the hematocrit and the higher the concentration of high molecular weight plasma proteins, the higher the forces necessary to disperse the aggregates.

The deformability of the cells is abolished after discreet changes in its internal and external milieu (pH, state of hydration, ATP content, membrane integrity, solubility of cell content and plasma environment) and can therefore be easily affected by a great number of factors as well as by inherent abnormalities of cell metabolism (13). These recent findings have strongly supported the idea that "rheological" factors play a role in tissue injury. However, it should be noted that many microrheological abnormalities evade detection in traditional viscometers (13). In order to understand possible rheological factors in the natural history (aetiology and pathogenesis) of the response to injury, one has to go through three stages:

1) quantification of the flow properties of blood elements under shear, followed by an analysis of their flow behaviour under well-defined artificial flow conditions in vitro.

2) a description of the complex natural flow conditions in the microcirculation of various organs as defined by dimensions, architectural arrangement of the different classes of vessels as well as by a quantification of the forces of flow.

3) predictions about the in vivo flow behaviour of the complex fluid "blood" (normal or abnormal) in the complex vascular beds under varying flow conditions.

The present report is concerned with the first stage; quantification of rheological properties in vitro and the effect of physical and chemical factors upon it. To this end special methods had to be developed which allowed generalized predictions about hemodynamic consequences of the observed abnormalities.

METHODS

The effect of direct thermal and chemical stress on the rheology of human venous blood (anticoagulated with EDTA, 1,5mg/ml) was studied by measuring viscosity in two Wells-Brookfield viscometers (14) with cone-plate as well as Couette systems. The sensitivity of the two varied between 336 dyn/cm and 6737 dyn/cm (full torque) and allowed adaptation to wide changes in apparent blood viscosity. Both apparent blood viscosity and relative apparent blood viscosity were plotted as a function of the shear stress, the latter value ($\eta_{blood}/\eta_{plasma}$) quantifies the effect of red cells on blood viscosity, provided the hematocrit is kept constant (45% in our experiments).

The deformability of red cells was measured by determining the flow rate of 10% red cell suspensions (divided by the flow rate of plasma alone) through pores of 5 μm in diameter. (Nuclepore Sieves, General Electric, Pleasantown Calif.) The pore length was 11 μm, the pressure head was 5 cm of H_2O, the resulting shear stress at the pore wall corresponds to values estimated in nutrient capillaries (542 dyn/cm^2).

The computed values of relative flow rate measured the effects of red cells on plasma flow in geometric and dynamic models of nutrient capillaries, the diameter of which is equal in size or smaller than the diameter of red cells.

Red cell aggregation was studied in viscometric flow in a cone plate chamber in which the shear stresses necessary to keep red cell aggregates dispersed are determined (8). The transition of erythrocytes from the aggregated to the deformed state makes blood maximally opaque, because of their randomisation. Aggregated as well as deformed red cells increase the transparency of blood. The shear rate (and thence shear stress) at which this transition occurs is measured photometrically and is used as an indicator of aggregate dispersion which evades detection by viscometry alone (12).

The effects of hypo- and hyperthermia were studied by executing the rheological experiments after appropriate temperature equilibration. The effects of pH and chemical irritants were tested after dialysing the blood samples (which were kept continuously stirred) against the respective agents or buffers. The agents used by Rickert and Regendanz (11) in their study of in vivo stasis were tested, the concentrations and exposure times that these authors employed were simulated as closely as possible (they are listed in table II).

RESULTS AND DISCUSSION

As previously shown, lowering the temperature increases the viscosity of blood. This effect is more pronounced at low rates of shear than at high ones (Fig 1A). When plotting the relative apparent viscosity as a function of shear stress, the reason for this behaviour becomes evident: η_{rel} rises to a much smaller extent at high shear stresses (where cell deformation determines viscous resistance) than at low ones (where cell aggregation predominates). Independent measurement of these microrheological variables corroborated this. At low temperatures, especially below 20° C, the tendency to aggregation is greatly enhanced. The shear stresses necessary to keep red cell aggregates dispersed are up to 10 times higher than under normothermic conditions (Fig 2). Quantification of the red cell deformability shows that in passing through 5 µm filter pores, the absolute flow rate of 10% red cell

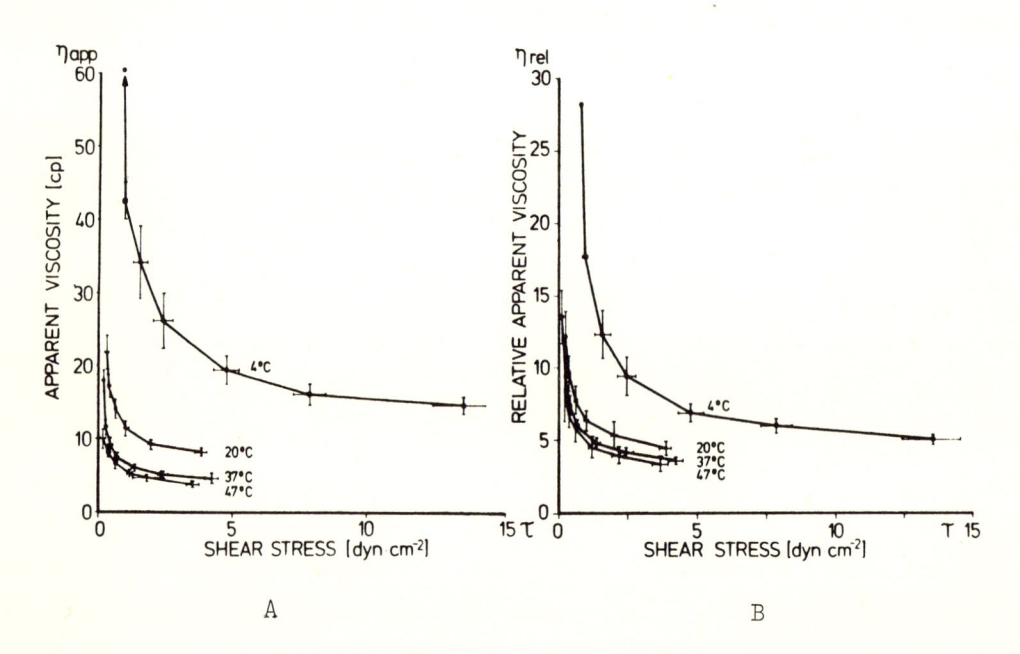

Fig 1. Influence of hypothermia on viscous behaviour of human blood. A) Apparent viscosity as a function of shear stress: elevated values at high and low flow. B) Relative apparent viscosity as a function of shear stress (= measurement of red cell effects on viscosity): elevated values only at low flow.

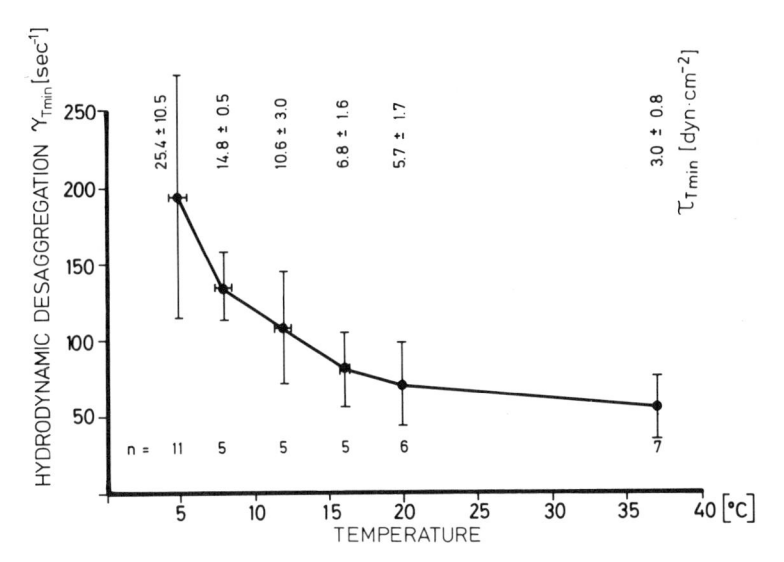

Fig 2: Effect of hypothermia on shear resistance of red cell aggregates. After lowering temperature below 20°C, the shear stresses necessary to disperse red cells are increased.

			control		
Temp. °C	4	20	37	47	49
$\overset{\circ}{V}_{pl}$ (μl·sec^{-1})	22.1±7.9	51.6±8	76±18	113±25	73.4±6
$\overset{\circ}{V}_{susp}$ (μl·sec^{-1})	14±4	37±7	51±8	72±11	0
$\overset{\circ}{V}_{rel}$	0.66±.13	0.75±.07	0.69±.08	0.64±.05	0

Table I: Effect of temperature on red cell deformability. ($\overset{\circ}{V}_{rel}$, 5 μm pore diameter, 10% Hct, pore wall shear stress 542 dyn/cm^2, n = 8). Between 4 and 47°C, plasma and suspension flow rate increase, $\overset{\circ}{V}_{rel}$ stays equal. At 49°C, cells do not pass pores.

suspensions falls, but the relative flow rate (Table I) remains virtually unaffected down to 4°C. At least in the present temperature range, there is no evidence that the red cells specifically lose their deformability or fluidity to a greater extent than plasma. In this temperature range, red cell aggregation primarily limits the fluidity of blood, especially if hypothermia is accompanied by a loss of driving force. During actual freezing and after thawing, when the cell undergoes gross physical damage, there is a permanent loss of fluidity.

Hyperthermia affects blood rheology in an entirely different fashion. At temperatures between 37 and 47°C, the viscosity falls at all shear rates, aggregation and deformability are unaffected (Fig 1A & B, Table I). However, as soon as temperatures are elevated above 49°C, the deformability is severely impaired so that the cells can no longer pass 5 μm pores; they behave like solid particles. The cell rigidification and plasma precipitation at 49°C and above greatly increase viscosity but make precise measurements impossible.

	%	min	η_{rel} (115 sec^{-1})	$\dot{V}_{rel}$
controls (n = 9)	-	-	3,57±0.50	0,64±0.50
silver nitrate	0.1	10	12,4	0.00
tannic acid	1.0	3	5,92	0,00
ammonia sol.	1.0	3	3,07*	0.00
iodine sol.	1.0	3	10,2	0,00
mustard oil	10.0	10	5,92	0,00
camphor oil	25.0	30	4,19	0.09
turpentine oil	-	30	5,45	0,00

*hemolysis

Table II: Effect of chemical irritants on blood viscosity (η_{rel}) and deformability ($\dot{V}_{rel}$) of red cell suspensions. The concentration of the irritants in the dialysing bath and exposure times (min) are as described by Rickert and Regendanz (11).

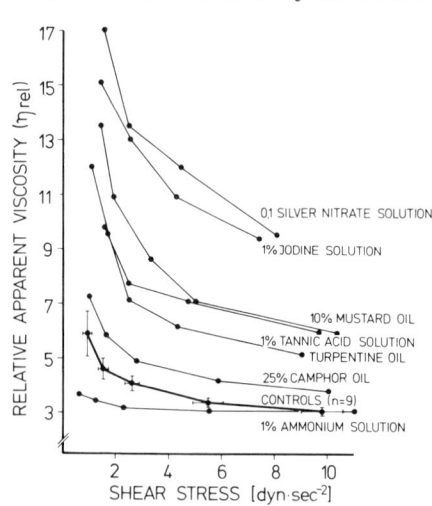

Fig. 3: Influence of chemical irritants on relative apparent viscosity of whole blood (Hct 37° vc., 45% Hct). The viscous effects of a given volume fraction of cells are grossly increased after chemical alteration.

Chemical irritants, (Table II) have a predominant effect on cellular deformability: the relative flow rates through 5 μm pores fall to zero after reaction with all these substances. In other words, when altered by these agents, the cells in the nutrient capillaries are likely to interrupt perfusion through these and all downstream vessels through cellular rigidification. In larger vessels as well the viscosity is likely to be increased. With the exception of ammonia (which caused rapid hemolysis), all chemical irritants substantially increased relative viscosity at all shear stresses (Fig. 3). Where, however, the relative viscosity is increased only up to 4 times in suspensions, the viscous resistance of single cells in models of microvessels is infinity.

Once stasis has occurred in vivo, the blood cells are immobilized in an anoxic environment in which lactacidosis is likely to develop. With falling pH, the cells also lose their flexibility, as first described by Murphy (10). Under better defined conditions of pore flow, deformability is already decreased at pH below 7 and is completely abolished at pH below 6. In these experiments, there is a considerable scatter among different blood donors in the response of their cells to acidosis. A number of clinically healthy subjects repeatedly showed cell rigidification at pH 7.2. It should be noted that with the present technique rather high shear stresses are applied (542 dyn/cm^2). When the forces are lower, the effect of pH on deformability is likely to be much more pronounced. In the capillaries during stasis, the shear stresses are, of course, very small.

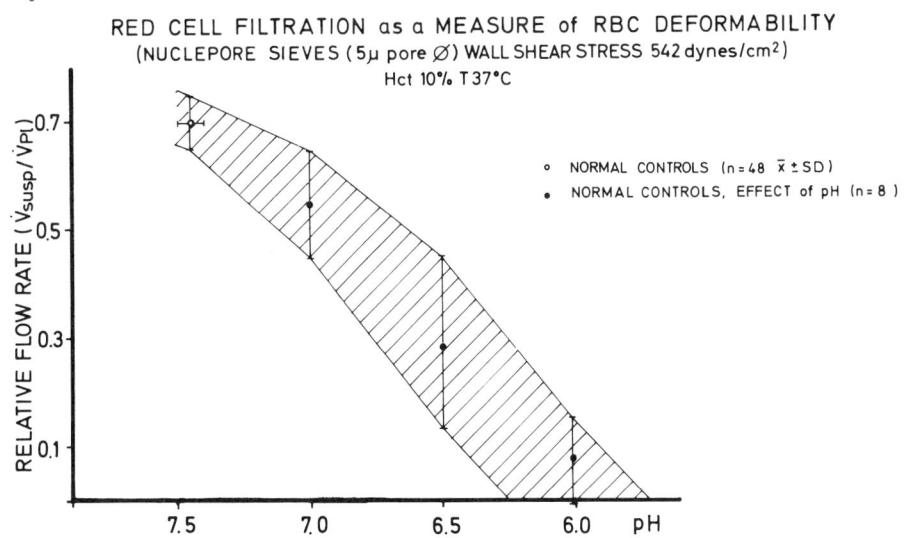

Fig 4: Effect of pH on red cell deformability: relative flow rate of 10% red cell suspensions through 5 μm pores. Shear stress at pore wall 542 hyn/cm^2.

CONCLUSIONS

The present microrheological experiments have shown that chemical and physical factors are able to abolish the fluidity of the erythrocyte, a property essential for the normal perfusion of the microvasculature. This effect can be due to two mechanisms:

1) Loss of fluidity through structural rigidification due to changes in the erythrocyte membrane, its intracellular contents or due to abnormal interaction between membrane and plasma or membrane and cell content. All chemical irritants used in the classical studies of experimental pathologists as well as hyperthermia and acidosis cause this loss of cell deformability. The etiology of the cellular defect awaits clarification. Its pathogenic consequences must be considered in the analysis of the response to injury. For simple geometrical reasons it is likely that these alterations interfere with normal capillary perfusion, which is possible only through continuous erythrocyte deformation. It is obvious that a block in capillary perfusion will reduce flow in the post-capillary venules as well.

2) Loss of fluidity due to functional rigidification through immobilization of red cells in large aggregates (or agglutinates) at low shear rates. This is a normal property of blood at low flow states and is intensified by lowering temperature, irrespective of changes in the plasma protein composition which occur in response to injury. Due to normal as well as pathologically intensified red cell aggregation, the viscosity in prestatic flow increases exponentially. The hemodynamic consequences of shear dependent red cell aggregates and concomitant reversible increases in viscosity are exceedingly complex and only partly understood. However, there is general agreement (2,12) that the postcapillary venules are the vessels in which the shear rates are most likely to fall below a critical level at which aggregation occurs and below which viscosity starts to increase. This increase in viscosity may reach infinity and is therefore capable of maintaining prolonged stasis. This postcapillary curtailment of flow also affects the capillaries upstream. The assumption of rheological causes for venular stasis, although not yet quantitatively established, is well supported by unequivocal qualitative observations. A number of consequences of venular stasis, such as leukocyte margination, red cell diapedesis, exudation, transudation and hemoconcentration are likewise affected by rheological behaviour of erythrocytes and deserve further study.

REFERENCES

1. Braasch, D. & Gössling, G. (1966) Pflüger's Arch. 289:1

2. Chien, S. In: Hemodilution, theoretical basis and clinical application. K. Messmer and H. Schmid-Schönbein. Basel-New York (Karger) 1972.

3. Cohnheim, J., (1867) Virchow's Arch. 40:1.

4. Fahraeus, R. In: The Biochemical response to Injury, p. 161, H.B. Stoner & C.J. Threlfall, Oxford (1960)

5. Fung, Y.C. (1966) Fed. Proc. 25:1761.

6. Gelin, L.E. & Zederfeldt, B. (1961) Acta chir. Scand. 122:336.

7. Klose, H.J., Volger, E. & Schmid-Schönbein, H. (1972) Pflüger's Arch. 333, 126.

8. Knisely, M.H., In: Handbook of Physiol. Sect. 2, Vol. III. pp. 2249. Ed. W.F. Hamilton and P. Dow, Washington D.C. 1965.

9. Letterer, E. In: Handbuch der allgem. Pathologie VII Springer Berlin, 1967.

10. Murphy, J.R. (1967) J. Lab. Clin. Med. 69:758.

11. Rickert, G. & Regendanz, P. (1921) Virchow's Arch. 231:1.

12. Schmid-Schönbein, H. & Wells, R.E. (1971) Ergebn. der Physiol. 36:145.

13. Schmid-Schönbein, H., Wells, R.E. & Goldstone, J. (1971) Biorheology 7, 227:234.

14. Wells, R.E., Jr., Denton, R. & Merrill, E.W. (1961) J. Lab. Clin. Med. 57:646.

Supported by grants from the Deutsche Forschungsgemeinschaft.

BLOOD RHEOLOGY AFTER HEMORRHAGE AND ENDOTOXIN

S. Chien, S. Usami, J. Dellenback and V. Magazinovic

Laboratory of Hemorheology, Department of Physiology
College of Physicians and Surgeons, Columbia University
New York, N.Y. 10032, U.S.A.

Since blood viscosity increases markedly as the shear rate is reduced (1-6), the elevation of blood viscosity may conceivably play a role in the pathogenesis of low flow states. In this paper the factors regulating blood viscosity in different flow conditions will be considered and the possible role of rheological changes in endotoxin and hemorrhagic shock in the dog will be analyzed.

FACTORS DETERMINING BLOOD VISCOSITY

Blood is a non-homogeneous system and its viscosity is higher than that of plasma because the suspended cells reduce the volume through which the streamlines of plasma can pass (7). The volume available for the plasma streamlines is reduced not only by the true cell volume but also by the volume of plasma immobilized hydrodynamically by the cells and the sum of these two can be regarded as the effective cell volume (7). The immobilization of plasma depends on the degree of axial asymmetry of the suspended particles, as well as the temporal and spatial distribution of the axial orientation. Thus, the volume and plasma immobilized and the effective cell volume decreases with cell deformation (7, 8) and generally increase with cell aggregation (7, 9). Since an increase in shear force causes cell deformation (10,11,12) and cell disaggregation (13), the ensuing reduction in effective cell volume forms the micro-rheological basis of the shear thinning behaviour of blood viscosity (7,14) (Fig. 1).

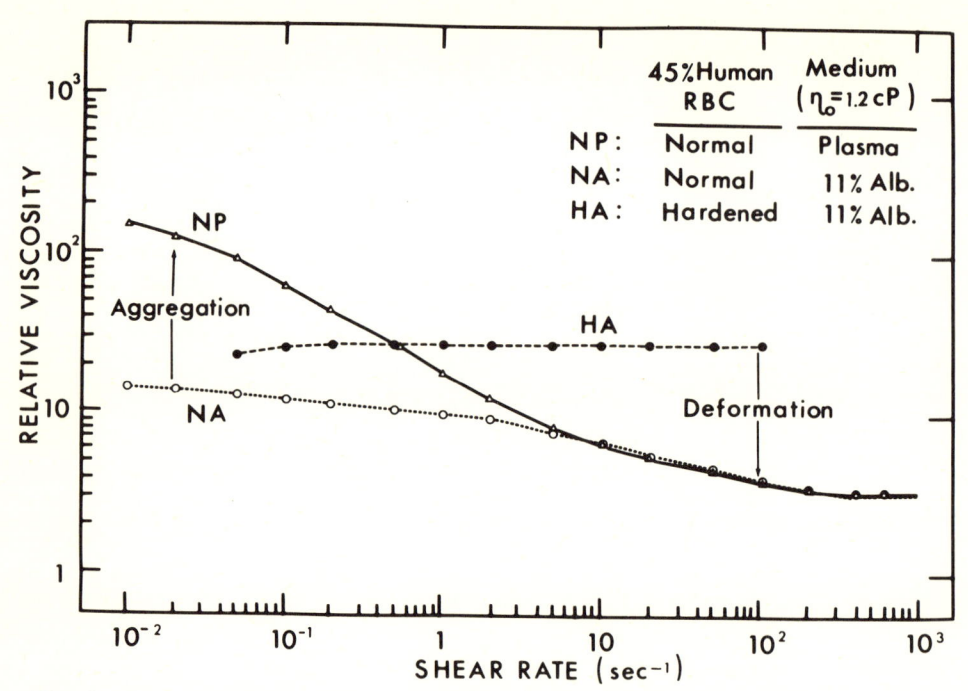

Fig 1 Showing the roles of erythrocyte deformation and erythrocyte aggregation in affecting the blood viscosity - shear rate relation. Hardened RBC were prepared by fixation of RBC in 1% glutaraldehyde solution.

The above considerations indicate that the relative viscosity of the blood depends on the true cell concentration as well as the effectiveness of the cells in reducing the volume of plasma participating in the streamlines. This effectiveness of the cells is a function of shear deformation and shear disaggregation. Blood viscosity, being the product of plasma viscosity and the relative viscosity, is therefore determined by the following 4 factors, which are not entirely independent of one another:

    A. Plasma viscosity
    B. Cell concentration
    C. Cell deformation
    D. Cell aggregation

The plasma viscosity is independent of shear rate (15) and, at a given temperature, is a function primarily of the plasma protein concentration. The red cells generally form the bulk of the cell concentration which can be determined by centrifugal packing and appropriate correction for plasma trapping (16).

The red cell is composed of a fluid (17,18) surrounded by a membrane which can undergo considerable uni-axial stretching without a significant area extension (19,20,21). This ability of the red cell to deform results from an excess surface area in relation to volume and from the flexibility of the membrane (19,22). The viscosity of the internal fluid is a function of the concentration and the physico-chemical properties of hemoglobin (23). Membrane flexibility can be affected by its chemical composition and metabolic state (24, 25). In addition to the intrinsic deformability of the cell, the extent of deformation also depends on the external conditions, including the shear rate, geometric limitations and the external fluid viscosity (14).

Aggregation of red cells results from the bridging of cell surfaces by long-chain plasma proteins, e.g. fibrinogen (9,26). Aggregation results when the surface adsorption force of the bridging macromolecules exceeds the electrical repulsive force between cell surfaces and the mechanical shearing force (26). At a given shear condition, cell aggregation depends on the aggregating effectiveness of the plasma proteins as well as the aggregation tendency of the red cells (14).

SHEAR RATES IN THE CIRCULATION AND INFLUENCE OF NON-NEWTONIAN VISCOSITY ON TRANSCAPILLARY FLUID SHIFT

Since blood viscosity at given concentrations of red cells and plasma proteins shows a striking dependence on shear rate, it is of interest to estimate the shear rates in the circulation. The circulatory system is composed of a complicated network of distensible vessels in which pulsatile flow occurs. Therefore, the shear rate should vary with space and time. The calculation of intravascular shear rates is made difficult by the non-Newtonian behaviour of the blood, in addition to the complexity of the vascular architecture. Nevertheless, one may obtain a rough estimate of the mean shear rate from the ratio of mean velocity to the mean radius ($4\bar{v}/r$ in Fig 2) from anatomical data (27,28). The highest shear rate is found in the capillaries (4), where the passage of red cells through narrow channels necessitates shear deformation. The lowest shear rate

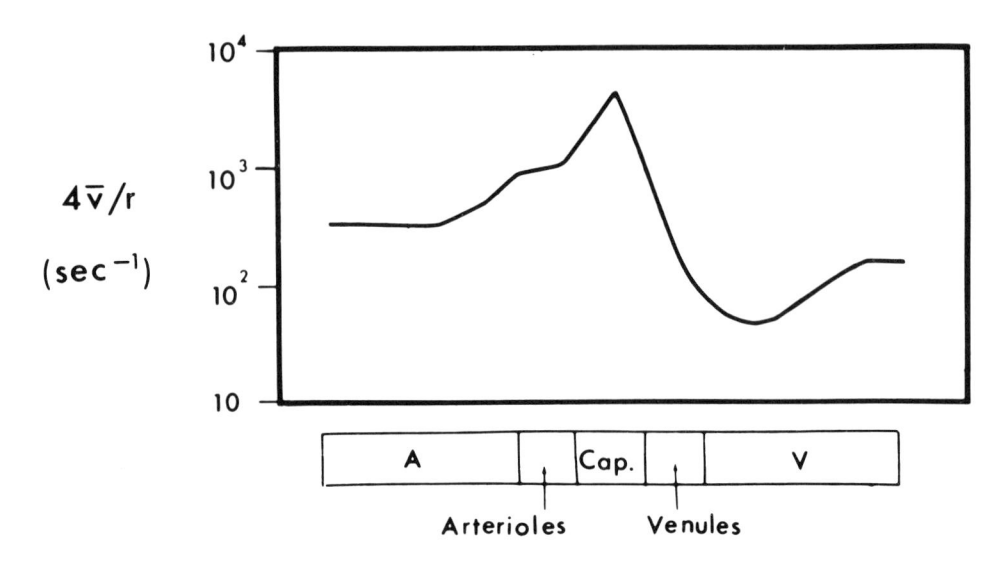

Fig 2 Variations in the parameter $4\bar{v}/r$, which gives a rough estimate of the shear rate, in various parts of the normal circulation of the dog (modified from 4). Note the high shear rates in capillaries and the relatively low shear rates in the venules and small veins.

is found in the postcapillary venules and small veins, but even there the normal mean shear rate is probably of the order of 50 sec^{-1}, a value at which shear deformation and shear disaggregation are essentially complete (7). Under normal conditions the precapillary shear rate (A_N in Fig 3) and the postcapillary shear rate (V_N) are such that the viscosity values are comparable in these segments. With a fall in blood flow,

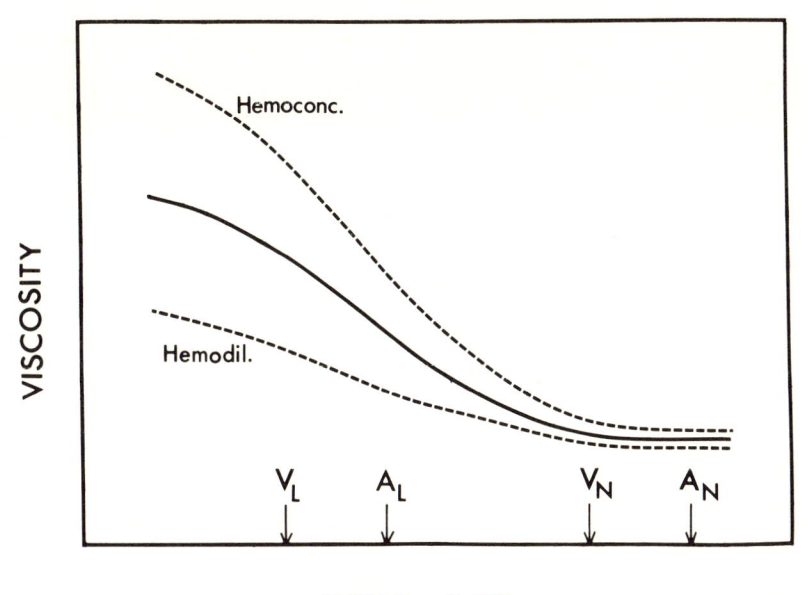

Fig 3 Showing the disproportionate influence of flow rate on viscosity of postcapillary (V) and precapillary (A) segments. Solid line indicates the relation between viscosity and shear rate for normal blood. According to Fig 2, the normal shear rate in the postcapillary segment (V_N) is lower than that in the precapillary segment (A_N). In low flow states, the shear rates in these segments are reduced to V_L and A_L respectively. The broken lines above and below the solid curve indicate the relation obtained when the shear dependence of blood viscosity is altered by hemoconcentration and hemodilution respectively. Because of the difficulty in determining the exact shear rate values in the circulation, the diagram is not quantitative and no scales are given for the co-ordinates.

even if the blood composition remains unchanged, the viscosity values would increase as a result of the reduction of the precapillary and postcapillary shear rates to A_L and V_L respectively in Fig 3. Because of the non-linear relation between viscosity and shear rate and the lower shear rate existing

in the postcapillary segments, low flow states would be
accompanied by an increase in the post-/pre-capillary viscosity
ratio. If the arterial and venous pressures are not excessively
low, this tendency for the post-/pre-capillary resistance ratio
to rise may increase capillary pressure and transcapillary fluid
loss. Such rheologically induced fluid loss would be minimized
if the slope of the viscosity-shear rate curve is reduced by
hemodilution and it would be enhanced if the slope becomes
steeper due to hemoconcentration (Fig 3). From this point of
view, hemoconcentration in low flow states could initiate a
positive feedback mechanism through rheological changes, leading
to transcapillary fluid loss and progressive deterioration of
the circulation (Fig 4). Data on endotoxin and hemorrhagic shock
in normal and spenectomized dogs will be used to illustrate the
influence of viscosity changes on: (a) flow resistance and
(b) transcapillary fluid shifts.

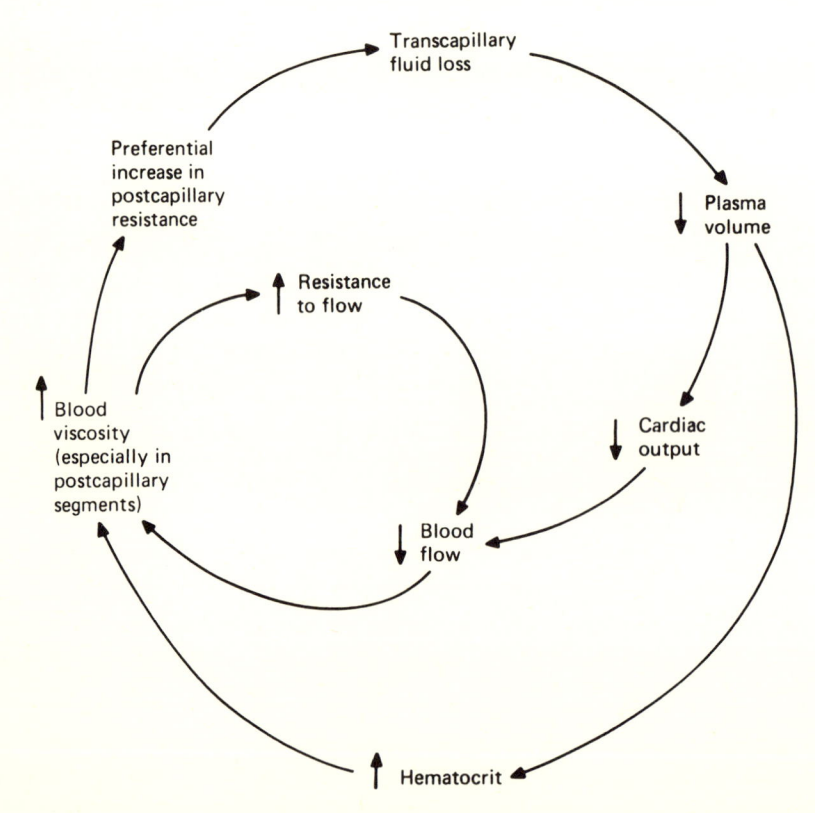

Fig 4 A diagram showing the possible role of positive feedback
mechanisms involving rheological changes in impeding blood flow
through the microcirculation (from ref. 4).

RHEOLOGICAL CHANGES IN ENDOTOXIN SHOCK

The hemodynamic effects of E. coli endotoxin (Difco, 3 mg/kg, i.v.) were studied in 30 dogs under pentobarbital anesthesia (4,29,30). Ten of these dogs had been chronically splenectomized. Representative experiments are compared in Fig 5.

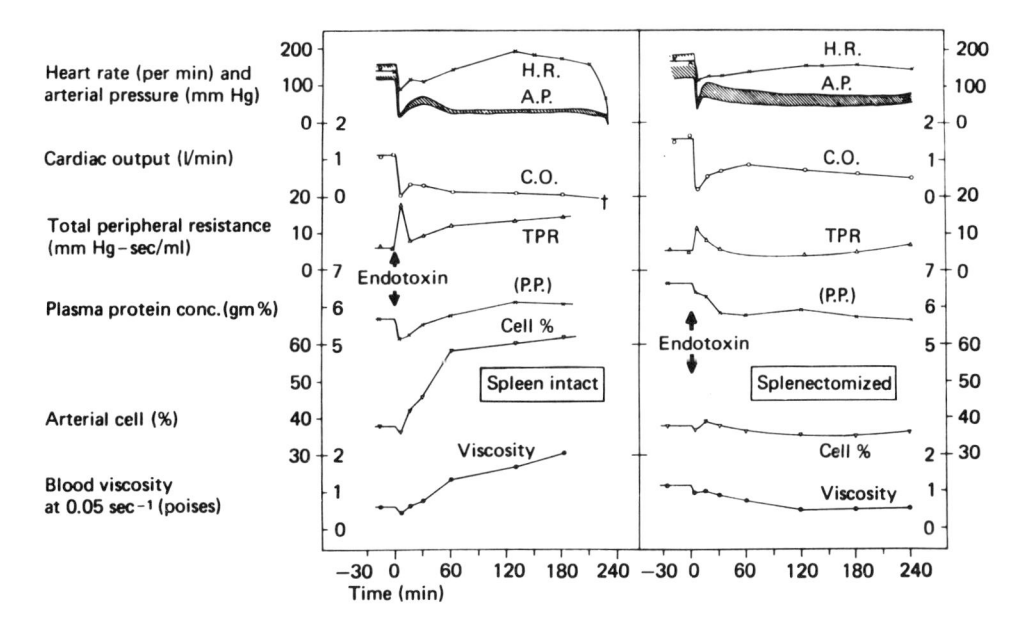

Fig 5 Comparison of hemodynamic changes in endotoxin shock between a dog with spleen and a splenectomized dog (from ref. 29). Note the increase in blood viscosity (at 0.05 sec^{-1}) in the dog with spleen and the decrease in the splenectomized dog. These are accompanied by parallel changes in arterial cell percentage and plasma protein concentration. The secondary rise in TPR in the dog with spleen follows a time course similar to that of the increase in viscosity.

The initial responses to endotoxin were similar in both groups and consisted of precipitous decreases in arterial pressure and cardiac output and a marked increase in TPR. Approximately 30 min after endotoxin, the cardiac output of the splenectomized dog showed a considerable recovery, and the TPR decreased. In contrast, the dog with spleen had secondary declines in cardiac output and arterial pressure, and an increase in TPR. This secondary increase in TPR occurred at a time when the hemotocrit (cell % in Fig 5) and blood viscosity increased progressively. These results indicate that the postendotoxin release of cell-rich blood from the spleen increased blood discosity and flow resistance, resulting in progressive curtailment of tissue perfusion and circulatory deterioration. Eight of the 20 dogs with spleen died within 6 hr after endotoxin. In the splenectomized dog the hematocrit did not increase, and the viscosity decreased as the plasma protein concentration was reduced. The low viscosity in the splenectomized dog was accompanied by a lower TPR than the dog with spleen and better tissue perfusion. All 10 splenectomized dogs survived the 6 hr period of the experiment.

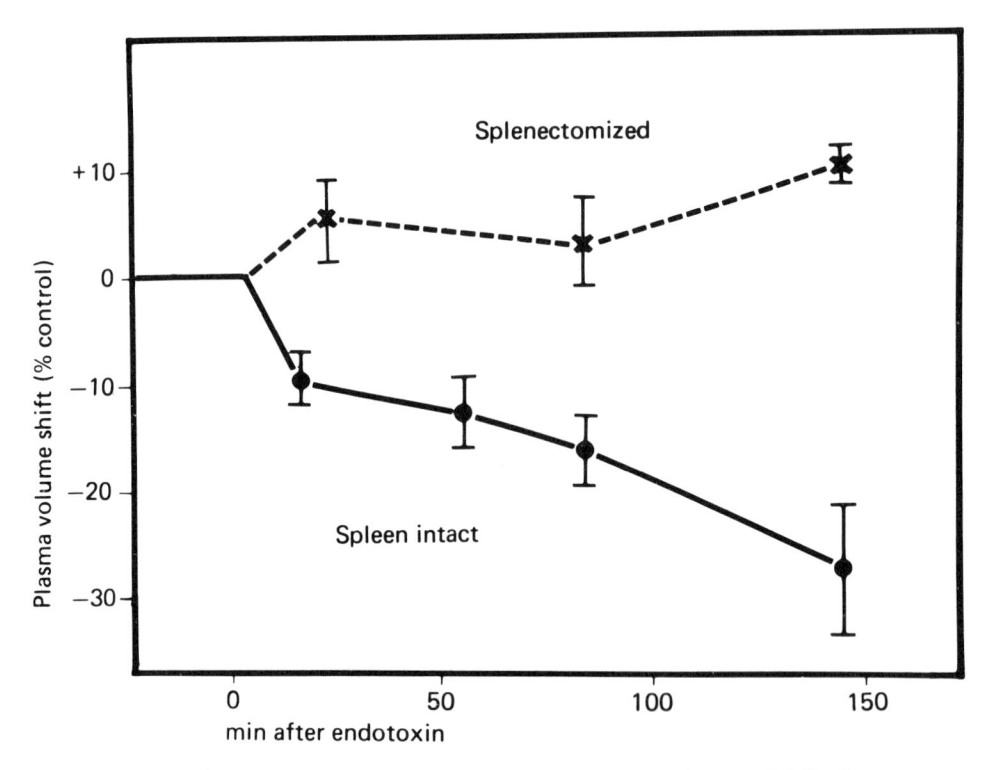

Fig 6 Effect of endotoxin on plasma volume shift in splenectomized dogs and dogs with spleen.

According to the concept developed above, rheological changes not only affect flow resistance, but also transcapillary fluid transfer and hence the plasma volume. The plasma volume was measured with I^{131} labeled serum albumin. In splenectomized dogs, endotoxin caused an increase in plasma volume, indicating transcapillary fluid influx (Fig 6). In contrast, there was a marked decrease of plasma volume in the dog with spleen, indicating a transcapillary fluid loss. These results can be explained in terms of the blood viscosity-shear rate relation in the precapillary and postcapillary segments (Fig 3). The influence of endotoxin on blood viscosity-shear rate relation is shown in Fig 7. In the splenectomized dog, the postendotoxin viscosity curve was lower than the control. Therefore, the

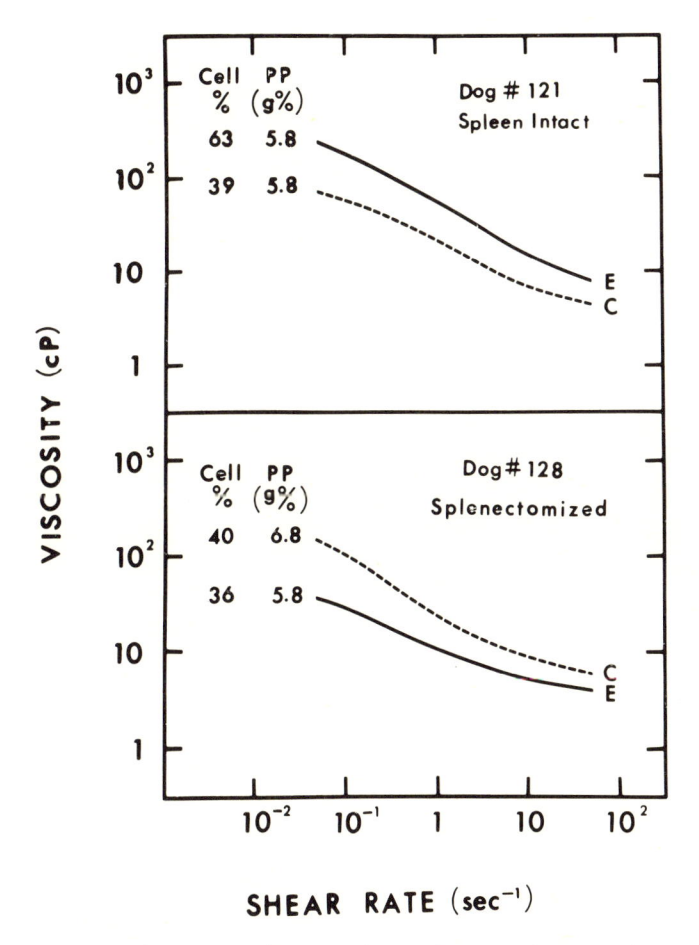

Fig 7 Effect of endotoxin on viscosity - shear rate relation in a dog with spleen (upper panel) and splenectomized dog (lower panel).

post-/pre-capillary resistance ratio was not significantly elevated by the rheological changes, and it may actually decrease due to sympathetically induced precapillary constriction, resulting in fluid influx. In the dog with spleen, however, the elevation of the viscosity curve coupled with a reduction in shear rate caused an increase in post-/pre-capillary resistance ratio and transcapillary fluid loss. Among the dogs with spleen, those which did not survive had significantly greater fluid loss and higher viscosity than the survivals. One of the possible sites of fluid loss is the intestine. The plasma loss in late endotoxin shock was associated with a progressive increase in intestinal lymph flow (I in Fig 8A).

The results in endotoxin shock suggest that the release of cell-rich blood from the spleen induces fluid loss by rheological mechanism. This further increases the hematocrit and raises the

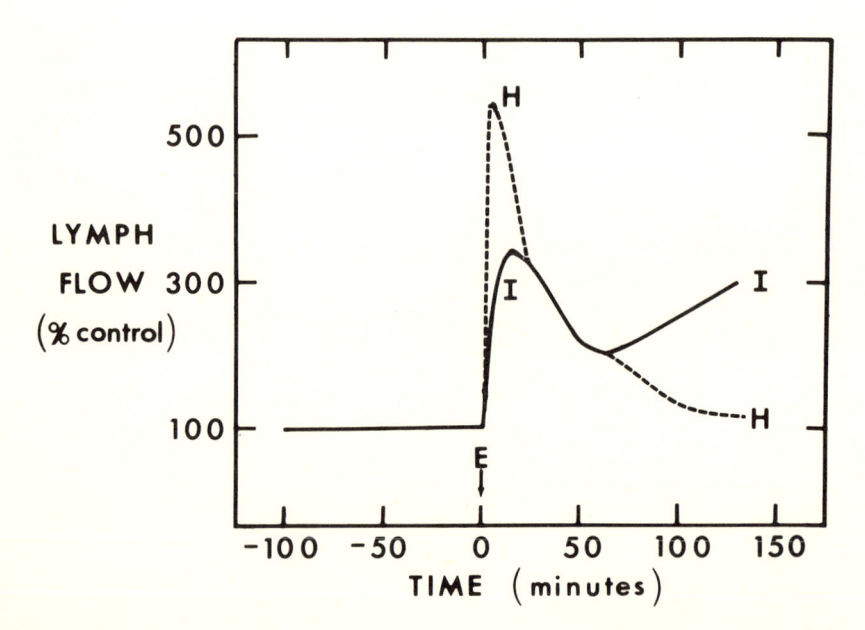

Fig 8 Effects of endotoxin on flow rates of hepatic lymph (H) and intestinal lymph (I).

RHEOLOGICAL CHANGES IN HEMORRHAGIC SHOCK

viscosity, thus perpetuating a positive feedback mechanism which contributes to the deterioration of the animal (Fig 4).

Hemodynamic changes in hemorrhagic shock were studied in 26 beagle dogs without general anesthesia (4). Eleven of the dogs had been chronically splenectomized. After heparinization, hemorrhagic hypotension of approximately 40 mmHg (or 30% of control arterial pressure) was produced. This degree of hypotension was maintained by the further removal or return of blood for 3 hr when the shed blood was returned (Fig 9). The dogs with spleen had a

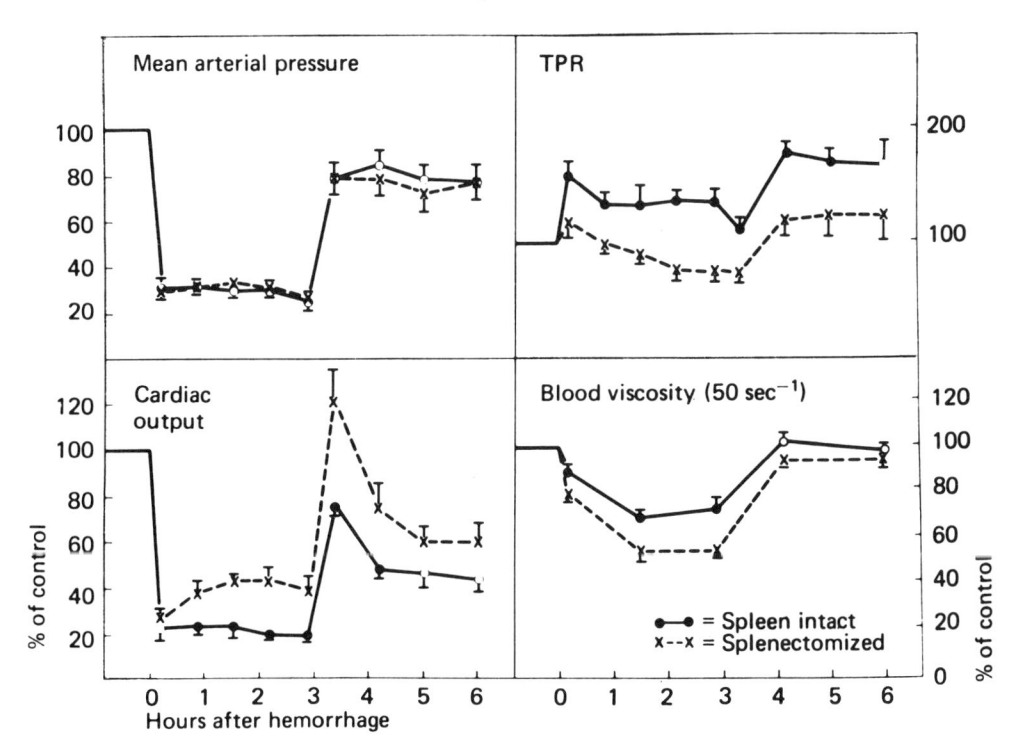

Fig 9 Changes in mean arterial pressure, cardiac output, total peripheral resistance (TPR) and blood viscosity (at a shear rate of 50 sec^{-1}) in hemorrhagic shock in unanesthetized beagles. The shed blood was retransfused at the end of 3 hr. The circles joined by solid lines represent the mean results on 15 dogs with spleen and the crosses joined by broken lines represent the mean results on 11 splenectomized dogs. Horizontal bars represent standard errors of the mean. Student t test is used to evaluate the significance of difference between these two groups, and solid circles are used for the dogs with spleen whenever the P value is less than 0.05 (from ref. 4).

larger control blood volume than the splenectomized dogs and they suffered a larger bleeding volume for the same degree of hypotension. At the same arterial pressure, the dogs with spleen had a lower cardiac output than the splenectomized dogs and hence a higher TPR. This difference in TPR was accompanied by higher viscosity values in the dog with spleen. The lower viscosity values in the splenectomized dogs were associated with a decrease in TPR during most of the oligemic period.

During oligemia, decreases in arterial and venous pressures, together with the sympathetically induced precapillary constriction, reduced the capillary pressure and caused transcapillary fluid influx (31,32). Therefore, the hematocrit and plasma protein concentration decreased during oligemia (Fig 10). In early

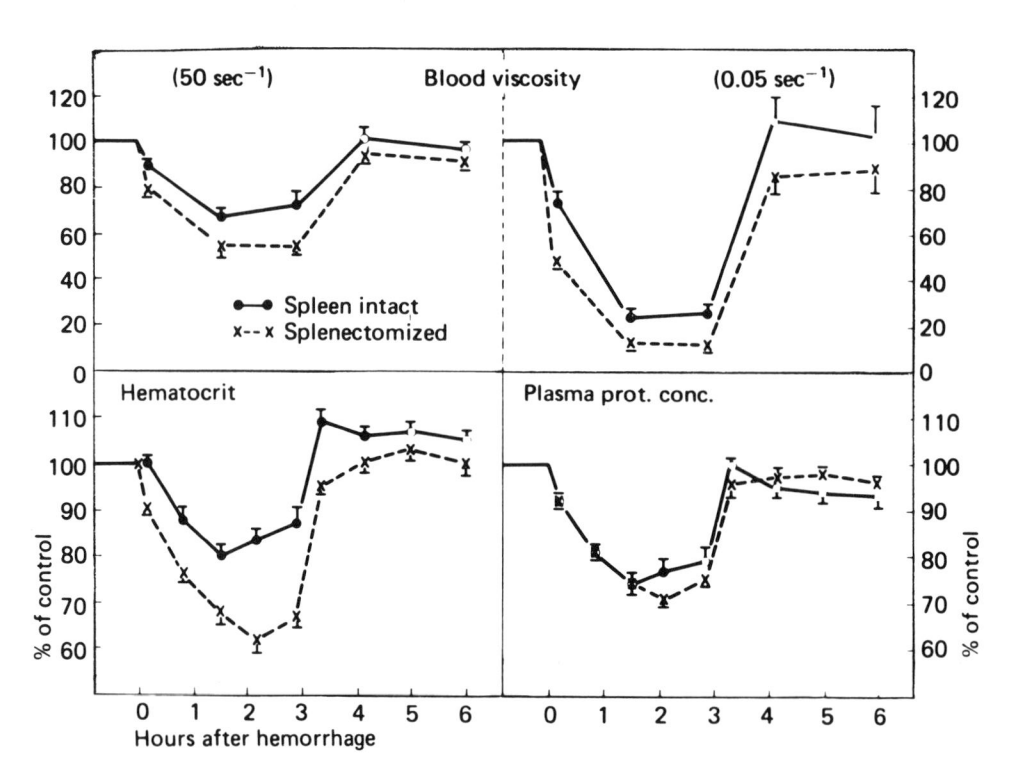

Fig 10 Changes in blood viscosity (50 and 0.05 sec^{-1}), hematocrit and plasma protein concentration in hemorrhagic shock in unanesthetized beagles. See Fig 9 for explanations of symbols (from ref. 4).

oligemia the decrease in plasma protein concentration was almost identical in the splenectomized dogs and in those with spleen, indicating a similar degree of fluid influx. As a result of the release of cell-rich blood by the spleen, however, the dogs with spleen showed higher hematocrit and hence higher viscosity values than the splenectomized dogs (Fig 10). According to the concept developed above, the lower viscosity curve in the splenectomized dogs should result in a lower post-/pre-capillary resistance ratio and a greater fluid influx. Evidence for such a concept is provided by direct measurement of the plasma volume (Fig 11). Early in oligemia the increase in plasma volume was similar in the 2 groups, but as oligemia proceeded the plasma volume was much better maintained in the splenectomized dogs. This difference in transcapillary fluid influx between the splenectomized dog and the dog with spleen is shown in the lower panel of Fig 11, and can be correlated with the difference in

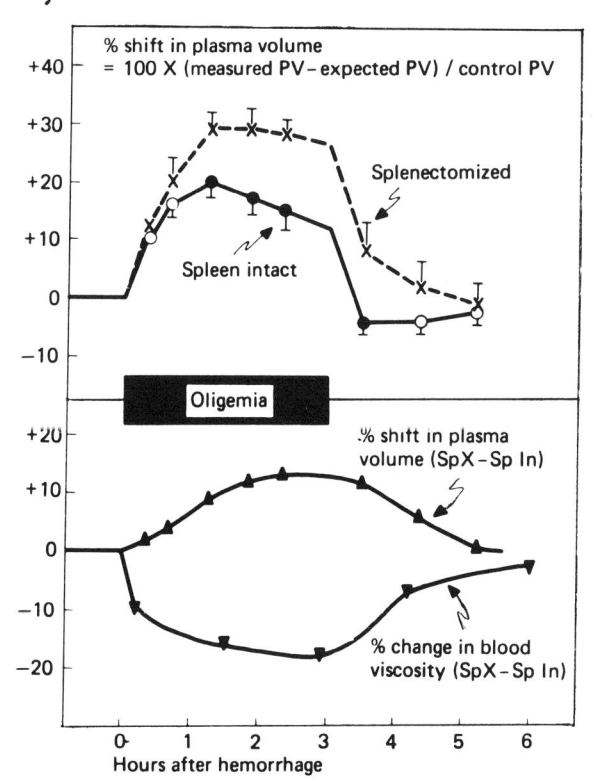

Fig 11 Upper panel: Effects of hemorrhagic shock on shift of plasma volume in unanesthetized beagles. Note the larger transcapillary influx in splenectomized dogs during late oligemia. See Fig 9 for explanation of symbols.

Lower panel: Differences between splenectomized dogs and dogs with spleen in plasma volume shift (upright triangles) and blood viscosity at 50 sec-1 (inverted triangles) (from ref. 4).

blood viscosity. Thus the lower blood viscosity in the splenectomized dogs was associated with not only a lower flow resistance, but also a larger transcapillary fluid influx.

A comparable mortality (approximately 1/2 died) was observed in splenectomized dogs and dogs with spleen bled to the same degree of arterial hypotension. Since a larger volume was bled from the dog with spleen, it follows that they actually had a greater tolerance to hemorrhage if volume removal rather than the level of hypotension were used as the criterion. Therefore, although the release of cell-rich blood from the spleen in hemorrhagic shock tends to impede transcapillary fluid influx, it exerts an overall beneficial effect. There are at least 2 possible explanations for this. First, the volume expansion effect of the splenic reservoir may compensate for the lesser fluid influx. Second, the maintenance of a higher hematocrit (approximately 40% in late oligemia) than that attained in the splenectomized dogs (approximately 25%) increased the oxygen carrying capacity and favoured O_2 transport. The lower O_2 capacity in the splenectomized dogs was indeed accompanied by a higher cardiac output (Fig 9) at the same survival rate as the dogs with spleen. Therefore, these results suggest that the critical factor in surviving hemorrhage is the rate of O_2 transport or delivery to tissues.

GENERAL DISCUSSION

The results on endotoxin shock and hemorrhagic shock in the dog indicate that rheological changes significantly affect flow resistance and transcapillary fluid transfer. In both forms of shock, the splenectomized dogs had lower values for hematocrit, blood viscosity and flow resistance and they showed a greater tendency for fluid influx than the dogs with spleen. The influence of the spleen on survival, however, is different in these two forms of shock. In endotoxin shock, the fundamental disturbance is a reduction in flow without an initial loss of blood volume, and the release of cell-rich blood from the spleen further aggravates the rheological disturbance to flow. Therefore, O_2 transport is reduced by the decrease in flow, despite the increase in O_2 capacity, resulting in the detrimental effect of the spleen in endotoxin shock. In hemorrhagic shock, the initial disturbance is loss of blood volume, and the O_2 capacity is greatly reduced by transcapillary fluid influx. The release of splenic blood tends to expand the blood volume and increase the O_2 capacity. These beneficial effects of splenic contraction outweigh the detrimental flow effects, resulting in an increase in tolerance to hemorrhagic shock. The present study illustrates the interaction between the two components of O_2 transport in influencing survival. In endotoxin shock, the hematocrit is high (approximately 60% in dogs with spleen and 40% in splenectomized dogs) and rheological influences on flow resistance and fluid shifts become the limiting factor. In hemorrhagic shock, the hematocrit is low in the oligemic phase (approximately 40% in the dogs with spleen and 25% in the splenectomized dogs), and alteration in O_2 capacity assumes critical importance. These findings are in agreement with the concept of an optimum hematocrit for O_2 transport, with both the blood flow and O_2 capacity taken into account (33,34).

In this analysis of the rheological changes in shock, emphasis has been placed on the comparison between splenectomized dogs and dogs with spleen. The difference between these two groups has been attributed to the release of cell-rich blood by the spleen. The possibility that the canine spleen may play other roles in shock (4) should be further studied. It is conceivable that the splenic release of formed elements other than erythrocytes (platelets, leucocytes) may contribute to the difference observed.

The present investigation on hemorrhagic and endotoxin shock indicate that rheological changes assume a much greater role in the dog with spleen than in the splenectomized dog. Therefore, these rheological changes would not be very important in species without a large, contractile spleen. Indeed, our studies on the monkey in hemorrhagic and endotoxin shock (unpublished observation) have shown that the rheological changes are closer to those seen in the spenectomized dog than in the dog with spleen. The results of hemodynamic measurements by several investigators (35-38) on various forms of shock in man are summarized in Table 1. In hemorrhagic shock and shock due to infection, the hematocrit is below normal and blood viscosity is probably reduced. Despite signs of intense vasoconstriction, the measured TPR in hemorrhagic shock is often unchanged, indicating that the vasoconstriction is balanced by the decrease in viscosity, as in the splenectomized dogs in the present study. In Table 1, the only form of shock with significant hemoconcentration is burns. Therefore, the type of rheological disturbance seen in endotoxin shock in the dog might be operative in burn shock in man. At a comparable level of low cardiac output in various forms of shock, the burnt patient shows the highest arterial pressure and hence the highest TPR, probably reflecting the superimposition of elevated viscosity on vasoconstriction. Our studies on the dog would indicate that the balance between the rheological influence on flow and the O_2 capacity should be considered in transfusion or fluid replacement therapy in human patients in shock.

In the present analysis, emphasis has been placed on the behaviour of the whole blood viscosity, which is regulated by many

TABLE 1 CIRCULATION IN SHOCK IN MAN*

	BV (L/m^2)	Hct (%)	BP (mmHg)	CO (L/min/m^2)	TPR (vs. normal)	RAP
Normal	2.9	45	120/80	3.5		
Hemorrhagic shock	1.6	32	90/50	1.8	= or ↑	↓
Burn shock	1.8	50	110/70	1.8	↑	↓
Cardiogenic shock	2.5	47	85/70	1.8	↑	↑
Shock due to infection	2.7	42	95/55	1.8	↑	=

*Summary of measurements by several groups of investigators (35,36,37). Numbers represent only general trend. Adapted from Ref. 38.

factors. Further experimental studies are needed to investigate the factors affecting cell deformation and cell aggregation in low flow states. Recent studies by Braasch (39) and by Folkow's group (40) have shown an alteration of red cell deformability in hemorrhagic shock. This represents another important rheological change in shock which deserves further investigation.

SUMMARY

The factors regulating blood viscosity at different flow conditions have been analyzed. Considerations of the shear rates in the circulation and the shear dependence of blood viscosity indicate that decrease in flow and increase in viscosity may lead to transcapillary fluid loss. In hemorrhagic shock and endotoxin shock in the dog, blood viscosity measured in vitro have been correlated with flow resistance and fluid shifts determined in vivo. In both forms of shock, the release of cell-rich blood from the spleen is associated with increases in blood viscosity and flow resistance and a tendency to fluid loss. In endotoxin shock these rheological changes are detrimental to survival. In hemorrhagic shock, however, the release of splenic blood may improve O_2 transport and survival by expanding the blood volume and increasing the O_2 capacity. A comparison of these findings in the dog with data on primates indicates important species differences. Nevertheless, the present study offers some insight into the possible role of rheological factors in the pathogenesis and management of injury in man.

REFERENCES

1. Gelin, L.-E. (1956) Acta Chir. Scand. Suppl. 210:1.

2. Dintenfass, L. (1962) Circulation Res. 11:233.

3. Wells, R.E., Jr. (1964) New Engl. J. Med. 270:832 and 889.

4. Chien, S. (1969) Advances in Microcirculation 2:89.

5. Merrill, E.W. (1969) Physiol. Rev. 49:863.

6. Schmid-Schönbein, H. & Wells, R.E., Jr. (1971) Ergeb. der Physiologie 63:146.

7. Chien, S. (1970) Science 168:977.

8. Chien, S., Usami, S., Dellenback, R.J. & Gregersen, M.I. (1970) Am. J. Physiol. 219:136.

9. Chien, S., Usami, S., Dellenback, R.J. & Gregersen, M.I. (1970) Am. J. Physiol. 219:143.

10. Chien, S., Usami, S., Dellenback, R.J. & Gregersen, M.I. (1967) Science 157:827.

11. Schmid-Schönbein, H. & Wells, R. (1969) Science 165:288.

12. Goldsmith, H.L. (1971) Biorheology 7:235.

13. Schmid-Schönbein, H., Gaehtgens, P & Hirsch, H. (1968) J. Clin. Invest. 47:1447.

14. Chien, S., Usami, S & Jan, K.M. (1971) The Symposium on Flow - Its Measurement and Control in Science and Industry. (ed. R.B. Dowdell), Pittsburgh, 1971.

15. Chien, S., Usami, S., Taylor, H.M., Lundberg, J.L. & Gregersen, M.I. (1966) J. Appl. Physiol. 21:81.

16. Chien, S. & Gregersen, M.I. In: Physical Techniques in Biological Research, edited by W.L. Nastuk, Vol 4: p 1 Academic Press, New York, 1962.

17. Cokelet, G.R. & Meiselman, H.J. (1968) Science 162:275.

18. Schmidt-Nielsen, K. & Taylor, C.R. (1968) Science 162:274.

19. Fung, Y.C. (1966) Federation Proc. 25:1761.

20. Hochmuth, R.M. & Das, N.M. (1970) Am. Soc. Mech. Engrs. Preprint 70-WA/BHF-12.

21. Skalak, R. in Biomechanics: Its Foundations and Objectives (ed. by Y.C. Fung), p. 457, Prentice-Hall, Englewood Cliffs, N.J. 1971.

22. Fung, Y.C.B. & Tung, P. (1968) Biophys. J. 8:175.

23. Chien, S., Usami, S. & Bertles, J.F. (1970) J. Clin. Invest. 49:623.

24. Weed, R.I. (1970) Am. J. Med. 49:147.

25. LaCelle, P.L. (1970) Seminars Hematol. 7:355.

26. Chien, S., Luse, S.A. Jan, K.M., Usami, S., Miller, L.H. & Fremount, H. (1971) Proc. 6th Europ. Conf. Microcirc. Karger, Basel, p. 29-34, 1971.

27. Green, H.D. In: Medical Physics, (edited by O. Glasser), Vol. 2, p. 228, Year Book Publishers, Chicago, Ill. 1944.

28. Wiedman, M.P. (1963) Circulation Res. 12:375.

29. Chien, S., Chang, C., Dellenback, R.J. Usami, S. & Gregersen, M.I. (1966) Am. J. Physiol. 210:1401.

30. Chien, S., Dellenback, R.J. Usami, S., Treitel, K. Chang, C. & Gregersen, M.I. (1966) Am. J. Physiol. 210:1411.

31. Chien, S. (1967) Physiol. Rev. 47:214.

32. Haddy, F.J. & Scott, J.B. (1968) Physiol. Rev. 48:688.

33. Crowell, J.W., Bounds, S.H. & Johnson, W.W. (1958) Am. J. Physiol. 192:171.

34. Crowell, J.W., Ford, R.G. & Lewis, V.M. (1969) Am. J. Physiol. 196:1033.

35. Cournand, A., Riley, R.L., Bradley, S.E., Breed, E.S., Noble, R.P. Lauson, H.D. Gregersen, M.I. & Richards, D.W. (1943) Surgery 13:964.

36. Freis, E.D., Schnaper, H.W., Johnson, R.L. & Schreiner, G.E. (1952) J. Clin. Invest. $\underline{31}$:131.

37. Udhoji, V.H., Weil, M.H., Sambhi, M.P. & Rosoff, L. (1963) Am. J. Med. $\underline{34}$:461.

38. Chien, S. & Gregersen, M.I. In Medical Physiology (edited by V.B. Mountcastle) 12th edition, p 262 C.V. Mosby, St. Louis, 1968.

39. Braasch, D. (1970) Abstr. VI. Conf. Microcirc. p. 23, Aalborg, Denmark.

40. Baeckström, P., Folkow, B., Löfving, B., Kovach, A.G.B. & Öberg, B. (1970) Abstr. VI. Conf. Microcirc. p. 25, Aalborg, Denmark.

ACKNOWLEDGMENTS

This investigation was supported by U.S. Public Health Service Research Grant HE-06139 from the National Heart and Lung Institute, by U.S. Army Medical Research and Development Command Contract DA-49-193-MD-2272, by the Scaife Family Charitable Trust in Pittsburgh, Penn. and by the Pharmacia Lab., Inc.

Dr. Magazinovic was U.S.P.H.S. National Institutes of Health International Postdoctoral Fellow. Present address: Department of Physiology, Veterinary Faculty, University of Sarajevo, Sarajevo, Yugoslavia.

RHEOLOGY OF RED CELL SUSPENSIONS IN

EXPERIMENTAL DEHYDRATION

F. Torres and L.-E. Gelin

Department of Surgery I, University of Göteborg

Sahlgrenska Sjukhuset, Gothenburg, Sweden

INTRODUCTION

Dehydration is a serious clinical condition which in itself can deleteriously complicate the course of a basic disease such as in burns, intestinal obstruction, vomiting, diarrhea, and heatstroke. The hemodynamic effects of dehydration are mainly due to decreased blood volume and hemoconcentration (Calcagna and Rubin, 1951; Senay and Christensen, 1965). Hemoconcentration has several rheological consequences: increased hematocrit, increased protein concentration and increased osmolarity all of which will affect blood viscosity.

Increased viscosity of blood is known even in case of slightly increased blood volume and arterial blood pressure to decrease venous return (Gelin et al., 1968) and tissue perfusion (Litwin et al., 1965), especially in organs which are nourished by a postcapillary flow system (Fajers and Gelin, 1959) such as the portal flow of the liver, the postglomerular flow of the nephron and the portal flow of the anterior lobe of the pituitary gland.

Since the relationship between dehydration with its associated phenomena and the viscous properties of blood has not been delineated experiments on the rheologic alterations of blood in dehydration were initiated using an evaporation method of blood in vitro.

MATERIAL AND METHODS

400 ml of blood were drawn from 23 healthy blood donors into glass bottles containing 100 ml ACD solution and stored at 4° for

3-10 days at the blood bank of Sahlgrenska sjukhuset. The viscosity and hematocrit of this blood was determined after storage and after careful mixing at the time of study. The ACD blood was then centrifuged in plastic tubes at 2500 rpm for 15 minutes = 1050 g/units to separate plasma and cells.

Plasma was removed by slow suction and divided into three aliquots. One aliquot was used for control and the other two were prepared by evaporation. The viscosity protein concentration and protein pattern and osmolarity of the different plasma samples were analysed.

Evaporation of plasma. The two aliquots of plasma for evaporation were first weighed with an accuracy of 1 mg using a Mettler balance and then frozen for 3 hours at -20°. The frozen aliquots were placed in a vacuum freeze-dryer model 1 PTD Edward Alta Vuato for 20 hours of gradual total evaporation. The aliquots were reweighed. To reconstitute the dry plasma to a desired degree of rehydration i.e. 90, 80, 70 or 50% of original value either distilled water or saline was added. The aliquots were then reweighed to confirm the amount of fluid added. The values given in results are according to the weights after final redilution.

Blood cells. The packed cell portion after centrifugation was washed twice in saline and then used for resuspensions in the plasma samples so as to produce red cell suspensions with hematocrits varying between 5 and 70. From each suspension hematocrit determination was done in duplicate.

Hematocrit was determined in capillary tubes after centrifugation at 12,000 i.p.m. for 5 minutes = 8000 g/units. The capillary tube method was compared to the Wintrobe method and no significant difference was noted. The number of red cells per mm^3 of suspension was determined in an automatic particle counter Celloscope (AB Lars Ljungberg, Stockholm, Sweden).

Protein concentration of plasma was determined according to the Folin phenol reagent method of Lowry, 1951.

Plasma protein pattern was analysed electrophoretically using the Beckman Microzone method and is accounted for in Table 1.

Osmolarity of plasma samples was determined with a freezing point depression technique using Advanced Osmometer Model 66 31 LAS.

Viscosity of whole blood, plasma samples and reconstituted suspensions of red cells in the different plasmas were determined at 37° using a Wells-Brookfield Microviscometer (Brookfield

TABLE I

ACD Plasma samples	g/100ml Total Protein	Percent of total protein					
		Albumin	α_1	α_2	β	γ	Fibrinogen
				Globulin			
Normal	5.0	55.3	2.5	8.0	12.3	10.7	11.2
100% Reconstituted	5.0	56.7	2.5	8.3	12.0	8.8	11.7

Engineering Laboratories Inc., Staughton, Mass., U.S.A.) at the shear rates of 6, 12, 30 and 60 rotations per minute corresponding to 23, 46, 115 and 230 inverse seconds (Gelin, 1961; Wells, 1964). The viscosity data were plotted in 3 ways: viscosity expressed as centipoise against shear rate in inverse seconds, viscosity against hematocrit and viscosity against number of red cells.

RESULTS

Viscosity of ACD Blood and Plasma

The viscosity values obtained on the stored ACD blood and plasma from 23 blood donors are given in Fig. 1 as mean values with total deviations. The hematocrits and plasma concentrations are lower than in normal blood due to the added ACD solution. They represent the initial values for the experiments reported here.

Effect of Plasma Dehydration on Plasma Viscosity and Protein Pattern

The freeze dried plasma diluted to 90, 80, 70 and 50 per cent of the original ACD plasma were taken for viscosity determination and analysed for protein concentration, as well as for electrophoretic pattern and osmolarity. One aliquot of dried plasma was reconstituted with distilled water to 100 per cent of its original weight. This fully reconstituted plasma should be equivalent to the initial values for viscosity, protein concentration, electrophoretic pattern and osmolarity. The slight deviations from original values observed thus include all the errors during the handling of the different samples.

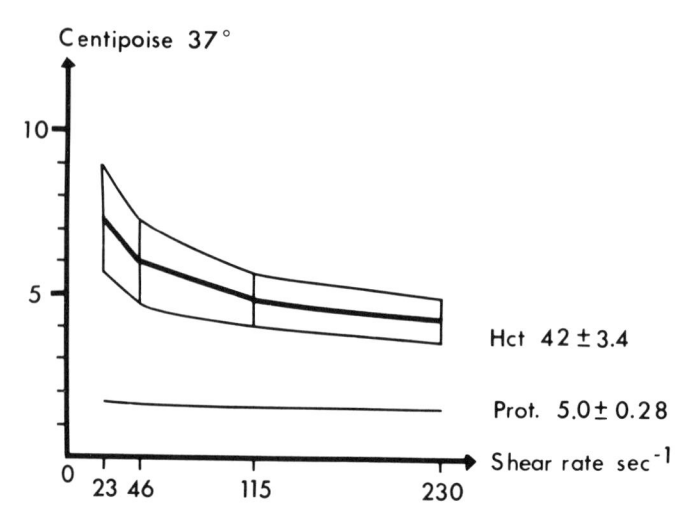

Fig. 1. Viscosity of ACD-blood and plasma.

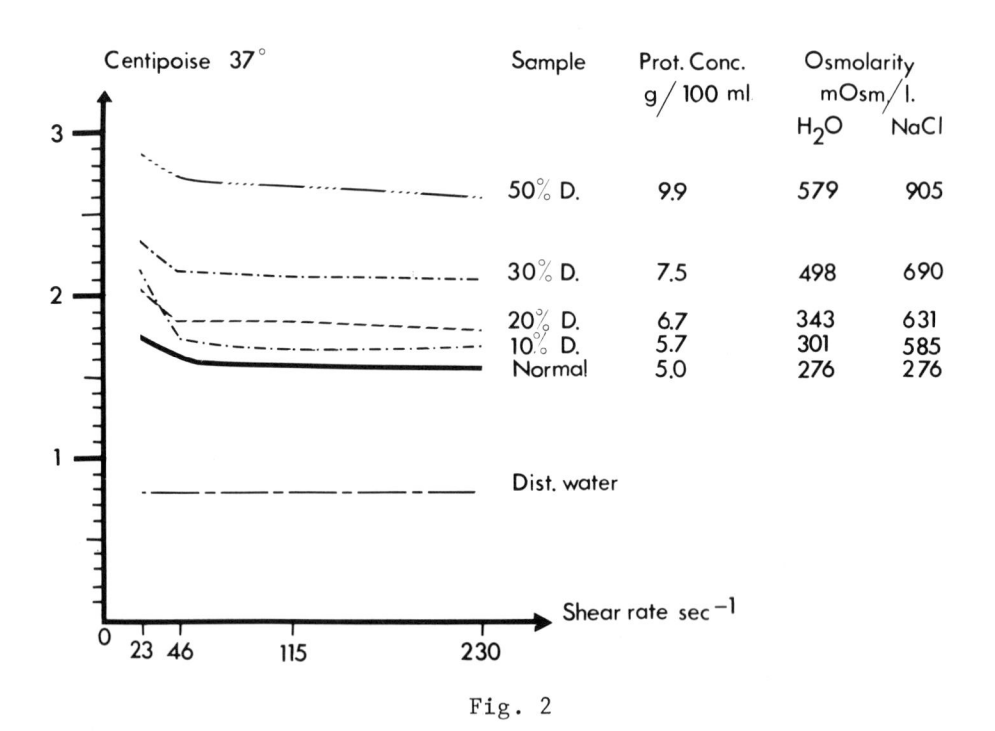

Fig. 2

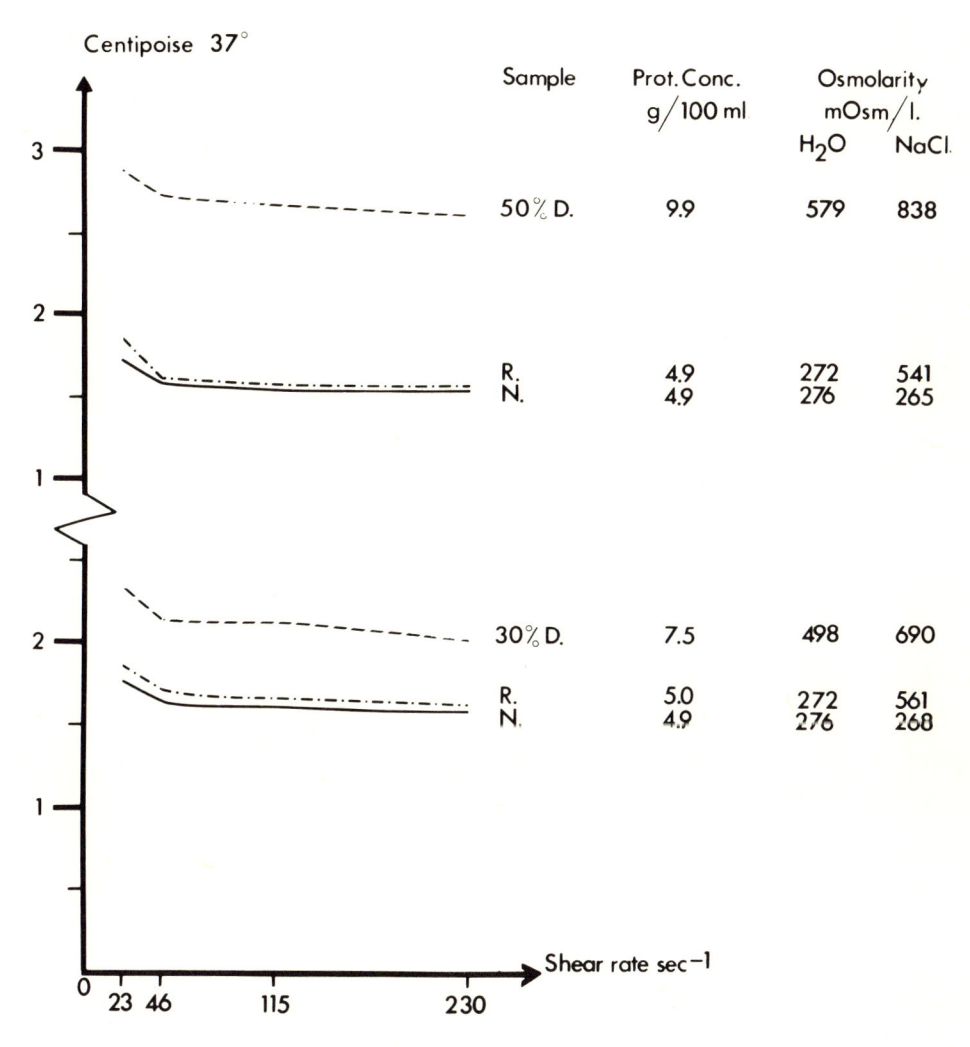

Fig. 3. Viscosity of dehydrated and reconstituted ACD plasma.

The data are given in Table I and Fig. 2 as mean values of five experiments for each group of dehydration.

The viscosity and osmolarity of the original plasma (N-plasma), the dehydrated plasma (D-plasma) and the fully reconstituted plasma (R-plasma) are given in Fig. 3 for experiments on 30 and 50 per cent dehydration.

As apparent from Table I and Fig. 2 and 3 dehydration of plasma caused a significant increase of the plasma viscosity paralleling the raise in protein concentration. Wide variations in plasma osmolarity did only produce minimum influence on plasma viscosity. The experimental procedure induced no significant alterations of the electrophoretic pattern.

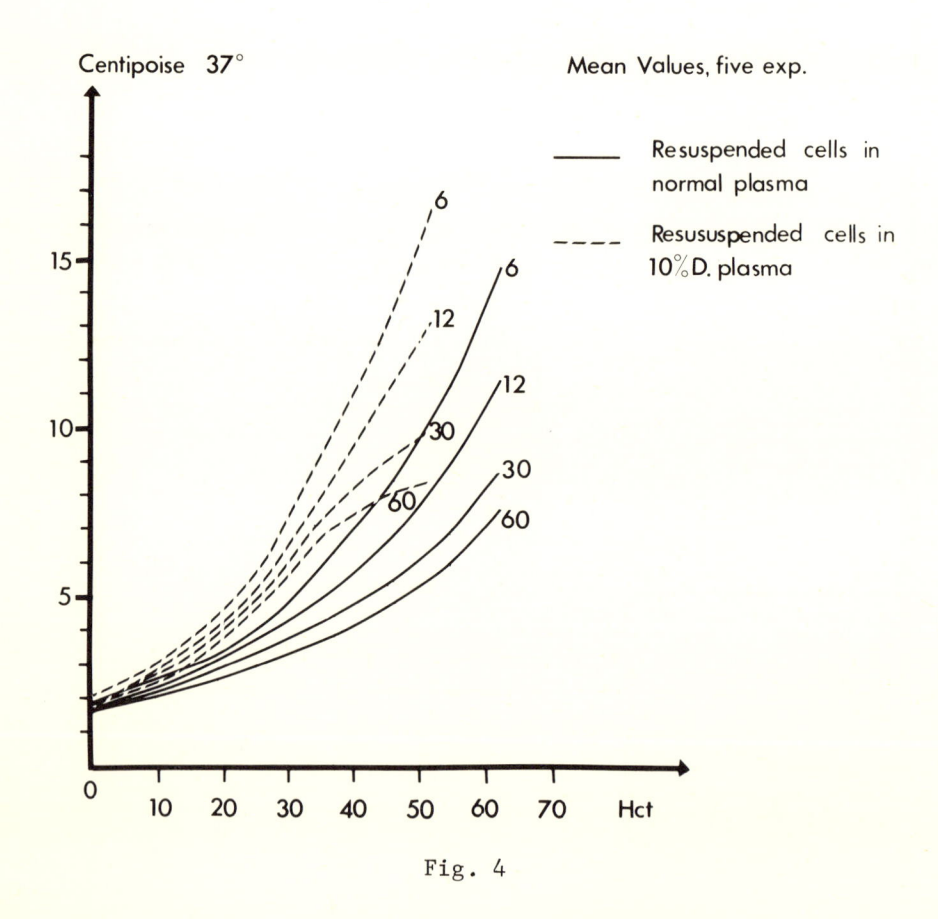

Fig. 4

Effect of Plasma Dehydration on the Viscosity of Red Cell Suspensions

The plasmas of varying degree of dehydration were used as vehicles for resuspension of red cells. The cells were added to the vehicle so as to produce suspensions with hematocrits, increasing from 5 to 70. The viscosity value of such individual suspension was plotted against hematocrit. The results given are mean values of five experiments for each different vehicle, i.e., 10, 20, 30, and 50 per cent dehydration in the Fig. 4, 5, 6, and 7 and after full reconstitution with distilled water in Fig. 8. All values are compared with the values obtained from resuspensions of the cells in their original plasma.

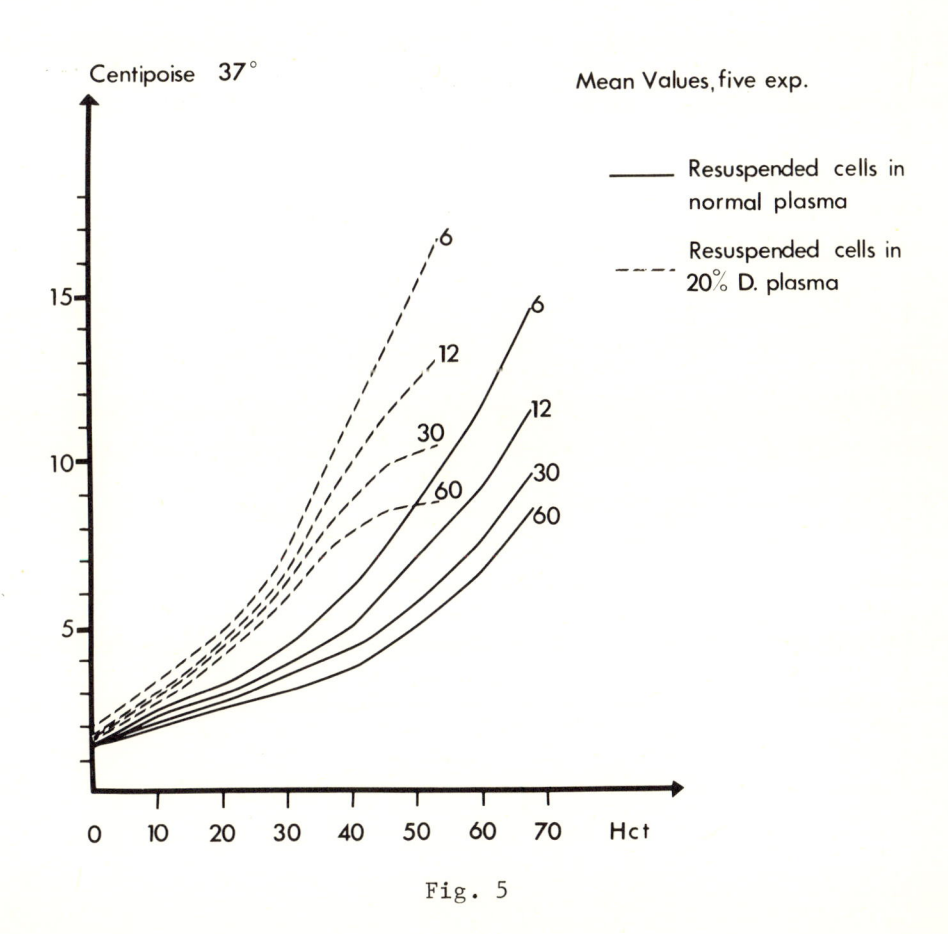

Fig. 5

Fig. 6

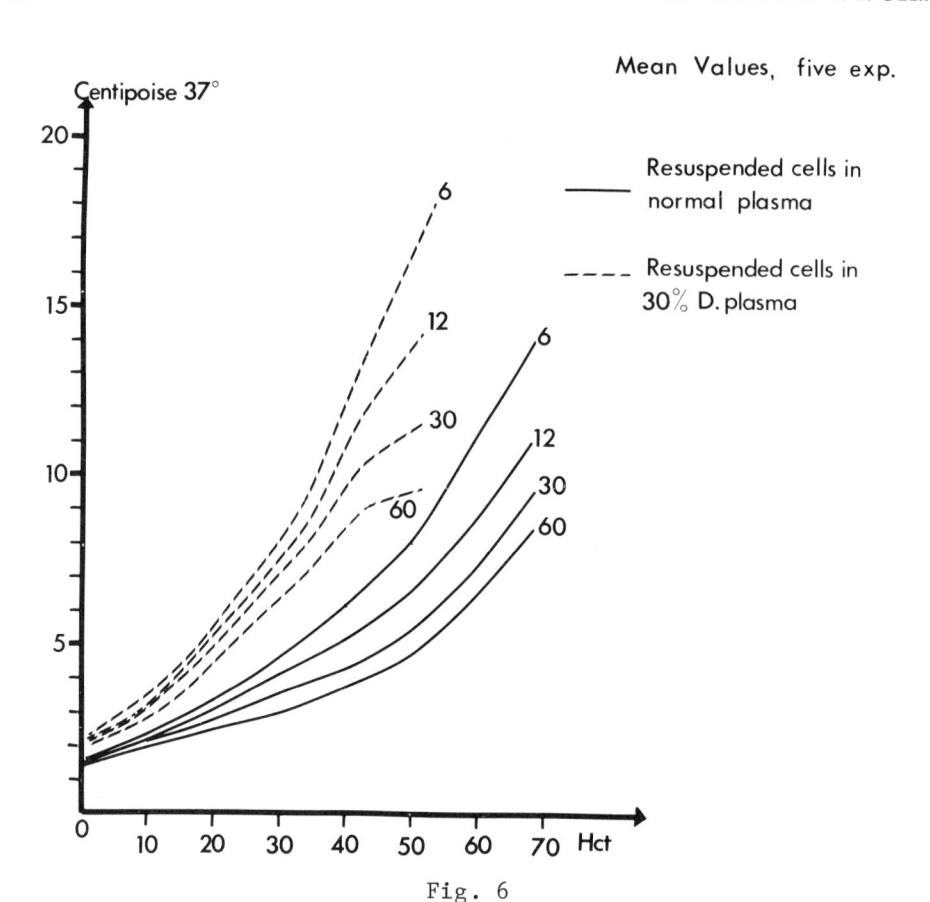

From the figures appears that with increasing degree of dehydration the viscosity of the vehicle increased. When the cells were resuspended in the different vehicles the viscosity of the suspensions increased with increasing hematocrits and decreasing shear rates and more so the higher the degree of dehydration of the plasma. It is also evident from the Figures 4, 5, 6, and 7 that the higher the degree of of dehydration the more marked was the increase of suspension viscosities even at low hematocrits. At hematocrits above 40 the viscosity curves tended to level off with increasing hematocrits at the higher but not at the lower shear rates.

Effect of Vehicle Osmolarity on the Viscosity of Red Cell Suspensions

In order to elucidate the influence of plasma osmolarity on the cell size and the viscosity of red cell suspensions the viscosity data were plotted both against hematocrit and number of cells.

The data from a comparative study after reconstitution to 50 per cent of the dried plasma with distilled water and with saline is given in Table II. The suspension viscosities at the shear rates of 23, 46, and 230 inverse seconds are given in Fig. 9, 10, and 11 both against hematocrit and against number of cells.

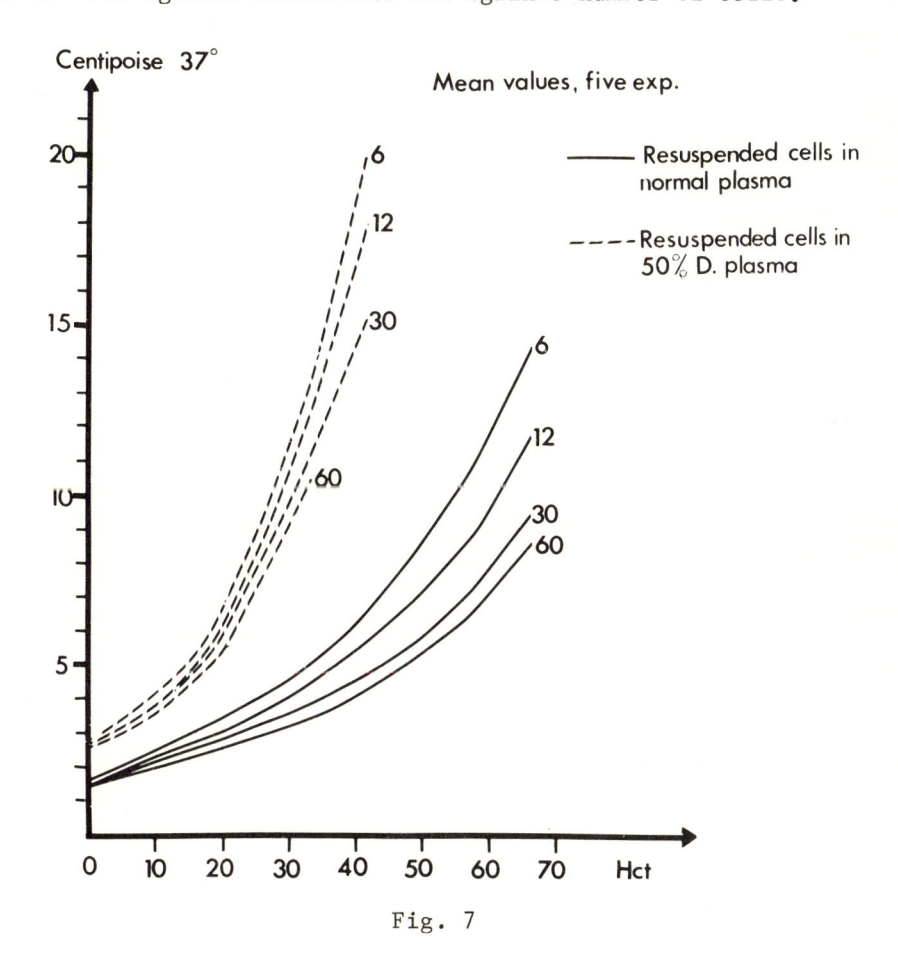

Fig. 7

Fig. 8

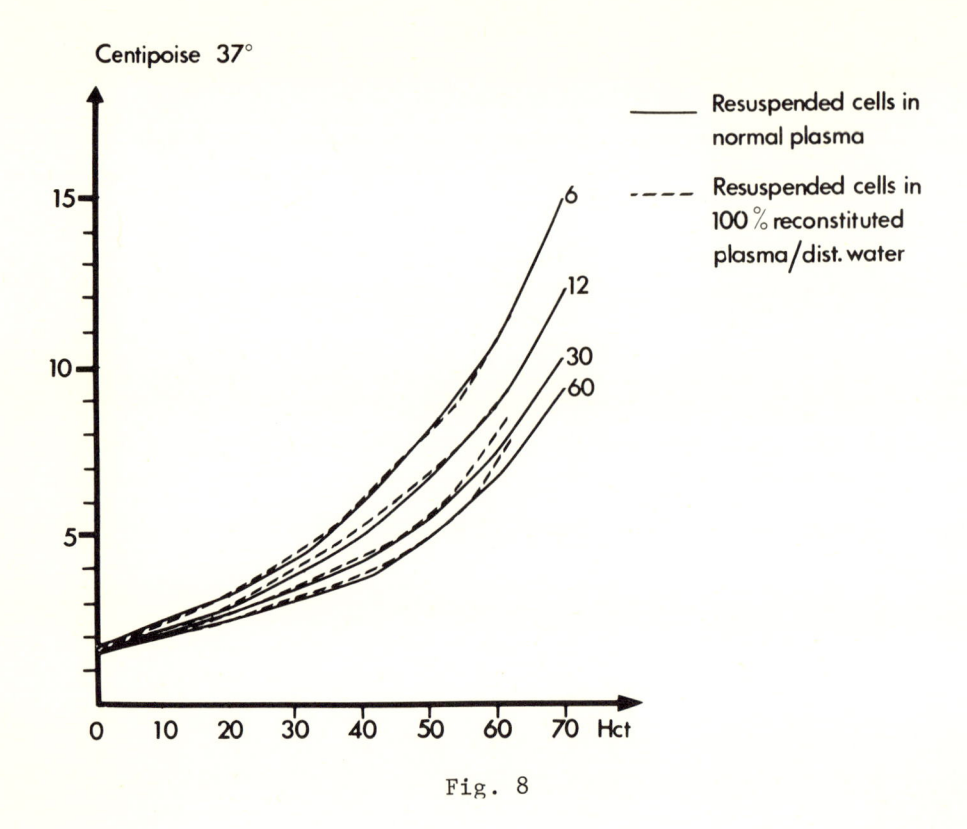

From Table II appears that the number of cells per mm³ packed cells will increase with an increase of plasma osmolarity. The vehicle osmolarity was about 270 m/osm for the original ACD-plasma, 580 m/osm for 50 per cent reconstituted plasma with distilled water and 900 m/osm for 50 per cent reconstituted plasma with saline.

The shape of the viscosity curves demonstrates that full reconstitution with distilled water restored the viscous properties to initial property both when related to hematocrit and to number of cells. The shape of the viscosity curves was different in dehydration with a levelling off at higher shear rates with increasing hematocrits and increasing number of cells.

TABLE II
Viscosity of Suspensions of Cells in Plasmas of Different Hydration and Osmolarity

Vehicle	Viscosity (sec^{-1}) 23	46	115	230	Hct in sample	Nr. of cells per mm^3 in sample	Nr. of cells per ml packed cells	Prot. Conc. g/100 ml	Osmolarity m Osm/l
Normal plasma	1,7	1,6	1,6	1,6	0	-	-	5,0	276
	2,5	2,2	2,1	2,0	10	0,88	8,8		
	3,1	2,8	2,6	2,5	20	1,76			
	4,3	3,8	3,3	3,1	32	2,81			
	6,2	5,0	4,2	3,7	41	3,60			
	8,5	6,8	5,6	5,0	51	4,48			
	10,7	8,8	7,3	6,7	60	5,28			
	15,0	12,3	10,2	9,3	70	6,16			
50 % Dehydrated plasma/H$_2$O	2,9	2,7	2,7	2,6	0	-		9,9	579
	3,5	3,3	3,2	3,1	6	0,76	12,8		
	4,7	4,3	4,2	3,8	12,5	1,6			
	5,3	5,2	5,0	4,7	19	2,4			
	6,8	6,6	6,3	6,1	26	3,3			
	9,2	8,6	7,7	7,2	34	4,3			
	11,8	10,2	8,5	7,6	43	5,5			
	13,8	11,2	8,7	7,6	51	6,5			
50 % Dehydrated plasma/NaCl	3,0	2,7	2,7	2,6	0	-		9,9	905
	3,8	3,7	3,6	3,6	5,5	0,78	14,2		
	4,5	4,3	4,1	4,1	11	1,56			
	5,6	5,5	5,2	5,1	17	2,4			
	7,1	6,8	6,5	6,2	23,5	3,3			
	9,7	9,3	9,0	8,7	29,5	4,2			
	13,1	12,5	11,3	10,5	37	5,3			
	19,3	17,7	14,7	11,5	44	6,2			
Dehydrated plasma Fully reconstituted with distilled water	1,7	1,6	1,6	1,6	0			4,9	272
	2,3	2,1	2,0	2,0	9	0,81	9,0		
	3,0	2,6	2,5	2,3	17,5	1,57			
	4,1	3,7	3,2	3,0	27	2,4			
	5,3	4,7	4,0	3,5	36	3,2			
	7,5	6,0	4,8	4,3	45	4,1			
	8,8	7,5	6,0	5,3	54	4,86			
	11,5	9,3	7,6	7,0	62	5,6			

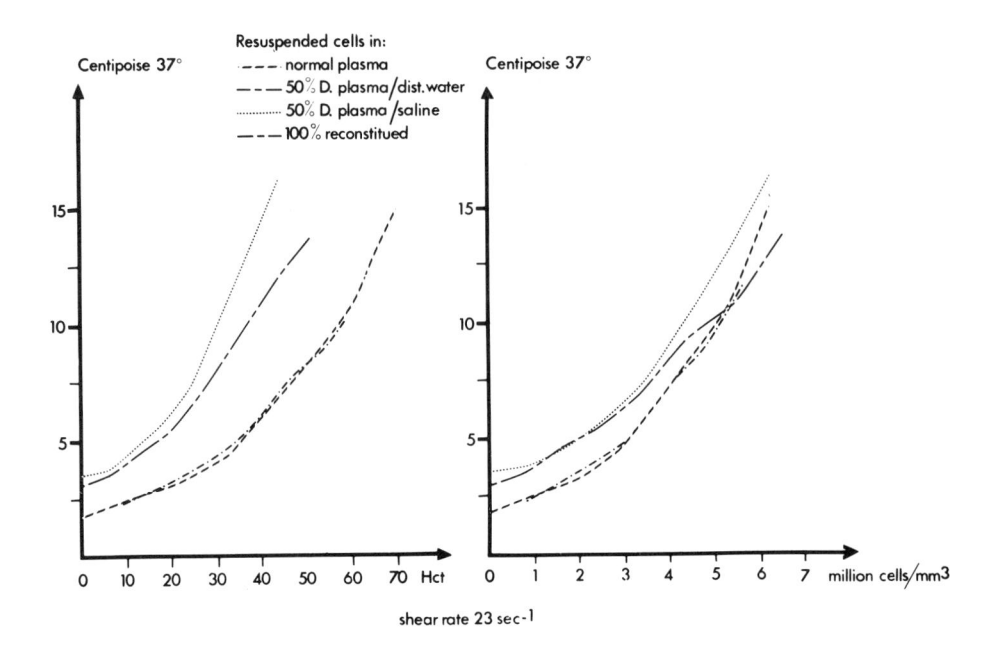

Fig. 9

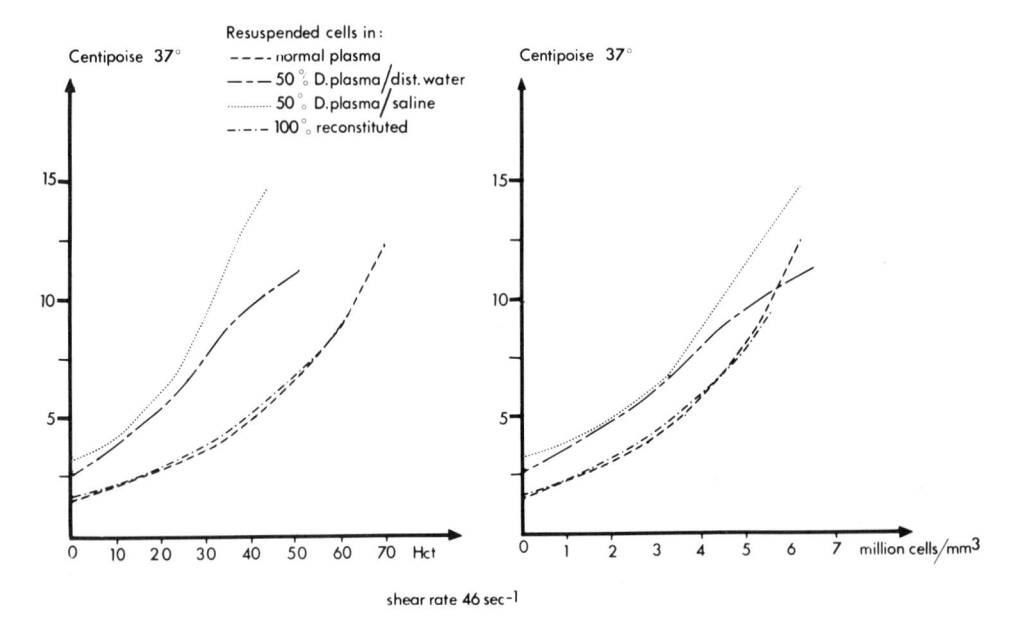

Fig. 10

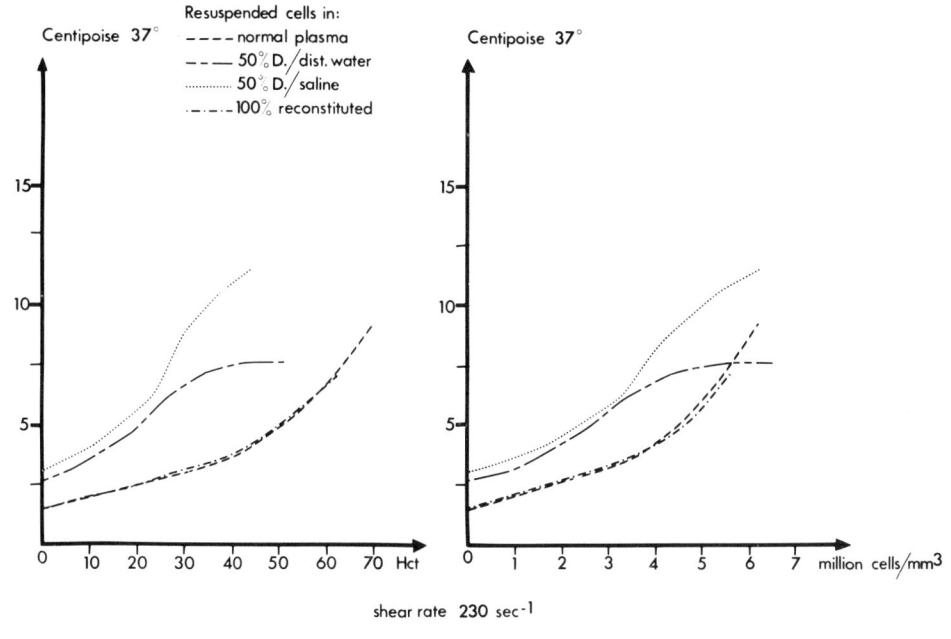

Fig. 11

DISCUSSION

Dehydration of plasma by a freeze drying procedure increased the viscosity of the plasma and of the suspension of cells. Full reconstitution of freeze dried plasma with distilled water restored the viscosity of plasma and the viscosity of cells suspensions to initial values, which leads to the conclusion that the freeze-drying procedure did not seriously invalidate the viscosity data despite some minor alterations were noted in the electrophoretic plasma protein pattern.

Plasma Viscosity increased with increasing degree of dehydration which paralleled the simultaneously increasing concentrations of protein, while even wide variations in osmolarity only very slightly affected plasma viscosity. One important observation was the shear rate dependence of plasma with increasing degree of dehydration. In our measurements normal plasma in contrast to water has a slight shear rate dependence in vitro which is in disagreement with some other investigators. The changes in viscous properties obtained at increasing degree of dehydration stresses the non-Newtonian characteristics when the concentration of proteins increased. The influence of a relative increase in the concentrations of the more viscous proteins, i.e., α- and β-globulins and fibrinogen, will become more important as dehydration proceeded and thereby emphasize the non-Newtonian characteristics of plasma. Senay and Christensen, 1965, observed in clinical dehydration due to heat not only an increased protein concentration but also a lowering of the albumin/globulin ratio.

Suspension viscosity. When red cells were resuspended in plasmas with different degrees of dehydration the suspension viscosity increased progressively with the degree of dehydration and much more than expected from the viscosity increase of the plasmas. Comparing the data on suspension viscosities according to volume of cells and to number of cells will show that the suspension viscosity is more dependent on the number than on the volume of cells which is evidenced in Fig. 9, 10, and 11. The shape of these curves demonstrates that full reconstitution with distilled water restored the suspension viscosity to initial values both when related to volume and to number of cells and that the slope of the viscosity curve with increasing hematocrit was steeper than with increasing number of cells at any given shear rate.

The levelling off of the curves at the higher hematocrits and shear rates in dehydration indicates a change in the viscous properties of the cells more than of the suspension. This will mean that the interaction between the particles and the vehicle becomes less significant than the interaction between the particles themselves. From the data in Table II is apparent that the number of cells per mm^3 packed cells increased with dehydration and with osmolarity.

Increased plasma osmolarity will affect cell size and density, i.e., the cell shrinks and becomes rigid. The decrease in cell size means an increased number of particles for any given volume per cent of cells. According to Einstein the viscosity of a suspension is related to volume concentration of particles independent of their size as long as the concentration is low. The data observed here show a closer relation to the number of cells than to the volume per cent of cells. The present data therefore suggests that the suspension viscosity depends more on the friction between each cell and the suspending medium than on the volume they occupy in relation to the flow stream at low shear rates. At higher volume concentration of cells and higher shear rates on the other hand the leveling off of the viscosity curves indicate less shearing forces between the individual cells which then behave more like solid rigid masses than as individual particles. The forces which then will regulate the viscous properties of the cell masses originate from changes of the internal fluidity of the cells as emphasized by Dintenfass, 1962. This rigidity of cells which is one consequence of increased osmolarity therefore should be prevented and treated in clinical dehydration by dilution with water and not with electrolyte solutions to restore normal fluidity of blood.

SUMMARY

In vitro experiments performed on ACD blood to study the effects of variations in water content and osmolarity on the viscous properties of plasma and of red cell suspensions showed that

1. Plasma viscosity increased with dehydration in proportion to protein concentration.

2. Non-Newtonian behavior of plasma was evident with increasing degree of dehydration.

3. Osmolarity did not significantly influence the viscous properties of plasma but markedly the viscous properties of cell suspensions.

4. Increased plasma osmolarity altered the relationship between cell volume and cell number.

5. Increased osmolarity increased the suspension viscosity by affecting cell size, density and shape.

6. In dehydration the viscosity of blood is closer related to the number of cells than to the volume of cells.

REFERENCES

Calcagno, P. L. and Rubin, M. I.: Effect of dehydration produced by water deprivation, diarrhea, and vomiting on renal function in infants. Pediatrics, 17:329-339, 1951.

Dintenfass, L.: Considerations of the internal viscosity of red cells and its effect on the viscosity of whole blood. Angiology, 13:333-344, 1962.

Fajers, C.-M., and Gelin, L.-E.: Kidney-, liver-, and heart-damages from trauma and from induced intravascular aggregation of blood-cells. An experimental study. Acta Path. Microbiol. Scand. 46:97, 1959.

Gelin, L.-E.: Studies in anemia of injury. Acta Chir. Scand. Suppl. 210, 1956.

Gelin, L.-E.: Hematorheological disturbances in surgery. Acta Chir. Scand. 122:287-323, 1961.

Litwin, M. S., Bergentz, S.-E., Carlsten, A., Gelin, L.-E., Rudenstam, C.-M., and Söderholm, B.: Hidden acidosis following intravascular red blood cell aggregation in dogs. Ann. Surg. 161:532-538, 1965.

Lowry, O. H., et al.: Protein measurement with the Folin phenol reagent. S. Biol. Chem. 193:265-275, 1951.

Senay, L. C. Jr., and Christensen, M. L.: Changes in blood plasma during progressive dehydration. J. Appl. Physiol. 20:1136-1140, 1965.

Wells, R. E.: Rheology of blood in the microvasculature. New England J. Med. 270:832-839, 889-893, 1964.

REGULATION OF INTERSTITIAL FLUID VOLUME AND PRESSURE

Arthur C. Guyton, Aubrey E. Taylor, Harris J. Granger and W. Harry Gibson

Department of Physiology and Biophysics, University of Mississippi School of Medicine, Jackson, Mississippi

Recently the structure of the interstitial fluid spaces has been greatly clarified, and with the clarification has also come, we believe, a clear understanding of the mechanisms by which both interstitial fluid volume and pressure are regulated.

Structural Features of the Interstitial Spaces Important to Understanding Interstitial Fluid Volume and Pressure Regulation. Though 1/6 of most tissues is composed of interstitial fluid, this fluid is generally held tightly in place in the form of a gel. The reticulum of this gel is composed mainly of hyaluronic acid, but this is cross-linked slightly with collagen fibers to form a meshwork in which the interstitial fluids are trapped. The presence of this gel reduces the mobility of fluid in the tissues by many thousands to more than a million fold, but it hardly affects the individual molecular diffusion of small molecules such as glucose, urea, and so forth, their diffusion rates being at least 95 per cent as great as the corresponding diffusion rates in free water.

Because of several properties of the interstitial gel, it tends to imbibe fluid. The imbibition pressure of typical interstitial gel (as exemplified by Wharton's jelly from the umbilical cord) is approximately -7 mm. Hg, which means that fluid is actually sucked into the gel from the free fluid phase with approximately this force. However, after only a small amount of fluid has been sucked into the gel, the gel has already lost some of its imbibition pressure; in fact, it loses all of this imbibition pressure by the time it has absorbed a volume of fluid equal to 30 to 50 per cent of its original volume.

A very small amount of normal interstitial fluid, probably less than one per cent, is in the free fluid phase (that is, not gelled) and can move freely in the interstitial spaces. In contrast, in edema the free fluid volume is often the major proportion of the interstitial fluid volume (1).

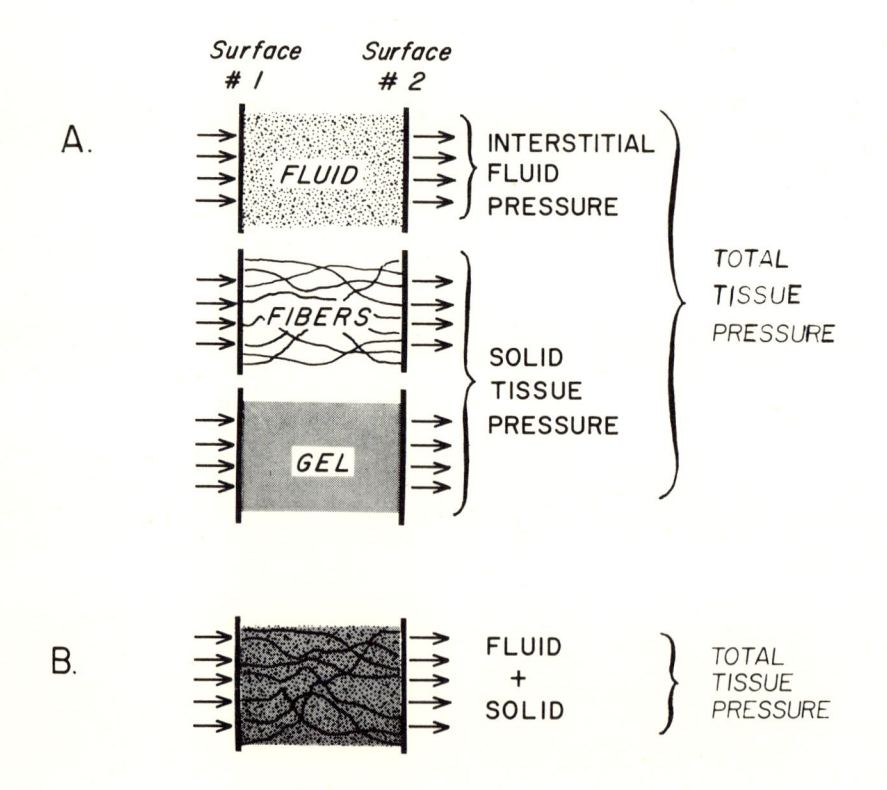

Fig. 1. A: Transmission of pressure separately through fluid, fibers and gel. B: Transmission of pressure through all the media simultaneously. (Reprinted by permission from: Guyton et al.: (1971) Physiol. Rev. 51:527.)

TRANSMISSION OF PRESSURES IN THE INTERSTITIAL SPACES:

Figure 1A illustrates three different methods by which pressure can be transmitted from one intra-tissue surface to another, such as from one cell surface to the surface of an adjacent cell. These three methods are by way of (1) fluid between the two surfaces, (2) fibers between the two surfaces, and (3) gel between the two surfaces. Figure 1B illustrates transmission of pressure between the two surfaces by all three interstitial elements at the same time. The fibers and the gel both represent solid structures because both of them exhibit solid elastic coefficients. On the other hand, the fluid is a hydrodynamic factor. Therefore, interstitial pressure can be divided into two different types: <u>interstitial fluid pressure</u> and <u>solid tissue pressure.</u> And <u>total tissue pressure</u> is the sum of both of these.

It is important to make these distinctions between types of interstitial pressure because it is fluid pressure that determines movement of fluid through the tissues, through capillary membranes, and into lymph channels. On the other hand, solid tissue pressure contributes to such effects as compression of blood vessels, distortion of cells that touch each other and so forth, but does not play a role in movement of fluids (2).

<u>Measured Interstitial Fluid Pressure.</u> Several new methods have been developed in the past few years which are believed to measure interstitial fluid pressure in contradistinction to total tissue pressure which has been measured by most previously used methods. In general, these methods utilize some implantable device that is left in the tissues for a considerable period of time until the fluid in the tissues comes to equilibrium with fluid in the device; then the pressure in the fluid of the device is measured. One such embodiment of this principle is an implanted perforated hollow capsule in which pressure is measured inside the fluid cavity of the implanted capsule.

The fluid found in the hollow space of the perforated capsule is free fluid, and the pressure measured in this space in most tissues averages about −6 mm. Hg. In other words, the pressure is slightly subatmospheric, and it tends to pull the surfaces of adjacent cells together all of the time. On the other hand, the fibers of the tissue spaces as well as the gel in these spaces exert solid tissue pressure to hold the surfaces apart. Therefore, when the interstitial fluid pressure is −6 mm. Hg, the solid tissue pressure (averaged over the cell's surfaces) calculates to be about +7 mm. Hg. And the total tissue pressure is the sum of these two, or approximately +1 mm. Hg (3).

<u>Recoil Pressure of the Gel</u>. The negative pressure of the fluid in the interstitial spaces is balanced by a positive recoil pressure of the gel reticulum—that is, a spring-like tendency for the reticulum to expand into a larger volume. It is this recoil nature of the reticulum that is mainly

responsible for transmission of solid tissue pressure by the gell. When free fluid pressure is decreased, fluid is pulled from the gel, but even slight removal of this fluid increases the recoil pressure of the gel reticulum to prevent further fluid loss. Because of this, the fluid volume of the gel can be changed only moderately by changes in free fluid pressure surrounding the gel (4).

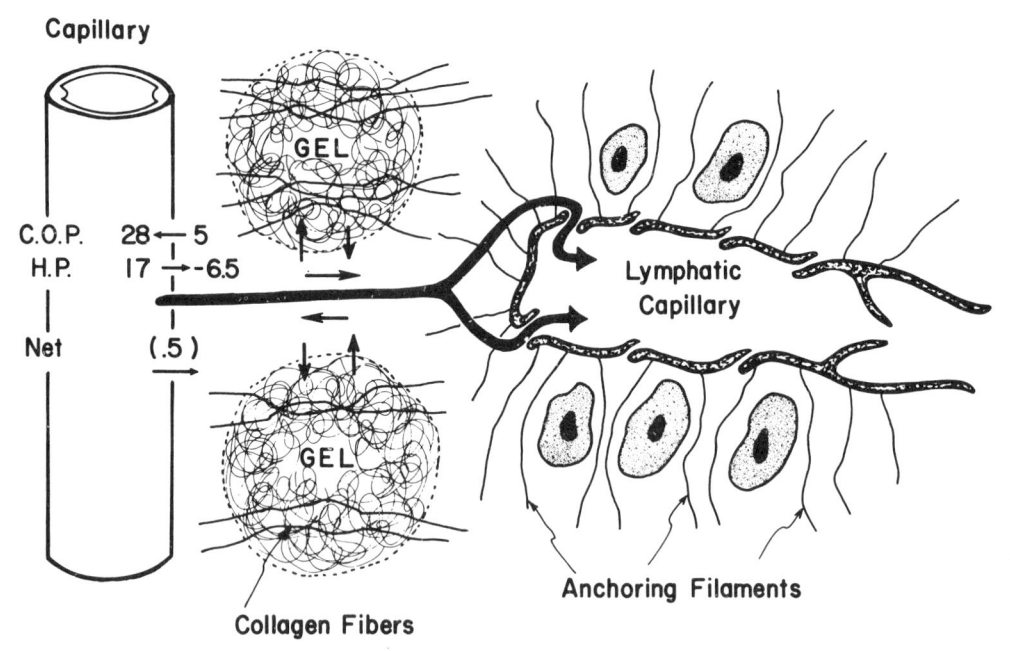

Fig. 2. Structure of the terminal lymphatic capillary and its surrounding tissues, illustrating a valve mechanism at the junctures of the endothelial cells. (Reprinted by permission from: Guyton; in Sodeman and Sodeman: Pathologic Physiology, 5th Ed., Saunders, Philadelphia).

FUNCTION OF THE LYMPHATIC SYSTEM:

Recent studies of lymph vessel function demonstrate two important features that have not generally been taught in the past. First, when an animal is not anesthetized, essentially every lymph vessel in its body undergoes rhythmical contraction and thereby pumps lymph along the lymph channels. Second, the openings into the terminal capillaries of the lymphatic system are constructed in such a way that the endothelial cells act as valves to allow fluid entry into the lymph capillaries but not to allow fluid flow backwards out of the capillaries. This construction is illustrated in Fig. 2, showing that each endothelial cell overlaps the adjacent cell, forming a flap-valve.

Also, the outer surfaces of the endothelial cells are held tightly to the surrounding tissues by anchoring filaments. Therefore, when the tissue expands, the lymph capillary also expands, and the internal edges of the endothelial cells flap inward, opening the spaces between the endothelial cells and allowing fluid to flow to the interior of the lymph capillary. Upon compression of the tissue, the flapping edges of the endothelial cells close over the openings, and the fluid can leave the capillary only by flowing into the larger lymph vessels and thence back into the blood. Therefore, any motion in the tissues whatsoever will cause the terminal lymph capillaries first to trap free fluid and, second, to pump this fluid on up the lymph vessels. This mechanism, therefore, acts as a scavenger mechanism to remove any free fluid from the tissue spaces as it is formed (5).

Relationship Between Free Fluid Pressure and Lymph Flow. Normally, lymph flow from most tissues is extremely slight, less than 2 ml from the entire body each minute. However, when the interstitial fluid pressure rises above its normal value of about -6 mm. Hg, lymph flow also increases markedly. As the free fluid pressure approaches atmospheric pressure (zero pressure), the lymph flow increases an average of 20 fold. (6).

AUTOMATIC FEEDBACK MECHANISM FOR REGULATING INTERSTITIAL FLUID PRESSURE:

We can now use the above principles to explain how an automatic feedback control mechanism occurs in the interstitial spaces to control interstitial fluid pressure. When the free fluid pressure rises even slightly above its normal level of approximately -6 mm. Hg, the rate at which lymph flows from the tissue spaces increases markedly (as much as 20 fold as the pressure approaches zero pressure). This rapid lymph flow causes rapid loss of protein-rich fluid (2 per cent concentration in the average tissue) from the tissue spaces. This is replaced by protein deficient fluid (0.1 to 0.2 per cent concentration) filtering from the capillaries. Therefore, the protein concentration becomes reduced in the tissue spaces, consequently also reducing the tissue fluid colloid osmotic pressure. This allows the plasma colloid osmotic pressure to overbalance the capillary pressure, thereby promoting rapid osmosis of fluid from the tissue spaces into the blood. In the usual subcutaneous tissue, about 90 to 95 per cent of the fluid is absorbed by the osmotic mechanism while 5 to 10 per cent flows into the lymph vessels. Yet, this 5 to 10 per cent is extremely important because it carries protein from the spaces, thus setting conditions for the osmotic mechanism to function. Because of fluid loss by these two different routes, (a) through the lymphatics and (b) by osmosis, the interstitial fluid pressure falls back once again to approach the normal value of -6 mm. Hg (7).

AUTOMATIC FEEDBACK MECHANISM FOR REGULATING INTERSTITIAL FLUID VOLUME:

Essentially the same mechanism as that described above for regulating interstitial fluid pressure also regulates interstitial fluid volume. That is, when the fluid volume rises above normal, the fluid pressure also rises above normal, lymphatic flow increases, protein concentration falls, and the remainder of the excess fluid is then absorbed into the capillaries. Thus, the interstitial fluid volume is returned back toward normal (7).

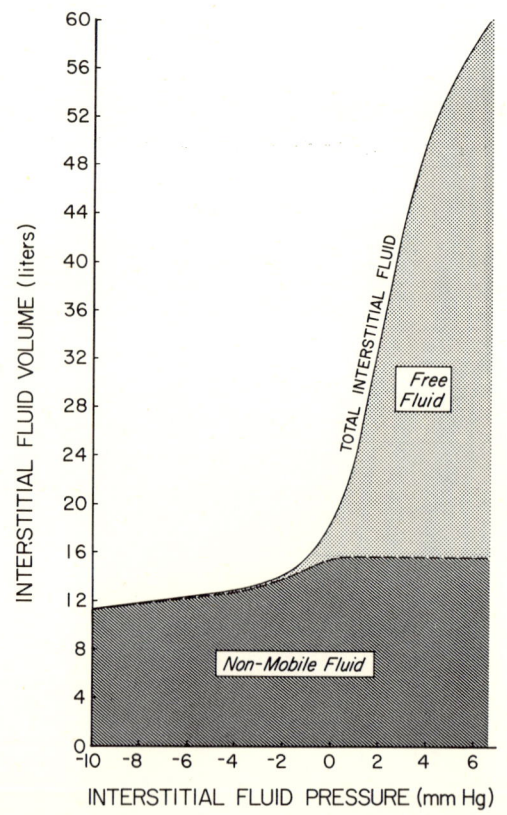

Fig. 3. Effect of interstitial free fluid pressure on total interstitial fluid volume, gel fluid volume (non-mobile fluid), and interstitial free fluid volume. (Reprinted by permission from: Guyton et al. (1971) Physiol. Rev. 51:527).

DEVELOPMENT OF EDEMA:

From the above considerations one can now understand very readily the changes that take place in the tissue fluids during the onset of edema. These are illustrated in Fig. 3, which shows the relationship between interstitial free fluid pressure and the volumes of free fluid and of gel fluid (non-mobile fluid). Note that so long as interstitial free fluid pressure is subatmospheric (less than zero in the figure), there is almost no free fluid. As the pressure approaches zero, the gel fluid at first expands about 30 to 50 per cent, but as the pressure rises above atmospheric pressure essentially no additional fluid is imbibed by the gel. Therefore, the gel volume remains constant thereafter, even though the free fluid volume continues to increase vastly.

When there are great excesses of free fluid, it can flow readily through the tissues, explaining the well known phenomenon of fluid mobility in edematous tissues. In fact, fluid can even flow from one breast to another as a patient is rolled in bed, allowing the dependent breast to become edematous while the higher breast becomes non-edematous (8).

<u>Hydrostatic Conditions in the Circulation and Lymphatic System that Predispose to Edema.</u> The conditions in the circulation and lymphatic system that cause edema are already so well known that they need only to be mentioned here. These include any factor that will cause excess filtration of fluid from the capillaries into the tissue spaces or poor return of fluid to the circulation, such as

(1) increased capillary pressure,
(2) decreased plasma colloid osmotic pressure,
(3) increased permeability of the capillaries,
(4) lymph vessel blockage.

Lymph vessel blockage causes edema mainly because of failure to remove proteins from the tissues. The protein concentration rises markedly and the normal osmosis of fluid from the interstitial spaces into the blood capillaries is then blocked by the resultant high colloid osmotic pressure of the tissue spaces (9).

SUMMARY

The normal mechanisms of the circulation and lymphatic system are geared to keep the tissues almost completely "dry" of free interstitial fluid. The basic mechanism that leads to this event is the pumping capability of the lymph vessels. Lymph flow causes removal of protein rich fluid from the tissue spaces, thereby decreasing the tissue colloid osmotic pressure. When this has been achieved, additional excess free fluid is absorbed by osmosis into the blood capillaries. This mechanism normally keeps the interstitial free fluid volume at an almost zero level and keeps the interstitial free fluid pressure at a normal value of about -6 mm. Hg.

Under normal conditions, almost all the interstitial fluid volume is comprised of interstitial gel. The gel imbibes fluid from the free fluid of the tissues. At very low free fluid pressures, gel volume decreases; at higher free fluid pressures, gel volume increases. When the free fluid pressure rises above atmospheric pressure, the gel imbibes no additional free fluid; therefore, large pockets of free fluid then begin to develop in the tissues to form edema.

REFERENCES

1. Ogston, A.G. & Sherman T.F. (1961) J. Physiol. (London) 156:67.

2. Guyton, A.C., Granger, H.J. & Taylor, A.E. (1971) Physiol. Rev. 51:327.

3. Guyton, A.C. (1963) Circ. Res. 22:399.

4. Flory, P.J. (1953) Principles of Polymer Chemistry. Cornell Press.

5. Casley-Smith, J.R. (1967) In: Progress in Lymphology, edited by A. Ruttiman, N.Y.: Hafner. p. 348.

6. Gibson, H., Gaar, K.A. Jr., & Taylor, A.E. (1970) Fed. Proc. 29:319.

7. Guyton, A.C. & Coleman T.G. (1968) Ann. N.Y. Acad. Sci. 150:537.

8. Guyton, A.C., Scheel K.W. & Murphree D. (1966) Circ. Res. 19:412.

9. Rusznak, I., Foldi, M., & Szabo, G. (1960) Physiology and Pathology. New York, Pergamon Press, p. 350.

Supported by research grants-in-aid from the National Heart Institute.

INFLUENCE OF HISTAMINE ON TRANSPORT OF FLUID AND PLASMA PROTEINS
INTO LYMPH

Eugene M. Renkin and Reginald D. Carter

Department of Physiology and Pharmacology, Duke
University Medical Center, Durham, North Carolina
27710, U.S.A.

It has been proposed that materials are transported across capillary walls by diffusion and ultrafiltration through small pores of 35-45 Å radius, a few large pores or "leaks" over 200 Å in radius and by pinocytosis in vesicles of 250 Å internal diameter (Landis & Pappenheimer, 1963; Mayerson, 1963; Winne, 1965). The small pores account for the exchange of water and low MW solutes, and contribute to the transport of larger molecules up to the size of serum albumin, effective radius 35.5 Å. Larger pores or vesicles are required for the transport into lymph of molecules larger than this. Fig. 1 illustrates the contribution of the small pore and large pore or vesicular

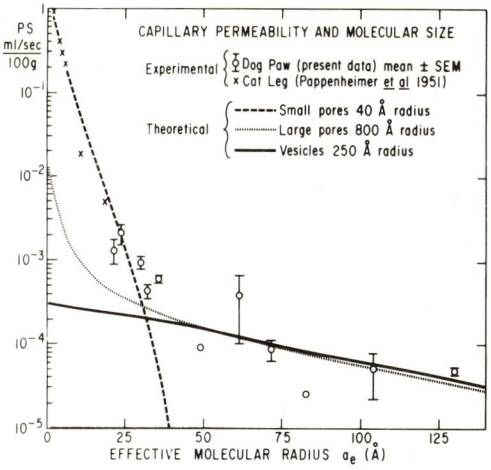

Fig. 1. Permeability and molecular size for serum albumin and graded dextrans in the dog's paw calculated from lymph flows and lymph/plasma concentration ratios (Garlick & Renkin, 1970).

mechanisms to the transport of serum albumin and of dextran molecules of graded molecular radii from 24 to 130 Å in the dog paw (Garlick & Renkin, 1971). The ordinate represents permeability-surface area product for the pore systems, or volume clearance for the vesicles.

Garlick and Renkin (1970; Renkin & Garlick, 1970) argued that under conditions of normal permeability, transport of molecules larger than 40 Å must be by endothelial cell vesicles and not through large pores or leaks, because at high lymph flows, the lymph/plasma concentration ratio for serum albumin fell to about 0.1, substantially less than would be expected for filtration through pores wide enough to account for the passage of larger molecules. According to this view, about half the serum albumin appearing in lymph crosses the capillary wall by the small pores, half by the endothelial vesicles. For larger molecules, like gamma globulins (55 Å effective radius), all must go by the vesicles.

Histamine is known to increase permeability to plasma proteins and other large molecules (see Rusznyak, Foldi & Szabo, pp. 309ff), and it has been proposed that it acts by opening large gaps between the endothelial cells in venous capillaries or venules (Majno and Palade, 1961). However, Diana et al (1971) who used the method of Pappenheimer et al. (1951) to evaluate capillary pore size and pore area in the dog's hind leg reported that intra-arterial histamine produced no increase in pore radius, only an increase in total area. We thought it would be interesting to examine the effects of this substance on blood-lymph transport of plasma proteins and other molecules in the dog paw relation to their molecular size, to see if we could characterize its action on the different transport mechanisms which it might influence.

METHODS

We collected lymph from pre-nodal lymphatics of both hind legs of anesthetized dogs. These drain principally the toe pads of each paw, approximately 24 g of tissue. The legs were kept passively in motion to promote lymph flow. Collection periods were usually 30 min, at the mid-point of which arterial blood samples were taken. Heparin dried on the collecting tubes prevented coagulation. Lymph flow was calculated from sample weight and time. In many experiments a narrow dextran fraction of mean MW 110,000 (Pharmacia D-110) was infused according to a schedule which kept its arterial plasma level nearly constant (see Garlick & Renkin, 1970). Lymph/plasma concentration ratios for dextran and endogenous plasma proteins were determined under control conditions, after s.c. injection of 0.01 ml Ringer's solution into each toe pad of one leg and after s.c. injection of 1 to 25 µg of histamine phosphate in 0.01 ml Ringer's into each toe pad. Protein concentrations were measured by the Folin method (Lowry et al., 1951), albumin and globulin fractions after Na_2SO_4 precipitation. Plasma proteins were also separated according to molecular size by gel filtration on Sephadex G-200 (Granath & Kvist, 1967) and according to electrophoretic mobility and molecular size by disc electrophoresis on 6.5% polyacrylamide gels (Davis, 1964) fixed and stained with Ponceau S. Variable affinities of individual proteins for the Folin reagent or the stain cancel out in calculating lymph plasma ratios. Dextran was measured by the anthrone method (Roe, 1954), either directly or after gel chromatography of plasma or lymph samples.

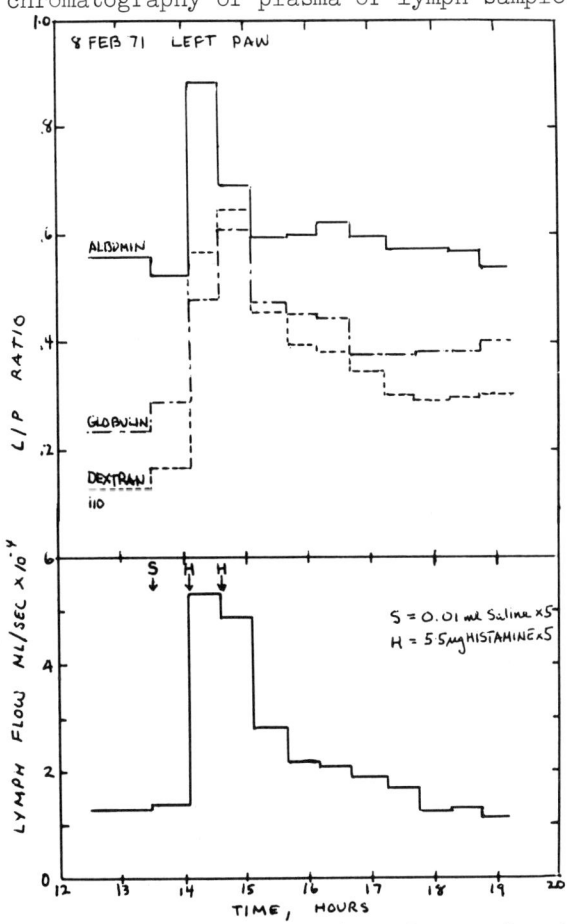

Fig. 2. Influence of subcutaneous histamine on lymph flow and lymph/plasma concentration ratios in the dog paw. Albumin and globulin fractions of lymph and plasma were separated with saturated Na_2SO_4.

RESULTS

Subcutaneous histamine in all doses studied increased lymph flow and lymph/plasma concentration ratios of proteins and dextrans. Fig. 2 shows results of an experiment in which 2 consecutive injections were given 30 min apart. These were preceded by an injection of Ringer's solution which had no effect. After 27.5 µg histamine, lymph flow increased 5-fold, and concentration ratios for albumin, globulin and Dextran-110 increased. The elevation of lymph flow was sustained during the 30 min collection period after each injection, but fell to the control 2 hr after the second injection. The L/P ratio for serum albumin also returned to its control level, but the ratios for higher molecular weight constituents remained above their controls 4 hr later. In other experiments, similar increases in lymph flow and lymph/plasma concentration ratios were maintained during repeated injections at 30 min intervals for as long as 4 hr. We observed an effect with as little as 1 µg histamine per pad (5 µg per paw) which was graded to a maximum at 5 or 10 µg per pad.

Lymph/plasma ratios for plasma globulins and D-110 increased more than that for serum albumin. We calculated capillary permeability-surface area products (PS) for the different constituents of the lymph from lymph flow (L) and lymph plasma concentration ratio (R) by the formula (Renkin, 1963; Johnson, 1966)

$$PS = \frac{LR}{1 - R} \qquad (1)$$

In most instances PS for serum albumin was increased 10-20 fold after histamine, PS for the larger molecules 20 to 100 fold. Fig. 3 shows graphs of PS for albumin, globulin and Dextran-110 for the experiment illustrated in Fig. 2 and for another experiment in which histamine was given over 2 hr. In all our experiments histamine reduced the selectivity of the capillary wall for proteins of different molecular size.

In control lymph, the predominant protein components are serum albumin and alpha globulins. Beta and gamma globulins, macroglobulins and fibrinogen are much reduced. Histamine brings the protein distribution of lymph closer to that of plasma. Fig. 4 shows tracing of disc electrophoresis scans of plasma, control lymph and histamine lymph. The ordinate is protein concentration (dye) adjusted to make the height of the albumin peak (a) the same in all three (the actual albumin L/P ratios were 0.30 for control lymph and 0.45 for histamine lymph at a

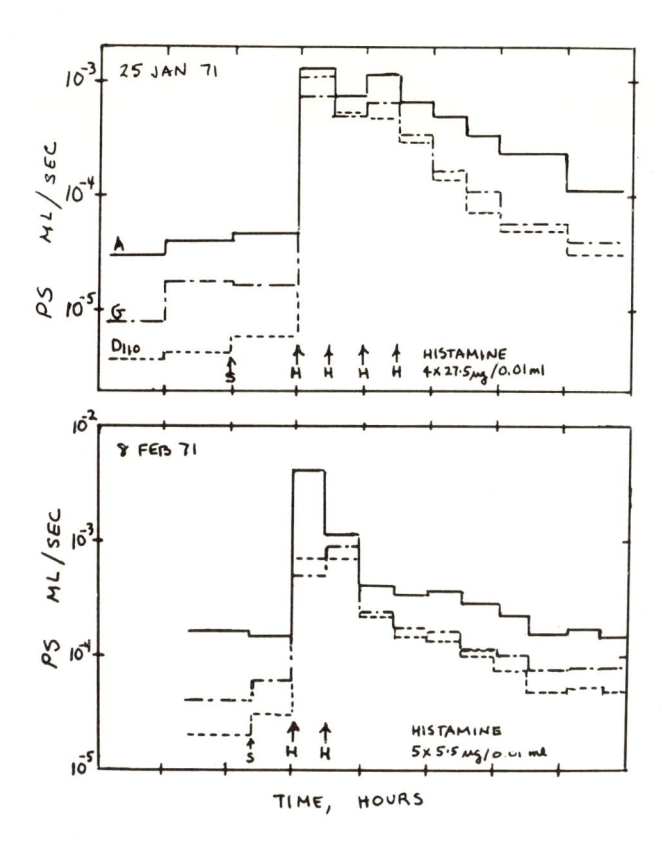

Fig. 3. Calculated permeability-surface area products (PS) after histamine in two experiments.

flow 5 times control). The Postalbumin (PA) peak 1 is an α globulin of MW 45,000 to 50,000, PA 2 and 3 are larger alpha and beta globulins, MW 100,000 to 200,000. TF is transferrin (MW 90,000) plus another globulin of MW 150,000. Post-transferrin (PTF) 1 and 2 represent the gamma globulins, MW 160,000, PTF 3 contains fibrinogen, MW 340,000 and macroglobulins, MW 800,000 and above.

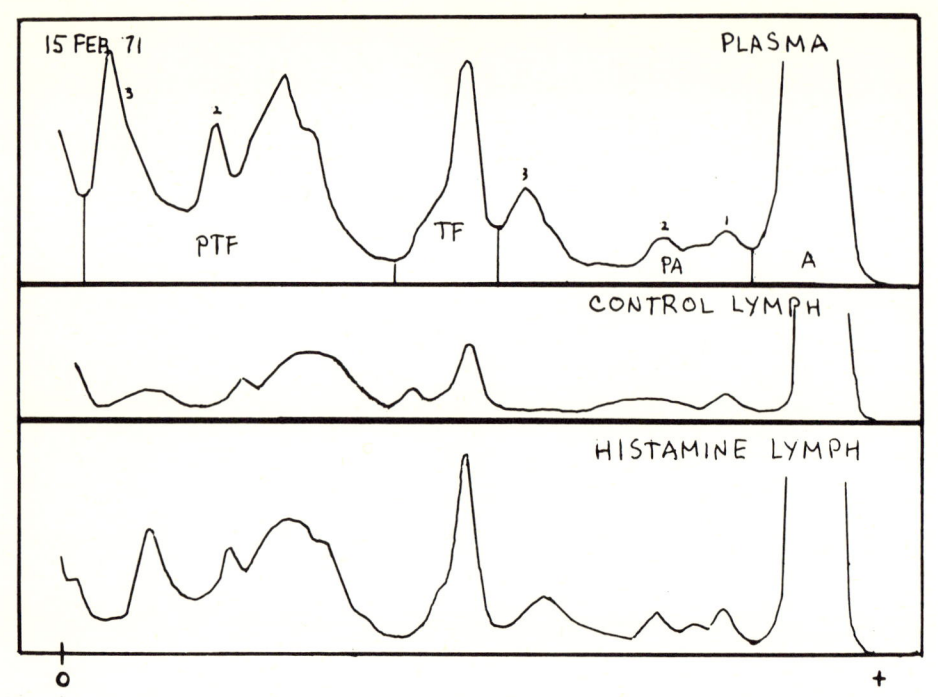

Fig. 4. Disc electrophoresis scans of plasma, control lymph and histamine lymph on 6.5% polyacrylamide, stained with dye. Abscissa: electrophoretic mobility, origin = 0. Ordinate: relative concentration of protein (dye), value for albumin (A) set equal in each run. PA-1, alpha-globulin, MW 45-50,000, PA-2,3 beta globulins, MW 100-200,000, TF transferrin, MW 90,000 and globulin of MW 150,000, PTF-1,2 gamma globulins, MW 160,000, PTF-3, fibrinogen, MW 340,000 and macroglobulins, MW 800,000.

Fig. 5. Gel chromatographic separation of proteins of plasma, control lymph and histamine lymph on Sephadex (R) G-200. Calibration scales for partition coefficient (K_d) and effective molecular radius a_e were determined with known dextran and protein standards. Peak I contains macroglobulins and fibrinogen. This peak is much reduced in two plasma samples which had clotted before chromatography. Peak II is mainly globulins, Peak III albumin, alpha and beta globulins.

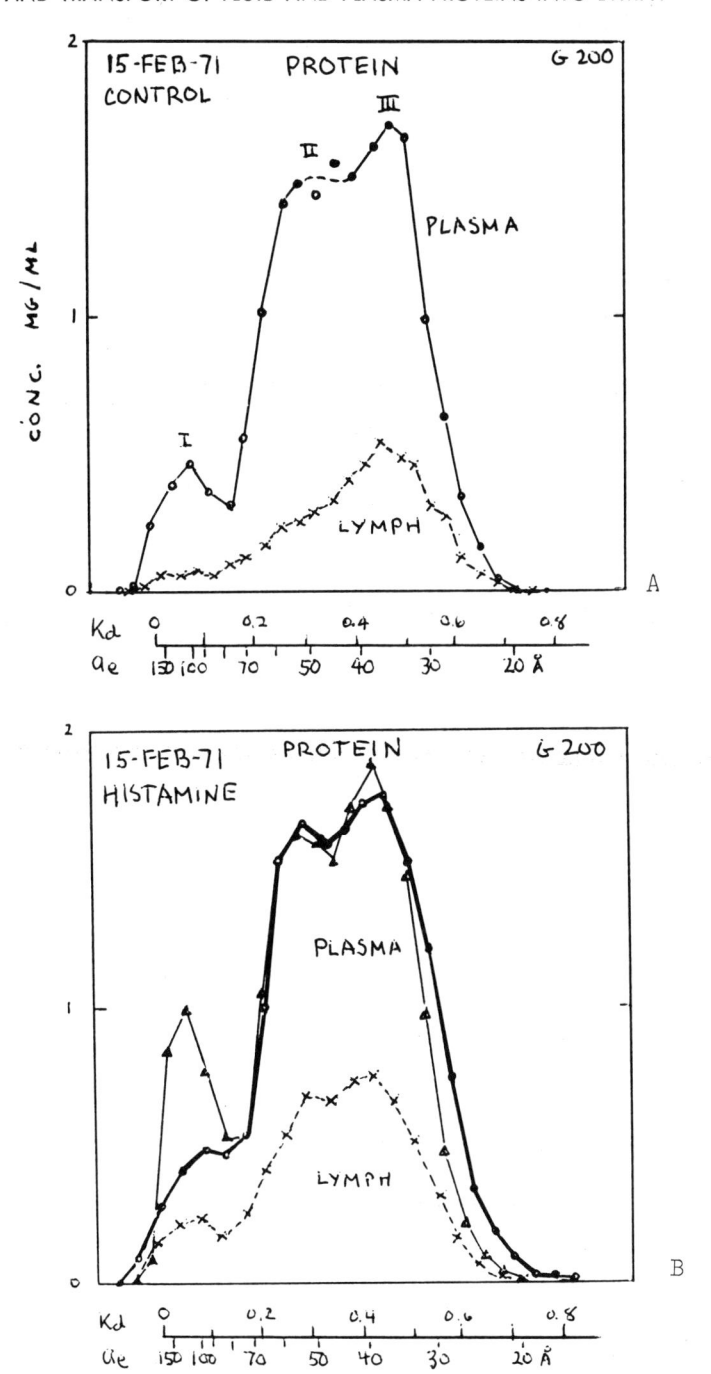

The same shift in protein composition of lymph relative to plasma is shown in Fig. 5 by gel-chromatographic separations of plasma, control and histamine lymph. The 3 peaks for plasma contain I-macroglobulins and fibrinogen (in 2 of the samples shown, the fibrinogen had been lost through clot formation), II-globulins of intermediate MW, III-serum albumin, low MW alpha globulin and transferrin. A calibration for molecular size is given below each graph. Resolution of individual proteins is not possible with this technique. Control lymph shows progressively lower ratios for the 3 peaks with increasing molecular size. After histamine, concentration ratios still fall with increasing MW, but to a lesser degree.

Fig. 6 shows gel chromatographic separations for Dextran-110 in the same samples of plasma and lymph. The plasma curves reflect the molecular size distributions of the preparation we used. Its peak lies between protein peaks I and II (indicated) and corresponds to a molecular radius of 70 Å, corresponding to an average MW of 110,000. In control lymph there is very little dextran, and most of this consists of smaller than average molecules. In histamine lymph, the larger molecules are present in substantial amounts.

In contrast to the action of histamine, vasodilators, such as methacholine, which increase lymph flow do so without increasing protein or dextran concentration ratios, or altering the distribution of these substances with respect to molecular size.

Fig. 6. Gel chromatographic separations of dextrans in the same samples as Fig. 5. Pharmacia Dextran-110 was infused i.v. The positions of the plasma protein peaks are indicated for comparison.

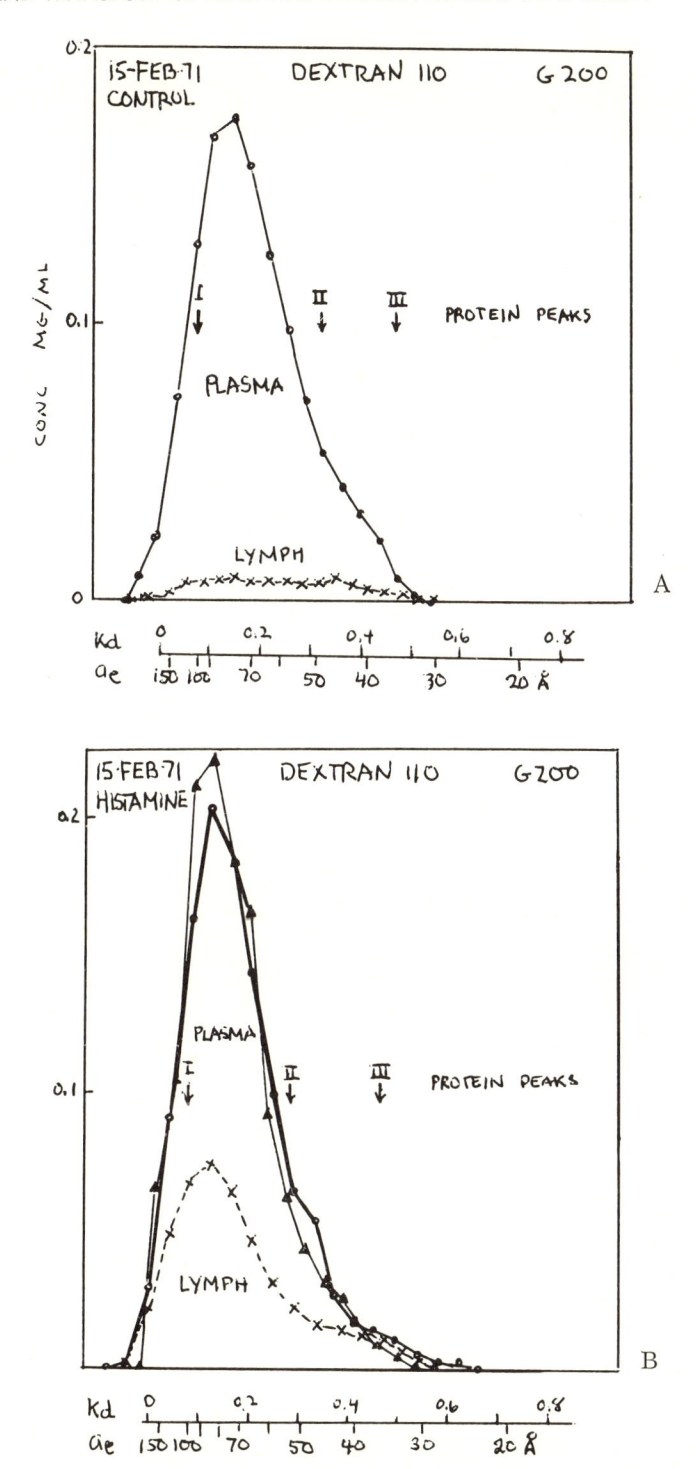

DISCUSSION

If we accept the capillary endothelium as the barrier to large-molecular transport between plasma and lymph and as the site of histamine action, here are four mechanisms by which histamine might act.

1. It might increase the number of the small pores, but not their size (radius 35-45 Å). This is equivalent to increasing the area of exposed capillary surface.

2. It might increase the size of the small pores moderately, allowing larger molecules to pass through them, still retaining molecular-size selectivity.

3. It might cause large openings to appear which allow the unrestricted passage of the largest macromolecules of plasma. Such openings would be readily visible in electron micrographs.

4. It might enhance the rate of vesicular transport of fluid and macromolecules.

The first mechanism, increase in the number but not in the size of the small pores may occur with histamine as it does with other vasodilators, but it cannot contribute to the increase in transport of molecules substantially larger than serum albumin or to the decrease in molecular selectivity. To test which of the 3 remaining mechanisms might be effective, we carried out experiments in which lymph flow was further increased during the action of histamine by elevating venous pressure to 30-40 mm Hg. If there are large openings in the capillary wall capable of allowing unrestricted passage of plasma proteins, we should expect whole plasma to pour into the lymph and raise all L/P ratios. Fig. 7 shows the result of one of these experiments: when venous pressure was increased, lymph flow increased, and L/P ratios for albumin, globulin and Dextran-110 fell (there was a lag of one collection period due to the expanded interstitial fluid volume of the paw). The results were essentially the same in all experiments. The ratios fell more for the larger molecules. These observations confirm that the capillary wall is still capable of restricting the passage of

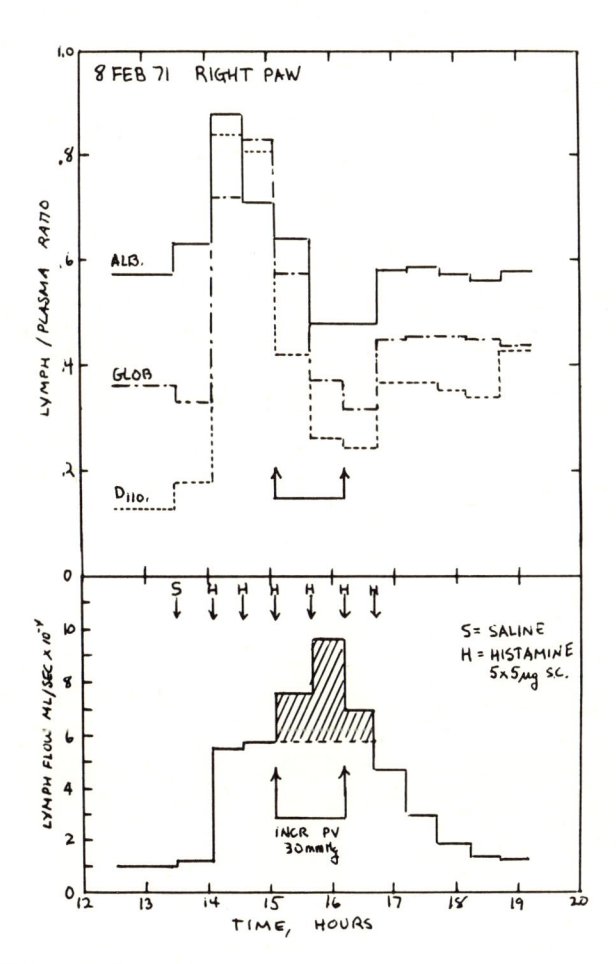

Fig. 7. Effect of raising venous pressure during histamine action. Albumin and globulin were separated with saturated Na_2SO_4. See text for discussion.

large molecules after histamine, and thus rule out the presence of large, non-restrictive openings except possibly as a transient phase lasting a few min after the first injection. The sustained action of histamine must be due to the presence of moderately enlarged openings in the capillary wall, still of molecular dimensions, or to increased vesicular transport. We can evaluate the contribution of these two mechanisms by comparing the experimental relation of PS to molecular radius before and after histamine with a set of theoretical curves for diffusion through pores of various sizes and for transport in 250 Å radius (internal) vesicles at different clearance rates. These are shown in Fig. 8. The pair of curves indicated by the heavy lines

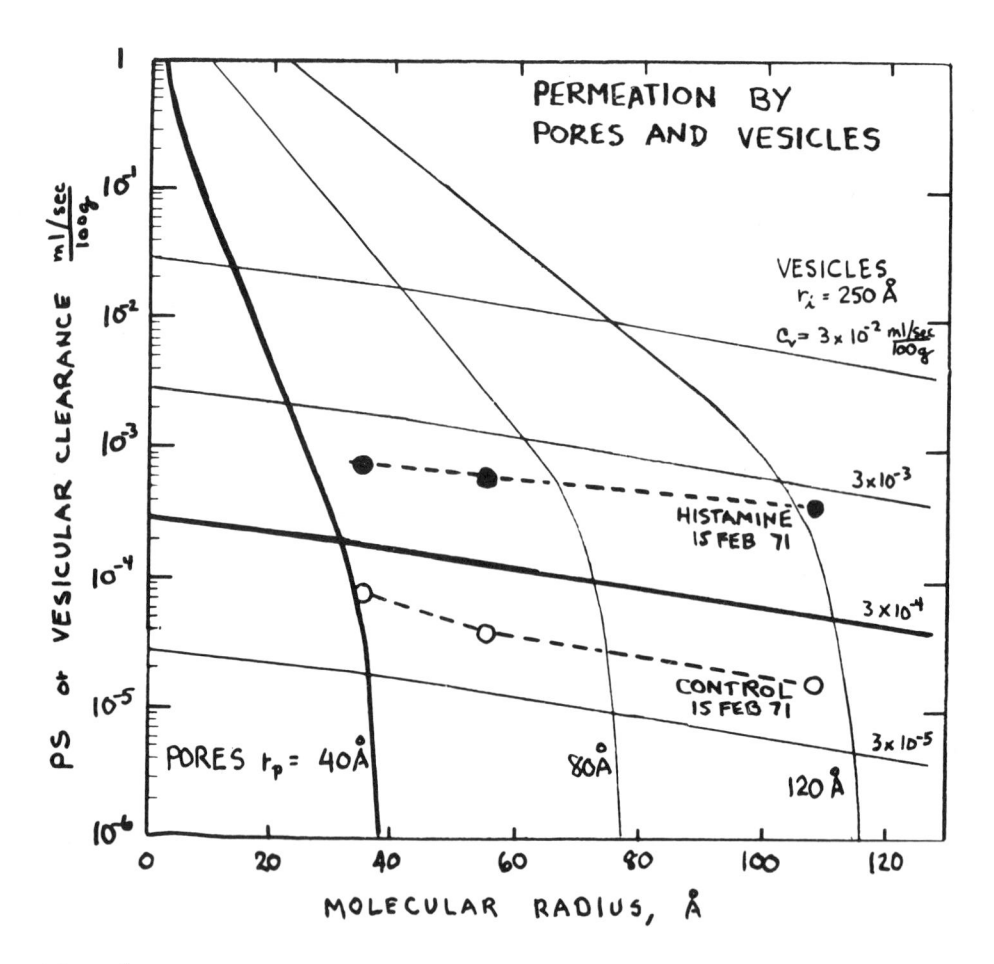

Fig. 8. Permeability and molecular size in relation to capillary transport processes. The solid lines are theoretical curves

for cylindrical pores of the dimensions indicated (constant number) and for spherical vesicles 250 Å internal radius with bi-directional volume flux (C_o, clearance) as indicated. To get total transport for any pore size and vesicular clearance, add the appropriate curves. The pair shown by heavy lines corresponds to those in Fig. 1. For details of the calculation, see Garlick & Renkin, 1970.

represent the hypothetical pore size and vesicular transport rate for the normal dog paw according to Garlick and Renkin (1970), which were shown in Fig. 1. Total transport follows the pore curve down to its intersection with the vesicle curve, and then curves follow the latter (it is the sum of the two). The open circles connected by a dotted line are control PS values for albumin, gamma globulin and fibrinogen/macroglobulin in the experiment of 15 Feb, other aspects of which have been illustrated in Fig. 3, 4 and 5. The points for the larger molecules lie on a vesicular transport curve with a clearance of 1×10^{-4} ml/sec x 100 g, one-third the mean value for the earlier experiments, but within their range. The albumin point lies somewhat higher due to the contribution of the 40 Å pores to its transport. The PS values after histamine are shown as filled circles connected by a dotted line. All are elevated, and fall close to a vesicular transport curve of clearance 2×10^{-3} ml/sec x 100 g, an increase of 20-fold. At this level, vesicular transport of albumin is 10 times that through the small pores. There is no indication that pore size has increased. If pore radius after histamine were 80 Å, for example, PS values for albumin and for gamma globulin would have to lie on the curve for 80 Å pores, substantially higher than their observed values.

Our conclusion is that the sustained action of histamine on the capillaries of the dog's paw consists mainly of an increase in the rate of vesicular transport, and that the opening of large pores or gaps in the endothelium can only be a transitory phase.

REFERENCES

Davis, B.J. (1964) Ann. N.Y. Acad. Sci. 121:404.

Diana, J.N., Long, S & Yao, H. (1971) Fed. Proc. 30:661 (abstract).

Garlick, D.G. & Renkin, E.M. (1970) Am. J. Physiol. 219:1595.

Granath, K.A. & Kvist, B.E. (1967) J. Chromatog. 28:69.

Johnson, J.A. (1966) Am. J. Physiol. 211:1261.

Landis, E.M. & Pappenheimer, J.R. (1963). In Handbook of Physiology, Sect. 2, Circulation. Amer. Physiol. Soc. Washington, D.C. Vol. II, Chapt. 29, p.961.

Lowry, O.H., Rosebrough, N.J., Farr, A.L. & Randall, R.J. (1951) J. Biol. Chem. 193:265.

Majno, G. & Palade, G.E. (1961) J. Biophys. Biochem. Cytol. 11:571.

Mayerson, H.S. (1963) In Handbook of Physiology, Sect. 2, Circulation. Amer. Physiol. Soc., Washington, D.C., Vol. II, Chapt. 30, p.1035.

Pappenheimer, J.R., Renkin, E.M. & Borrero, L.M. (1951) Am. J. Physiol. 167:13.

Renkin, E.M. (1964) Physiologist 7:13.

Renkin, E.M. & Garlick, D.G. (1970) Microvasc. Res. 2:392.

Roe, J.H. (1954) J. Biol. Chem. 208:889.

Rusznyak, I., Foldi, M. & Szabo, G. (1967) Lymphatics and Lymph Circulation, Physiology and Pathology, 2nd ed. Oxford: Pergamon Press, p.303 ff.

Winne, D. (1965) Pflügers Arch. 283:119.

This work was supported by U.S. Public Health Service Grant HE-10936.

The authors wish to express their appreciation for the technical assistance of Mrs. Diane Sams and Mr. Ezzat S. Mishriky.

Fig. 1 is reproduced with the permission of the Editors of the American Journal of Physiology.

EFFECT OF VASOCONSTRICTORS ON LYMPH OXYGEN TENSION IN NORMO-
AND HYPOVOLEMIA

T. Barankay, S. Nagy, G. Horpácsy and A. Petri

Institute of Experimental Surgery, University Medical
School, Szeged, Hungary

The effects of various vasoconstrictor agents at the
microcirculatory level have been debated for many years. As
these agents are still rather widely used in the therapy of
various forms of shock, it seems important to acquire information
about their effect on tissue oxygenation. Tissue hypoxia has
a key role in the pathogenesis of shock, there are, however, few
data available on the level of tissue oxygen tension because of
methodological difficulties.

We have shown previously (NAGY et al. 1969) that
measurement of lymph pO_2 provides good information on the level
of tissue oxygen tension in hemorrhagic shock without causing
tissue destruction at the site of measurement which can be very
extensive with tissue electrodes and could change local oxygen
tension. In the present experiments the effect of some
vasoconstrictors on lymph oxygen tension was examined under
normo- and hypovolemic conditions.

Methods

The experiments were performed on mongrel dogs under
sodium pentobarbital anesthesia. Using intermittent positive
pressure ventilation a right thoracotomy was made and the
thoracic duct was cannulated in the caudal part of the right
chest with a stainless steel tube. This was connected to the
flow through a measuring cuvette of a Clark type polarographic
platinum electrode maintained at 37°C. Oxygen tension of
lymph was measured continuously with a Beckman Model 160
Physiological Gas Analyzer and recorded on a Beckman Dynograph

recorder together with mean arterial blood pressure measured with a Statham transducer.

The course of normovolemic experiments was as follows: after a 30 min control period the various vasoconstrictors in saline solution were infused intravenously during a period of 20-30 min using an infusion pump. The speed of the infusion was 0.15 ml/min.

In the hypovolemic experiments the animals were bled to a mean arterial blood pressure of 50 mm Hg. This pressure was maintained for 30 min when bleeding was stopped. A further 30 min was allowed to pass during which blood pressure and lymph pO_2 increased and stabilized at a certain level. Infusion was begun at this time. The following vasoconstrictors were examined: norepinephrine, phenylephrine, metaraminol, angiotensin and a synthetic vasopressin analog, ornithin-8-vasopressin (POR-8).

The results were analyzed by the method of Wilcoxon (1945) using tables by Wilcoxon et al. (1963). A 5 per cent probability level was selected as the cut off point between significant and nonsignificant results. The percentile method (Herrera, 1958) was used for the estimation of scattering.

Results

The experimental results are summarized in the tables. Values are shown before the infusion, in the 10th min of infusion and 10 min after the infusion. It can be seen (Table I) that after norepinephrine lymph oxygen tension rises significantly ($P < 0.05$), while there is a decrease after angiotension (Table II) and POR-8 (Table III). This is also statistically significant ($P < 0.05$). Postinfusion values in the norepinephrine and angiotensin treated groups are close to control values. Ten min after the infusion of POR-8 lymph oxygen tension did not rise to the control value as this agent has a sustained effect.

The recording from a typical norepinephrine and angiotensin experiment is shown in Fig. 1.

Tables IV and V show the effect of phenylephrine and metaraminol on lymph pO_2. No significant changes were observed with either drug.

The effect on lymph pO_2 of norepinephrine, angiotensin and POR-8 during hypovolemia is shown in Tables VI, VII and VIII respectively. As in normovolemia norepinephrine caused a statistically significant (P<0.05) rise of lymph pO₂ while angiotensin and POR-8 decreased it, also significantly (P<0.05). A typical recording of the effect of norepinephrine and angiotensin in hypovolemia is shown in Fig. 2.

Phenylephrine (Table IX) and metaraminol (Table X), just as in normovolemia, had no effect on lymph pO₂ in hypovolemia either.

Table I. Effect of norepinephrine (2µg/kg/min) on lymph oxygen tension (mm Hg) in 6 normovolemic dogs

	Control period	Infusion period	Postinfusion period
Mean	22.0	28.0	17.0
Median	21.5	28.5	16.0
20th percentile	17.4	24.6	10.4
80th percentile	28.0	31.2	24.6

Table II. Effect of angiotensin (2µg/kg/min) on lymph oxygen tension (mm Hg) in 6 normovolemic dogs

	Control period	Infusion period	Postinfusion period
Mean	25.0	14.0	23.0
Median	23.0	11.0	26.0
20th percentile	16.8	8.0	12.6
80th percentile	40.0	23.0	33.2

Table III. Effect of POR-8 (0,015 µg/kg/min) on lymph oxygen tension (mm Hg) in 6 normovolemic dogs

	Control period	Infusion period	Postinfusion period
Mean	31.8	15.1	22.5
20th percentile	19.4	5.1	8.8
80th percentile	44.4	28.2	39.2

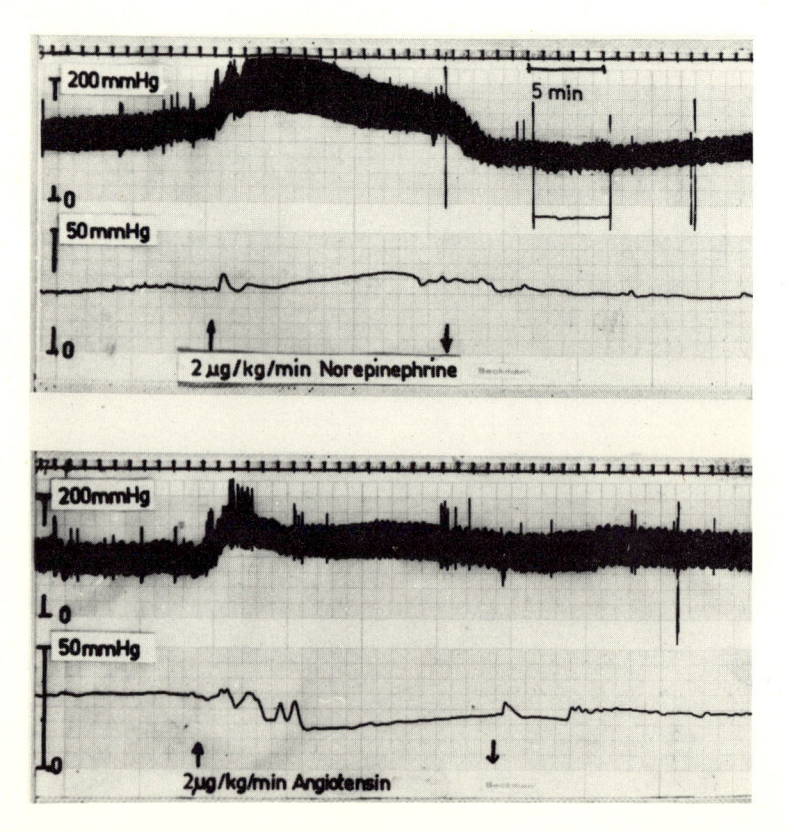

Fig. 1. Effect of norepinephrine and angiotensin on arterial blood pressure and lymph pO$_2$ in normovolemia. (Recordings from typical experiments).

Table IV. Effect of phenylephrine (5µg/kg/min) on lymph oxygen tension (mm Hg) in 6 normovolemic dogs

	Control period	Infusion period	Postinfusion period
Mean	22.0	22.4	25.7
Median	19.5	18.5	25.0
20th percentile	14.4	11.6	12.9
80th percentile	34.2	35.4	34.6

Table V. Effect of metaraminol (1µg/kg/min) on lymph oxygen tension (mm Hg) in 6 normovolemic dogs

	Control period	Infusion period	Postinfusion period
Mean	36.0	36.0	34.0
Median	40.0	38.0	34.5
20th percentile	25.4	26.6	24.6
80th percentile	43.8	43.6	43.2

Table VI. Effect of norepinephrine (2µg/kg/min) on lymph exygen tension (mm Hg) in 6 hypovolemic dogs

	Control period	Infusion period	Postinfusion period
Mean	15.0	22.0	14.0
Median	16.5	21.0	13.5
20th percentile	7.0	17.0	7.0
80th percentile	23.4	29.4	21.4

Table VII. Effect of angiotensin (2µg/kg/min) on lymph oxygen tension (mm Hg) in 6 hypovolemic dogs.

	Control period	Infusion period	Postinfusion period
Mean	11.0	8.0	11.0
Median	8.0	6.0	13.0
20th percentile	5.0	4.0	3.0
80th percentile	24.0	15.0	20.0

Table VIII. Effect of POR-8 (0,015 µg/kg/min) on lymph oxygen tension (mm Hg) in 6 hypovolemic dogs

	Control period	Infusion period	Postinfusion period
Mean	24.0	10.0	12.3
20th percentile	7.0	3.2	7.0
80th percentile	34.0	25.0	21.0

Table IX. Effect of phenylephrine (5 µg/kg/min) on lymph oxygen tension (mm Hg) in 6 hypovolemic dogs

	Control period	Infusion period	Postinfusion period
Mean	21.0	21.0	21.0
Median	19.5	22.0	20.0
20th percentile	12.8	12.4	12.2
80th percentile	33.8	29.2	32.0

Table X. Effect of metaraminol (5µg/kg/min) on lymph oxygen tension (mm Hg) in 11 hypovolemic dogs

	Control period	Infusion period	Postinfusion period
Mean	11.8	12.7	13.4
Median	10.0	10.0	9.0
20th percentile	2.0	2.6	2.0
80th percentile	21.2	25.6	24.4

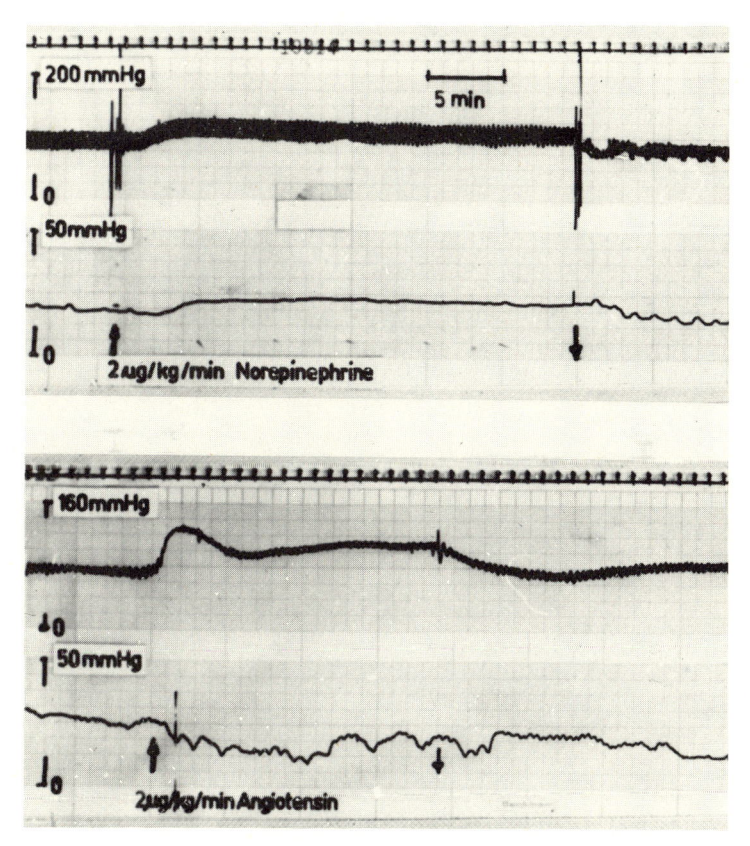

Fig. 2. Effect of norepinephrine and angiotensin on arterial blood pressure and lymph pO$_2$ in hypovolemia. (Recordings from typical experiments).

Discussion

These experimental results support the view that different vasoconstrictors have different effects on the microcirculation (Altura and Hershey, 1967; Cohen et al. 1970; Ericsson, 1971; Folkow et al. 1961; Greenway & Lawson 1966; Mellander & Johansson 1968; Messmer et al. 1968). By constricting mainly precapillary vessels angiotensin and the vasopressin analog influence tissue oxygen supply unfavourably. In the recent experiments of Järhult (1970) on cat skeletal muscle it was found that angiotensin was an effective constrictor of precapillary resistance vessels and precapillary "sphincters" but a poor constrictor of the main capacitance vessels.

In the case of norepinephrine constriction of the capacitance vessels simultaneously with the resistance vessels occurs and, at the same time, cardiac output and venous return increase (Emerson 1966; Laks et al. 1971), and conditions favourable for O_2 supply for the tissues can prevail.

It is interesting that identical blood pressure elevations together with an increase, a decrease, or, as in the metaraminol and phenylephrine treated groups, no change in lymph pO_2 could be observed.

In summary it can be said that various vasoconstrictors have different effects on the microcirculation with respect to its main function: the supply of O_2 to the tissues. Their effects do not change qualitatively in hypovolemia.

REFERENCES

Altura B.M. & Hershey, S.G. (1967) Angiology 18:428.

Cohen, M.M., Sitar, D.S., McNeill, J.R. & Greenway, C.V. (1970) Amer.J.Physiol. 218:1704.

Emerson, T.J., Jr. (1966) Amer.J.Physiol. 210:933.

Ericsson, B.F. (1971) Acta chir.scand. Suppl. 414.

Folkow, B., Johansson, B. & Mellander, S. (1961) Acta physiol. scand. 53:99.

Greenway, C.V. & Lawson A.E. (1966) J. Physiol. (Lond.) 186:579.

Herrera, L. (1958) J. Lab.clin.Med. 52:34.

Jarhult, J. (1971) Acta physiol. scand. 81:315.

Laks, M. Gallis, G. & Swan, H.J.C. (1971) Amer.J.Physiol. 220:171.

Mellander, S. & Johansson, B. (1968) Pharmacol.Rev. 20:117.

Nagy, S., Barankay, T. & Tárnoky, K. (1969) Acta physiol. Acad. Sci.hung. 35:87.

Wilcoxon, F. (1945) Biometrics 1:80.

Wilcoxon, F., Katti, S.K. & Wilcox, R.A. (1963) Critical values and probability levels for the Wilcoxon rank sum test and the Wilcoxon signed rank test American Cyanamid Co. and Florida State University Pearl River and Tallahassee.

TRANSVASCULAR FLUID FLUXES DURING SEVERE, PROLONGED CANINE

HEMORRHAGIC OR ENDOTOXIN HYPOTENSION

George J. Grega and Francis J. Haddy

Department of Physiology, Michigan State University
East Lansing, Michigan 48823

This paper is an attempt to summarize our findings related to transvascular fluid fluxes in circulatory shock states. One aim was to determine if fluid filtration into tissue (especially skeletal muscle) during severe, prolonged hypotension is an important determinant of irreversibility in circulatory shock states resulting from blood loss or endotoxin. Experiments were conducted on mongrel dogs anesthetized with Na pentobarbital. The forelimb was selected as the test organ since it largely comprises skin and skeletal muscle. Since total body soft tissue mass is mainly composed of these two tissues, fluid fluxes in skin and skeletal muscle would be of paramount importance in shock states. Transvascular fluid fluxes in collateral-free, innervated, naturally perfused canine forelimbs were inferred from changes in organ weight and total and segmental (large artery, small vessel, large vein) vascular resistances (1-4). Brachial venous outflow was used as an index of skeletal muscle blood flow and cephalic venous outflow was used as an index of skin blood flow.

Hemorrhage. The effects of 25 (n=10), 50 (n=17), or 60 (n=28) % blood volume depletion (total blood volume assumed to be 8% body weight) on forelimb weight, blood flows, vascular pressures, and vascular resistances were investigated (1,2). The bleeding rate was adjusted so that systemic pressure remained above 35 mm/Hg during the bleeding period; the duration of the hypovolemic period was 4 hr. In all animals forelimb weight decreased rapidly initially and then more slowly throughout the remainder of the hypovolemic period. Total forelimb weight loss was least in animals subjected to 25% blood volume depletion and most marked in those animals subjected to 60% blood volume depletion. Forelimb total and segmental (large artery, small vessel, large vein) vascular resistances increased abruptly in response to arterial hemorrhage; vascular resistances then either further increased or were largely maintained during the remainder of the hypovolemic period. The rise in total forelimb vascular resistance resulted from intense constriction of both skin and skeletal muscle vessels. The rapid forelimb weight loss (0-60 min) was associated with marked increases in total and segmental vascular resistances. This suggests that mean vessel caliber decreased and, inferentially, that forelimb intravascular blood volume also decreased. Hence, the rapid weight loss is largely attributable to a decreased forelimb vascular volume due to intense vasoconstriction. Clearly a decreased intravascular blood volume cannot account for all the weight loss. In animals subjected to 50 or 60% blood volume depletion the total forelimb weight loss exceeded the weight of the blood volume normally contained in the forelimb. Forelimb blood volume is approximately 4g/100g forelimb (5); with 60% blood volume depletion the weight loss after 4 hr of hypovolemia was 7g/100g. Hence, some of the forelimb weight loss must have represented extravascular fluid reabsorption. In fact, the weight loss during the last 3 hr of the hypovolemic period is largely attributable to extravascular fluid reabsorption. Resistances in the large vein and small vessel segments were relatively constant during the last 3 hr of the hypovolemic period. Since these two vascular segments together contain over 90% of the total vascular volume it is assumed that forelimb mean vessel caliber and inferentially, intravascular blood volume was reasonably constant during this time. Weight loss in the face of a relatively constant intravascular blood volume must largely be due to extravascular fluid reabsorption. This weight loss pattern persisted even in 8 of the 28 animals subjected to 60% blood volume depletion that died during the late hypovolemic period owing to the severity of the hemorrhage. The weight loss was similar in collateral-free (brachial and cephalic veins cut or intact) and in completely intact forelimbs. Therefore, the forelimb weight loss cannot be attributed to trauma associated with extensive surgery or to bypassing the effects of the right atrium and the veins between the elbow and the right atrium on capillary pressure in the preparation in which the brachial and cephalic veins were cut. All arterial and venous pressures were substantially reduced throughout the hypovolemic period suggesting that the fluid influx may have resulted from a fall in capillary hydrostatic pressure (P_c). A fall in Pc could have resulted from a rise in precapillary resistance and a fall in aortic pressure despite an increase in postcapillary resistance. Furthermore, these data are not inconsistent with a hemorrhage-induced increase in microvascular permeability to plasma proteins. If microvascular permeability to plasma proteins increased, however, the fall in the transmural colloid osmotic pressure gradient must have been counteracted by a greater fall in the transmural hydrostatic pressure gradient since the forelimb weight loss continued unabated.

Endotoxin. The forelimb preparation was also used to study transvascular fluid fluxes in endotoxin shock (3). Mongrel dogs anesthetized with Na pentobarbital were given either 2(n=10) or 5(n=10) mg/Kg purified E. coli endotoxin, I.V. The endotoxin was suspended in 20 cc of saline and infused over 10 min. Forelimb weight, vascular pressures, blood flows and vascular resistances were followed for 4 hr afterwards. Both doses of endotoxin produced an initial rapid weight loss (0-10 min) and then a slower weight loss during the remainder of the hypotensive period (10-240 min). The weight loss was largest in those receiving the high dose. The rapid weight loss (0-10 min) was associated with increases in forelimb total and segmental vascular resistances. This suggests that mean vessel caliber decreased and that forelimb intravascular blood volume also decreased. Thus, the rapid weight loss is largely attributable to a decreased vascular capacity due to vasoconstriction. The slow weight loss (10-240 min) was associated with further increases in forelimb skin segmental vascular resistances and decreases toward control in forelimb skeletal muscle vascular resistances; the net effect was a fall in total forelimb resistance toward control. This suggests that mean vessel caliber was increasing with an increase in forelimb intravascular blood volume. A weight loss in the face of an increasing vascular capacity must be attributed to extravascular fluid reabsorption. All arterial and venous pressures were substantially reduced throughout the hypotensive period suggesting that the fluid influx may have resulted from a fall in Pc. A fall in Pc would occur if the rise in precapillary resistance and the fall in aortic pressure overwhelmed the effects of the rise in postcapillary resistance. These findings do not support the concept that endotoxin shock greatly increases microvascular permeability to plasma proteins. If microvascular permeability to proteins did increase, it must be assumed that the fall in transmural colloid osmotic gradient was counteracted by a greater fall in the transmural hydrostatic pressure gradient since the forelimb weight loss persisted. Hinshaw and Owens (5) studied transvascular fluid fluxes during prolonged endotoxin hypotension in a similar forelimb preparation. They also failed to find fluid filtration into the forelimb.

Catecholamines. Many investigators have attributed the alleged fluid filtration in hemorrhagic or endotoxin shock to catecholamine-induced venoconstriction at a time when precapillary resistance wanes owing to local regulation. The resulting decrease in the pre/postcapillary resistance ratio is said to increase Pc above plasma colloid osmotic pressure causing net fluid filtration. To test this hypothesis the effects of prolonged infusions (3 hr duration) of high doses of norepinephrine (1.5 (n=10) or 3.0 (n=10) µg base/kg/min i.v.) or epinephrine (3.0 (n=10) or 6.0 (n=10) µg base/Kg/min, i.v.) on forelimb transvascular fluid fluxes was also investigated (4). As in the previous experiments, the forelimb weight, vascular pressures, blood flows and vascular resistances were followed. Epinephrine and norepinephrine produced sustained increases in forelimb total and segmental vascular resistances and net decreases in forelimb weight. Vascular resistances in skin increases proportionately more than vascular resistances in skeletal muscle. Systemic pressure increased markedly initially but rapidly waned falling below control by the end of the infusion period. The weight loss from min 0-60 was largely due to a decreased intravascular blood volume subsequent to intense vasoconstriction. The weight loss during the last 2 hr of the infusion period was associated with a falling systemic pressure and nearly constant total forelimb small vessel and large vein resistances. The nearly constant resistance in the capacitance vessels suggests a reasonably steady mean vessel caliber with a nearly constant forelimb blood volume. A weight loss in the face of a nearly constant forelimb vascular capacity is suggestive of extravascular fluid reabsorption. These data clearly fail to provide evidence for extensive net catecholamine-induced fluid filtration into skin and skeletal muscle during the prolonged infusion of either epinephrine or norepinephrine while systemic pressure is at or below control. Since both hemorrhagic and endotoxin shocks are associated with hypotension, catecholamine-induced fluid filtration into skin and skeletal muscle would not be expected in these circulatory shock states.

Histamine. Other investigators have suggested that the alleged fluid filtration in shock is partly due to an increased microvascular permeability to plasma proteins. As protein leaves the vascular system, the transmural colloid osmotic pressure gradient is said to fall below the transmural hydrostatic pressure gradient resulting in fluid filtration. The change in microvascular permeability to plasma proteins is frequently attributed to the presence of histamine or bradykinin which are known to be released in a variety of circulatory shock states. However, because locally administered histamine clearly produces marked edema formation in doses which do not elicit systemic effects, it should not be implied that histamine or bradykinin-induced fluid filtration would necessarily be expected in hypotensive shock states. Microvascular permeability to plasma proteins and the pre-postcapillary resistance ratio are the determinants of the direction and rate of transvascular fluid fluxes during local administration of vasoactive agents. When the whole body participates in response to a circulatory stress or to a vasoactive agent, aortic pressure and right atrial pressure must also be considered as important determinants of Pc and, hence, transvascular fluid fluxes. Changes in these two variables may magnify, nullify, or reverse the effects of changes in microvascular permeability and the pre/postcapillary resistance ratio on transvascular fluid fluxes. This prompted a study to determine the effect of histamine on transvascular fluid fluxes when all 4 (aortic pressure, precapillary resistance, postcapillary resistance, right atrial pressure) indirect determinants of Pc were operant. The effects of intravenous infusions of histamine (60 µg base/min I.V. for 4 hr) on forelimb weight, vascular pressures, blood flows, and vascular resistances were determined (n=10). Histamine produced large sustained reductions in systemic pressure and decreases in forelimb blood flows. Total skeletal muscle vascular resistance increased slightly with time but total skin vascular resistance was significantly decreased during the entire infusion period, the net effect was that total forelimb vascular resistance failed to increase relative to control throughout the infusion period. This suggests that the mean vessel caliber and the intravascular blood volume remained at least at control levels during the entire infusion period. Forelimb weight, however, decreased markedly during this time. A large weight loss in the presence of a reasonably constant blood volume must be attributed to extravascular fluid reabsorption. Hence, these data clearly demonstrate that extravascular fluid reabsorption may still occur despite high doses of histamine, and also fail to support the hypothesis that histamine-induced fluid filtration would likely occur in circulatory shock states.

Control animals. A series of control animals anesthetized with Na pentobarbital was also studied (4). Forelimb weight and systemic pressure failed to change significantly from control during a 5 hr observation period (n=7). Segmental vascular resistances increased slightly but significantly with time. However, the spontaneous rise in segmental vascular resistances in the control animals could only account for a fraction of the increase in segmental vascular resistances in the experimental animals. Hence, the changes in forelimb weight, systemic pressure, and segmental vascular resistances must largely be attributed to the intervention (hemorrhage, endotoxin, histamine, or catecholamines) and not to deterioration of the preparation with time.

Summary

These data fail to provide evidence for fluid filtration into skin and skeletal muscle during severe, prolonged hemorrhagic or endotoxin hypotension and also fail to support the hypothesis that catecholamine or histamine-induced fluid filtration are likely to occur in hypotensive shock states. We are presently studying the effects of transfusion and over-transfusion on forelimb transvascular fluid fluxes to determine if fluid filtration might occur during the post-transfusion period.

REFERENCES

1. Schwinghamer, J.M., Grega, G.J. & Haddy, F.J. (1970) Am. J. Physiol. 219:318.

2. Grega, G.J., Schwinghamer, J.M. & Haddy, F.J. (1971) Circulation Res. 29:691.

3. Weidner, W.J., Grega, G.J. & Haddy, F.J. (1971) Am. J. Physiol. 221:1229.

4. Grega, G.J. & Haddy, F.J. (1971) Am. J. Physiol. 220:1448.

5. Baker, C.H. (1969) Am. J. Physiol. 216:368.

6. Hinshaw, L.B. & Owens, S.E. (1971) J. Appl. Physiol. 30:371.

CAPILLARY FILTRATION RATE IN CAT HIND LIMB MUSCLE DURING HEMORRHAGE

Susanna Biró and A.G.B. Kovách

Experimental Research Department, Semmelweis Medical

University, Budapest, Hungary

Muscle tissue comprises about 40 per cent of the body weight in mammals. It appears that in the development of irreversible hemorrhagic shock changes in skeletal muscle microcirculation may be critical.

Studying the changes of fluid transport in skeletal muscle microcirculation we were interested in the following:

1. Characteristic changes in the capillary fluid transfer during the course of mild and severe hemorrhage.

2. The mode of action of Butylsympathon (p-oxy-butylamino-acetophenon-hydrochloride) an agent with shock protecting effect (Kovách et.al. 1966, unpublished),(Thuránszky et.al. 1965) on the changes of fluid transfer occurring during hemorrhage.

Changes in capillary filtration coefficient during hemorrhage and after treatment with Butylsympathon were studied. This coefficient is known to indicate the actual state of the precapillary sphincters and thus the number of capillaries open to flow and exchange at any given moment (Mellander et.al. 1963).

Methods. Experiments were performed on 55 cats anesthetized intraperitoneally with Chloralose and Urethane (50 mg/kg and 200 mg/kg). No artificial respiration was used unless the condition of the animal seemed to deteriorate rapidly. Isovolumetric plethysmographic method described by Kjellmer (1965) was applied. Briefly, the calf muscles of one of the hind limbs of the cat were isolated so that the only remaining connections were the popliteal artery and vein and the femoral and sciatic nerves. The paw was excluded from the circulation by tight ligatures, the femur drilled open and the bone marrow cavity plugged. Heparin (1000 IE/kg) was given in order to prevent blood coagulation.

The following parameters were recorded: 1.) arterial blood pressure from the controlateral femoral artery with mercury manometer; 2.) hind limb blood flow from the cannulated popliteal vein - venous outflow method - the blood being immediately reinfused into the opposite femoral vein; 3.) volume of the isolated calf muscles placed in a plethysmograph with temperature control. Capillary filtration coefficient (CFC) was calculated from the changes in volume recorded after increasing venous outflow pressure by 10-15 mmHg. The response consists of two components and the CFC was calculated from the second, slower slope of the recording.

Hemorrhagic hypotension was induced by means of the pressure bottle technique of Engelking and Willig. The animals were bled via the carotid artery to a blood pressure level of 50-60 mmHg for 60 minutes (first group) for 120 minutes (second group) and the third group was exposed to standardized hemorrhagic shock where the arterial blood pressure was maintained at a level of 55 mmHg for 90 minutes and at 35 mmHg for another 90 minutes. At the end of the bleeding periods the shed blood was reinfused and all parameters were recorded for one hour after reinfusion.

Butylsympathon was given slowly intravenously in a dose of 15 mg/kg to normotensive cats (fourth group) and to animals exposed later to standardized hemorrhagic shock (fifth group).

Results. Effects of hemorrhagic hypotension. During hemorrhagic hypotension to 50-60 mmHg for 60 or 120 minutes parallelly to a drop in blood pressure and blood flow there were no changes or there was a slight increase in CFC values. After reinfusion all parameters returned to the control level (Fig.1).

Effects of hemorrhagic shock. In figure 2 changes observed during standardized hemorrhagic shock are demonstrated. During the first period of bleeding to 55 mmHg the changes in blood flow and CFC are similar to those observed during hypotension of 60 and 120 minute duration. Entirely different changes occurred however, during the second bleeding period to 35 mmHg. The blood flow decreased to 20 per cent and the CFC measured in this period was 56 per cent ($P < 0,001$). After reinfusion of the shed blood none of the measured parameters reached the initial level, the blood pressure was 68 per cent, the blood flow 70 per cent and the CFC 56 per cent of the prebleeding values 60 minutes after reinfusion.

Effects of Butylsympathon in normotensive animals. Changes in arterial blood pressure, hind limb blood flow and capillary

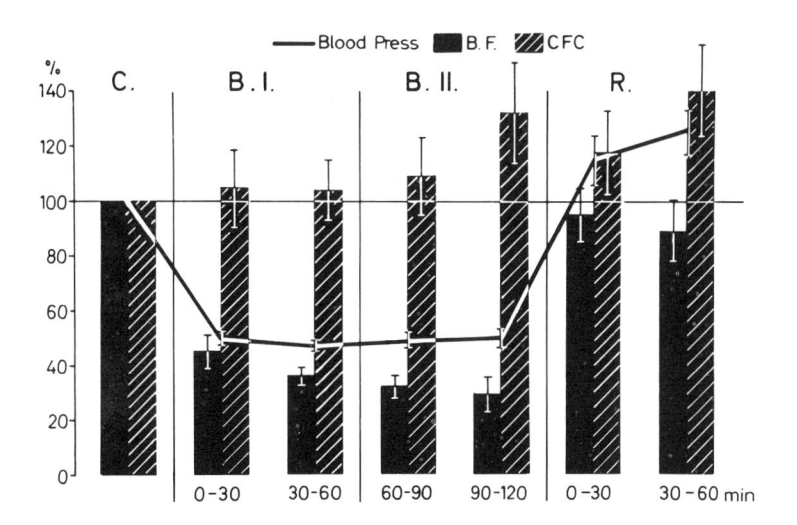

Fig.1. Effects of 120 min hemorrhagic hypotension on blood pressure, blood flow and capillary filtration coefficient.
N.B. In all Figs. the vertical lines on the columns indicate the standard errors of the means.

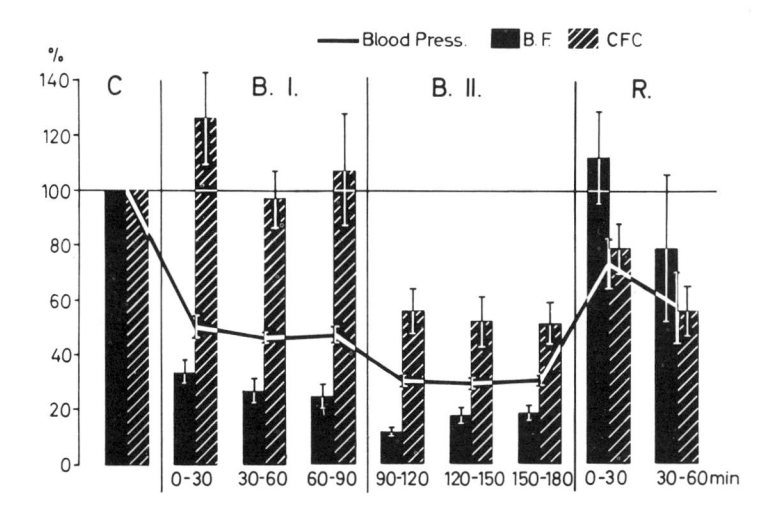

Fig. 2. Effects of standardized hemorrhagic shock on blood pressure, blood flow and capillary filtration coefficient.

filtration coefficient before and after administration of Butylsympathon are demonstrated in figure 3. The arterial blood pressure and the hind limb blood flow decreased to about 80 per cent during the first 60 minutes following the administration of the drug while there was a 35 per cent increase ($0.05 < P < 0.02$) in CFC values. The peripheral resistance increased to 110 per cent 15-30 minutes after the onset of the drug action and then returned to the initial level. All measured parameters returned to their initial values 90 minutes after the drug administration.

Effects of Butylsympathon during hemorrhagic shock. Figure 4 demonstrates changes in arterial blood pressure, hind limb blood flow and CFC during standardized hemorrhagic shock in cats pretreated with Butylsympathon (15 mg/kg i.v.).

In these animals decrease in blood flow was 47 per cent during the first bleeding period and only 25 per cent during bleeding to 35 mmHg, after reinfusion it was 57 per cent of the initial value.

The most important difference comparing to the non-treated animals was observed in the CFC values during the second bleeding period. The capillary filtration coefficient was 0.028 ml/min/Hgmm/100 g, (149 per cent, $0.02 < P < 0.01$) at the beginning of this period and did not fall below the control value (0.019/ml/min/Hgmm/100 g) either during bleeding or after reinfusion.

Changes in peripheral resistance during hemorrhagic shock in treated and non-treated cats are illustrated in Figure 5. There is a great increase in resistance up to 300 per cent in the non-treated animals while the resistance in the cats pretreated with Butylsympathon did not differ significantly from the control values during bleeding periods.

The bleeding volume was much higher in the non-treated animals (Figure 6) the maximal value being 37 ml/kg during the second period of bleeding while in the pretreated group at all intervals the shed blood was significantly less, the maximal value being 13 ml/kg during the first bleeding period.

Discussion. We have found a severe diminution of the capillary filtration coefficient during the second bleeding period of standardized hemorrhagic shock. Earlier Mellander (1963) demonstrated an approximately 300 per cent increase in CFC values during hemorrhage. The apparent contradiction can be explained by some differences in the experimental conditions such as (1) in our experiments the calf muscles were totally isolated while Mellander measured the CFC values of the muscles and skin together; (2) in Mellander's experiments unilateral adrenalectomy had been performed and the other gland was ligated while in our experiments the adrenals were left intact; (3) in our experiments the sym-

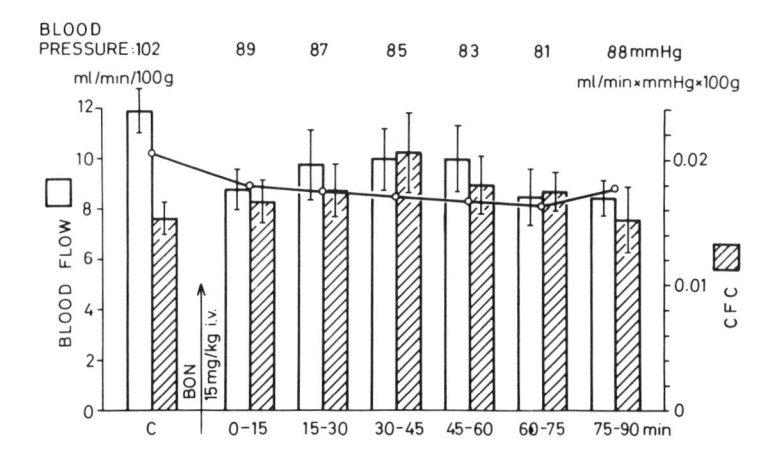

Fig. 3. Effects of Butylsympathon on blood pressure, blood flow and capillary filtration coefficient in normotensive cats.

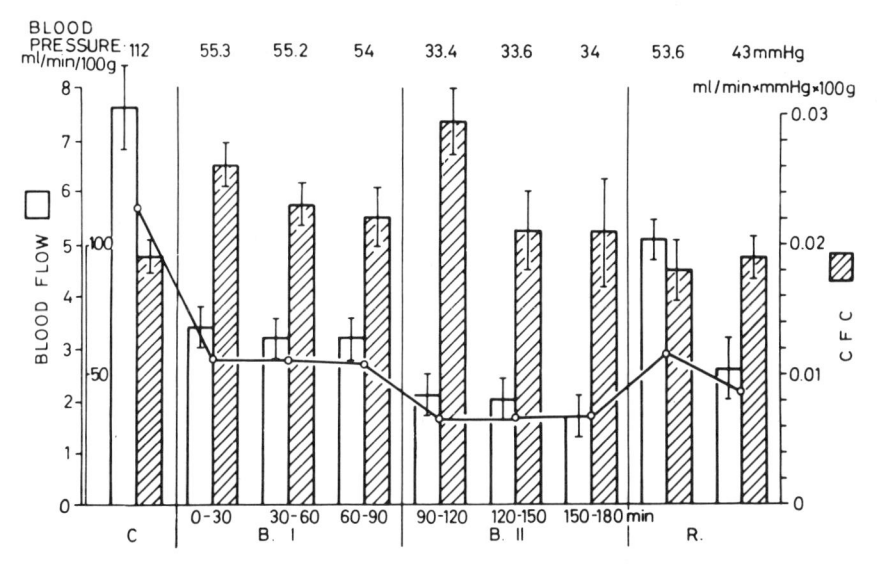

Fig. 4. Effects of Butylsympathon on blood pressure, blood flow and capillary filtration coefficient during standardized hemorrhagic shock.

pathetic innervation remained intact while Mellander cut the sympathetic trunk; (4) as to the methodology of hemorrhagic shock the 30-35 mmHg level seems to be critical and in Mellander's experiments the blood pressure was maintained at or above 40 mmHg.

According to the decrease in CFC values during the second bleeding period of hemorrhagic shock it seems obvious that the increased sympathetic activity combined with severe hemodynamic alterations can influence the tone of the precapillary sphincters and this may lead to a severe and irreversible diminution of CFC observed in our experiments. This explanation seems to be reasonable considering that neither 60 nor 120 minute hypotension to 55 mmHg level, nor the first period of bleeding in the course of hemorrhagic shock had led to any reduction in CFC.

Butylsympathon increased the CFC values in anesthetized cats and prevented the 50 per cent diminution of the capillary filtration coefficient in the second phase of standardized hemorrhagic shock. This drug was found to exert a protective action in the course of irreversible hemorrhagic shock in dogs and had a potent ganglionic blocking effect acting also at the medullary level. It is supposed that the protective action of the drug is due to its sympathetic blocking activity viz. decreasing the tone of the precapillary sphincters it prevents the severe and irreversible diminution of CFC during the second bleeding period of hemorrhagic shock.

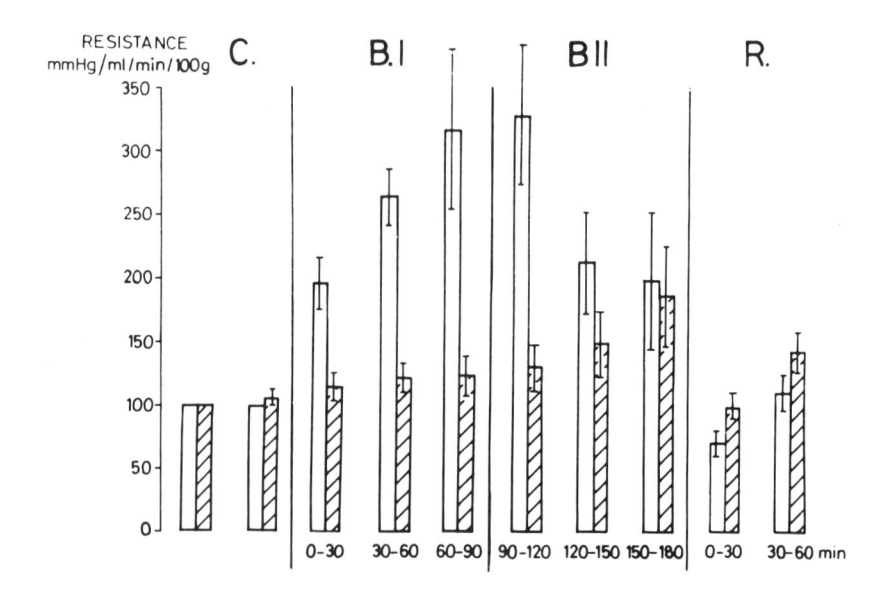

Fig.5. Peripheral resistance of hind limb of cats with and without Butylsympathon. Treated animals - hatched columns.

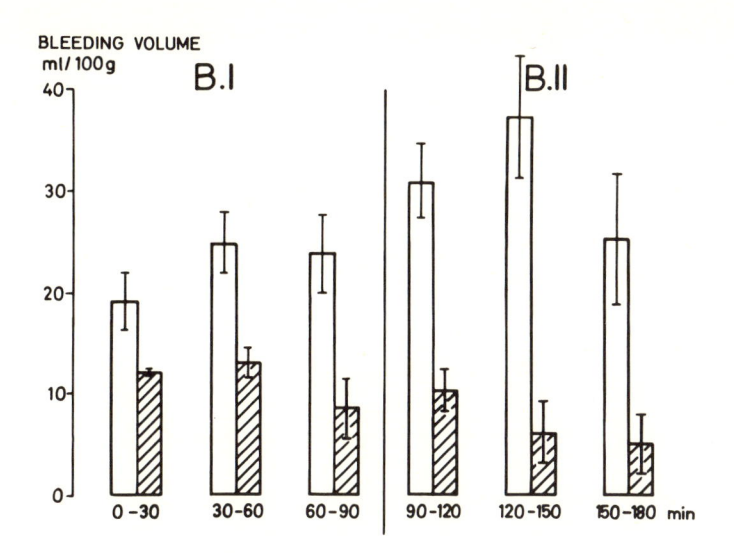

Fig 6. Bleeding volume during standardized hemorrhagic shock with and without Butylsympathon. Treated animals - hatched columns.

REFERENCES

Kjellmer, I. (1965) Acta physiol. scand. $\underline{63}$:450.

Kovách, A.G.B. (1966) unpublished.

Mellander, S. & Lewis, D.H. (1963) Circ. Res. $\underline{13}$:105.

Thuránszky, K., Szabó, A.K., Bálint, G. & Madarász, I. (1965). MTA V. Orvostud. Oszt. Közl. $\underline{15}$:379.

THE ADENOHYPOPHYSIAL RESPONSE TO STRESS AND ITS HYPOTHALAMIC CONTROL

S.M. McCann, L. Krulich, P. Illner, M. Quijada,
K. Ajika and C.P. Fawcett

Dept. of Physiology, University of Texas
Southwestern Medical School, Dallas, Texas

In this communication we will review the participation of the anterior pituitary gland in the stress response and concentrate in particular on research in this area from our own laboratory. The well-known participation of the neurohypophysis and its hormone, vasopressin, in the stress response will not be considered.

Adrenocorticotropin (ACTH). ACTH was the first anterior pituitary hormone found to be discharged in response to noxious stimuli of a variety of types (1). The release of ACTH takes place within a minute after the application of the stressful stimulus and continues until it is removed. Almost any alteration in the internal or external environment of the organism if sufficiently pronounced, can produce a release of ACTH. Although most such stimuli are physical in nature, such as hemorrhage, fracture, or surgical operation, emotions can also elicit a discharge of ACTH and it can even follow exposure to heat or cold (2). Afferent pathways concerned with ACTH release appear to pass up the spinal cord in the lateral spinothalamic tract and from the cord into the brain stem (3). It appears likely that impulses in the ascending pathways enter the reticular formation and traverse this to the ventral hypothalamus (4).

Electrical stimulation in other regions of the brain in addition to the hypothalamus can also bring about ACTH secretion. One of these areas is the amygdala. Thus, it is probable that emotional influences are transmitted to the ACTH-controlling centers in the hypothalamus via the limbic system (4).

It is also clear that some inhibitory influences impinge on the hypothalamic controlling centers since isolation of the hypothalamus from the rest of the brain using the knife developed by Halasz leads to augmentation of ACTH release (5).

It is now generally accepted that the stimulatory hypothalamic influence over ACTH release is mediated by a corticotropin releasing factor(s) (CRF) (6). This factor is released into the hypophyseal portal capillaries in the median eminence and is transported to the anterior lobe via the hypophyseal portal system of veins. Even though CRF was the first hypothalamic-releasing factor to be discovered, its chemical nature is still elusive. It appears to be dissimilar from vasopressin and on the basis of enzyme inactivations it is probably a small polypeptide. Whether or not there is more than one CRF as originally claimed (7) remains to be determined (6).

Although vasopressin was the first substance found to possess corticotropin-releasing activity (8), it is now clear that it is not the major corticotropin-releasing factor. In fact, its actions in increasing ACTH release are so complex that they have generated considerable research and controversy for over 15 yr. It is now fairly clear that exogenous vasopressin can evoke ACTH release by a variety of means. First, the injection of large doses of vasopressin alters blood pressure which may cause the activation of ACTH release by stimulating CRF discharge. Second, it appears that, at certain doses, vasopressin can stimulate ACTH release by a direct action on the pituitary gland in view of the fact that the polypeptide is active in animals with hypothalamic lesions which eliminate nonspecific stress responses, if injected directly into interstices of the pituitary gland, if injected into the arterial supply of a pituitary tumor, or if incubated with pituitaries _in vitro_ (6).

Recently, data have been presented to suggest that vasopressin may, in addition, potentiate the response to CRF. The principal evidence for this is the fact that microinjection of vasopressin into the anterior lobe in doses which are too small to evoke ACTH release potentiates the response to systemically administered CRF. Systemic administration of subthreshold doses of vasopressin similarly potentiates the response to CRF and the effect of vasopressin is not reproduced by other vasoactive substances (9).

The animal with hereditary diabetes insipidus which lacks vasopressin and the animal with posterior lobectomy which also suffers from vasopressin deficit appear to have a subnormal release of ACTH in response to stress (9, 10, 11). This would be consistent with a role for vasopressin in potentiating the response to CRF. Conversely, dehydration to evoke endogenous vasopressin secretion also potentiates the response of the rat pituitary to CRF (9).

Lastly, extremely large pharmacological doses of vasopressin may act directly on the adrenal gland to evoke corticoid release (12).

Adrenal corticoids, released in response to ACTH, feedback on the hypothalamic pituitary unit to inhibit the release of ACTH. The weight of evidence at the present time suggests that this feedback probably occurs both at the hypothalamic and at the anterior pituitary level. For example, Chowers et al (13) have shown that corticoids can modify the content of CRF in the hypothalamus pointing to a hypothalamic site for the feedback. On the other hand, it has been clearly shown by several groups that corticords can at least partially block the response to CRF which indicates a site of action at the pituitary level (14).

In addition to the feedback action of corticoids, there is little doubt that large doses of ACTH can feedback to suppress ACTH release, a so-called "autofeedback" action (15). Whether this effect operates at physiological levels of ACTH remains to be established.

Growth hormone. The development of a radioimmunoassay for human growth hormone led relatively recently to the discovery that stress elevates plasma growth hormone in man (15a). This also occurs in the monkey and bioassay by the tibia test revealed depletions of pituitary growth hormone in rats exposed to stress (16) which led to the conclusion that growth hormone release followed stress in the rat as well. The development of radioimmunoassay for rat growth hormone, however, failed to confirm the existence of stress-induced growth hormone release in the rat. On the contrary, stress was found to reduce growth hormone levels in this species (17). This discrepancy between immuno and bioassay results remains to be resolved and it is hard to explain by postulating nonspecificity of the tibia assay. In addition, growth hormone is released in response to certain "metabolic" stimuli such as hypoglycemia or the infusion of amino acids (15a).

The afferent pathways by which stress activates growth hormone release have not been studied; however, there is now good evidence for a hypothalamic controlling mechanism as in the case of ACTH. On the basis of both stimulation and lesion studies, it would appear that the ventromedial nucleus is the region most concerned with growth hormone release (18). As in the case of ACTH, hypothalamic control over growth hormone secretion is mediated by means of hypothalamic factors. A growth hormone releasing factor was discovered in extracts of hypothalamus on the basis of the ability of these extracts to deplete pituitary growth hormone in the rat or to increase growth hormone secretion by pituitaries incubated in vitro as determined by bioassay using the tibia test (19). The ability of these extracts to release growth hormone both in vivo (20) and in vitro (21) in the rat has now been confirmed by immunoassay but here again a discrepancy still exists between immuno and bioassay in that the depletions of pituitary growth hormone observed by bioassay have not been reproduced by immunoassay. Growth hormone releasing factor has been purified and separated from other releasing factors and prepared in highly purified form (22, 23).

During routine screening of fractions from Sephadex columns for GRF activity in vitro, Krulich discovered that certain fractions reduced release of GH from the pituitary rather than augmenting it (24). This activity could be further purified (25) and was given the name growth hormone inhibiting factor (GIF). In recent work in our laboratory, the GRF and GIF activity of sheep hypothalamic extract has been followed using immunoassay of media from pituitaries incubated in vitro. The existence of both GRF and GIF activities has been confirmed by this method (21).

As part of a study of the localization of various hypothalamic releasing and inhibiting factors in the hypothalamus, the distribution of GRF and GIF in frozen sections of the hypothalamus has been studied. The sections have been cut along the frontal, horizontal and sagittal planes. Although the results are complex, it appears that GRF is localized mainly to the region of the ventromedial nucleus which would be consistent with the earlier stimulation and lesion studies. On the other hand, GIF was localized only to the median eminence region. There was minimal activity in a region lying at the border between the anterior hypothalamic and preoptic areas. It would appear likely that the neurons which secrete GRF have their cell bodies in the region of the ventromedial nucleus with axons which project to the median eminence, there to release GRF into the hypophyseal portal vessels (21). The distribution of CRF in the hypothalamus has received scant attention although in one report it was found only in the median eminence (26).

In the case of GH, autofeedback of the hormone to inhibit its own release may be quite important since there is no target hormone feedback. It has been shown that exogenous GH can lower pituitary GH content if given systemically (27) or if implanted in the hypothalamus (28). In the monkey exogenous GH blocks the response to insulin-induced hypoglycemia (29).

Prolactin. In addition to ACTH and growth hormone, it is now necessary to add prolactin to the list of hormones which are released in response to stress. Early evidence for this was the development of pseudopregnancy in rats following stress and the ability of stress to induce lactational changes in the mammary gland of estrogen-primed rats. Grosvenor *et al* (30) demonstrated that stress could induce a depletion of pituitary prolactin. Very recently with the development of a radioimmunoassay for rat prolactin, Neill (31) has observed that stress can produce a very rapid elevation in plasma prolactin in the rat. We have recently observed that the exposure to two minutes of ether anesthesia elevates plasma prolactin by a factor of three in ovariectomized rats (32). Nembutal anesthesia can block the stress-induced elevation of prolactin and lower the levels of the hormone to pre-stress (decapitated) levels (32).

The secretion of prolactin by the adenohypophasis is also under hypothalamic control and the predominate control appears to be inhibitory since lesions in the median eminence which interrupt this control or grafting the pituitary to a site distant from the hypothalamus, led to a rapid and sustained elevation in plasma prolactin level (33). The addition of hypothalamic extracts can inhibit the release of prolactin from pituitaries incubated *in vitro* (34) and their injection either intravenously (35) or into a cannulated hypophyseal portal vessel leads to a prompt lowering of plasma prolactin levels (36). The prolactin inhibiting factor (PIF) has been purified and separated from some of the other hypothalamic factors (37).

Increasing evidence suggests that prolactin secretion is under dual hypothalamic control by a prolactin releasing factor (PRF) as well as a prolactin inhibiting factor (38). In recent work from our laboratory, the localization of PRF and PIF in the hypothalamus has been studied by the technique described above for growth hormone. PIF activity was localized to the dorsolateral part of the preoptic area in the region corresponding to the lateral preoptic nucleus (39). Median eminence and a narrow medial basal portion of the preoptic area showed on the other hand prolactin releasing activity suggesting the presence of a PRF.

Other anterior pituitary hormones. The possible participation of MSH in the stress response is controversial (40, 41). There has been little evidence to support the view that gonadotropins play a significant role in the stress response. Ajika, in our department, has shown that etherization for two minutes can lead to rapid elevations in immunoassayable FSH and LH in plasma of ovariectomized rats (32). The response to ether could be blocked by Nembutal anesthesia in the case of LH whereas the elevation of FSH was not inhibited by Nembutal. Earlier attempts in the rat to demonstrate stress effects on gonadotropins were probably masked by the use of ether anesthesia (42). It will be of interest to study further the possible participation of gonadotropins in the stress response.

Conclusions. It is now clear that the adenohypophyseal response to stress is not limited to ACTH but also involves the discharge of growth hormone, prolactin, and possibly FSH and LH as well. The afferent pathways involved in these responses have only been studied in the case of ACTH but it is tempting to speculate that the same pathways may be utilized in mediating the response of the other adenohypophyseal hormones to stress as well. Afferent impulses from the brain stem, limbic system and possibly cortex converge on the hypothalamic controlling centers which regulate the release of the various pituitary hormones. This leads to modification in the rate of discharge of the hypothalamic releasing and inhibiting factors into the hypophyseal portal vessels to bring about the stress-induced release of the adenohypophyseal hormone in question. The utility to the organism of this mass discharge of adenohypophyseal hormones remains to be elucidated.

REFERENCES

1. Sayers, G. (1950) Physiol. Rev. 30:241.

2. Chowers, I., Hammel, H.T., Eisenman, J., Abrams, R.M. & McCann, S.M. (1966) Am. J. Physiol. 210:606.

3. Gibbs, F., (1969) Am. J. Physiol. 217:78.

4. Ganong, W.F. In the Hypothalamus, ed. L. Martini, M. Motta & F. Fraschini, Academic Press, New York, 1970, p. 313.

5. Halasz, B., In Frontiers in Neuroendocrinology, ed. W.F. Ganong and L. Martini, Oxford, London, 1969, p. 307.

6. McCann, S.M. & J.C. Porter (1969) Physiol. Rev. 49:240.

7. Guillemin, R., (1964) Rec. Prog. in Hormone Res. 20:89.

8. McCann, S.M. & J.R. Brobeck, (1954) Proc. Soc. Exp. Biol. and Med. 87:318.

9. Yates, F.E., Russell, S.M., Dallman, M.F., Hedge, G.A., McCann S.M. & Dhariwal, A.P.S. (1971) Endocrinology 88:3.

10. McCann, S.M., Antunes-Rodrigues, J., Nallar, R. & Valtin, H. (1966) 79:1058.

11. Smelik, P.G., Gaavenstroom, J.H., Konijnendijk W. & de Wied, D. (1962) Acta Physiol. Pharmacol. Neerl 11:20.

12. Hilton, J.G., Scian, L.F., Westermann, C.O. & Kruesi, O.R. (1959) Science 129:971.

13. Chowers, I., Conforti, N. & Feldman, S. (1967) Neuroendocrinology 2:193.

14. Russell, S.M., Dhariwal, A.P.S., McCann S.M. & Yates, F.E. (1969) Endocrinology 85:512.

15. Vernikos-Danellis, J. & Trigg, L.N. (1967) Endocrinology 80:345.

15a Glick, S.M. (1969) In Frontiers in Neuroendocrinology, ed. W.F. Ganong & L. Martini, Oxford, London, p. 141.

16. Krulich, L. & McCann, S.M. (1966) Proc. Soc. Exp. Biol. and Med. 122:612.

17. Takahashi, K., Daughaday, W.H. & Kipnis, D.M. (1971) Endocrinology 88:909.

18. Frohman, L.A. & Bernardis, L.L. (1968) Endocrinology 86:305.

19. Deuben, R.R. & Meites, J. (1966) Endocrinology 74:408.

20. Frohman, L.A., Moran, J.W., Yates, F.E. & Dhariwal, A.P.S. (1971) Fed. Proc. 30:198.

21. Krulich, L., Illner, P. Fawcett, C.P., Quijada, M. & McCann, S.M. (1971) Proceedings of the 2nd International Symposium on GH, Milan, Italy, in press.

22. Dhariwal, A.P.S., Krulich, L., Katz, S.H. & McCann, S.M. (1965) Endocrinology 77:939.

23. Schally, A.V., Sawano, S., Arimura, A., Barret, S.F., Wakabayashi, I & Bowers, C.Y. (1969) Endocrinology 84:1493.

24. Krulich, L., Dhariwal, A.P.S. & McCann, S.M. (1968) Endocrinology 83:783.

25. Dhariwal, A.P.S., Krulich, L. & McCann, S.M. (1969) Neuroendocrinology 4:282.

26. Vernikos-Danellis, J. (1964) Endocrinology 75:514.

27. Krulich, L. & McCann, S.M. (1966) Proc. Soc. Exp. Biol. and Med. 121:1114.

28. Katz, S.A., Molitch, M. & McCann, S.M. (1969) Endocrinology 85:725.

29. Sakuma, M. & Knobil, E. (1970) Endocrinology 86:890.

30. Grosvenor, C.E., Nallar, R. & McCann S.M. (1965) Endocrinology 76:883.

31. Neill, J.D. (1970) Endocrinology 87:1192.

32. Ajika, K., Kalra, S.P., Fawcett, C.P. Krulich, L & McCann, S.M. (1971) Endocrinology 90: 707.

33. Bishop, W., Krulich, L., Fawcett, C.P. & McCann, S.M. (1971) Proc. Soc. Exp. Biol. Med. 136:925.

34. Talwalker, P.K., Ratner, A. & Meites, J., (1963) Am. J. Physiol. 205:213.

35. Watson, J.T., Krulich, L. & McCann, S.M. (1971) Endocrinology, in press.

36. Kamberi, I.A., Mical, R.S. & Porter, J.C. (1971) Endocrinology 88:1288.

37. Dhariwal, A.P.S., Grosvenor, C.E., Antunes-Rodrigues, J. & McCann, S.M. (1968) Endocrinology 82:1236.

38. Nicoll, C.S., (1970) In Hypophysiotropic Hormones of the Hypothalamus, Assay and Chemistry, ed. J. Meites, Williams and Wilkins, New York, p. 115.

39. Krulich, L., Quijada, M. & Illner, P. (1971) Program of the 53rd Meeting of the Endocrine Society, p. 83.

40. Kastin, A.J., Arimura, A., Viosia, S., Barrett, L. & Schally, A.V., (1967) Neuroendocrinology 2:200.

41. Taleisnik, S. & Orias, R. (1965) Am. J. Physiol. 208:293.

42. Ramirez, V.D. & McCann, S.M. (1965) Proc. Soc. Exp. Biol. and Med. 118:169.

Research in the authors' laboratory was supported by NIH grant No. AM10073 and by grants from the Ford Foundation and Texas Population Crisis Foundation.

THYROID FUNCTION AFTER INJURY

Roger Kirby and Ivan D.A. Johnston

University Department of Surgery, Royal Victoria
Infirmary, Newcastle upon Tyne, England

The participation of the thyroid gland and its hormonal products, thyroxine (T_4) and triiodothyronine (T_3) in the response of the body to major injury has long been the subject of research and speculation. Hanbury (1959), summarizing the available data at that time, stated that any general conclusions as to the function of the thyroid in the postoperative state would be tenuously drawn from the work reviewed. Since then investigation of thyroid physiology has proved a fertile field of research and there has been much progress in the development of techniques for measuring free thyroid hormones, the understanding of thyroid hormone transport mechanisms and the introduction of sensitive immunoassays for detecting thyrotrophic hormone.

Some of the features of the post injury period, notably the increases in O_2 consumption and metabolic expenditure suggest the effect of thyroid hormone. In addition, thyroxine is known to stimulate protein synthesis and to increase the incorporation of amino acids into protein and may thus be important in tissue repair.

Most previous studies have utilised standard tests of thyroid function repeated after operation. These have the disadvantage of large day-to-day variations and are known to be affected by other hormones, some of which are increased in the postoperative period e.g. cortisol and catecholamines depress the thyroid uptake of radioactive iodine which is reduced after operation (Johnston et al.,1965). Studies of serum thyroxine levels after trauma as measured by P.B.I. and B.E.I. have yielded varying results, some authors finding increases

(Franksson et al, 1959) whilst others have found no change (Engstrom & Markardt, 1954). Serum levels, however, depend upon the rate of degradation as well as the rate of synthesis and may therefore not be representative. Oppenheimer et al. (1967) found that contrary to what might be expected, operation decreased the fractional removal of $I^{131}T_4$ from plasma in the absence of a diminished degradative clearance. They suggested the possibility of either decreased disposition of $I^{131}T_4$ via the faecal route or redistribution of thyroxine between vascular and extra-vascular stores. Blomstedt (1965), studying the peripheral degradation of I^{131} thyroxine, found a significant increase in the radioiodide in the urine and concluded that there is a postoperative increase in the peripheral degradation of thyroxine.

Current concepts of thyroxine physiology emphasise the importance of free hormone. After secretion, thyroxine becomes rapidly bound to serum proteins and only a small fraction (0.05%) of the total hormone remains free. There is, however, considerable evidence to suggest that the concentration of free hormone is the most reliable determinant of the thyroid status of an individual; this fraction is able to penetrate cells and exert a metabolic effect whereas the protein-bound hormone acts largely as a buffering reservoir. Binding occurs to 3 individual serum proteins, thyroxine-binding globulin (TBG) which is normally the major carrier, thyroxine-binding prealbumin (TBPA) and albumin, and the relationship between thyroxine bound to individual protein sites and free thyroxine can be described in terms of a multiple equilibrium system conforming to mass action principles. A decrease in a binding protein will thus lead to a redistribution of T_4 and an increase in free hormone. The concentration of TBPA is reduced in many situations of stress, including trauma, and it has been postulated that TBPA may act as a short-term regulator of the delivery of free hormone to the tissues. Recently the relationships between blood thyrotrophin levels and the free thyroxine concentration, have been studied in man under normal physiological conditions (Lemarchand-Beraud & Vannoti, 1969). The formation and release of TSH is regulated by the hypothalamic thyrotrophin-releasing factor (TRF) and by the feed-back mechanism of the pituitary-thyroid axis (Brown-Grant, 1960). The variations in free hormone are in an inverse relationship with the blood TSH level and the free T_4 level seems to be one of the main factors regulating the feed-back mechanism which acts on the TSH secretion. However, it appears that this is not the only factor since variations in TSH occur in the presence of a normal free T_4 concentration in certain physiological conditions and the metabolic activity of thyroxine in the cell and cellular binding factors may be significant in the regulation of TSH secretion. Investigating anterior pituitary function during surgical stress, Crane Charters, Odell and Thompson (1969) found no significant intra-operative changes in the concentration of TSH although there was a small, questionably significant, decrease in TSH on the first postoperative day.

The present investigation was designed to study thyroid function after injury in a group of comparable subjects undergoing a standard surgical operation, utilising some of the current parameters of measurement.

Material and Methods

Ten endocrinologically normal male patients undergoing elective upper abdominal operation (vagotomy & pyloroplasty) were studied, and measurements made before and after operation of the free thyroxine concentration, the percentage free triiodothyronine, the thyroxine-binding proteins, protein bound iodine and plasma thyrotrophin (TSH). All operations were conducted under a general anaesthetic; induction with thiopentone, and maintenance with nitrous oxide, oxygen and halothane. Venous blood samples were allowed to clot at room temperature and the serum separated after centrifugation within 3 hr of withdrawal. Grouped serum samples were stored at -10°C. pending analysis.

Thyroxine Binding Studies

The maximal binding capacities of TBG and TBPA were determined by a modification of the electrophoretic technique of Launay (1966). Electrophoresis was performed in a Shandon cell using Oxoid cellulose acetate strips and a TRIS-Boric Acid-EDTA buffer at pH 9.0. I^{131} labelled l-thyroxine was obtained from the Radiochemical Centre, Amersham, and used within 2 weeks of despatch. Stock solutions of stable l-thyroxine were prepared weekly by dissolving the Na salt in 50% aqueous propylene glycol. The 3 stock solutions were made up in such a way that addition of 2 μl quantities of the stable thyroxine to each serum sample produced a final concentration of 200, 600 and 800 μg T_4 per 100 ml. Four x 100 μl aliquots of a serum sample were then labelled with I^{131} thyroxine to a concentration of 2.0 μg 1100 ml and the 2μl samples of the respective stock solutions were added to three of these. Duplicate 5 μl samples were then submitted to electrophoresis, the strips dried and the distribution of labelled thyroxine between the binding proteins determined using an Actigraph III automatically recording scanner and a manual planimeter (Fig 1). It has been determined that TBPA is saturated by a T_4 concentration of

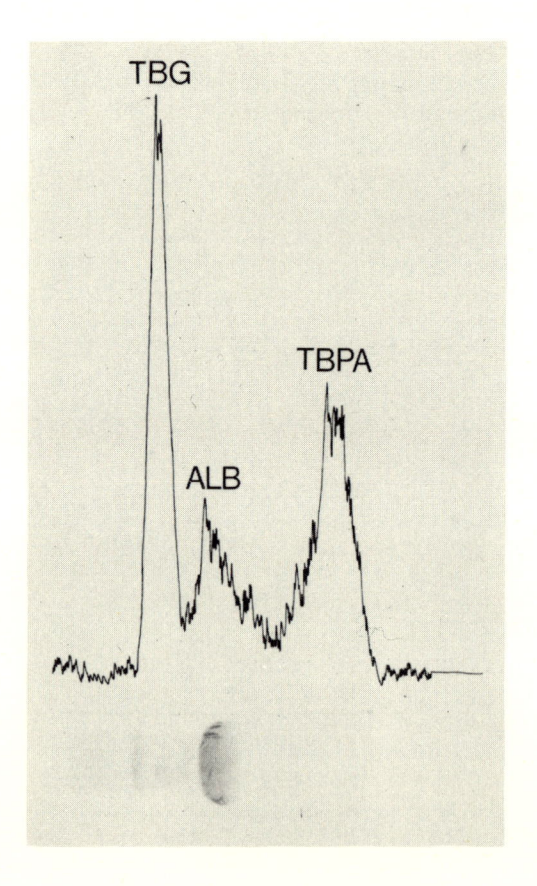

Fig. 1. Trace drawing of the distribution of ^{131}I-Thyroxine between the serum proteins as determined by cellulose acetate electrophoresis in Tris-boric and EDTA buffer at pH 9.0.

600 µg/100 ml and the mean of the duplicate 600 and 800 µg estimations was taken as the maximal binding capacity of TBPA. Similarly, TBG is saturated by a T_4 concentration of 200 µg/100 ml. Each group of 8 sera from one patient were analysed in a 4 day period using the same buffer, tracer and stock solutions.
All sera were utilised as soon as possible after collection and in no case kept longer than 4 weeks, during which time storage had no effect on T_4 binding capacities. The overall error of the method was ±12%.

Equilibrium Dialysis

Free thyroxine was estimated by a technique based on that originally described by Oppenheimer et al (1963) of equilibrium dialysis of diluted serum. I^{125} labelled l-thyroxine was added to serum samples in quantities sufficient to increase the endogenous concentration of T_4 by 1 µg/100 ml. After an equilibration period of 30 min, 0.5 ml of labelled serum was diluted in 12.0 ml of phosphate buffer, (pH 7.4, ionic strength 0.15) and 1.5 ml aliquot of diluted serum pipetted into a dialysis sac prepared from 8/32" Visking tubing. Each dialysis sac was then suspended in 7.0 ml phosphate buffer, sealed and incubated at 37°C on a rotating platform for 18 hr, during which time equilibrium is attained. Aliquots were then taken from inside and outside the dialysis sac and contaminating iodide eliminated by the addition of outdated plasma and trichloroacetic acid precipitation. The precipitates were assayed for radioactivity using a Nuclear Chicago well type scintillation counter. The percentage free thyroxine and the concentration of free thyroxine in the system were calculated by the formula of Oppenheimer et al. (1963).

All estimations of free T_4 were performed in duplicate and all sera from each patient were analysed as a batch on the same day to eliminate possible variation due to incubation temperature, buffer constitution, radioactive impurities, etc. A standard, using pooled serum of known free thyroxine content was included in each series of specimens analysed. In a preliminary investigation of the reproducibility of the method using sera from 11 healthy volunteers, the mean (± S.D.) free thyroxine level was 0.046 ± 0.004% and the mean free thyroxine concentration was 5.29 ± 1.02 (SD) x 10^{-11} moles per litre.

A similar dialysis technique was used to measure the percentage free triiodothyronine. I^{125} labelled T_3 was added to serum samples to a concentration of not more than 0.20 µg 1100 ml following which dilution, dialysis and TCA precipitation was carried out as with the T_4 estimations. The mean free T_3 in 11 healthy volunteers was 0.256% i.e. approx 5 times that of free T_4.

TSH Immunoassay

Serum TSH levels were measured by a modification of the double antibody method (Raud & Odell 1969), with guinea pig anti-human TSH in the presence of human chorionic gonadotrophin and using rabbit anti-guinea pig gamma-globulin as second antibody. National Pituitary Agency human TSH was used for iodination by the method of Hunter and Greenwood (1962), iodinated human TSH being purified by a modification of the Quso method of Berson and Yalow (1968). Medical Research Council human TSH Stardard A was used for reference. The Serum Protein Bound Iodine was estimated by the Technicon Auto-Analyser technique of Riley and Gochman (1964).

RESULTS

Estimates for the maximal binding capacities of TBPA and TBG are to some extent dependent upon the method used for their

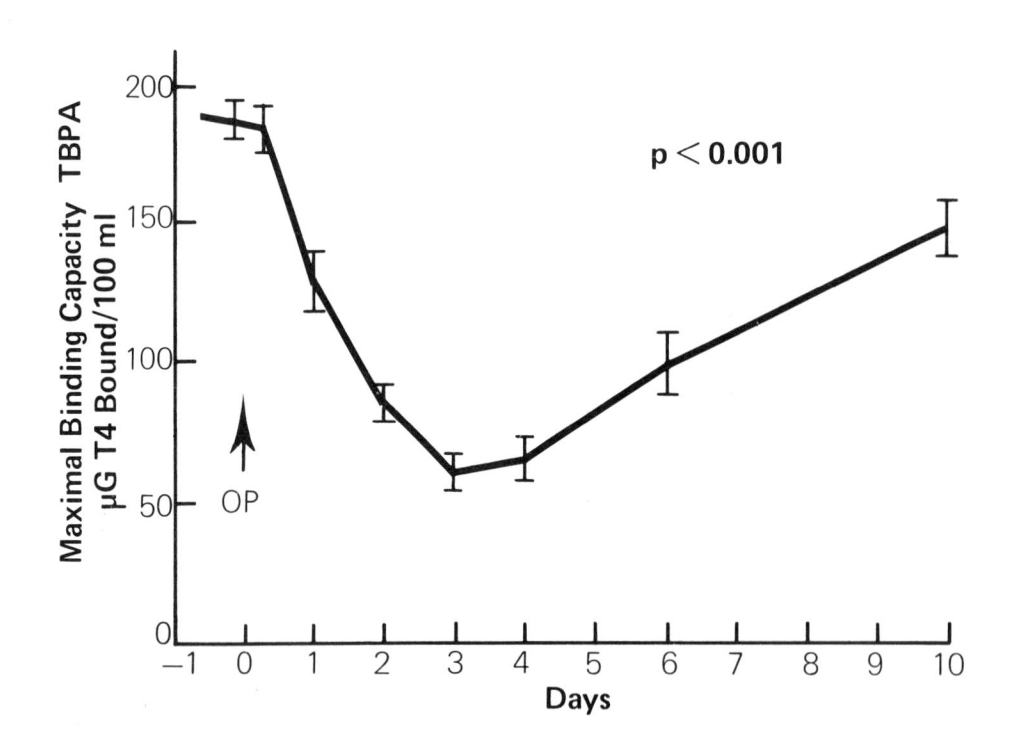

Fig. 2

determination. The normal range of TBPA by this particular technique is 120 to 250 µg 1100 ml. The maximal binding capacities of TBPA of the patients used in this study ranged between 123 and 220 µg 1100 ml with a mean of 172 µg 1100 ml. Following operation TBPA fell to 35% of the preoperative level on the third postoperative day with no demonstrable change in TBG (Fig 2). Levels were almost back to normal on the tenth postoperative day. Postoperative complications such as respiratory tract or wound infections tended to prolong the depressions of TBPA. Oppenheimer et al. (1967) showed that TBPA is a protein with a rapid turnover and a half-life of only 44 hr, and the fall in its serum concentration in the postoperative period is compatible with a sudden cessation in synthesis of this protein.

This acute decrease in TBPA would be expected to lead to an increase in the proportion of free hormone and an increase in the free thyroxine concentration. Measurement of the free thyroxine concentration which is the product of the percentage of free thyroxine and the PBI showed an increase of 65% above the pre-

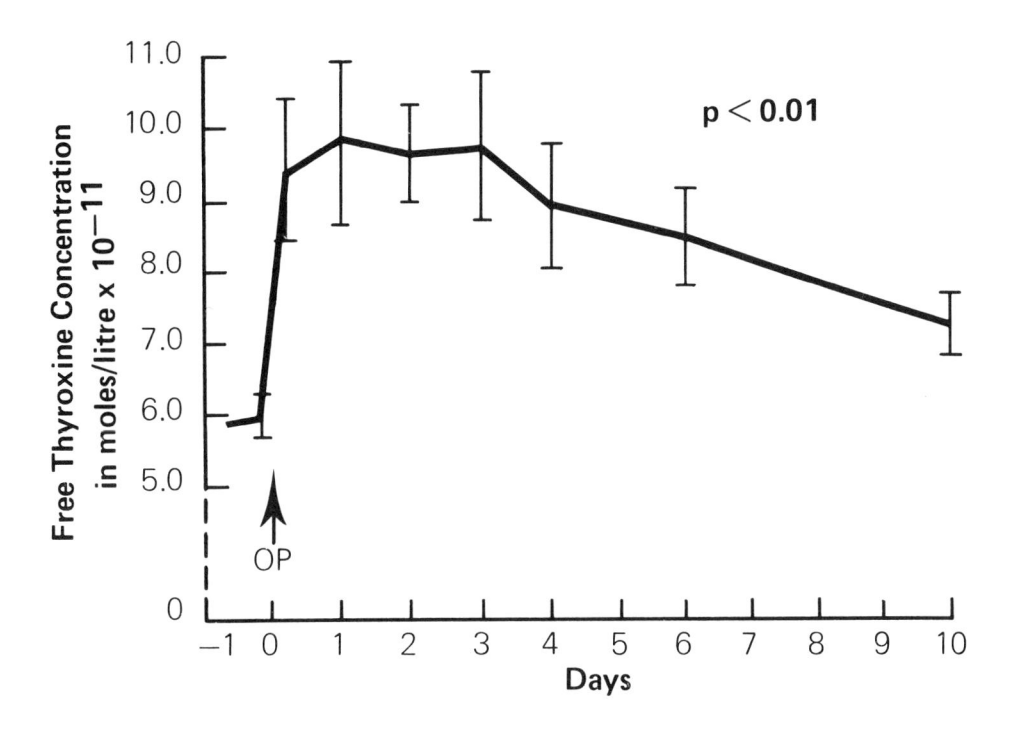

Fig. 3

operative level on the first postoperative day with significantly increased levels for at least 6 days. The first 2 samples were taken 2 hr before and 3 hr after operation, showing the rise to occur rapidly following trauma. As with TBPA the presence of postoperative complications tended to prolong the increases in free thyroxine and to delay the return to normal levels (Fig. 3).

The measurement of total serum triiodothyronine concentration is at present a detailed technique available in only a few laboratories. Nevertheless, it was felt to be of interest to measure percentage free T_3 before and after trauma. T_3 has a much lower serum concentration than T_4 but due to its more rapid turnover and its greater biological potency, it accounts for about half the biologically effective secreted hormone. These studies indicated a sharp rise in free T_3 to values 19% above the pre-operative level on the third postoperative day (Fig. 4).

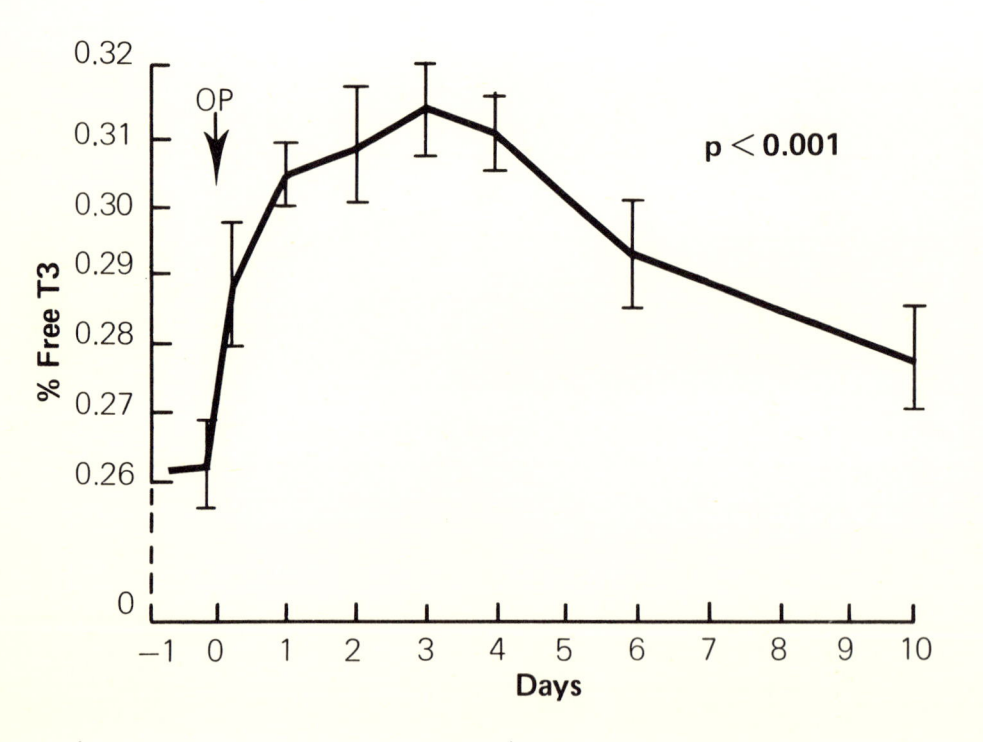

Fig. 4

EFFECT OF SURGICAL OPERATION ON PROTEIN BOUND IODINE

MEAN OF 10 PATIENTS ± S.E.

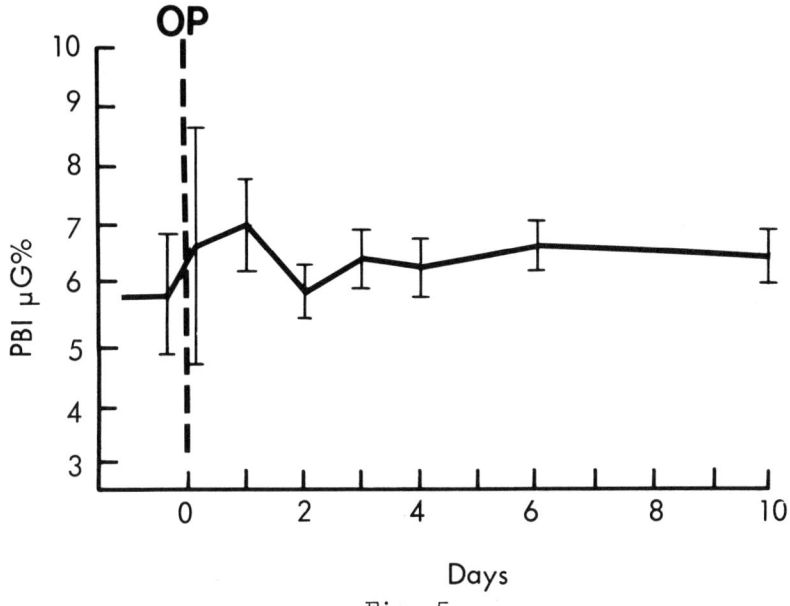

Fig. 5

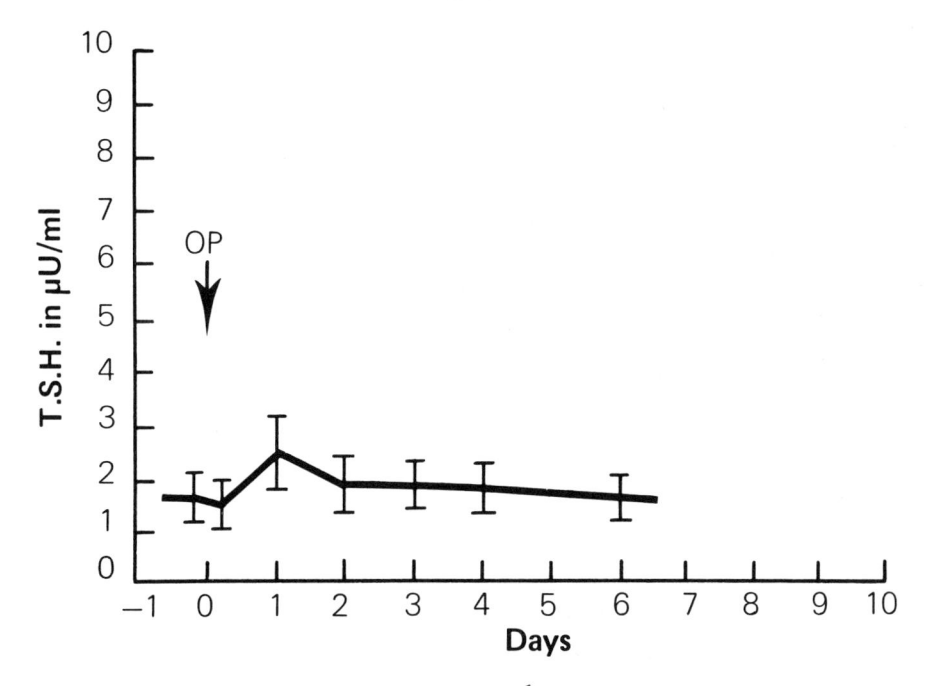

Fig. 6

A transient rise in total thyroxine was demonstrated on the first postoperative day but this was not of statistical significance (Fig 5). There was no significant increase in serum thyroid stimulating hormone levels after operation. The normal level by this method is 0 to 6 I mμ/ml. A fall in TSH could not however be excluded (Fig 6).

DISCUSSION

A constant pattern of change is apparent from these observations. There are relatively prolonged increases in tissue active thyroid hormones following abdominal operation in man, and these variations occur in the presence of unchanged levels of total thyroid hormone and in the absence of increased concentrations of thyrotrophic hormone. Although changes in PBI and TSH concentrations are not necessarily proportional to their secretory rates, it is unlikely that there is an increased secretion by the thyroid gland after trauma.

The maximal binding capacities of TBPA and TBG are assumed to be proportional to their protein concentrations and this has been shown to be so with TBPA (Oppenheimer et al. 1966). The depression in TBPA levels following trauma occurs at the same time as the increases in free hormone, although in this study there was a non-significant correlation between TBPA and free thyroxine. The degree and purpose of changes in plasma proteins following trauma is still unclear but it is known that the depression in TBPA following surgery is due to an acute reduction or complete inhibition of synthesis of a protein with a short half life (Socolow et al, 1965; Oppenheimer & Bernstein, 1965). How important this phenomenon is in the alteration in free hormone is uncertain. Theoretical considerations suggest that such a reduction in TBPA levels would produce an increae in free thyroxine and an increase in free triiodothyronine due to displacement of T_3 molecules from TBG by T_4 molecules, which are preferentially bound. Thus TBPA would be ideally situated to participate in a dynamic physiological mechanism for the peripheral control of the delivery of free hormone to the tissues. There are, however, objections to this hypothesis and results of radiothyroxine turnover studies after operation suggest that TBPA cannot be the only factor controlling the rate of T_4 turnover in the postoperative period. Since the liver is known to accumulate rapidly up to 30% of an injected dose of radioactive thyroxine (Cavalieri & Searle, 1966), it may be important following trauma in a redistribution of T_4 between vascular and extravascular stores.

Various components of the operative procedure were investigated to assess which factors were of importance in the production of the changes in TBPA and free thyroxine. Although the effect of general anaesthesia has not been evaluated in detail, 2 patients were given a general anaesthetic for performance of aortography, involving minimal tissue trauma, and neither showed changes in TBPA or free thyroxine. The induction of minor "stress" situations sufficient to increase the activity of the hypothalamo-pituitary-adrenal axis (insulin hypoglycaemia and production of a pyrogenic reaction) also failed to produce a response. Tissue trauma is probably the most important factor in initiating these changes as there is a variation in the degree of response with the severity of trauma. Fig 7 shows the change in percentage free thyroxine with varying degrees of surgical trauma. Rectal resection leads to more prolonged increases presumably due to the extensive tissue damage. By contrast inguinal herniorrhaphy gives a milder response with an apparent delay in onset.

Free fatty acids are known to affect free thyroxine levels. They compete with thyroxine for specific T_4-binding sites and the increase in free fatty acids which occurs after operation (Truman Mays, 1970) may therefore be of significance in the genesis of the changes in free hormone. Studies in which the free fatty acid concentration was raised in vitro, however,

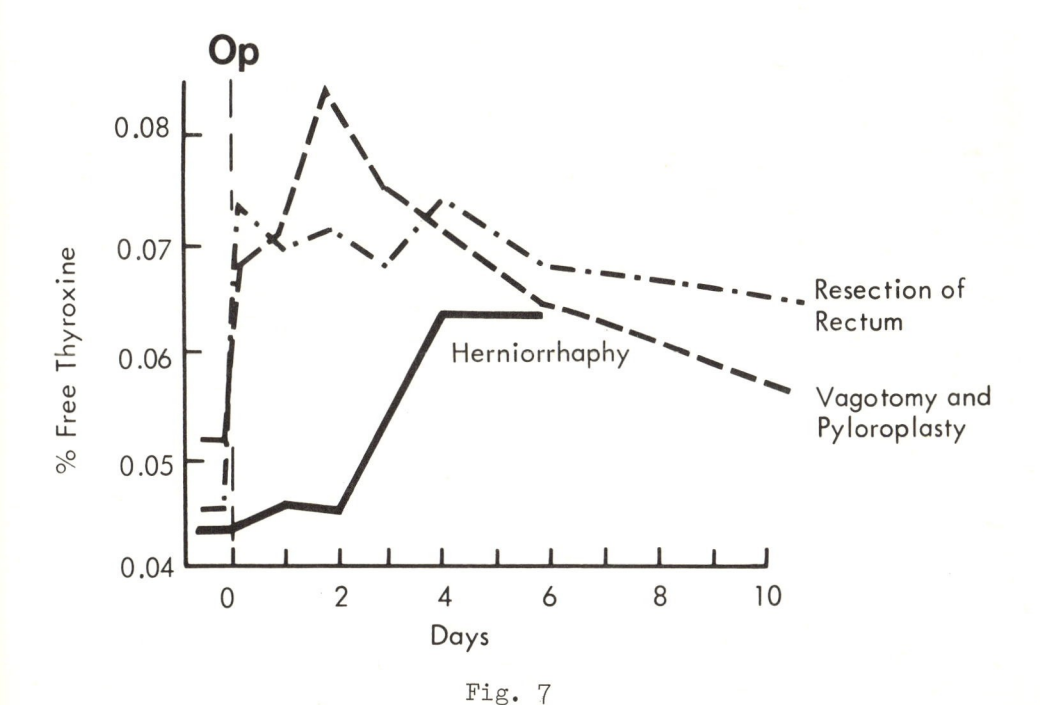

Fig. 7

EFFECT OF ADDITION OF FATTY ACIDS ON FREE THYROXINE IN POOLED SERUM

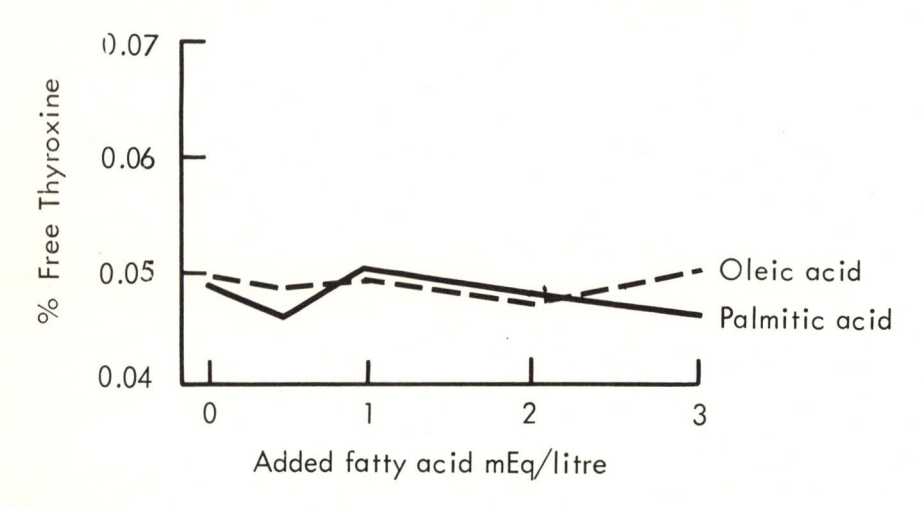

Fig. 8

demonstrated no change in either free T_4 or free T_3 (Fig 8). In addition physiological variation in FFA concentration produces no change in PBI or free T_4 in vivo (Braverman, 1969). Free fatty acids do not therefore appear to be contributing to these changes following surgery.

SUMMARY

Elective abdominal operation in man produced a significant increase in the free thyroxine concentration and the percentage free triiodothyronine. There were no detectable changes in PBI and no increases in thyrotrophic hormone. It is suggested that thyroid hormones may play an important role in post-injury metabolism accompanying tissue repair.

REFERENCES

Bernstein, G., Hasen, J. & Oppenheimer, J.H. (1967) J. clin. Endocr. 27:741.

Berson, S.A. & Yalow, R.S. (1968) J. clin. Invest. 47:2725.

Blomstedt, B. (1965) Acta chir. scand., 130:424.

Braverman, L.E. (1969) J. clin. Invest. 48:878.

Brown-Grant, K. (1960) The pituitary gland, Vol. 2. Butterworth: London.

Cavalieri, R.R. & Searle, G.L. (1966) J. clin. Invest. 45:939.

Crane Charters, A., Odell, W.D. & Thompson, J.C. (1969) J. Clin. Endocr. 29:63.

Engstrom, W.W. & Markardt, B.J. (1954) J. clin. Invest. 33:931.

Franksson, C., Hastad, K. & Larsson, L.G. (1959) Acta chir. scand. 264:118.

Hanbury, E.M. (1959) Metabolism, 8:904.

Hunter, W.M. & Greenwood, F.C. (1962) Nature, 194:495.

Johnston, I.D.A. & Bell, J.K. (1965) Proc. roy. Soc. Med. 58:1017.

Launay, M.P. (1966) Canad. J. Biochem. 44:1657.

Lemarchand-Beraud, Th. & Vannott, A. (1969). Acta endocr. 60:315.

Oppenheimer, J.H., Squef, R., Surks, M.I. & Hauer, H. (1963) J. clin. Invest. 42:1769.

Oppenheimer, J.H. & Bernstein, G. (1965) In Cassano, C. and Andreoli, M. (eds.): Current topics in thyroid research, Academic Press, New York, p. 674.

Oppenheimer, J.H., Martinez, M. & Bernstein, G. (1966) J. Lab. clin. Med. 67:500.

Raud, H.R. & Odell, W.D. (1969) Brit. J. Hosp. Med. 2:1366.

Riley, M. & Gochman, V. (1964) In Technicon Symposium: Automated Analytical Chemistry, p. 62. Chertsey, Surrey, Technicon Inst.

Socolow, F.L., Woeber, K.A., Purdy, R., Holloway, M.T. & Ingbar, S.H. (1965) J. clin. Invest. 44:1600.

Truman Mays, E. (1970) J. Surg. Res. 10:315.

INSULIN SECRETION DURING HEMORRHAGIC SHOCK

M. Vigaš, R.E. Haist, F. Bauer and W.R. Drucker

Institute of Experimental Endocrinology, Slovak Academy of Sciences, Bratislava, and Departments of Physiology and Surgery, University of Toronto, Toronto, Ontario

Hemorrhage as a stress stimulus activates the endocrine system which results in an increase of circulating catecholamines (7), glucocorticoids and ACTH (8). These hormones induce various metabolic changes including hyperglycemia which regularly appears in the early period of hemorrhagic shock in well-fed animals (17), being predominantly mediated by catecholamines, whereas glucocorticoids contribute to its intensity and persistence (9). The hyperglycemia acts as a physiological stimulus for insulin secretion, but in the presence of an increased blood catecholamine level the secretory response of the islets is blocked (1,14). The activation of the endocrine system with resulting metabolic changes may increase the demand for insulin and its absolute or relative deficiency may seriously jeopardize survival in shock.

The level of circulating insulin depends not only on its secretion by beta cells, but also on its uptake and degradation by tissues. Decreased perfusion rate during hypovolemia impairs a considerable part of the functional circulation in the liver and thus the degradation processes in the liver cells may be decreased. A decreased insulin degradation prolongs the presence of the hormone in the circulation and increases its level. Moreover, the augmentation of the insulin concentration in the peripheral blood may result, in part, from a reduction of the blood volume in shock.

A high percentage of animals in shock perish when the blood glucose level is low (5). However, the role of insulin in the onset of hypoglycemia in terminal shock remains to be defined, since the administration of glucose at this stage may prolong life in experimental animals (12).

Thus, the further elucidation of the role of insulin in the regulation of metabolism during hemorrhagic shock may be not only of theoretical, but also of practical importance. The aim of present experiments was to elucidate some of the problems related to insulin secretion during shock and its regulatory effect on blood glucose in the dog.

Well-fed mongrel dogs of both sexes weighing 12 to 20 kg were used. Following the induction of anesthesia with Nembutal (20 mg/kg body weight) and endotracheal intubation, cannulae were placed into both femoral arteries and veins and used to record the blood pressure, to obtain samples of blood and to inject the drugs. Four ml sodium heparin (1:1000) were administered to the animal immediately after cannulation of the vessels. After anesthesia, but before hemorrhage, two blood samples were taken to obtain control values for glucose and insulin.

Two different types of hemorrhagic shock were used. In the first type (10 dogs) the blood pressure was reduced to 50 mm Hg and maintained at this level for 4 hr. The shed blood was then reinfused and blood glucose and insulin levels were measured for the next 75 min.

In one dog two catheters were inserted to obtain additional samples from the portal vein and from the hepatic vein.

In the second type of experiment 17 dogs were subjected to hemorrhage with a reduction of arterial blood pressure to 50 mm Hg and subsequently to 30 mm Hg, this pressure being maintained until the dog died. After different time intervals necessary for the deterioration of cardiovascular function in individual dogs, small amounts of previously withdrawn blood were returned in order to prevent a fall in mean arterial blood pressure below 30 mm Hg. When the animals received back, from the reservoir, a volume of blood equivalent to 1% of body weight (1% uptake), 85% of them ultimately died even though all the blood remaining in the reservoir was rapidly reinfused at this time. In this study the time of 1% uptake was selected as the time to infuse 1 g glucose per Kg body weight during 3 min. The glucose was infused in 9 animals to determine whether the pancreatic islets remained capable of responding to the stimulatory effect of hyperglycemia. Another 8 animals received no glucose and served as controls.

Samples of whole blood were analysed for glucose (4) and 1 ml aliquots of plasma were frozen for subsequent analysis of immunoreactive insulin by chromatoelectrophoresis (18).

RESULTS

The first type of experiment with a shorter and more moderate hypovolemia showed an increase of blood glucose level following the hemorrhage, attaining its peak at the third hour; then it began to decline. Reinfusion of the shed blood caused a further decrease in glycemia. The changes of plasma insulin level correspond to changes of glycemia after the hemorrhage and also after the reinfusion (Fig. 1).

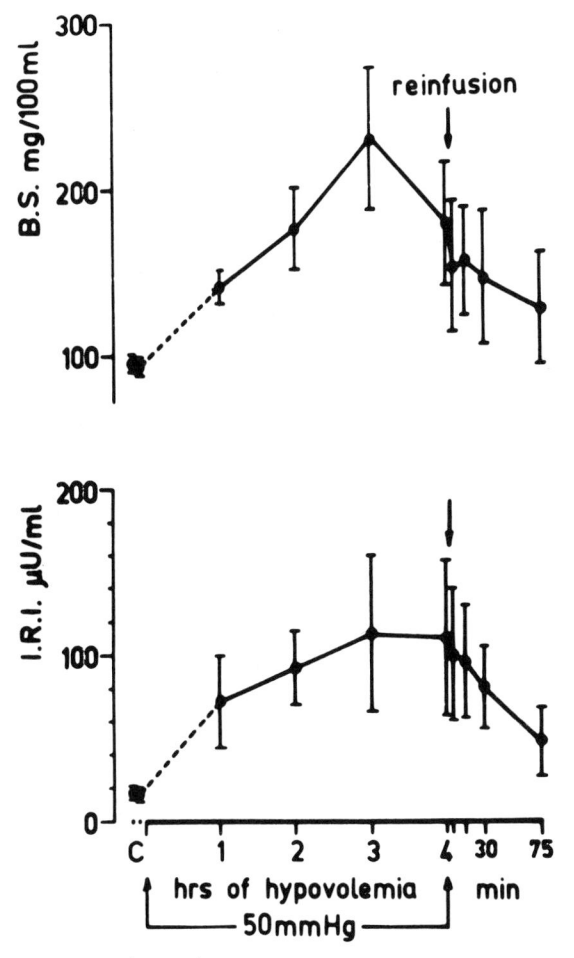

Fig. 1. Blood sugar (B.S.) and immunoreactive insulin (I.R.I.) levels in 10 dogs. Mean arterial blood pressure lowered to 50 mm Hg for 4 hr, followed by reinfusion of the shed blood. (Mean ± S.E.)

The highest glucose concentration was found in the blood flowing from the liver and the lowest from the splanchnic area, drawn from the portal vein. In contrast, the highest insulin level was found in blood from the portal vein since this originates in part directly from the pancreas. The insulin level was substantially decreased after its passage through the liver. The difference in insulin concentration before and after its passage through the liver remained relatively constant throughout the experiment (Fig. 2).

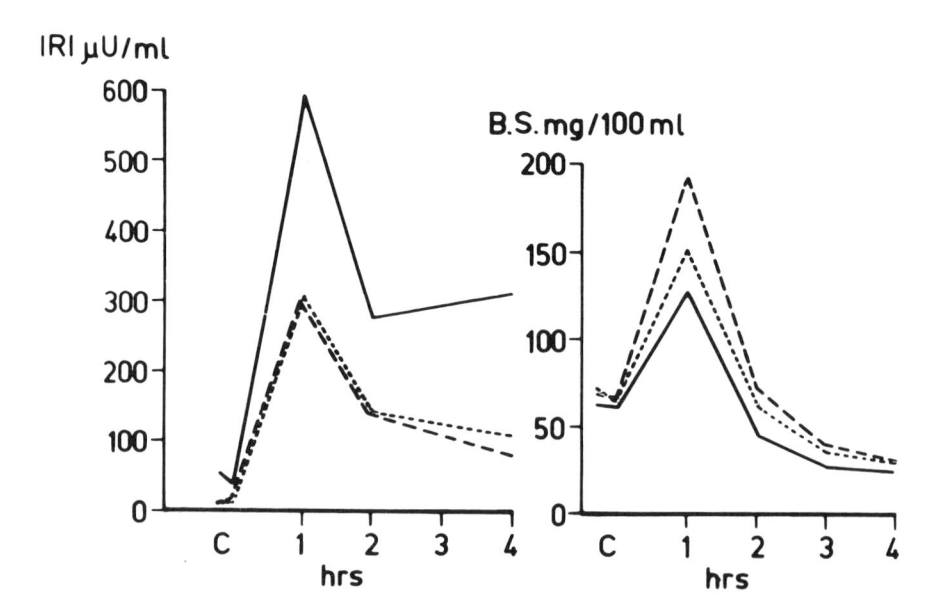

Fig. 2. Immunoreactive insulin (I.R.I.) and blood sugar (B.S.) levels in one dog during hemorrhagic shock (blood pressure 50 mm Hg). Blood samples were taken from aorta (dotted line), portal vein (solid line) and hepatal vein (interrupted line).

During the experiment in which blood pressure was lowered to 30 mm Hg the hyperglycemic reaction and increase of circulating insulin level were greater than in the first type of experiment. During the period of shock compensation, the courses of blood sugar and insulin run roughly in parallel again. In late shock, after 1% uptake, a different pattern of blood sugar and insulin levels was observed following glucose administration. Blood sugar rose abruptly, but the regular rapid decline of its concentration, normally brought about by dilution, failed to appear. One hour after glucose injection its value was unchanged but then it declined slowly. At the end of the experiment the mean blood sugar values in such dogs that had survived the shock for the longest period, were at the starting level.

The insulin concentration rose abruptly following the injection of glucose. The highest levels being attained 2 hr after glucose injection. Later, the glycemia returned to approximately its pre-injection values. Further decrease was very slow; the final insulin values being still higher than the maximum attained before glucose injection (Fig. 3).

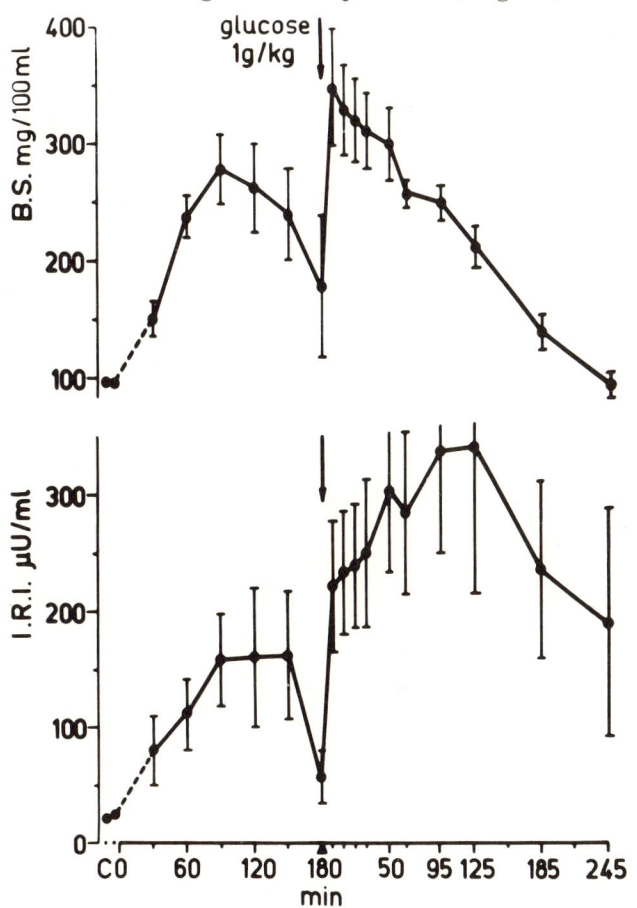

Fig. 3. Blood sugar (B.S.) and immunoreactive insulin (I.R.I.) levels in 9 dogs. Mean arterial blood pressure 30 mm Hg. The glucose was administered, when dogs have taken back the volume of blood equal to 1% of body weight from the reservoir (Mean ± S.E.).

In control animals that did not receive glucose at the 1% uptake point, the decline in both glucose and plasma insulin continued throughout the decompensatory phase of shock.

On average, the dogs which received glucose survived significantly longer than the dogs not receiving glucose (439 min. versus 257 min.).

DISCUSSION

The results of these experiments on hemorrhagic shock in dogs do not show any decrease of the insulin secretion response. In early hemorrhagic shock insulin levels correlate well with changes in blood sugar values and the percentage of insulin degraded in the liver in this stage of shock is the same as reported for the intact organism by Rieser (16). Moreover, the renal arteriovenous difference of insulin in the same model of shock was also the same as in control studies (11).

The injection of glucose in late shock results in an increase in the insulin level. The response of beta cells to glucose stimulation is still well preserved. The administered glucose seems to be inadequately distributed within the glucose space and persists for a long time at high concentration in the circulating blood. However, the cause of its insufficient utilization is not a deficiency of insulin, but an inadequate perfusion rate through the tissues (17). A persistence of blood sugar at high level is a permanent stimulus for insulin secretion: its blood level further increases, although the glucose concentration gradually diminishes. From these results we may suppose that a high level of circulating insulin itself does not exert any inhibitory effect on its secretion from beta cells. At the terminal stage, when blood sugar is again at its original value, the blood insulin remains at high levels. It may be presumed that some liver disorder (2) and an inadequate perfusion in tissues are also responsible for the inhibited degradation of the circulating hormone in late shock.

Our experiments confirmed the favourable effect of glucose administration in the terminal stage; dogs that received glucose survived much longer than those that had not.

Our findings differ from these reported by Halmagyi et al.(6) on sheep, and Cerchio et al. (3) on baboons. Halmagyi found poor correlation between glucose and insulin levels in sheep during hemorrhagic shock. This finding agrees with the fact that insulin secretion in ruminants is regulated by the concentration of volatile fatty acids in blood (10); beta cells in sheep are markedly less sensitive to glucose. Cerchio et al. (3) found that the insulin level in baboons in hemorrhagic shock declines at the same time as the blood sugar rises. Even tolbutamide administration failed to affect the level of circulating insulin. We have no plausible explanation for this species difference. There may be perhaps a different islet reaction to endocrine situations and pancreatic hypovolemia in the dog. ACTH, which is increased in peripheral blood in hemorrhagic shock (8), results in an increased release of insulin (13). Rappaport et al. (15) found that mechanical restriction of blood flow rate through the pancreas resulted in a higher concentration of insulin in the effluent blood.

SUMMARY

Plasma immunoreactive insulin levels were studied in two models of hemorrhagic shock which differed in mean arterial blood pressure, e.g. 50 or 30 mm Hg, respectively. In early stage of hemorrhagic shock the typical rise of blood glucose level was associated with a significant increase in plasma insulin. Insulin degradation in the liver was not changed in the early period of shock.

Insulin response to intravenous glucose administration was well preserved in late shock. The decrease of plasma insulin in this stage of shock, when some essential physiological functions started to deteriorate, was significantly slower than that of glucose. Thus, the impaired degradation rate of insulin could participate in the terminal stages of shock.

The decreased glucose utilization during hypovolemia is a consequence of limited perfusion in tissues. Insulin response to glucose is not altered by hemorrhage in dogs.

REFERENCES

1. Altszuler, N., Steel, R., Rathged, I. & De Bodo, R.C. (1967) Am. J. Physiol. 212:677.

2. Blair, O.M., Stenger, R.J., Hopkins, R.W. & Simeone, F.A. (1968) Lab. Invest. 18:172.

3. Cerchio, G.M., Moss, G.S., Popovich, P.A., Butler, E. & Siegel, D.C. (1971) Endocrinology 88:138.

4. Drucker, W.R., Schlatter, J. & Drucker, R.P. (1968) Surgery 64:75.

5. Engel, F.L. (1952) Ann. N.Y. Acad. Sci. 55:381.

6. Halmagyi, D.F.J., Gillett, D.J., Lazarus, L. & Young, J.D. J. Trauma 6:623.

7. Huma, D.M. Federation Proc. 20; suppl. 9, 87.

8. Kuzela, L. & Mikulaj, L. (1966) Klin. Med. 21:298.

9. Levenson, S.M., Einheber, A. & Malm, O.J. Federation Proc. 20: suppl. 9, 99.

10. Mayhew, D.A., Wright, P.H. & Ashmore, J. Pharmacol. Rev. 21:183.

11. McCormick, J.R., Sonksen, P.H., Soeldner, J.S. & Egdahl, R.H. (1969) Surgery 66:175.

12. Moffat, J.G., King, J.A.C. & Drucker, W.R. (1968) Surgical Forum 14:5.

13. Oshawa, N., Kuzuya, T., Tanioka, T., Kanazawa, Y., Ibayasha, H. & Nakao, K. Endocrinology 81:925.

14. Porte, D., Graber, A.L., Kuzuya, T. & Williams, R.H. (1966) J. Clin. Invest. 45:228.

15. Rappaport, A.M., Davidson, J.K., Kawamura, T., Lin, B.J., Ohira, S., Ziegler, M., Coddling, J.A., Henderson, J. & Haist, R.E. (1970) Abstract of papers, Regional Congress of Int. Union of Physiol. Sci., Brasov, Romania, pp.436.

16. Rieser, P. (1967) Insulin, membranes and metabolism. The Williams and Wilkins Co., Baltimore, p. 54.

17. Seligman, A.M., Frank, H.A., Alexander, B. & Fine, J. (1947) J. Clin. Invest. 26:536.

18. Yalow, R.S., & Berson, S.A. (1960) J. Clin. Invest. 39:1157.

THE EFFECT OF PERSISTING HYPOVOLEMIC SHOCK ON PANCREATIC OUTPUT OF INSULIN

W.R. Drucker, B.L. Gallie, T.S. Lau, G. Farago,
R.A. Levene & R.E. Haist

The Departments of Surgery and Physiology and the
Institute of Medical Sciences University of Toronto,
Toronto, Ontario

Study of the basic biological response of the body to an acute and prolonged deficiency of blood flow provides insight into the mechanisms of homeostasis and information necessary for rational therapy. Physiological changes define successive stages of homeostasis: compensation, equilibrium and deterioration. The relation of the biochemical changes, however, to the continued effectiveness of compensatory mechanisms is not clear. Initial metabolic responses reflect the combined effect of decreased peripheral perfusion and hormonal changes induced by acute circulatory insufficiency. In time the energy reserves of the body become depleted and homeostasis deteriorates. It is unlikely that any one change is critical but rather a series of complex interrelated events ultimately result in loss of compensation for a persisting reduction in peripheral blood flow. The problem for investigators today is to determine to what extent metabolic and endocrine alterations are involved in the loss of tolerance or compensation for shock (6).

Recently, because of its vital role in energy metabolism, considerable interest has centered on insulin during shock. In our previous studies the concentration of insulin was found to rise in the peripheral venous blood shortly after the mean arterial blood pressure was reduced to 50 mmHg by hemorrhage (4). This rise was associated with an increase in the level of blood glucose. As hypovolemia continued, the elevated blood levels of both glucose and insulin declined. It was recognized, however, that peripheral blood levels are at best an approximation of the pancreatic secretion of insulin. Opportunity to study the pancreatic response directly during shock was provided by the development of a method to isolate venous drainage from a segment of pancreas in vivo (19). This technique made it possible to determine blood flow and insulin output from the pancreas in relation to peripheral blood concentration of insulin and glucose.

METHODS

Well nourished mongrel dogs (15-20 kg) fasted 18 hr were anaesthetized with intravenous nembutal (25 mg/kg). Following endotracheal intubation, femoral arteries and veins were cannulated for monitoring the mean arterial blood pressure (MABP), collecting samples and bleeding into a reservoir.

The head and uncinate process were divided from the body of the pancreas. All venous connections to the duodenum and the inferior pancreatic - duodenal vein were ligated. The distal end of the superior pancreatic - duodenal vein was cannulated and a "chocker" placed at the proximal end near its entry into the portal vein. Venous flow rate from this segment of pancreas was determined by timed collections of blood (30-120 sec) in a graduated cylinder and corrected for plasma flow using the hematocrit. The concentration of insulin was determined in the collected blood. Four thousand units of sodium heparin were given intravenously after the surgical preparation was completed.

Two groups of animals were studied. In one group of 8 animals the mean arterial blood pressure (MABP) was reduced by hemorrhage over a 10 min period to 50 mm Hg and maintained at this level for 40 min. The MABP was then reduced by a second 10 min hemorrhage to 30 mmHg and this pressure maintained by periodic withdrawal or return of blood as necessary for a total period of hypovolemia of 180 min. Pancreatic venous insulin concentration, pancreatic venous flow rate, peripheral (vena caval) insulin concentration and arterial blood glucose concentration were determined twice prior to hemorrhage, 30 min after MABP reached 50 mmHg, 30 min after MABP reached 30 mmHg and at hourly intervals thereafter.

A second group of 8 animals was subjected to all surgical procedures but not hemorrhage. In this control group of animals similar determinations as in the first group were carried out for a total study period of 6 hr.

Serial assays for plasma insulin were performed by a radio-immuno assay technique (10). Blood glucose was determined by a modification of the inverse colorimetric technique with potassium ferricyanide in the Technicon Autoanalyzer (12). The isolated portion of pancreas was excised and weighed after completion of the study. After blood flow was converted to plasma flow using the hematocrit, pancreatic insulin output was calculated per gram of wet weight of pancreatic tissue.

$$\frac{\text{Pancreatic plasma flow (ml/min)} \times \text{Plasma insulin conc (u/ml)}}{\text{Wet weight of pancreas (g)}} = \text{u/min/g pancreas}$$

In a supplementary study to relate pancreatic blood flow to cardiac output 6 dogs were anaesthetized and subjected to hemorrhage to a MABP of 50 mmHg for a total period of 3 hr. Cardiac output was determined by the cardio green dye dilution technique (16), prior to hemorrhage and 15, 45 and 150 min after the MABP was stabilized at 50 mmHg.

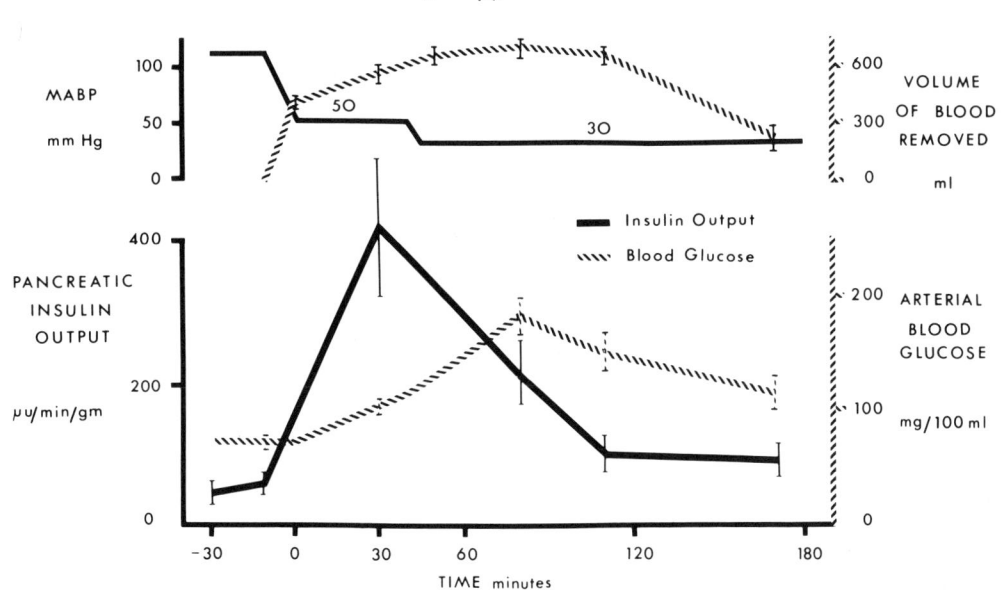

Fig. 1. Mean arterial blood pressure (MABP) and mean values with standard error of mean of the volume of blood removed, pancreatic output of insulin and arterial blood glucose plotted before and at intervals during hypovolemia.

RESULTS

Pancreatic insulin output, peripheral insulin concentration and arterial blood glucose did not change in the 6 hr following isolation and cannulation of a segment of the pancreas in 8 control anaesthetized normotensive animals.

The mean volume of blood removed from animals during the three hours of hypovolemia is illustrated in Fig. 1. While there was variation among animals, the mean maximal bleed-out volume occurred within 80 min after the onset of hypovolemia and within 20 min following stabilization of the MABP at 30 mmHg. Physiological deterioration, as shown by the need to return shed blood in order to maintain a stable MABP was seen in all animals to the extent that an uptake of blood equal to 1% of body weight occurred within the 3 hr period of hypovolemia (Fig. 1). Previous studies indicated that animals die subsequent to the return of all shed blood if they are allowed to remain in hypovolemia until they have reached the point of "1% uptake".

Hyperglycemia developed in all hypovolemic animals. Arterial blood glucose (mg/100ml) rose from a control of 72 ± 9 (mean $\pm$ SEM) to a maximum of 181 ± 33 within 80 min after hemorrhage and then fell to reach a level of 113 ± 27 after 170 min of hypovolemia (Fig 1). Pancreatic insulin output ($\mu u/min/g$) rose abruptly with hemorrhage from the pre-shock level of 53 ± 16 to a peak value of 418 ± 97 within 30 min after hemorrhage and fell to 95 ± 24 after 170 min of hypovolemia. 30 min after the onset of hypovolemia, when the pancreatic output of insulin had reached its maximum level, the arterial blood glucose had increased only 31 mg/100ml (Fig 1).

Peripheral venous insulin concentration rose more slowly than the pancreatic output of insulin (Fig 2). Plasma insulin concentration ($\mu u/ml$) rose from control levels of 18 ± 3 to 312 ± 88 during the initial 110 min of hypovolemia and continued to rise at a slower rate thereafter reaching 390 ± 126 after 180 min of hypovolemia. Most animals followed beyond this time were found to have a falling level of plasma insulin.

The pancreatic venous flow (ml/min/100g pancreas) fell from 44.3 ± 5.4 to 15.7 ± 2.4 when the MABP was reduced to 50 mmHg. Further reduction of MABP to 30 mmHg caused a continuing slight

EFFECT OF HYPOVOLEMIC SHOCK ON PANCREATIC OUTPUT OF INSULIN 191

Insulin Output during Hypovolemia

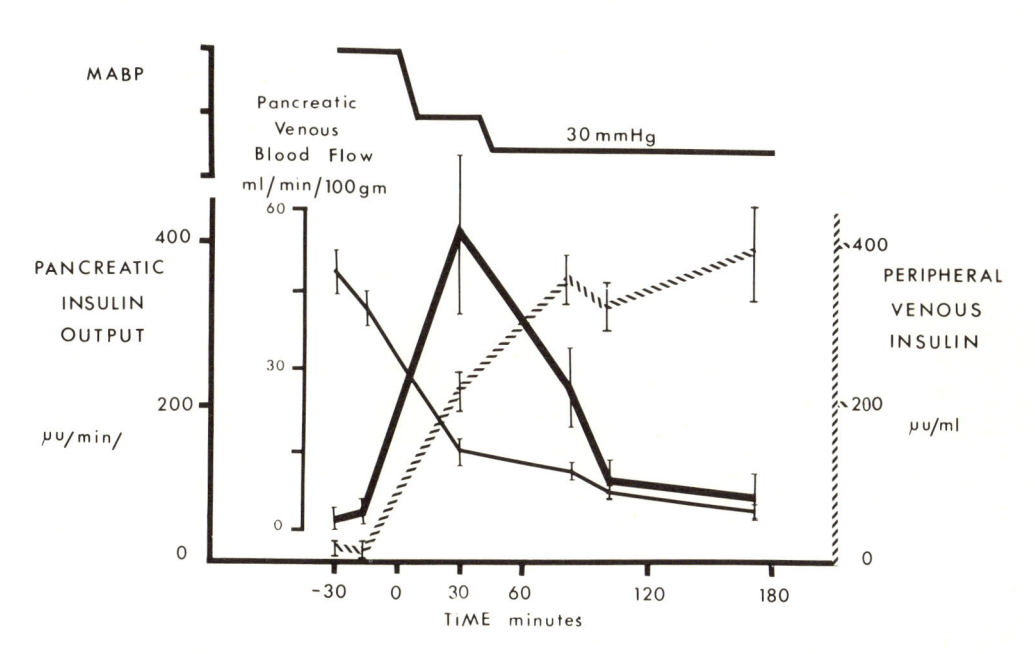

Fig 2. Mean arterial blood pressure (MABP) and mean values with standard error of mean of pancreatic output of insulin, pancreatic blood flow and peripheral plasma concentration of insulin plotted before and at intervals during hypovolemia.

decline in pancreatic venous flow to an average rate of 4 ml/min/100g during the next 2 hr of hypovolemia. However, in 6 animals maintained at a MABP of 50 mmHg for a 3 hour period, the pancreatic blood flow remained within the range of 11-16 ml/min/100g (Table 1). The percent fall in pancreatic venous flow (75%) observed in these dogs remained proportional to the decline in cardiac output (79%) during the period that the MABP was maintained at 50 mmHg by hypovolemia (Table 1).

TABLE 1

Pancreatic venous flow and cardiac output before and at intervals after hemorrhage sufficient to maintain mean arterial blood pressure (MABP) at 50 mmHg.

		Pancreatic venous flow ml/min/100g	Cardiac output l/min
Normotensive control		51.2 ± 5.8	1.38 ± 0.17
Hemorrhagic Shock (MABP at 50 mmHg)	After (Minutes)		
	10	13.2 ± 0.9	0.30 ± 0.5
	30	11.2 ± 0.8	---
	45	---	0.28 ± 0.01
	60	16.8 ± 1.3	---
	120	16.3 ± 1.2	---
	150	---	0.26 ± 0.02
	180	14.8 ± 1.2	---

All values = mean ± SEM

DISCUSSION

All 3 successive stages of hemorrhagic shock were included in this study. The adequacy of the mechanisms compensatory for acute hemorrhage was indicated in the first stage of shock by the periodic need to remove blood to maintain a stable low MABP. After 80 min of hypovolemia a stage of equilibrium was reached when the MABP remained constant without change in the volume of shed blood. Decompensation of homeostasis indicative of the third stage became apparent when it was necessary to return shed blood at intervals in order to sustain the MABP at 30 mmHg (Fig 1).

The results of this study confirm previous observations in dogs that an acute reduction in circulating blood volume is associated with an increase in the concentration of insulin in peripheral blood (4). Analysis of the insulin content and the quantity of venous drainage from the pancreas at intervals during a prolonged period of hypovolemia permits a more precise estimation of insulin metabolism than was possible by previous analysis of peripheral blood levels of insulin. In no instance did the concentration of insulin in the peripheral venous blood fail to reflect a discharge of insulin from the pancreas. The rise and fall of peripheral blood levels of insulin, however, are inadequate reflections of either the temporal or quantitative aspects of the pancreatic release of insulin (Fig 2).

A completely satisfactory explanation for the observed increase in output of insulin from the pancreas shortly after the inception of shock is not possible on the basis of currently available information. A high concentration of insulin was found in pancreatic venous blood at a time when blood flow from the organ was markedly reduced. It is possible that the abrupt hemodynamic changes triggered the release of large quantities of insulin from stores in the β cells. A disruption of β cell integrity with indiscriminate spilling of preformed insulin into the venous blood would seem most unlikely in view of the temporal factors involved and because no histological lesion has been identified in the pancreas of animals subjected to hypovolemia of this severity and duration.

Ordinarily the pancreatic release of insulin is exquisitely sensitive to fluctuations in the levels of blood glucose (14). McCormick did not observe a rise in plasma insulin during hemorrhagic shock in dogs if hyperglycemia was prevented by prior adrenalectomy (15). Since less blood loss is required to reduce the blood pressure in adrenal medullary deficient animals it is possible that fewer hemodynamic alterations occurred than within the pancreas in the animals with intact adrenals who exhibited hyperinsulinemia with shock. It is unlikely that a rise in the level of blood glucose can be the only factor responsible for the initial pancreatic output of insulin observed in the present study since glucose was only slightly elevated at the time pancreatic output of insulin was maximal (Fig 1).

The rapid rise of insulin after the inception of shock tends to exclude the possibility that other circulating insulin secretagogues such as enteric hormones are responsible for the pancreatic response. Products of protein digestion, in particular the amino acids arginine and leucine, stimulate insulin output but these would be unlikely agents for the observed changes so shortly after the onset of hemorrhage in fasted animals. Free fatty acids, rapidly mobilized from fat depots by trauma, are known to stimulate pancreatic release of insulin (14) but when the MABP is reduced to the levels obtained in this study the plasma concentration of FFA declines (8). At present therefore, it must be considered that the pancreatic output of insulin occurs in response to an acute reduction of circulating blood volume by unknown mechanisms.

Factors responsible for the decline in pancreatic output of insulin in shock after the initial burst of activity may be viewed in terms of the two-phase response of insulin output in a normal pancreas.

Porte clearly demonstrated that epinephrine can prevent a rise in plasma insulin despite the presence of the hyperglycemia that it initiates (18). The mechanism of this effect is now known to involve inhibition of the formation of cyclic adenosine monophosphate (cAMP). It is well established that release of epinephrine is an integral part of the homeostatic response to shock and many investigators have demonstrated a marked rise in the plasma concentration of both epinephrine and norepinephrine during hypovolemia (2,13). Conceivably the initial high output of insulin during shock reflects the release of readily available insulin from pancreatic stores (phase 1). This release may be relatively insensitive to inhibition by epinephrine. Sustained insulin secretion requires insulinogenesis (phase 2). The combined influence of the more severe reduction of pancreatic blood flow when the MABP was reduced to 30 mmHg and a steadily rising plasma concentration of catecholomines could inhibit insulin synthesis. Under this circumstance continued output of insulin would depend on the relative influence on the pancreas of the factors which stimulate and inhibit synthesis of insulin. There is support for this concept from incomplete studies in animals bled to a MABP of 50 mmHg without further reduction in pressure during a 3 hr period. The initial brief high pancreatic output of insulin in these animals (phase 1) was followed by a sustained output of insulin at a lower level (phase 2) suggesting that insulinogenesis continues during moderate hypovolemia. This curve, while differing in its quantitative aspects, is similar to the two phase pancreatic output of insulin found in normovolemic animals stimulated by glucose (14). If the concept of a two-phase pattern of insulin secretion is applicable to pancreatic activity during shock, it would help to explain why an increase in plasma insulin is not found invariably during shock (5,11,17).

A careful study in awake baboons, in which acute hypovolemia failed to elicit a rise in plasma levels of insulin, may have avoided the pitfalls of an anaesthetic agent influencing metabolic processes at the price of permitting increased sympathetic activity (17). High epinephrine levels in plasma may have inhibited release of insulin (phase 2) during hypovolemia. In these animals elevation of the plasma level of glucose by eating prior to shock may have been sufficient to diminish the pancreatic stores of insulin to the extent of precluding a prompt rise of insulin levels in plasma (phase I) in response to the onset of hypovolemia. While the variable responses of insulin release with shock reported in the literature (1,4,5,11,17) may reflect species differences, Skillman (20) demonstrated in human volunteers a rise in plasma levels of insulin following a hemorrhage of 15% of measured blood volume and Dykes (7) observed significant elevations of insulin in the plasma of patients in cardiogenic shock. The severity of the reduction of blood pressure is undoubtedly a major determinant in insulin release as indicated by the almost complete cessation of pancreatic output of insulin when the MABP maintained at 30 mmHg (Fig 1) in contrast to a continuing although reduced output when the MABP remained at 50 mmHg (Table 1).

Whatever the precise mechanisms of inhibition of insulin output may be during shock, our previous studies suggest it is not absolute because infusion of a bolus of glucose (1 g/kg) late in the course of shock at the time of "1% uptake" caused an abrupt rise in plasma insulin (3). Presumably the reduced pancreatic blood flow in the order of 78% when the MABP is maintained at the level of 30 mmHg does not damage the pancreatic capacity to release insulin in response to a hyperglycemic stimulus. It would seem likely that the ability of hyperglycemia to stimulate the pancreas under these circumstances is dependent upon the rate and magnitude of rise of the level of blood sugar.

The continuing rise in levels of plasma insulin while both the pancreatic release of insulin and pancreatic blood flow are declining indicates that the metabolic clearance rate for insulin is decreased by shock. This may simply reflect the hemodynamic alterations of decreased volume of distribution of insulin and reduced perfusion of the sites responsible for clearance of insulin from the plasma. There are preliminary data which indicate the half-life of insulin is prolonged in shock (15). Some clearance of insulin does occur despite severe reduction in peripheral blood flow since the plasma levels of insulin ultimately decline as shock persists (4). The physiological significance of the alterations in pancreatic output of insulin during shock will ultimately be reflected in the availability of insulin to support energy metabolism. No data are available in the present study regarding the biochemical effectiveness of the insulin released during shock.

It is noteworthy, however, that the observed fall in pancreatic blood flow was proportional to the reduction in cardiac output (Table 1). This suggests that the pancreas does not participate in splanchnic vasoconstriction as part of a homeostatic response to redistribute blood to vital organs during hypovolemia. In fact, the 3 observations made during 150 min of hypovolemia when the MABP was maintained at 50 mmHg indicate that the pancreatic venous flow rose slightly while the cardiac output declined from 0.30 to 0.26 liters per min. This limited observation does not justify the conclusion that the pancreas is protected at the expense of ischemia in other tissues but it does suggest that pancreatic blood flow closely reflects the degree of reduction in cardiac output throughout the period of hypovolemia. When the MABP was reduced to 30 mmHg pancreatic blood flow continued to decline and the output of insulin virtually ceased. Should there be therapeutic difficulties in restoring cardiac output it is conceivable that an occasion might arise in which insulin insufficiency compromised homeostasis. Under such circumstances the exogenous administration of insulin would be a rational adjunct to the therapy of shock.

There are clinical data which indicate insulin therapy is beneficial in patients suffering from shock following severe burns (1) or myocardial infarction (7). Flear has found the administration of insulin assists in recovery from the "sick cell syndrome" which can be a consequence of shock (9).

SUMMARY

Based on the observations of this study and previous data a concept is advanced that the pancreatic output of insulin during shock is a two-phase system dependent upon similar mechanisms to those involved in the release of insulin from a normal pancreas in response to the stimulus of hyperglycemia. The initial output represents release of pancreatic stores of insulin in response to an acute reduction of circulating blood volume and is very little affected by factors that ordinarily inhibit insulin secretion. If the stores are depleted prior to shock a rise in pancreatic output of insulin will not occur. The second phase of insulin output depends on insulinogenesis which can occur during hypovolemia in response to hyperglycemia, but may be inhibited by severe reduction in pancreatic blood flow and increased levels of catecholamines in the blood. Whether insulin output occurs or not depends on the relative strengths of these opposing factors.

CONCLUSIONS

1. Hemorrhagic shock in dogs causes an abrupt and marked release of insulin from the pancreas at a time when pancreatic blood flow is falling.

2. The fall in pancreatic blood flow is proportional to the decline in cardiac output with hypovolemia.

3. The initial high output of insulin declines rapidly to pre-shock levels when hypovolemic shock is prolonged at a MABP of 30 mmHg.

4. The release of insulin occurred earlier than a rise in blood glucose and declined when hyperglycemia was present.

5. Concentration of insulin in peripheral venous blood rose more slowly than the pancreatic output and remained elevated for a considerable period following reduction of pancreatic insulin release.

6. An hypothesis is advanced of a two-phase system of insulin release during shock. An initial output of insulin will occur if pancreatic stores of insulin are intact. Continuing output of insulin during shock depends on the relative strength of factors that stimulate and inhibit insulinogenesis.

REFERENCES

1. Allison, S.P., Hinton, P. & Chamberlain, J.J. (1968) Lancet 2: 1113.

2. Bauer, W.E., Levene, R.A., Zechwieja, A., Lee, M.J., Menczyk, Z. & Drucker, W.R. (1969) Forum 20:9.

3. Bauer, W.E., Vigas, S.N.M., Haist, R.E., Levene, R.A. & Drucker, W.R. Fed. Proc. 28:507 (1969).

4. Bauer, W.E., Vigas, S.N.M., Haist, R.E., Levene, R.A. & Drucker, W.R. (1969) Surgery 66:80.

5. Carey, L.C., Lowery, B.D. & Cloutier, C.T. (1970) Annals of Surgery 172:342.

6. Drucker, W.R., Levene, R.A., Koven, I.H. & Gallie, B.L. Workshop sponsored by the Department of Surgery, University of Miami, School of Medicine and the Office of Naval Research May 10-11, 1971 (in press).

7. Dykes, J.R.W., Saxton, C. & Taylor, S.H. (1969) Brit. Med. J. 2:490.

8. Farago, G., Levene, R.A., Lau, T.S. & Drucker, W.R. Surgical Forum 1971 (In Press).

9. Flear, C. (1969) J. Surg. Res. 9:369.

10. Hales, C.N. & Randle, P.J. (1963) Biochem. J. 88:137.

11. Halmagyi, D.F.J., Gillette, D.J. Lazarus, L. & Young, J.D. (1966) J. Trauma 6:623.

12. Hoffman, W.S. (1937) J. Biol. Chem. 120:51.

13. Hume, D.M. (1958) Surgical Forum 8:111.

14. Levene, R.A. (1970) The New England Journal of Medicine 283:522.

15. McCormick, J.R., Lieen, W.M., Herman, A.H. & Egdahl, R.H. Surgical Forum 20:12.

16. Miller, D.E., Gleason, W.L. & McIntosh, H.D. (1962) J. of Lab. & Clin. Med. 59:345.

17. Moss, G.S., Cercio, G., Siegel, D.C., Reed, P.C., Cochin, A. & Fresquez, V. (1970) Surgery 68:34.

18. Porte, D., Graber, A.L., Kuzuya, T. & Williams, R.H. (1966) J. Clin. Invest. 45:228.

19. Rappaport, A.M., Ohira, S., Coddling, J.A., Empey, G., Kalnins, A., Lin, B.J. & Haist, R.E. (1971) Abstracts of Communications XXV International Congress of Physiological Sciences.

20. Skillman, J.J., Hedley-White, J & Pallotta, J.A. (1970) Surgical Forum 21:23.

Work supported in part by the Medical Research Council of Canada, Grant MA2599.

The authors express their appreciation for the technical assistance of Mrs. Siu Lo, Mr. Henry Chan and Mr. Frank Rechnagel, and the secretarial assistance of Miss Una MacDonald.

THE DECLINE IN PANCREATIC INSULIN RELEASE DURING HEMORRAGIC

SHOCK IN THE BABOON

G.S.Moss, G.Cerchio, D.C.Siegel, P.C.Reed, A.Cochin
and V. Fresquez

Department of Surgery and Medicine, University of

Illinois College of Medicine, The Abraham Lincoln

School of Medicine, and the Hektoen Research Institute

Hyperglycemia has been noted in man and animal during hemorrhagic shock. Studies in the dog demonstrated that this hyperglycemia is accompanied by elevated levels of serum immunoreactive insulin, suggesting increased insulin resistance (2). However, recent shock studies by Carey (3) in man, and from our laboratory in the baboon (6) have shown a different pattern. Insulin levels have been depressed despite hyperglycemia, suggesting suppression of insulin release rather than increased insulin resistance.

Most of these studies have relied on peripheral blood sampling for insulin assay. Since peripheral insulin levels represent an equilibrium between pancreatic release and peripheral uptake, it is not certain that peripheral insulin values correlate well with the level of insulin release from the pancreas. A more sensitive sampling site for assaying the insulin response to hemorrhagic shock should be the pancreatic venous outflow system--the portal vein. This study was therefore designed with the following objectives:

1. to determine the changes in portal vein insulin concentration during hemorrhagic shock.

2. to determine how accurately peripheral insulin levels predict portal concentrations.

METHOD

Five adult baboons of both sexes (9-18 kg body wt) were used. Approximately one week before a siliconized PE240 catheter had been implanted in the portal vein via the inferior mesenteric vein. The catheter was filled with heparin solution and buried under the skin of the abdomen. The operation was carried out under nembutal anesthesia through a left paramedian incision.

On the day of the study the animal was tranquilized with a small dose of Sernylan* (0.5 - 1.0 mg/kg). Under local anesthesia, the portal vein catheter was re-exposed and the heparin aspirated. Polyethylene catheters were positioned in the abdominal aorta and inferior vena cava via the femoral vessels. These catheters were used to measure arterial pressure and sample caval blood. An endotracheal tube was inserted and the animal was allowed to breathe spontaneously.

The animal was then placed in the prone position. Axillary supports were employed to ensure free excursion of the chest wall. The aortic catheter was connected to a Statham strain gauge transducer. Pressures were recorded by means of a Lexington A101 pre-amplifier and a Hewlett Packard hot stylus recorder.

Hypotension was induced by removing blood through the aortic catheter into a sterile plastic bag containing 75 ml acid citrate dextrose solution. Mean arterial pressure was reduced to 60 mm Hg over a 15 min period and maintained at that level for 2 hr. At the end of the first hour of shock, 1 g tolbutamide was injected into the aortic catheter in an attempt to examine the portal-caval insulin relationship at higher insulin levels. Paired simultaneous blood samples for insulin assay were collected from the portal vein and the inferior vena cava at baseline, during the induction of shock, and at regular intervals during the period of hypotension. Insulin was determined by the immunoassay method (4).

RESULTS

In Fig.1, the simultaneous resting portal and caval insulin concentrations are shown. The mean difference 26.0 ± 24.6 µM/ml is significant (P < 0.05 by paired $\underline{t}$ test). Note the large variance in the portal samples.

*I - (phenylcyclohexl) piperidine HCL, Parke, Davis & Co.

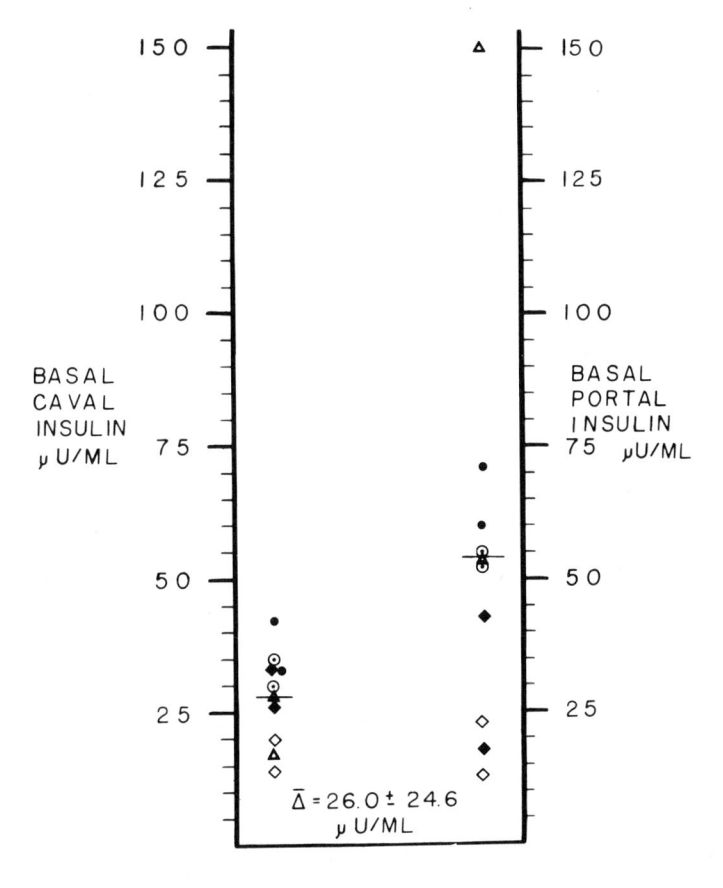

Figure 1.

The simultaneous basal caval and portal insulin concentrations are shown. Paired samples were drawn 5 min and 1 min before hemorrhage in each animal. Each animal is represented by a different symbol. Note the scatter in the portal concentrations.

The decline in mean value from baseline in portal insulin concentrations during shock is shown in Fig.2. This decline is significant in 6 of the 14 sample periods. Of greater interest is the observation that in all of the 14 sample periods without exception, the mean portal insulin value is less than the baseline. The probability of this occurring by chance is highly unlikely ($P < .005$ by sign test) indicating that a decline in portal insulin concentration should be an expected change in the baboon during shock. The injection of tolbutamide at the 60 min point had no measurable effect on the mean change from baseline in portal insulin concentrations.

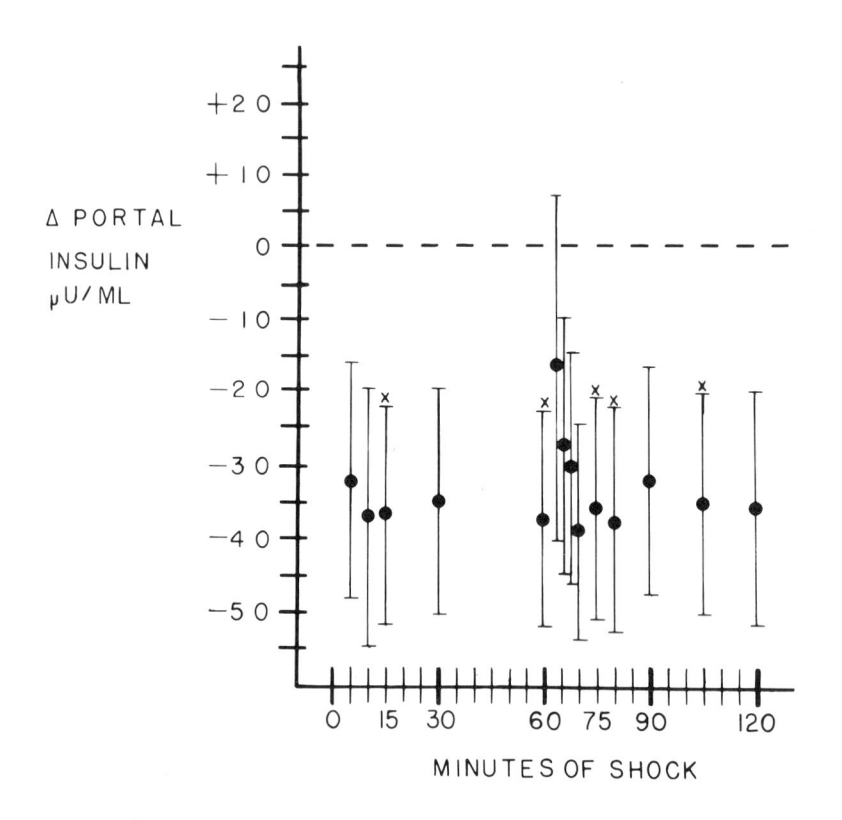

Figure 2.

Changes (Δ mean ± 1 SEM) from baseline in portal insulin concentrations during hemorrhagic shock. The asterisks indicate a significant decline from baseline.

In Fig.3, the decline in the mean value of caval insulin from baseline is shown. The changes are similar to that noted for portal insulin concentrations, the principal difference being a smaller variance in the caval measurement.

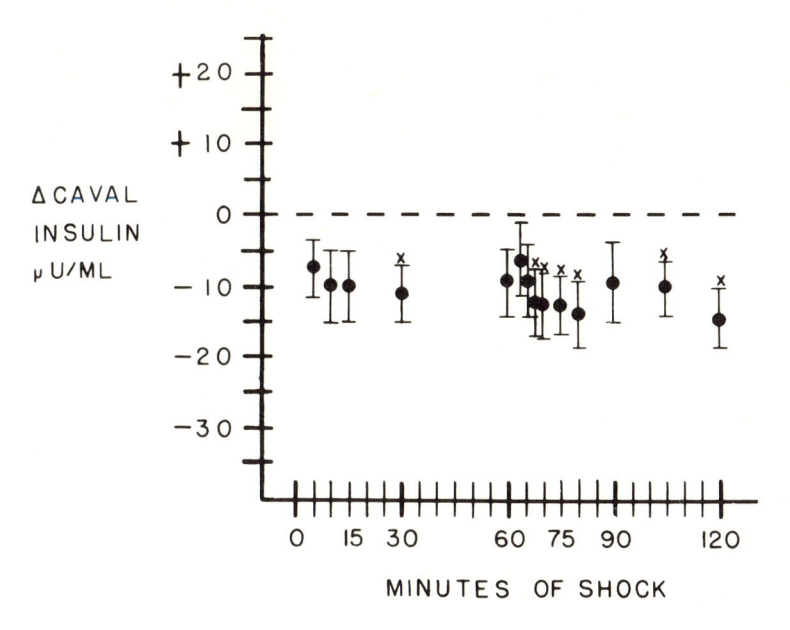

Figure 3.

Changes (Δ mean ± 1 SEM) from baseline in caval insulin concentrations during hemorrhagic shock. The asterisks indicate a significant decline from baseline.

The relationship between paired portal and caval insulin concentrations is shown in Fig.4. This relationship is linear, has a slope not different from one, and an intercept not different from zero. In view of the apparent scatter in the relationship, sources of variability were sought. Time was found to have no significant effect. Between animals effects were found to be significant as shown in Fig.5. When these animal effects were accounted for as covariates in the estimating equation, the correlation was substantially improved (R = 0.38 to 0.61). The slope remained indistinguishable from one.

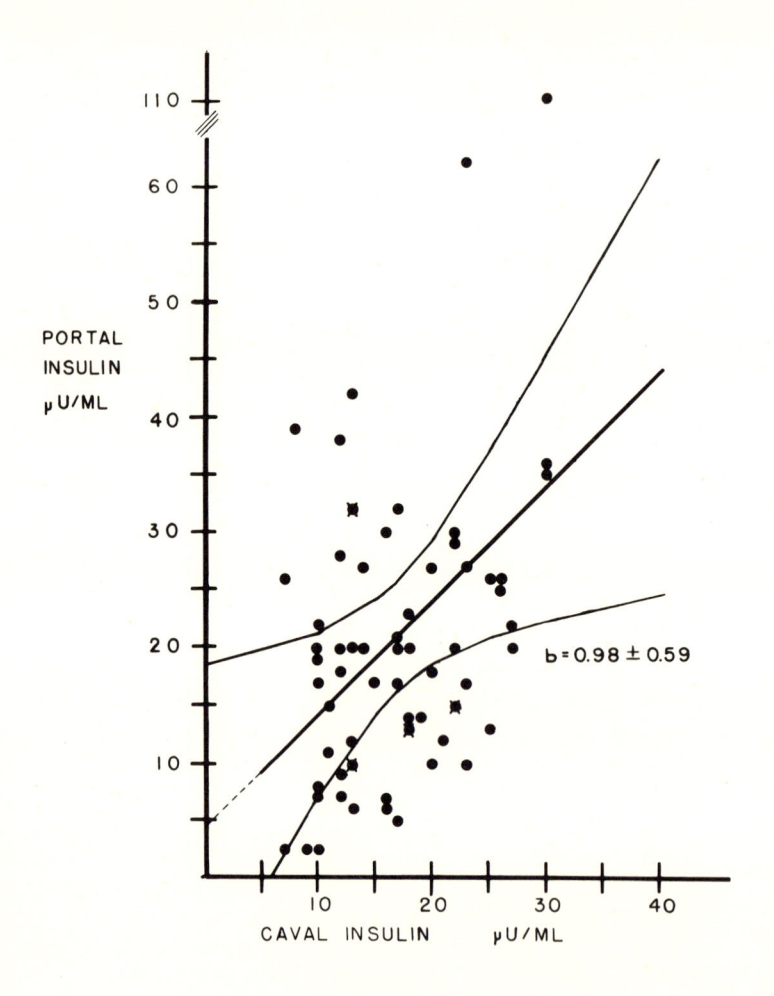

Figure 4.

Caval insulin concentrations are plotted against the corresponding portal values during the shock period. The symbol X indicates two values falling on the same point. The slope and the 99% confidence bands are shown. The estimating equation is:

Portal insulin = a + b . caval insulin

where:

a = 4.41 ± 10.65 (intercept ± confidence limits)
b = 0.98 ± 0.59 (slope ± confidence limites).

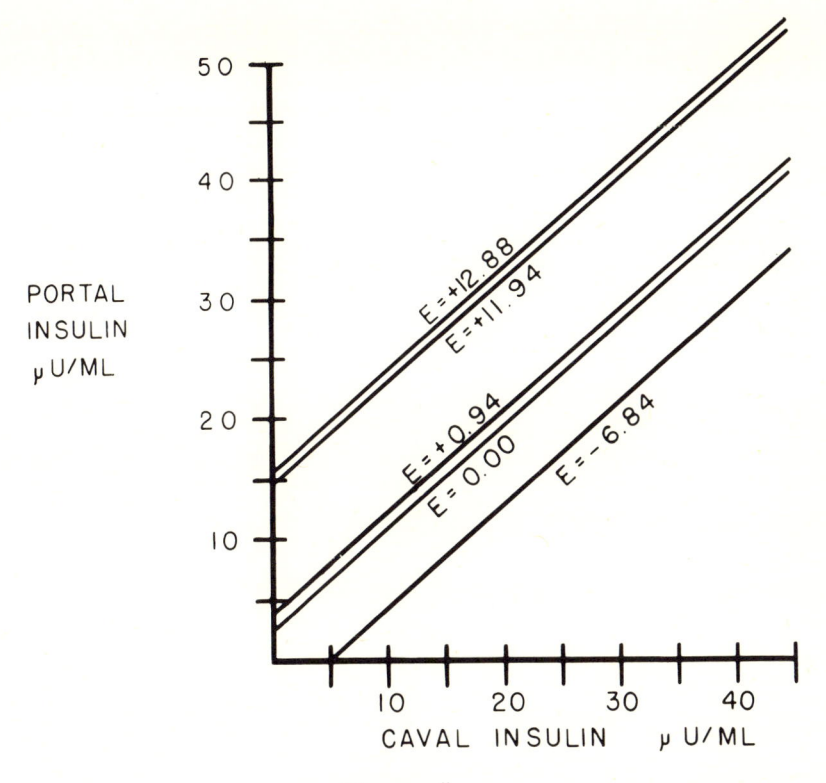

Figure 5.

The portal insulin versus caval insulin relationship during shock, where animal effect is taken into account. Each line represents the relationship between portal and caval insulin for each of the 5 animals.

The corrected estimating equation is:

Portal insul = a + b . caval insulin + E

where:

$a = 2.66$ (common intercept)
$b = 0.83 \pm 0.55$ (common slope $\pm$ 95% C.L.)
$E =$ animal effect.

Note that the slope for this equation is not different from the pooled slope (Fig.4), and is not different from a slope of one.

DISCUSSION

It is not suprising that resting portal insulin concentrations are higher than caval, since insulin released from the pancreas flows through the portal vein prior to uptake in the liver and dilution by systemic distribution. The wide variation in resting and shock portal insulin concentrations suggests that insulin release from the pancreas is intermittently augmented. In contrast, the smaller variance in resting and shock caval values emphasizes its peripheral equilibrium.

The decline in portal insulin concentrations during shock may have several explanations. The most plausible is a suppressed insulin response to the known hyperglycemia associated with hemorrhagic shock. The lack of response to tolbutamide supports this theory. Another explanation is that blood flow is increased in the portal vein during shock, thus producing a fall in insulin concentration. This explanation can be rejected since portal flow in fact declines during hemorrhagic shock (1). A third explanation might be that a major portion of insulin released during shock enters the circulation through a pathway other than the portal route. This seems highly unlikely in view of what is known of the physiology of the splanchnic bed. Thus we are left with the premise that insulin release is suppressed during hemorrhagic shock.

It has been postulated that this suppresion is related to the direct α-adrenergic action of epinephrine in the β-cell (3,6) by blocking the conversion of adenine triphosphate (ATP) to cyclic adenine monophosphate (cAMP)(5). Normally, in the β-cell, glucose is metabolized producing ATP. The ATP is converted to cAMP which is a prime stimulus for insulin release. Epinephrine, which is presumably produced in high concentrations in shock inhibits the conversion of ATP to cAMP. A diminished capacity to generate cAMP results in a suppressed insulin response to glucose. This inhibition, referred to as the α-adrenergic effect, reduces the stimulus to release insulin.

It has also been argued that the reduced insulin response during shock might be on the basis of reduced blood flow through the pancreas, rendering the β-cells hypoxic. Hypoxic cells produce reduced amounts of ATP and therefore have a diminished capability to release insulin. Against this theory is our previous observation in the baboon that within 5 min of the onset of slow hemorrhage, before a decline in aortic arch pressure can be detected, a significant fall in insulin concentrations can be seen (6). It is unlikely that a loss of blood volume insufficient to produce a decline in aortic pressure could seriously reduce pancreatic blood flow to the point where hypoxic effect occurs.

Finally, the data shows that a significant linear relationship exists between changes in caval and portal insulin concentrations can be expected to reflect a change in the portal concentration. However, because of the large variance in the relationship it is not possible to predict the exact portal insulin value if the caval is known.

SUMMARY

A study was carried out to determine if the previously described fall in peripheral insulin during hemorrhagic shock in baboons reflects a fall in pancreatic insulin release. Five adult baboons, with chronically implanted portal vein catheters, were subjected to 2 hr hemorrhagic hypotension (mean arterial pressure 60 mm Hg). Paired portal and caval insulin measurements demonstrated a decline in both portal and caval insulin levels during the entire shock period, and a significant correlation between caval and portal insulin levels during shock.

It is concluded that hemorrhage produces inhibition of pancreatic release of insulin, and that a fall in peripheral insulin levels reflects a diminished pancreatic insulin release.

REFERENCES

1. Abel, F.L., Waldhausen, J.A. & Selkurt, E.E. (1965). Amer.J. Physiol, 208, 256.
2. Bauer, W.E., Vigas, S.N.M., Haist, R.E. & Drucker, R.W. (1969) Surgery, 66, 80.
3. Carey, L.C., Lowery, B.D. & Cloutur, C.T. (1970). Ann.Surg., 172, 342.
4. Herbert, V.Lank, Gottlieb, C.W. & Blucker, S.J. (1966). J.Clin.Endocrinol., 25, 1375.
5. Levine, R. (1970). New Eng.J.Med., 283, 522.
6. Moss, G.S. Cerchio, G.M. Siegel, D.C., Popovich, P.A. and Butler, E. (1970). Surgery, 68, 34.

GLUCOSE HOMEOSTASIS IN THE POSTOPERATIVE STATE

J. Tiefenbrun, S. Finkelstein & W.C. Shoemaker

Mount Sinai School of Medicine, 5th Ave and 100th Street
New York, N.Y. 10029

The postoperative period is a unique and complex state caused by the controlled injury of surgery, anesthetic agents, and starvation. These factors produce profound physiologic and metabolic alterations which subject homeostatic mechanisms to stress. The glucose-insulin feedback system operates in an altered milieu of increased catecholamines and corticosteroids, liver glycogen depletion and changes in perfusion patterns. We have examined the response to insulin in the early and later postoperative periods to determine insulin clearance, hepatic glucose output and peripheral utilization of glucose.

METHODS

Mongrel Dogs (18-22 kg) in good physical condition were anesthetized with intravenous surital after 17-19 hr fasting. A laporatomy were performed through a midline incision and catheterization of the portal vein, hepatic vein, aorta and inferior vena cava was performed as previously described. Doppler ultrasonic flow probes were placed about the portal vein and hepatic artery. The catheters and leads were brought out through a separate stab incision. Normal saline was infused during the operative procedure. Animals were studied on arousal from anesthesia, early post-operative period (EPO), and in the unanesthetized state on the 7-14 post-operative day, late post-operative period (LPO). Animals studied in the late post-operative period had maintained their control weight and were studied after a 17-19 hr fast. Simultaneous blood samples were taken from each catheter at 2 to 4 control periods and at 5,10,15,30,60,90 and 120 min after injection of crystalline insulin (0.1 units/kg). The plasma was immediately frozen and analyzed for glucose and immune-reactive insulin. The hepatic arterial flow probes were calibrated by en-bloc excision of the arterial segment and direct perfusion with a Harvard pump. The portal venous probes were similarly calibrated after ligation of all entering veins and from the total hepatic blood flow calculated by the bromsulphapthalein clearance. Hepatic glucose output (HGO) was calculated from the transhepatic glucose concentration differences and the portal venous and hepatic arterial blood flow. Estimated extracellular glucose mass was calculated from the arterial plasma glucose concentration and the estimated extracellular volume based on body weight. Peripheral glucose uptake was calculated from the hepatic glucose output added to the instantaneous rate of change of the extracellular glucose mass at the time of sampling.

RESULTS

The levels of glucose and insulin in the control periods of both groups were identical. The mean glucose values for the early and late post-operative phases were 91 and 96 mg %. Insulin values were 12 and 13 units/ml respectively. There was no discernible difference in glucose levels until 30 min after insulin injection. LPO animals reached their maximum hypoglycemia of 64 mg % after 15 min and then rapidly returned to the baseline by 90 min. In many animals over-shooting followed by under-shooting was seen. Arterial glucose levels in the EPO group continued to decrease to a low of 35 mg % and then rose slowly and did not reach control values until 120 min after insulin. The differences between arterial glucose levels in the two groups after 15 min were statistically significant. There were significantly higher arterial insulin levels in the EPO animals in the 5 ($P<0.001$) and 10 ($P<0.05$) min periods, and resting levels were not reached until 90 min after injection. LPO animals, by contrast, had normal levels of arterial plasma insulin after 30 min.

Portal venous and hepatic arterial blood flow was higher in the LPO group. There were no significant changes in level or distribution of flow following insulin injection or hypoglycemia in either group.

EPO animals showed wide hepatic venous-portal and hepatic venous-arterial concentration differences when compared with the LPO group ($P<0.10$). These concentration differences were associated with the lower blood flow seen in the EPO. Portal venous glucose levels were the lowest in each time period showing net intestinal extraction of glucose.

The resting HGO in the EPO animals was 76 mg/min, more than 3 times the control values seen in the LPO. In the LPO, HGO increased above resting level following insulin and remained elevated until normoglycemia was restored; at 120 min there was a net decrease in HGO. By contrast, there was a significant decrease below control values in HGO 5 min after insulin injection. This was the first in a chain of under- and over-shooting of the baseline which continued until a consistent net increase in HGO occurred 30 min after insulin.

DISCUSSION

Following acute operative injury and anesthesia in the post-absorbtive state there is an increased hepatic glucose output, a delayed clearance of insulin, a lowered splanchnic blood flow, with increased transhepatic glucose differences. The initial response to insulin results in a fall in HGO followed by oscillations about the control level until a net effective increased glucose output is obtained. Following this period of oscillation, severe hypoglycemia develops. This can be contrasted to the smooth effective increase in HGO seen in the late post-operative state. These differences probably result from an altered humoral, biochemical and hemodynamic environment.

The present protocol was designed to test the time related response of glucose to an impulse function of insulin. The arterial glucose levels in the early post-operative period approximate an overdamped second order system. In the late post-operative period there is a tendency toward underdamping with late overshooting of the baseline. With less damping there is a faster reaction to stimuli with a tendency to oscillate about a given steady state level. In the EPO, although the overall arterial response is overdamped, in the first 15 min the hepatic glucose output varies in a positive and then negative direction until the appropriate response occurs. This ineffective compensation sets up the phase lag in the arterial glucose response curves between the LPO and EPO and allows the arterial glucose to continue to fall 15 min after insulin, resulting in a slower return of the glucose level to the baseline.

In these experiments, in the absence of changes in flow or changes in the distribution between PV and HA flow after insulin, and in the presence of wide changes in transhepatic concentration differences, hemodynamic alterations do not seem to be limiting factors. Liver glycogen depletion and increased catecholamines are well defined reactions to acute surgical stress. The well fed convalescent animal may primarily utilize stored glycogen to respond rapidly and effectively to hypoglycemia through neural mechanisms or small changes in adrenaline or glucagon. The acutely stressed animal has high ambient levels of cathecolamines already present and may synthesize the bulk of the hepatic glucose output through less efficient alternate pathways. The response to insulin in the EPO is similar to that reported after Eck fistula and portocaval transposition. Madison has suggested that with an HGO of approximately 54 mg/min, less than 40 % of the glucose was derived from glycogen stores. In the presence of increased substrate demands and less than optimal homeostatic control, the peripheral needs may not be met. Five min following insulin administration, despite a demonstrated need for 76 mg/min of glucose, only 45 mg/min was taken up in the periphery. Peripheral energy requirements may not have been satisfied.

The hepatic glucose output is 3 times greater in the EPO than in the LPO. This reflects a greater peripheral demand. The hepatic glucose output in the LPO ranged from 1-2 mg/kg/min. There was a 5-6 mg/kg/min output in the EPO. These LPO values approximate those seen in man. Although man and dog differ in metabolic rate, it can be calculated that a 70 kg. man in the acute post-operative period may require in excess of 20 g glucose per hr to meet peripheral demands. The surgical procedures can result in stress to the liver in its metabolic and synthetic pathways.

SUMMARY

These data suggest that the early postoperative period may be accompanied by an altered hepatic and peripheral response to insulin resulting in inefficient homeostatic compensations, which may result in substrate deprivation.

SOME IN VITRO AND IN VIVO EFFECTS OF A NEW PROSTAGLANDIN DERIVATIVE

B. David Polis, Anne Marie Grandizio, Edith Polis

Biochemistry Laboratory, U.S. Naval Air Development Center, Warminster, Pennsylvania, U.S.A.

On the simplistic premise that an anoxic-fatigue stress, like acceleration, could be defined in terms of energy demand under conditions of limited supply, we searched for molecular probes which would reveal or reflect those regulatory mechanisms pertinent to the bioenergetic pathways involved in adaptation to stress. It was our expectation that the exhaustion of adaptive events, and the onset of pathology would be presignaled by molecular changes which might afford a biochemical index or end point to stress tolerance. Such information would be useful also to amortize, pharmacologically, the energetic cost of a defensive reorganization against stress, and thereby enhance the survival of a crisis period.

Experiments with isolated particulate fractions from animal cells revealed marked changes in a specific phospholipid identified as phosphatidyl glycerol (G) that followed exposure to an acute stress like acceleration or a longer termed degenerative stress like x-irradiation (1). These stress induced changes in the phospholipid composition of tissues and their correlation with comparable changes in plasma phospholipids of the rat suggested an approach to stress induced chemical changes in humans.

The effects of both physical and psychic stress on human plasma phospholipids are shown in Figure 1. These are portrayed as three dimensional plots of the means ± two standard errors for G, phosphatidic acid (P) and phosphatidyl ethanolamine (PE). It is evident that in all the populations exposed to the various stresses of acceleration, sleep deprivation, combat flying, or the stress accompanying schizophrenia, there was a significant increase in G over the controls. Variations in other phospholipids made

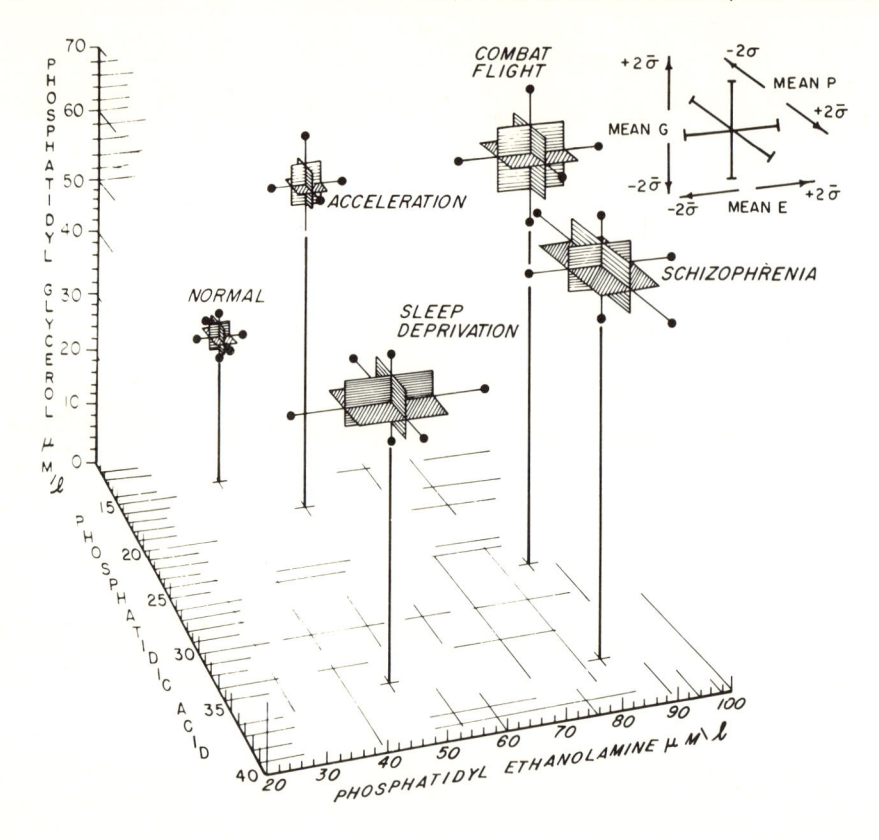

Fig. 1. Changes in Human Plasma Phospholipids Caused By Stress

possible the statistical discrimination of stressed populations from each other.

In all the stress reactions we studied, G was unique in the consistently elevated plasma levels which were common to all the stresses. In contrast, other phospholipids showed variable changes which facilitated a molecular characterization of the stress. These concentration shifts in individual phospholipids were not a direct consequence of variations in the total phospholipid content. Both increments and decrements of specific phospholipids were observed with no changes or even opposing changes in the levels of the total phospholipid content. Whether this represented a concentration effect in the output of a major regulatory factor or whether each phospholipid was uniquely controlled, the results implied the action of some brain centers which interpreted sensory inputs as "threats to survival" and in reacting, mobilized the phospholipids. That this hypothesis had some merit was indicated by the release of G from the brains of stressed humans.

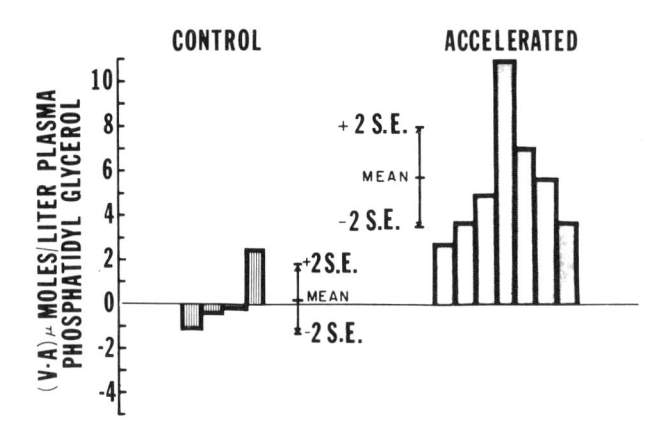

Fig. 2. Release of Phosphatidyl Glycerol by Stressed Human Brain

Figure 2 shows the results of collaborative studies with Dr. Martin Reivich on the differences in G between jugular venous and femoral arterial blood plasma from subjects under control conditions and after acceleration to grayout (2). It is apparent that there is indeed a significant release of G from the brains of the subjects after acceleration. The singularity of G is emphasized by the failure of all other phospholipid species to show any significant concentration change across the human brain after grayout.

Some indications of triggering factors for the phospholipid changes in stress can be obtained from the fact that the injection of various prostaglandins into rats caused major changes in G and lesser changes in other phospholipids that mimic the results obtained in stressed rats and humans. The increases in plasma G in the rat were accompanied by elevations of brain G. Although all four of the prostaglandins shown in Figure 3 caused elevations of plasma G, they differed in their effects on other phospholipids. Prostaglandin E_1 (PGE_1) caused a significant decrease in lecithin (Le) and total phospholipid while Prostaglandin $F_{1\alpha}$ ($PGF_{1\alpha}$) caused an increase in Le and total phospholipid. The changes in total phospholipid were nonsignificant for PGB_1 and PGB_x and the variation in Le less marked. Variable effects on the concentration changes of P were observed with the different prostaglandins. The greatest change was obtained with PGE_1. There also were marked differences in observable physiological response. PGE_1 injection was followed by severe lassitude and diarrhea so that the rats appeared visibly ill. This response was absent with the other prostaglandins. With PGB_x, the rats even appeared more lively and excitable. It was this prostaglandin derivative which showed the most dramatic in vitro effects on mitochondria and was studied in detail.

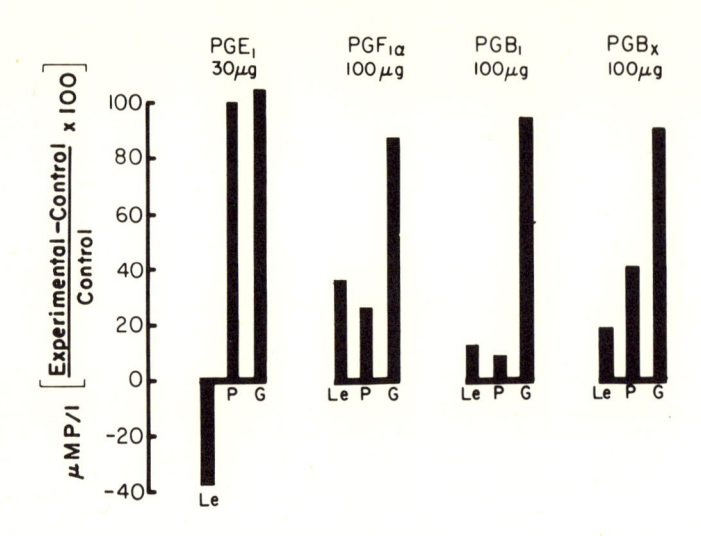

Fig. 3. Variation In Phospholipid Effects Correlated With Structural Changes In Prostaglandins

The cyclopentanone ring of PGE$_1$ has a hydroxyl group β to a keto group. Dehydration in dilute alkali then readily occurs and with rearrangement of the double bond PGB$_1$ is formed. These changes can be followed by the appearance of an absorption band at 278 nm. When the NaOH concentration is raised to 1 molar in 50% alcohol and the prostaglandin is heated at 65°, the 278 band disappears and two bands at 247 and 370 appear. These components react further to give a mixture that contains the active component PGB$_x$ and shows no well defined peak in the UV. When the base catalyzed reaction products are acidified and extracted into ether they can be resolved by thin layer chromatography into components which are distinguished by their fluorescence under long wave UV as well as by their mobilities. Purification of PGB$_x$ by elution of the orange band resolved by thin layer chromatography or in larger quantities by column chromatography on PVP eluted with heptane-alcohol gradients yielded the active component which had as an essential part of its structure an enolized β-diketone.

Using Warburg techniques for measuring oxygen uptake and hexokinase with glucose as a phosphate trap, mitochondria, aged for 4 to 5 days showed little or no phosphate esterification and comparatively low oxygen uptake with α-ketoglutarate and ADP or AMP as substrate even in the presence of serum albumin. With the addition of 0.1 micromole PGB$_x$ to the reaction, there was a marked recovery of both phosphorylation and oxygen uptake. Even with AMP as phosphate acceptor the PGB$_x$ channeled the reactions from dephosphorylation to phosphorylation with changes in P/O ratios from 0 to 1.5 (Table 1).

Table 1

Effect of PGB$_x$ on Phosphorylation Efficiency of Aged Mitochondria

Age (days)	Substrate	P esterified	O uptake	P/O
4	control	0.07 ± 0.8	2.1 ± 0.36	0.03
	ADP PGB$_x$	8.92 ± 0.45	8.0 ± 0.26	1.12
5	control	-0.1 ± 0.01	0.8 ± 0.04	---
	AMP PGB$_x$	5.8 ± 0.04	3.8 ± 0.20	1.53

To prove that PGB$_x$ actually reactivated the net synthesis of ATP, mitochondrial reactions were run using ADP as the acceptor for phosphate. The changes in the nucleotide composition were measured by chromatography of the deproteinized reaction on a pellicular ion exchange column. This procedure (3) can quantitatively separate mixtures of AMP, ADP and ATP in 10 µ liter samples with a sensitivity in the order of 0.1 nanomole of nucleotide. The net disappearance of inorganic phosphate was measured separately by colorimetry.

The data summarized in Table 2 shows the nucleotide distribution in the reactions. At zero time the reaction mixture contained essentially inorganic phosphate and ADP plus a small amount of AMP formed by the splitting of ADP. In the control reaction run for 20 min. there was actually an increase in Pi because of ATPase activity, while the ADP originally added was dismuted to form AMP and ATP by the adenylate kinase present. With the addition of PGB$_x$ to the reaction mixture there was clearly a shift to phosphorylation with a net synthesis of ATP corresponding to the decrease in inorganic phosphate.

The PGB$_x$ activation of oxidative phosphorylation was blocked by conventional inhibitors like oligomycin, dinitrophenol, dicumerol or pentobarbitol but with interesting concentration effects. At low levels of the inhibitor, a potentiation of the prostaglandin activation was observed. With increased concentrations of the inhibitor the effect was diminished and finally completely blocked (Figure 4).

Table 2
Nucleotide Distribution and P_i esterified in Control and PGB_x Activated Mitochondrial Reactions for Oxidative Phosphorylation

	μ Moles			
Reaction	AMP	ADP	ATP	P_i esterified
0 Time	0.22	4.55	<0.01	0
Control (20 min)	1.76	1.41	0.75	-0.53
PGB_x (20 min)	0.07	0.57	4.00	4.74

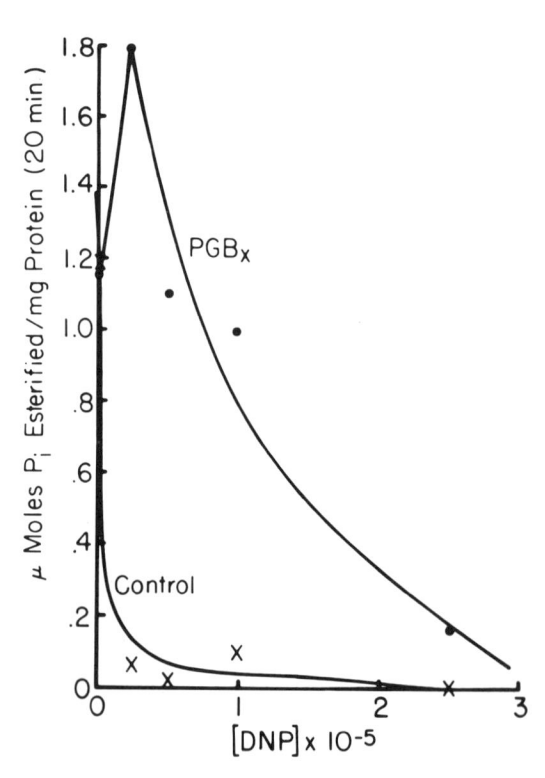

Fig. 4. Effect of PGB_x on DNP Inhibition of 24 Hour Mitochondria

Of the inhibitors of oxidative phosphorylation studied possibly the most important from the standpoint of its implications to stress was the interplay between Ca++ and PGB$_X$ to control mitochondrial phosphorylation. Recently the uptake of Ca++ in preference to phosphorylation has been detailed from a number of laboratories, especially those of Chance and of Lehninger (4,5). Figure 5 illustrates the activation curve for PGB$_X$ in a mitochondrial system inhibited with Ca++. In aged preparations the addition of Ca++ at low levels enhanced dephosphorylation. This increase in inorganic PO$_4$ was blocked as the Ca++ was increased (4x) with no net phosphorylation. With the addition of PGB$_X$, there was an activation of phosphorylation and a cancellation of the Ca++ effect over the Ca++ range studied. It would appear then that in the competition between Ca++ and PGB$_X$ for the direction of oxidative energy, an *in vitro* mechanism is available for the control of phosphorylation in mitochondria.

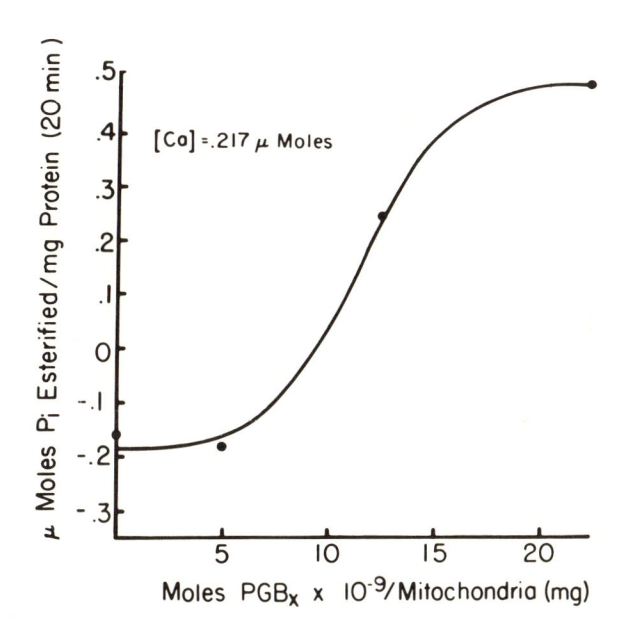

Fig. 5. Reversal of Ca Inhibition of Mitochondrial Phosphorylation by PGB$_X$

Any concern with problems of performance and survival in a stressful environment forces the recognition of the complex interplay between the animal and his internal and external environment. This interplay invokes intuitive and learned, involuntary and voluntary responses, designed to maintain a stable equilibrium essential for survival. For lack of better terminology the manifold reactions that constitute the response to a homeostatis displacement have been lumped into a catchall designation of stress. From a biochemical standpoint, using acceleration as a model, we interpreted stress as an increased demand for biological energy under conditions of limited supply and predicated our approach on the premise that common bioenergetic factors were operative in stresses of diverse etiology. Some new biochemical correlates of stress and some of their hormonal control factors have been demonstrated. These have been implicated through in vitro studies of mitochondria into primary energy transducing mechanisms. The studies with PGB_x offer a molecular approach to the channelling of biological energy to pathways that should enhance the tolerance to stress and survival of animals in a crisis period.

References

1. Polis, B. David, E. Polis, J. Decani, H. P. Schwarz, and L. Dreisbach. *Biochem. Med.*, 2(4), 1969.

2. Polis, B. David, M. L. Reivich, L. H. Blackburn, and D. P. Morris. *Aerospace Med. Assoc.*, 42nd Annual Meeting, April 26-29, 1971.

3. Shmukler, Herman W. *J. Chromatog. Sci.*, 8, 653(1970).

4. Rasmussen, Howard, Britton Chance, and Etsuro Ogata. *Proc. Nat. Sci.*, 53, pp:1069-1076 (1965).

5. Lehninger, Albert L. *Biochem. J.*, 119, 129-138 (1970).

NON-ESTERIFIED FATTY ACID (FFA) METABOLISM FOLLOWING SEVERE

HEMORRHAGE IN THE CONSCIOUS DOG

John J. Spitzer, Roslyn Wiener, Eugene H. Wolf

Hahnemann Medical College and Hospital Department of
Physiology and Biophysics Philadelphia, Pennsylvania,
19102

It is becoming increasingly evident that hypovolemic alterations involve at least two major areas of normal tissue physiology: the delivery of oxygen and oxidizable metabolites and the utilization of metabolites to release the energy needed for normal functioning of the tissues.

In the past, most of the investigative and therapeutic efforts have been aimed at the hemodynamic changes. More recently it has become apparent that multiple and marked alterations in intermediary metabolism accompany the development and may determine the irreversibility, of shock. Although a number of shock-related studies have attempted to characterize metabolic alterations under in vitro conditions (mostly on the cellular or subcellular level), there have been few efforts directed at the study of energy metabolism in the whole animal (or man), and most of these have dealt with carbohydrates (1,2,3,4). Much less information is available concerning changes of lipid metabolism during shock (5,6), although FFA are an important source of energy for most tissues (7). The present investigations were designed to provide more information on FFA metabolism during hemorrhagic hypotension in the conscious dog.

MATERIAL AND METHODS

The experiments were performed on 5 conscious mongrel dogs weighing from 12.7 to 25.0 kg. Food was withheld from the animals for 16-18 hr before the experiment. All procedures were performed under local anesthesia with 2% procaine (not containing epinephrine) and without administration of anticoagulants to the animals. A femoral artery was cannulated for monitoring blood pressure, sampling arterial blood and for inducing hemorrhage. A catheter was introduced into the pulmonary artery via a femoral vein for the removal of mixed venous blood samples, and kept patent by a slow saline drip. A constant infusion of a tracer dose of albumin-bound 1-C^{14}-palmitic acid was administered throughout the experiment via a polyethylene cannula in a superficial leg vein. Sixty to 120 min after the infusion was started, simultaneous arterial and mixed venous samples were taken into heparinized syringes. Processing of blood samples was begun immediately with all samples being kept at 4°C. Plasma volume (Evans blue) and cardiac output (indocyanine green) were then determined, followed by two more sets of control blood samples at 20 min intervals. Subsequently, the dogs were subjected to severe hemorrhage (47±3 ml of blood loss per kg body weight, or about 46% of the initial circulating blood volume). The first fraction of the removed blood was used for catecholamine assays (kindly performed by Dr. E.T. Angelakos). Forty to 60 min later the first pair of post-hemorrhagic blood samples were taken, following by two more at 20 min intervals. Plasma volume, cardiac output, and catecholamines were again estimated.

FFA were determined on duplicate aliquots, using the method of Dole and Meinertz (8). The lower phase washings of the titrated FFA (9) were counted in a Packard liquid scintillation spectrometer for estimation of FFA radioactivity. Blood O_2 and CO_2 were determined according to Van Slyke, $^{14}CO_2$ by the method of Passman et al. (10), plasma glucose by glucose oxidase (Worthington kit) and blood lactate by an enzymatic method (11). Cardiac output was measured by the Stewart-Hamilton technique, employing indocyanine green as the indicator.

The following calculations were used in obtaining the data (12):

FFA Flux (μmole/min/kg) =

$$\frac{\text{Radiopalmitate infused (dpm/min)}}{\text{Art. FFA specific activity (dpm/μmole) X dog wt. (kg)}}$$

Contribution of FFA oxidation to CO_2 production (%) =

$$\frac{\text{FFA oxidized (μmole/min/kg) X 17 X 100}}{\text{V-A } CO_2 \text{ difference (μmole/ml) X cardiac output (ml/min)}}$$

TABLE 1. Hemodynamic changes in the conscious dog following hemorrhage (N=5)

	Control	After Hemorrhage	Change	
Mean arterial blood pressure (mm Hg)	109 ±7	66 ±8	-43 ±8	(39.4%)
Heart rate (per min)	126 ±6	181 ±13	+55 ±11	(43.6%)
Cardiac output (ml/min)	2703 ±539	967 ±92	-1726 ±483	(63.8%)
Blood volume (ml)	1811 ±182	1070 ±96	-741 ±104	(40.9%)
Hematocrit (%)	38 ±3	29 ±2	-10 ±3	(26.3%)

Mean ± SE

TABLE 2. Changes in FFA turnover and oxidation in 5 conscious dogs after hemorrhage

Dog No.	Arterial FFA (μMole/ml) Before	Arterial FFA (μMole/ml) After	FFA specific activity (dpm/μMole) Before	FFA specific activity (dpm/μMole) After	FFA flux (μMole/min/kg) Before	FFA flux (μMole/min/kg) After	FFA oxidized (μMole/min) Before	FFA oxidized (μMole/min) After	$\frac{\text{Oxidation}}{\text{flux}} \times 100$ Before	$\frac{\text{Oxidation}}{\text{flux}} \times 100$ After	% contrib. of FFA to CO_2 prod. Before	% contrib. of FFA to CO_2 prod. After
1	0.235	0.436	5502	3400	13.1	22.3	2.5	1.5	20.2	6.4	10.2	10.4
2	0.457	0.788	1628	1316	25.3	31.3	5.7	10.1	23.6	33.1	28.3	67.3
3	0.344	0.526	2887	2431	9.8	11.7	1.5	0.8	15.8	6.8	10.7	7.1
4	0.411	0.953	5314	2881	21.2	38.9	1.9	2.9	9.2	7.4	13.8	15.1
5	0.594	0.861	2794	2164	23.1	30.7	5.6	2.4	24.2	7.7	14.0	12.2
Mean SE	0.402 ±0.061	0.703 ±0.106	3642 ±771	2438 ±350	18.4 ±3.0	26.9 ±4.6	3.4 ±0.9	3.5 ±1.7	18.6 ±2.8	12.3 ±5.2	15.5 ±3.4	22.4 ±11.3
Change SE	+0.302 ±0.068 (75.1%)		−1203 ±459 (33.0%)		+8.5 ±2.7 (46.2%)		0.1 ±1.3		−6.3 ±4.7		+6.9 ±7.9	

Per cent change is given in parenthesis when statistically significant.

TABLE 3. Correlation of arterial FFA with FFA flux and oxidation

	Control	After Hemorrhage
Arterial FFA conc. vs. FFA flux	y =36.9 X + 3.9 r =0.80	y=31.0 X +4.8 r=0.75
Arterial FFA conc. vs. FFA oxid.	y =9.6 X -0.3 r =0.64	y=3.8 X +1.03 r=0.25

Calculations based on data from individual samples

TABLE 4. Changes of O_2 consumption and CO_2 production in the conscious dog following hemorrhage (N=5)

	Control	After Hemorrhage	Change
Arterial O_2 (μMole/ml)	7.44±0.63	5.88±0.48	-1.56±0.45 (21.0%)
O_2 consumption (μMole/min/kg)	411±58	243±31	-167±34 (40.6%)
Arterial CO_2 (μMole/ml)	15.67±0.75	6.28±0.50	-9.39±1.18 (59.9%)
CO_2 production (μMole/min/kg)	386±75	287±37	-99±57

Mean ± SE---- Per cent change is given in parenthesis when statistically significant.

TABLE 5. Changes of arterial glycerol, glucose, and lactate in the conscious dog following hemorrhage (N=5)

	Control	After Hemorrhage	Change
Arterial glycerol (μMole/ml)	0.030±0.005	0.248±0.049	+0.218±0.047 (727%)
Arterial glucose (μMole/ml)	5.79±0.40	5.23±0.9	-0.47±0.90
Arterial lactate (μMole/ml)	1.264±0.226	8.415±0.339	+7.299±0.459 (577%)

Mean ± SE -- Per cent change is given in parenthesis when statistically significant.

RESULTS

Table 1 summarizes the observed hemodynamic changes following hemorrhage. The percentage decreases of mean arterial blood pressure and blood volume were similar, that of cardiac output was more marked, reflecting the diminished venous return. Heart rate increased considerably. The expected hemodilution is indicated by the diminished hematocrit.

Changes in various parameters of FFA metabolism for the 5 dogs are summarized in Table 2. Arterial FFA concentration and FFA flux increased considerably. There was no consistent change in total body FFA oxidation following hemorrhage; in 3 dogs it diminished, in the other 2 it increased. It can be further observed from Table 2 that 18.6% of the FFA flux was oxidized to CO_2. This rate of FFA oxidation contributed 15.5% to the total body CO_2 production. Neither of these values exhibited a statistically significant change following hemorrhage due to the great variability in the responses of the individual dogs.

The correlations of arterial FFA with FFA flux and with oxidation are indicated in Table 3. A significant correlation was found between the concentration of FFA and FFA flux in individual blood samples both before and after hemorrhage. A reasonably good correlation also existed between the arterial FFA concentration and FFA oxidation before, but not after hemorrhage.

Changes of blood O_2 and CO_2 content are shown in Table 4. Both parameters decreased, CO_2 more markedly than O_2. Total body O_2 consumption decreased in all dogs. CO_2 production diminished in 4 of the 5 animals. In each dog, the RQ value increased after hemorrhage.

Arterial concentrations of 3 other metabolites are shown in Table 5. Concentration of glycerol increased more than seven-fold, that of lactate almost six-fold. The changes of arterial glucose concentration were inconsistent.

The catecholamine concentration increased from 2.1 (±0.8) to 46.1 (±9.3) µg/l following hemorrhage.

DISCUSSION

These studies were designed to investigate alterations of lipid metabolism in the conscious animal subjected to major hemorrhage. It should be noted that they deal with the early stages of hemorrhagic hypotension; subsequent changes might well be qualitatively or quantitatively different.

It is noteworthy that there were differences in both hemodynamic and metabolic responses between the present series of conscious dogs and a previously studied group of animals given general anesthesia. The animals of the current series, when subjected to a very massive hemorrhage, amounting to almost half the blood volume, showed a marked increase in heart rate, while there was no change in the anesthetized animals which had suffered comparable blood loss (13). This can presumably be explained by better preservation of vascular reflexes in the conscious animals. A disparity in metabolic parameters between anesthetized and unanesthetized animals was also noted: the increase in plasma FFA and FFA flux observed in the present investigations did not occur in previous studies (13) conducted under Nembutal anesthesia. The difference between the two groups might be explained by a greater adipose tissue perfusion and catecholamine discharge in the conscious animals. It is important to point out that, despite the increased lipid mobilization (implied by the increased flux, FFA and glycerol concentration) in the present series of conscious animals, FFA oxidation was not increased. The direct correlation between flux and oxidation observed during the control period did not obtain after hemorrhage.

The dissimilar behaviour of fatty acid oxidation and total flux warrants an explanation. It might be attributed to limitation of tissue O_2 supply, although a decrease in lipid oxidation has been described in tourniquet shock, where tissue hypoxia did not occur (5).

The increased FFA flux without an increase in FFA oxidation suggests the possibility of enhanced esterification. There are suggestions from the data that increased triglyceride synthesis in fact might have occurred: (a) the greater percentage increase of plasma glycerol than of FFA values might indicate re-esterification of FFA moieties, (b) RQ's above 1.0 are consistently observed after hemorrhage (although the concurrent acidosis and lactacidemia might render this value less meaningful), (c) an increase in plasma insulin as previously observed in the course of hemorrhagic shock in dogs (4) would favour lipogenesis, (d) triglyceride synthesis might also be promoted by the combination of elevated blood lactate and catecholamines, the latter enhancing lipolysis, the former favouring re-esterification in adipose tissue (14,15). Increased triglyceride synthesis has been previously reported in both myocardium (16) and liver (17) after hypoxia or injury.

An enhanced channeling of FFA to triglyceride would tend to deprive the organism of this needed oxidizable substrate, leading ultimately to accelerated protein breakdown after exhaustion of the limited carbohydrate stores. This sequence of events might be implicated in the development of irreversibility of shock.

Given the limitation of lipid contribution to the total metabolism in the course of shock, attention must be focused on the utilization of other substrates. Such studies are now in progress in our laboratory.

SUMMARY

Changes of FFA metabolism were studied in 5 conscious dogs following severe hemorrhage. A continuous infusion of 1-^{14}C-palmitate was administered throughout the experiment. Blood samples were taken simultaneously from a systemic and the pulmonary artery both before and after the removal of 46% of the circulating blood volume. Arterial concentration and FFA flux increased following hemorrhage; however, FFA oxidation did not change consistently. A direct correlation between arterial FFA concentration and oxidation of this metabolite was observed before, but not after bleeding. Arterial glycerol and lactate increased greatly, glucose did not change consistently. Indirect evidence suggested increased lipid synthesis under these experimental conditions.

REFERENCES

1. Kinney, JM., Long, C.L. & Duke, J.H., (1970) In: Energy Metabolism & Trauma (R. Porter & J. Knight, eds.) pp. 103, Churchill, London, England.

2. Stoner, H.B., Heath, D.F., & Collins, O.M., (1960) Biochem. J. 76: 135.

3. Vigas, M., Hetenyi, G.J., & Haist, R.E., (1971) J. Trauma 11: 615.

4. Bauer, W.E., Vigas, S.N.M., & Haist, R.E. (1969) Surgery 66: 80.

5. Heath, D.F. & Stoner, H.B., (1968) Brit. J. Exptl. Path. 49: 160.

6. Kovach, A.G.B., Rosell, S., Sandor, P., Koltay, E., Kovach, E., & Tomka, N., (1970) Circul. Res. 26: 733.

7. Spitzer, J.J., & Spitzer, J.A., (1971) Progr. Biochem. Pharmacol. 6: 242.

8. Dole, V.P., & Meinertz, H., (1971) J. Biol. Chem. 235: 2595.

9. Gold, M. & Spitzer, J.J. (1964) Am. J. Physiol. 206: 153.

10. Passman, J.M., Radin, N.S., & Cooper, J.A.D., (1956) Anal. Chem. 28: 484.

11. Hohorst, H.J., (1963) In: Methods of Enzymatic Analysis (H.V. Bergmeyer, ed.) p. 266 Academic Press- New York.

12. Little, J.R., ioto, M. & Spitzer, J.J., (1971) Am. J. Physiol. 219: 1458.

13. Spitzer, J.A. & Spitzer, J.J. (1972) J. Trauma, in press.

14. Issekutz. B., Miller, H.I. & Rodahl, K., (1966) Fed. Proc. 25: 1415.

15. Fredholm, B.B., (1971) Acta Physiol. Scand. 81: 110.

16. Scheuer, J. & Brachfeld, N., (1966) Metab. 15: 945.

17. Lieber, C.S., (1967) In: Biochemical Factors in Alcoholism (R.P. Maickel, ed.) p. 167, Pergamon Press, Oxford.

These studies were supported by the U.S. Navy Themis Project and by the National Heart Institutes (grant HE 03130).

ALTERATIONS OF MITOCHONDRIAL STRUCTURE AND ENERGY-LINKED FUNCTIONS IN HEMORRHAGIC SHOCK AND ENDOTOXEMIA

Leena M. Mela, Leonard D. Miller, Leonardo V. Bacalzo, Jr. Kenneth Olofsson, and Raleigh R. White, IV

From the Shock and Trauma Unit, Hospital of the University of Pennsylvania and the Harrison Department of Surgical Research, School of Medicine, University of Pennsylvania, Philadelphia

Gross metabolic aspects of injury have been under intensive study for years and thus a lot of useful information about the metabolic effects of injury is available (1,2). On the other hand, until a couple of years ago, there were hardly any reports about specific cellular metabolic changes occurring during clinical or experimental shock reactions, such as the hypovolemic hemorrhagic shock or the normovolemic, primarily septic shock reaction, endotoxemia. Recently a few reports have appeared in the literature, indicating a block at the mitochondrial level of the cellular metabolic pathway (3-12).

The mitochondrial site of inhibition of energy transference has been studied to some extent but the exact biochemical mechanism of the inhibition is not known at the present time (9). The intracellular etiology of the shock-induced mitochondrial alterations is also still unresolved.

This paper presents our attempts to evaluate the correlation of the shock-induced alterations of the liver mitochondria with other intracellular changes in the liver (13). Information on such a correlation should prove helpful in searching for the cellular causes of mitochondrial damage during hemorrhage and endotoxemia.

METHODS

Experimental hemorrhage and endotoxemia were induced in male Sprague-Dawley rats (Charles River breed) weighing 120-200 g as described earlier (9,14). In the hemorrhage experiments the rats were bled to 30 mm Hg blood pressure and endotoxemia was induced by an intraperitoneal injection of an LD_{90} dose of E.coli endotoxin (Difco Laboratories).

Procedures for the isolation and assays of the liver mitochondria have been described (9,15).

Liver lysosomes were prepared from the supernatant fraction of the liver homogenates after the separation of the nuclear fraction and cell debris (at 480 g for 7 min), and the separation of the mitochondrial fraction (at 7700 g for 14 min), by centrifuging at 14500 g for 20 min.

Acid phosphatase was assayed by the method of Fiske and Subbarow (16) using β-glycerophosphate as substrate at pH 5.0. Tissue PO_2 was determined by a Pt tissue electrode with a tip diameter of 20 μm. The electrode was constructed according to Silver (17,18) and Lübbers (19). The tip of the electrode was membrane covered.

In vivo tissue hypoxia was induced by lowering the O_2 in the breathing air to 9% in exchange with N_2 (95% N_2, 5% CO_2 + air). PO_2 was monitored continuously during the experiment from the rat's hind leg muscle surface. A separate Ag-AgCl reference electrode was used outside the tissue.

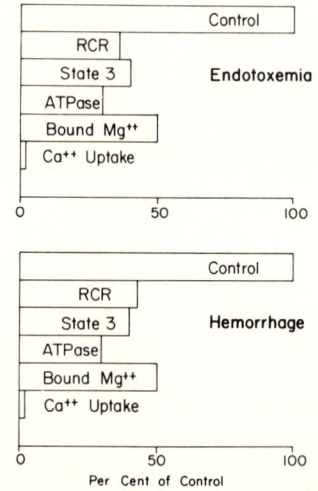

Fig. 1. A summary of the effects of 5 hr of endotoxemia or hemorrhage on rat liver mitochondria. The activities of the various mitochondrial parameters are presented as percentage of control.

RESULTS

Fig. 1 summarizes our earlier results on effects of 5 hr of endotoxemia or hemorrhage on rat liver mitochondrial functions (9). The various parameters characterizing mitochondrial functions are presented as a percentage of control. The figure shows that both endotoxemia and hemorrhage produce similar mitochondrial end results. The mitochondria exhibit loose coupling and thus low respiratory control ratios (RCR). State 3 respiratory activity becomes inhibited by about 60-70%, so does the uncoupler activated ATPase. The mitochondria lose about 50% of their bound Mg^{++}, and almost completely lose their capacity to transport Ca^{++} across the membrane and keep it in the mitochondria.

Lysosomal enzyme activity and mitochondrial function

Fig. 2 shows a comparative study of lysosomal enzyme activity and mitochondrial functions. Rat livers were removed

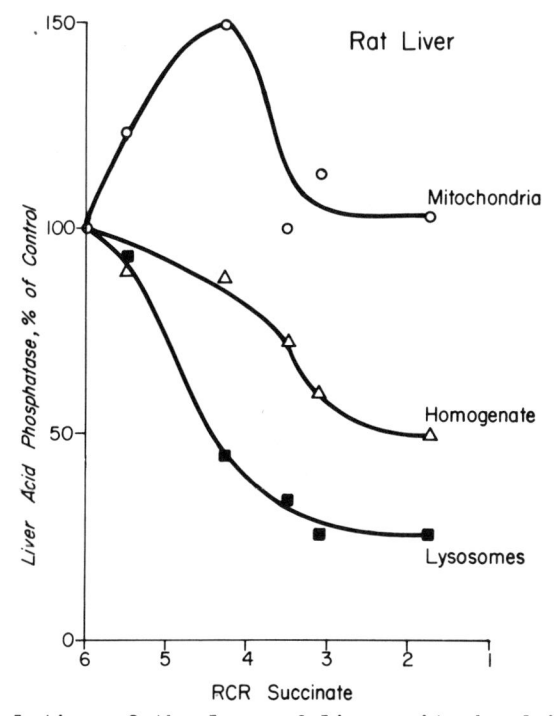

Fig. 2. Correlation of the lowered liver mitochondrial respiratory control, due to increasing exposure to endotoxin in vivo, to the lysosomal acid phosphatase activity measured in the homogenate, mitochondrial and lysosomal fractions of the same livers.

from 1-6 hr after endotoxemia. A sample of the homogenate was kept for lysosomal acid phosphatase assay, and both lysosomal and mitochondrial fractions were isolated and washed. The liver mitochondria were assayed for respiratory activity, ATPase, and electron transfer reactions. In Fig. 2, the respiratory control ratios of various preparations are used to indicate the severity of the mitochondrial defect caused by endotoxemia, and they are compared to the lysosomal acid phosphatase activity of the same livers. Acid phosphatase is used as an indicator of the lysosomal activity in these preparations. Fig. 2 shows that as the mitochondria become more severely damaged the lysosomal activity decreases in a sigmoidal fashion both in the lysosomal fraction and in the whole liver homogenate. In the mitochondrial fraction, there is an initial increase in the percentage activity of acid phosphatase. This initial increase suggests swelling of the liver lysosomes due to endotoxemia, and thus sedimentation at a lower centrifugal speed. Simultaneously, less activity sediments in the lysosomal fraction. After this phase there is a significant decrease in total amount of acid phosphatase activity in the liver, parallel with the mitochondrial impairment.

The results shown in Fig. 2 suggest disappearance of lysosomal enzyme activity from the liver during endotoxemia. The lysosomes possibly rupture and release the enclosed hydrolytic enzymes, which are then washed from the liver into the circulating plasma.

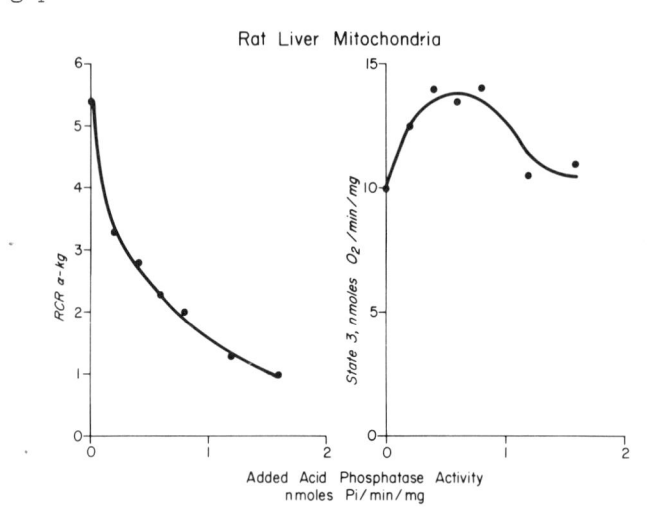

Fig. 3. <u>In vitro</u> effect of added lysosomal enzymes (activity indicated as acid phosphatase) on mitochondrial respiratory control (RCR) and State 3 O_2 uptake activity. Rat liver mitochondria were suspended at 1 mg/ml in 0.3 M mannitol-sucrose-20 mM Tris-Pi, pH 7.4 in the presence of 8 mM α-ketoglutarate as substrate and 520 μM ADP as phosphate acceptor (State 3). Broken lysosomes were added before the addition of substrate.

Assuming the release of lysosomal enzymes in the liver in vivo during endotoxemia we were concerned about their effects on mitochondrial functions, and studied this aspect in vitro. Figs 3 and 4 illustrate some results of these studies. In these experiments we used normal, isolated rat liver mitochondria, and added increasing amounts of broken lysosomes, measured as acid phosphatase activity, to the reaction mixture for the O_2 electrode and the ATPase assay. Fig. 3 shows the effect of these enzymes on the respiratory activity of the liver mitochondria using α-ketoglutarate as substrate. With increasing amounts of added lysosomal enzymes, the mitochondrial respiratory control ratios drop exponentially. This drop is due to increased State 4 respiratory activity and, thus, loose coupling, since the State 3 rates are not inhibited, as seen in the right hand diagram. The initial increase in State 3 respiratory activity is due to loose coupling, which tends to increase the ADP-induced respiratory rates slightly, to the level of the uncoupler activated respiratory rates. The fact that State 3 respiration is not inhibited by lysosomal enzymes, with any of substrates studied (α-ketoglutarate, glutamate + malate or succinate) separates the in vitro lysosomal effects from the in vivo effects of endotoxemia and hemorrhage on mitochondrial respiration into different categories. The uncoupler activated ATPase, however, as shown in Fig. 4, is inhibited by added lysosomal enzymes in vitro. The graph in Fig. 4 is a summary of 4 titrations of ATPase activity with the uncoupling agent FCCP, each point on the graph indicating the maximum ATPase activity in the presence of the amount of added lysosomal enzymes indicated on the diagram.

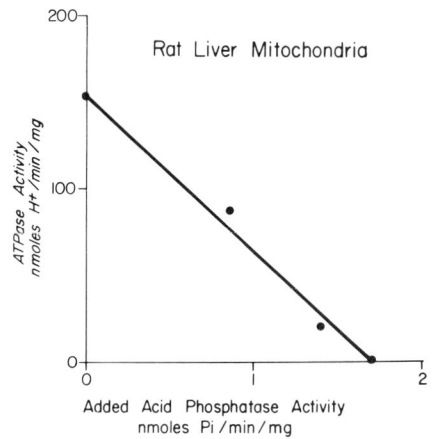

Fig. 4. In vitro effect of lysosomal enzymes on the uncoupler induced ATPase activity. Each point represents the peak of ATPase activity obtained by titration with increasing amounts of the uncoupler FCCP in the presence 1 mg/ml of mitochondrial protein 3 μM rotenone and 1.9 mM ATP in 0.3 M mannitol-sucrose, 10 mM Tris-Cl medium, pH 7.4.

Tissue hypoxia and mitochondrial function

Both endotoxemia and hemorrhage induce low tissue PO_2 values. Our surface O_2 microelectrode measurements showed that the skeletal muscle PO_2 dropped almost to zero after hemorrhage to 30 mm Hg blood pressure. Simultaneously, the liver PO_2 dropped by 50%. To induce similar levels of tissue hypoxia in vivo, in the absence of other variables such as hypovolemia or endotoxemia, we allowed anesthetized rats to breathe air with lowered O_2. Nine % O_2 in the respired air induced tissue hypoxia levels of 50-60% in skeletal muscle and 60% in the liver, as controlled by the surface O_2 microelectrodes. At various times up to 3 hr of hypoxia, liver mitochondria were separated and examined. The results on the respiratory activity are shown in Figs 5 and 6.

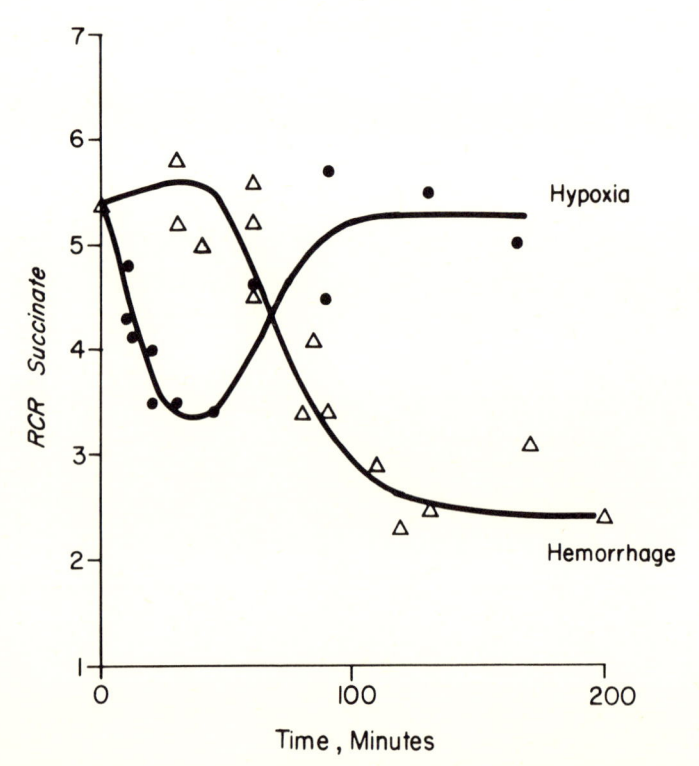

Fig. 5. Comparison of the effects of various durations of hemorrhage and hypoxia (9% oxygen) on mitochondrial respiratory control (RCR) using 8 mM succinate as substrate. Other experimental conditions were similar to Fig. 3. Each point indicates an individual animal experiment.

In Fig. 5, the effect of tissue hypoxia on mitochondrial respiratory control is shown as compared to the effect of hemorrhage. Each point indicates an individual experimental animal. Interestingly, the effects of hemorrhage and hypoxia are opposite. Within the first half an hour, hypoxia (9% O_2) induces decreased respiratory control. This is exclusively due to increased State 4 rates, because, as can be seen from Fig. 6, there is no inhibition of State 3 activity under these conditions. On the contrary, State 3 rates increased by about 100% during the first half hour of hypoxia (9% O_2). After this period, both the respiratory control and State 3 activities return to normal level and stabilize there, even after long exposure to hypoxia. It is also obvious from Fig. 6 that the response time of these mitochondrial alterations is dependent on the level of hypoxia.

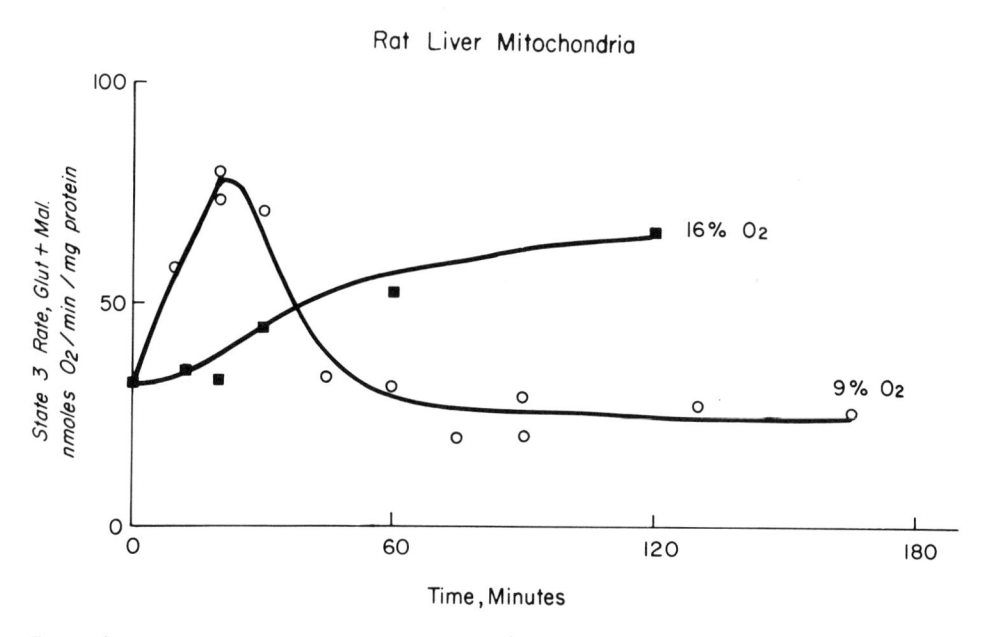

Fig. 6. Effect of tissue hypoxia (9 and 16% O_2) on mitochondrial State 3 respiratory activity using 8 mM glutamate and malate as substrates. Other conditions were similar to Fig. 5.

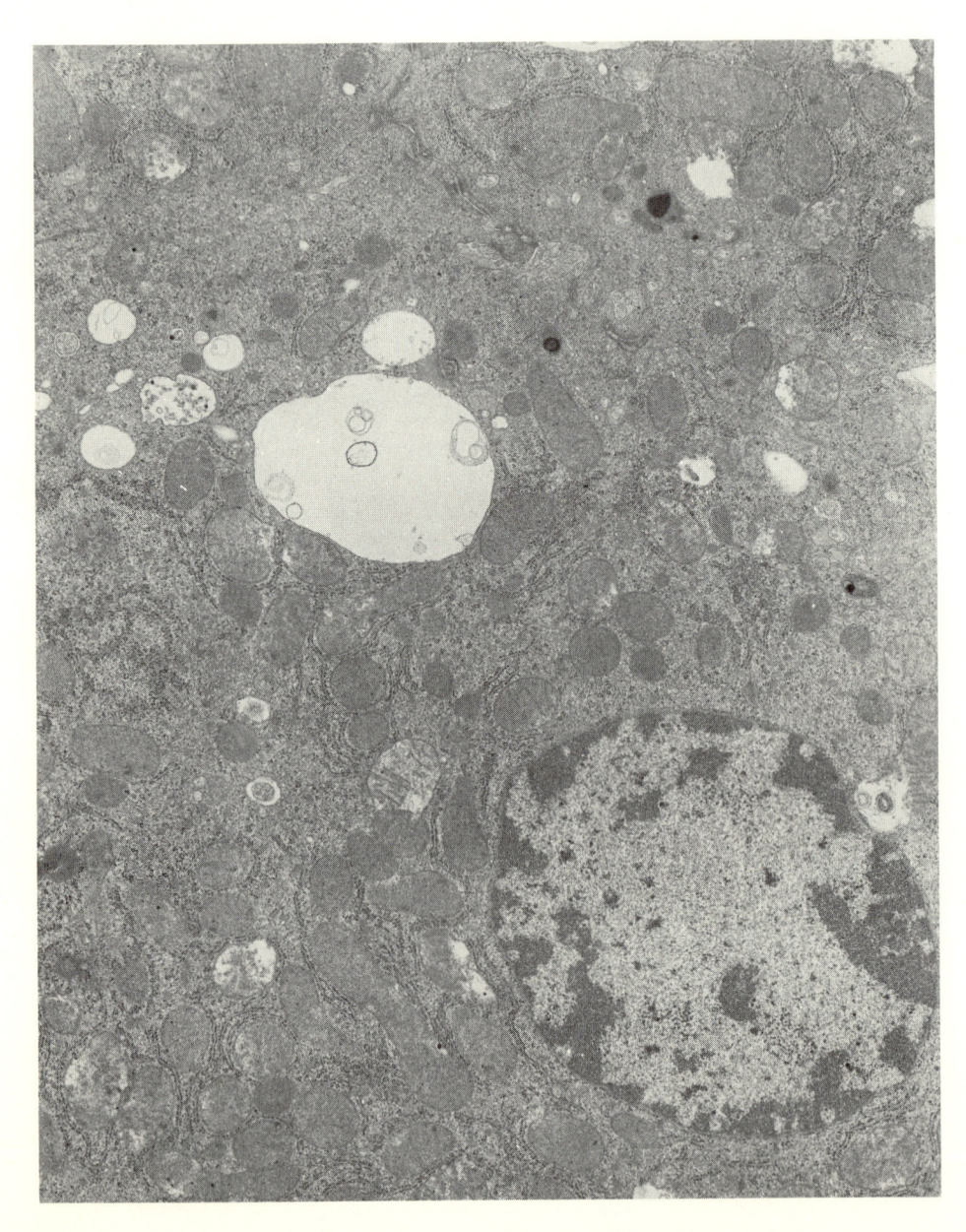

Fig. 7. Ultrastructure of the rat liver cell after 4 hr endotoxemia x 8250. (Reduced 35% for reproduction.)

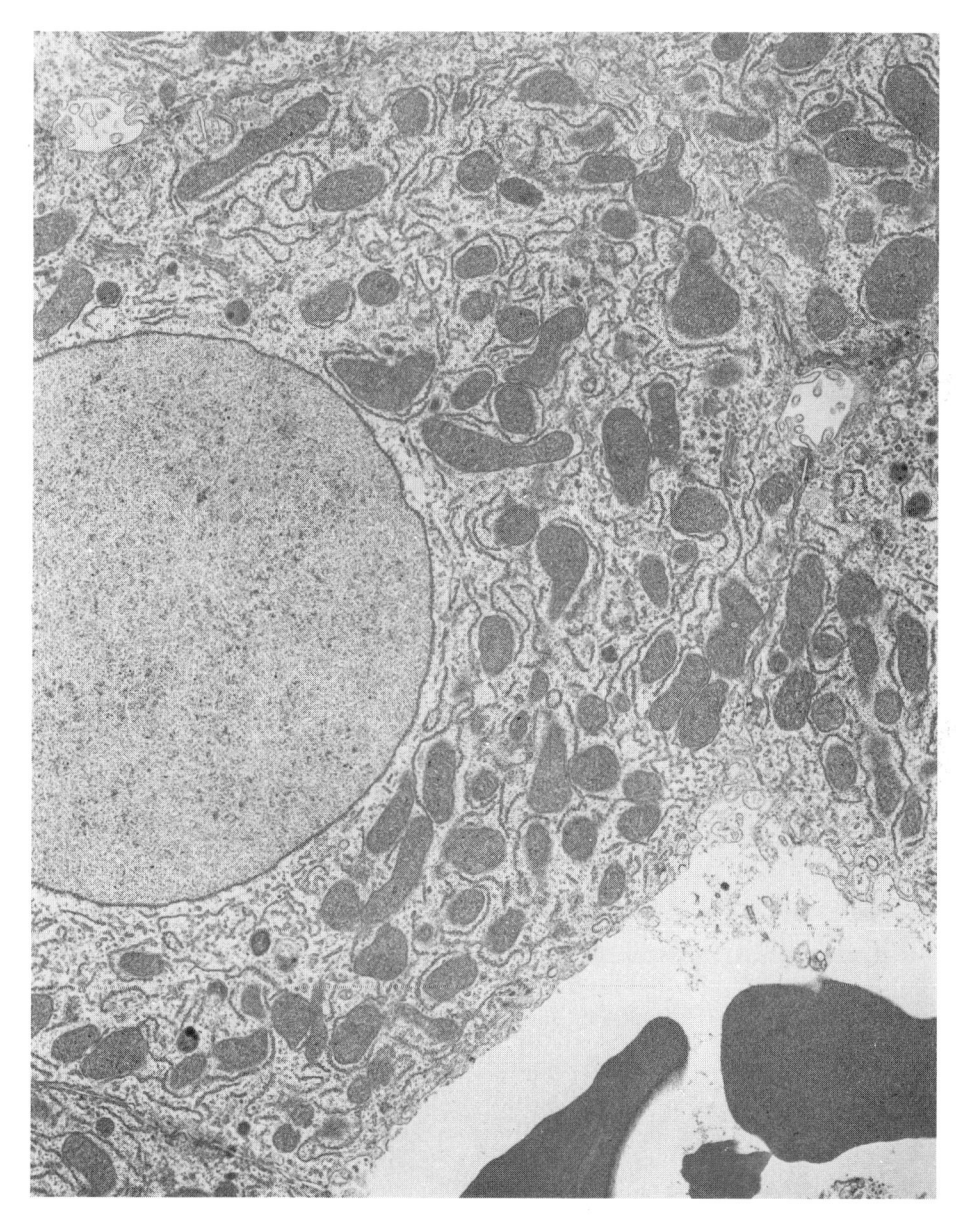

Fig. 8. Ultrastructure of the rat liver cell after 3 hr lethal tissue hypoxia (9% O$_2$) x 8250. (Reduced 35% for reproduction.)

Morphological Changes

Fig. 7 is an electronmicrograph of a rat liver 4 hr after an intraperitoneal injection of an LD$_{90}$ dose of endotoxin and shows generalized swelling of mitochondria. Many mitochondria exhibit matrical disintegration and patchy electron lucency. Also of note are several single-membrane-limited cytoplasmic vacuoles containing membranous debris. Many of these are the size of swollen mitochondria.

Fig. 8 is an electron microscopic picture, at the same magnification as Fig. 7, of a liver cell after 3 hr hypoxia (9% O$_2$). The mitochondria display normal configuration without swelling. No abnormal cytoplasmic vacuoles are seen. The cytoplasmic background, however, shows electron lucency which is caused by clusters of small vesicular blebs. Comparison of Figs 7 and 8 suggests very different morphologic responses of the liver cell to endotoxemia and hypoxia. In endotoxemia particularly the mitochondria appear damaged, while in hypoxia they preserve their intactness, although the cytoplasm exhibits diffuse swelling.

DISCUSSION AND CONCLUSIONS

Shock-induced mitochondrial alterations similar to these summarized here have also been reported by other investigators (3-5,7-8,10-12). Any intracellular etiologic factors causing these alterations have not been verified before. We believe that it is important to investigate these etiologic factors, because it is obvious that the intracellular alterations caused by endotoxemia and hemorrhage include more than just the mitochondrial damage. Other subcellular organelles are affected as well. By comparing the various subcellular alterations, it is possible to learn their relative importance.

The data reported clearly indicate that pure tissue hypoxia of the degree found during hemorrhage and endotoxemia, is not responsible for the mitochondrial damage. This is only feasible, since mitochondria are known to function normally, and even with maximal respiratory activity, at oxygen concentrations below 1µM(16). What role tissue hypoxia plays in connection with other factors that damage mitochondria, is not known.

The role of released lysosomal enzymes in damaging the mitochondria in hemorrhage and endotoxemia, is still somewhat controversial. Our data indicate that the lysosomal enzymes inhibit the uncoupler-activated ATPase as well as the cation transport activity. They also induce loose coupling of respiration and phosphorylation. However, the inhibition of State 3 respiratory activity, so clearly present in liver mitochondria isolated after hemorrhage or endotoxemia, cannot be reproduced in vitro by lysosomal enzymes. Here, it should be noted that the conditions under which these experiments were done, might not have been optimal to provide the lysosomal enzymes the intracellular environment necessary to reproduce the in vivo effects. Thus the question of any interaction between lysosomal and mitochondrial membranes during hemorrhage and endotoxemia remains open. Our experiments suggest, however, that the role of lysosomal enzymes, with other intracellular alterations, should be considered when attempting to solve the etiology of mitochondrial alterations in shock.

REFERENCES

1. Baue, A.E. (1968) Surg. Gyn. Obstet. 127:849.

2. Berry, L.J. In Microbial Toxins, Vol. V, Bacterial Endotoxins (Kadis, S., Weinbaum, G. and Ajl, S.J., Eds.) Academic Press, New York and London, 1971, p.165.

3. Schumer, W., Das Gupta, T.R., Moss, G.S. & Nyhus, L.M. (1970) Ann. Surgery 171:875.

4. Baue, A.E. & Sayeed, M.M. (1970) Surgery, 68:40.

5. DePalma, R.G., Harano, Y., Robinson, A.V. & Holden, W.D. (1970) Surg. Forum, 21:3.

6. Mela, L., Bacalzo, L.V., White, R.R. & Miller, L.D. (1970) Surg. Forum, 21:6.

7. Baue, A.E., Wurth, M.A. & Sayeed, M.M. (1970) Surg. Forum, 21:8.

8. Reed, R.C., Erve, P.R., DasGupta, T.K. & Schumer, W. (1970) Surg.Forum, 21:13.

9. Mela, L., Bacalzo, L.V. & Miller, L.D. (1971) Am.J. Physiol. 220:571.

10. Sayeed, M.M. & Baue, A.E. (1971) Am.J. Physiol. 220:1275.

11. Schumer, W., Erve, P.R. & Obernolte, R.P. (1971) Surg. Gyn. Obstet. 133:433.

12. Lavine, L., Harano, Y. & de Palma, R.G. (1971) Surg. Forum 22:5.

13. Mela, L., Olofsson, K., Miller, L.D., Bacalzo, L.V. & White, R.R. (1971) Surg. Forum, 22:19.

14. Bacalzo, L.V., Cary, A.L., Miller, L.D. & Parkins, W.M. (1971) Surgery, 70:555.

15. Mela, L., Miller, L.D., Diaco, J.F. & Sugerman, H.J. (1970) Surgery, 68:541.

16. Fiske, G.H. & Subbarow, Y. (1925) J. Biol. Chem. 66:375.

17. Silver, I.A. In: Oxygen Measurements in Blood and Tissues and Their Significance (Payne, J.P. and Hill D.W., Eds) Little, Brown and Company, Boston, 1966, p.135.

18. Cater, D.B. & Silver, I.A. In: Reference Electrodes, Theory and Practice (Ives, D.J.G. and Janz, G.J., Eds.) Academic Press, New York and London, 1961, p.464.

19. Lübbers, D.W. In: Oxygen Transport in Blood and Tissue (D.W. Lübbers, V.C. Luft, G. Theirs and E. Witzlab, Eds), Georg Thieme Verlag, Stuttgart, 1968, p.124.

Supported by the U.S. Public Health Service Grant GM-15001-05, and Career Development Award, 1 K04 GM 50318-01, for Leena M. Mela, M.D.

EFFECT OF HEMORRHAGIC SHOCK ON GLUCONEOGENESIS, OXYGEN CONSUMPTION AND REDOX STATE OF PERFUSED RAT LIVER[1]

A. G. B. Kovách[2], P. Sándor[3]

Johnson Research Foundation, Medical School, University of Pennsylvania, Philadelphia, Pa. 19104

Trauma or shock is known to produce metabolic changes, anaerobiosis, increased lactate production, a reduction in ATP stores and increase in inorganic phosphate (Kovách et al. 1952). Studies on depression of mitochondrial function in shock are inconclusive. Aldridge and Stoner (1960) found no change in the behaviour of liver mitochondria isolated from rats after limb ischemia. Levenson et al. (1961) reviewing this question concluded that no significant changes in the mitochondrial metabolism in shock could be demonstrated. Fonnesu (1960) demonstrated a partial inhibition of phosphorylation in liver of rats given S. typhi murium toxin. Hift and Strawitz (1961) found normal P/O ratios in liver mitochondria of dogs in severe hemorrhagic shock. The same observations were made by De Palma et al. (1970) in states where profound ultrastructural changes in mitochondria have been demonstrated (Holden et al. 1965; Strawitz and Hift, 1965).

Recently, Baue and Sayeed (1970, 1971), and Mela et al. (1970, 1971) reported progressive irreversible alterations in cellular and subcellular functions in the liver in hemorrhagic shock; these findings were confirmed in our laboratory (Kovách, 1969) in late terminal shock phases.

1 These studies were supported by a grant from John A. Hartford Foundation.
2 Senior Foreign Scientist of the National Science Foundation. Permanent address: Experimental Research Department, Semmelweis Medical University, Budapest, VIII. Üllői ut 78/a. Hungary.
3 Research Fellow, supported by the Heart Association of South Eastern Pennsylvania.

The question arose of the extent to which the demonstrated mitochondrial changes are only a secondary development occuring during the isolation procedure of the swollen and more sensitive mitochondria.

Because the study of metabolic control mechanisms in highly complex systems such as intact tissues has been limited in the past by the lack of suitable methods permitting continuous measurements, we adapted to our present studies the time-sharing surface fluorometer for the readout of intracellular oxidation reduction states of NADH and flavoprotein developed by Chance et al.(1969, 1971). This optical method has the distinct advantage of allowing the monitoring of cellular parameters without disruption of the tissue.

Methods and Procedure. 117 male albino rats of the Sprague-Dawley Strain, weighing 230-300 g were used. The animals were deprived of food for 24 hours before the experiment.

Hemorrhagic shock was induced during pentobarbital (50 mg/kg intraperitoneal) anesthesia, by a modified Wiggers procedure. The animals, after administration of heparin (700 U.S.P.unit/100 g) were bled through a cannula in the femoral artery into a polyethylene reservoir, until the arterial blood pressure reached 30 mmHg. The cannula was left open, and blood would be taken back or given up by the animal automatically to maintain this pressure. When the spontaneous re-uptake from the reservoir reached either 25% (early shock) or 50% (late shock) of the maximal bleeding volume, the rest of the blood was reinfused. One hour after reinfusion the liver perfusion was started. The other femoral artery was cannulated and connected via a transducer to a Grass polygraph for continuous monitoring of blood pressure.

Liver perfusion. (Fig.1). The perfusion apparatus and technique were based on those described by Scholtz and Bücher (1965). The portal vein and the vena cava inferior were cannulated and the liver was removed from the carcass. The liver, placed in a Lucite chamber which maintained the temperature at $37^{\circ}C$, was perfused with a pump (15 ml/100 g body wt/min) with an initial circulating fluid volume of 150 ml. The perfusion medium was Krebs-Henseleit bicarbonate buffer solution containing 4% bovine serum albumin (Fraction V. Sigma). The buffer was dialysed against two changes of Krebs-Henseleit bicarbonate solution and filtered through Millipore filters (0.45 µ porosity) before use. After saturation with an (95:5) O_2-CO_2 mixture, the pH was adjusted to 7.40.

Substrate for gluconeogenesis (lactate 10 mM) was added to the medium after perfusion was started or infused continuously at a rate sufficient to maintain an approximately constant concentration during perfusion. The adequacy of perfusion was judged from the maintenance of approximately linear rates of glucose production,

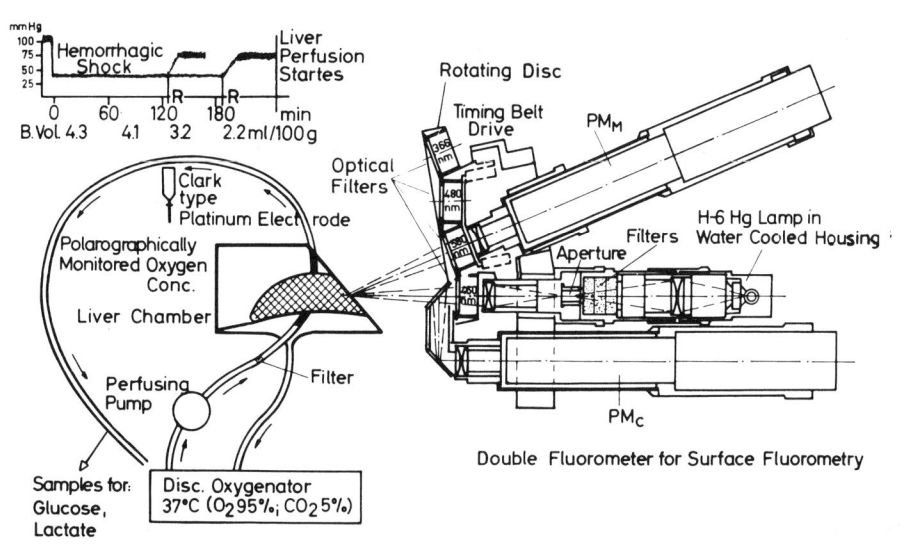

Fig. 1.: Rat liver perfusion. Perfusion fluid: 150 ml Krebs-Henseleit buffer, pH 7.40, containing 6 g Bovine serum albumin (Fraction V, Sigma; twice dialysed.

the physical appearance of the liver, and measurements of oxygen tension of the fluid leaving the liver.

Analytical techniques. Samples of perfusion medium (2 ml) were taken at 10 or 15 min intervals. Glucose was measured by glucose oxidase (Hugget, Nixon, 1957), lactate by lactate dehydrogenase (Bergmeyer, 1965).

Surface fluorometery. Chance et al. (1971) developed a double fluorometer for the simultaneous measurement of flavin and pyridine nucleotide fluorescence in suspensions of isolated mitochondria and for oxidation-reduction changes in perfused rat liver (Fig.1.)

It was shown that changes in fluorescence intensity excited on the liver surface by 366 nm light represent oxidation-reduction changes in both mitochondrial and extramitochondrial NAD system. A yellow fluorescence excited at 436 nm is predominantly due to certain oxidized mitochondrial flavoproteins.

RESULTS

Gluconeogenesis. A comparison of the rates of glucose production from lactate in perfused livers from normal, hemorrhagic shock (early and late), phenoxybenzamine (PBZ) (20 mg/kg i.p.)-treated normal,

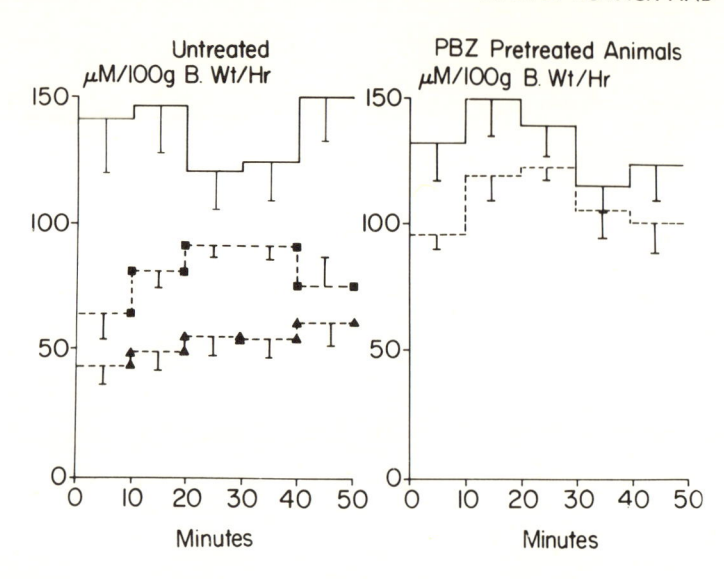

Fig.2.: Glucose production by perfused rat livers. ──── control, ■-----■ early shock (25% reuptake) ▲-----▲ late shock (50% reuptake), ----- 20 mg/kg i.p. phenoxybenzamine treated and bled for the same time period as the late shock group. The liver perfusion was started one hour after reinfusion of the shed blood.

and PBZ-treated hemorrhagic shock rats is shown in Fig.2. The rate of gluconeogenesis was 140+15 µM/100 g body wt./hr in normal animals without or with PBZ treatment. Both in the early shock (25% spontaneous reuptake) and in the late shock (50% reuptake) group the rate of gluconeogenesis was significantly reduced. In contrast the PBZ-pretreated and bled group (hypotension maintained at 30 mmHg for 190 min, equal to the late shock group) after the first 10 minute period was not significantly different from the normal PBZ-treated group. In all groups the rate of removal of lactate from the medium corresponded to the gluconeogenesis (glucose/lactate ratio of 1:2).

Oxygen Consumption. The oxygen uptake by perfused rat livers is demonstrated in Fig.3. In normal rats the oxygen consumption was 1322+32 µA/100 g body wt./hr Phenoxybenzamine had no effect on the control values. Early hemorrhagic shock did not decrease the oxygen consumption in perfused livers. In the late shock group one hour after retransfusion (preterminal phase) the oxygen consumption was significantly lowered 687+59 µA/100 g body wt./hr In the PBZ-pretreated group, when the rats were bled according to the late shock procedure, the oxygen consumption remained at the control level.

Fig.3.: Oxygen uptake by perfused rat livers. ——— control,
■ ----- ■ early shock, ▲ ------ ▲ late shock, ----- 20 mg/kg i.v.
PBZ-treated group. Bleeding lasted in the PBZ group for the same
time period as the late shock group.

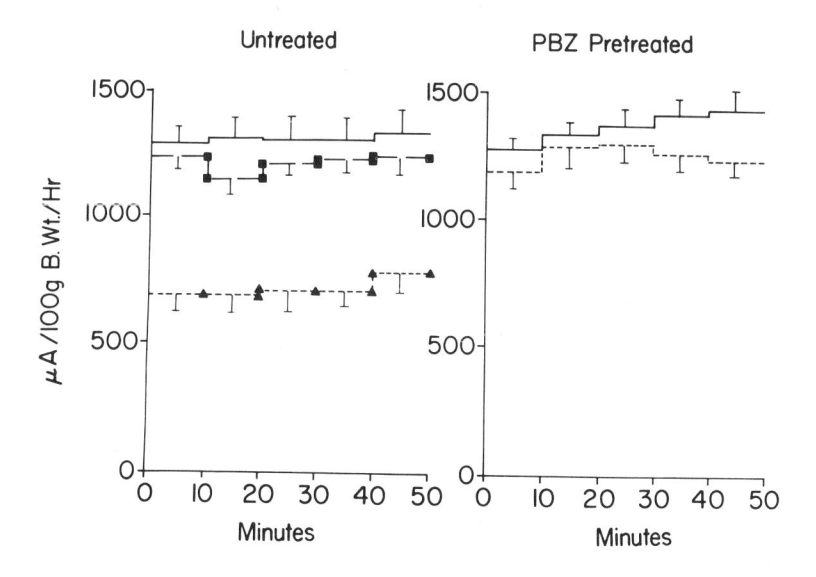

Oxidation-Reduction Changes of Flavin and Pyridine Nucleotides.
Reversible intracellular oxidation-reduction changes were obtained
when the perfused liver was subjected to a cycle of anoxia (see Fig.
4). After oxygen withdrawal the intensity of fluorescence excited
at 436 nm decreased, while that excited at 366 nm increased, representing a reduction of both flavin and pyridine nucleotides. The
changes indicate the maximal oxidation-reduction shift possible
from the aerobic steady state level. They were reversible after
restoration of the oxygen supply, indicating a reoxidation of the
coenzymes. 20 mM DNP increased oxygen consumption from 1346 to
1750 μA/100 g body wt./hr at the same time the pyridine and
flavoprotein steady states were reduced.

The flavin and pyridine nucleotide oxidation-reduction changes
of the perfused liver from previously shocked rats were considerably
different from the control, as can be seen in Fig.5. The anoxic reduction cycle was much smaller and the reoxidation kinetics after
oxygen supply were slower in the case of livers from shocked animals.
The pyridine nucleotide reduction cycle after anoxia was earlier
affected in shock than was the flavoprotein cycle. In later terminal shock states the pyridine nucleotide and flavoprotein of the
liver were converted almost completely to the reduced form, and
anoxia could reduce them no further.(Fig.6). The ATP content of

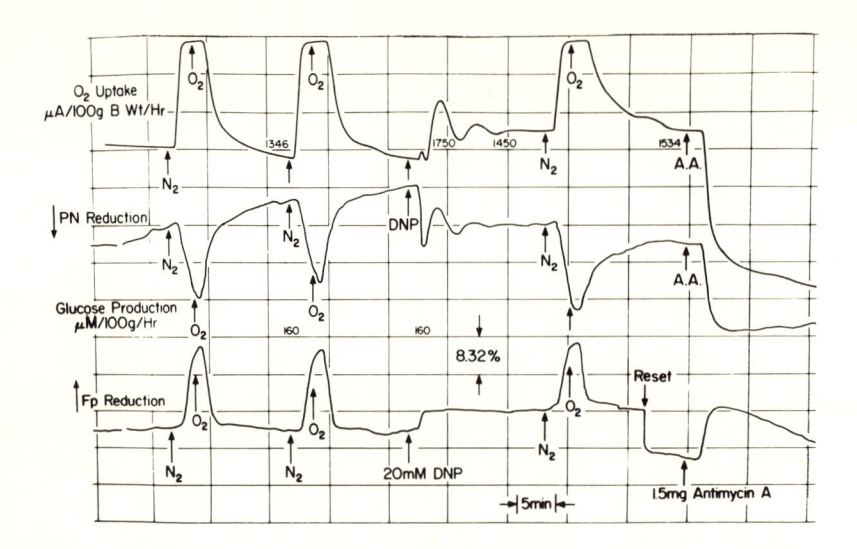

Fig. 4.: Oxidation-reduction changes of flavin and pyridine nucleotides in control perfused rat liver. O_2 uptake in $\mu A/100$ g body wt./hr Glucose production in $\mu M/100$ g/hr 3 minutes anoxic cycles as indicated with N_2.

the liver tissue in these cases was reduced significantly to very low levels.

Conclusion. Our results demonstrated that gluconeogenesis in the perfused liver from hemorrhagic shocked rats is significantly reduced in early shock states. According to the results of Williamson et al. (1970), gluconeogenesis of endotoxin shock rat liver is different from that of our hemorrhagic shock animals. Observation of the redox state of flavoprotein and pyridine nucleotides with surface fluorometry show that the steady state oxidation of pyridine nucleotides is partially reduced before any change occurs in flavoprotein oxidation-reduction steady state. These results suggest that cytosolic redox changes develop earlier than the mitochondrial defects. As shock progresses the steady state reduction gradually involves both flavoprotein, as well as pyridine nucleotide. These reductions can be demonstrated in states where the perfused liver oxygen consumption is normal. This suggests that the respiration is uncoupled. There is a fall in the ATP/ADP ratio.

According to the present results, progressive alterations in cellular and subcellular metabolic functions in the liver develop in vivo in late hemorrhagic shock. The cytosolic changes are demonstrated earlier than the mitochondrial defects. Pyridine nucleotide and flavoprotein are reduced because ATP potential is low;

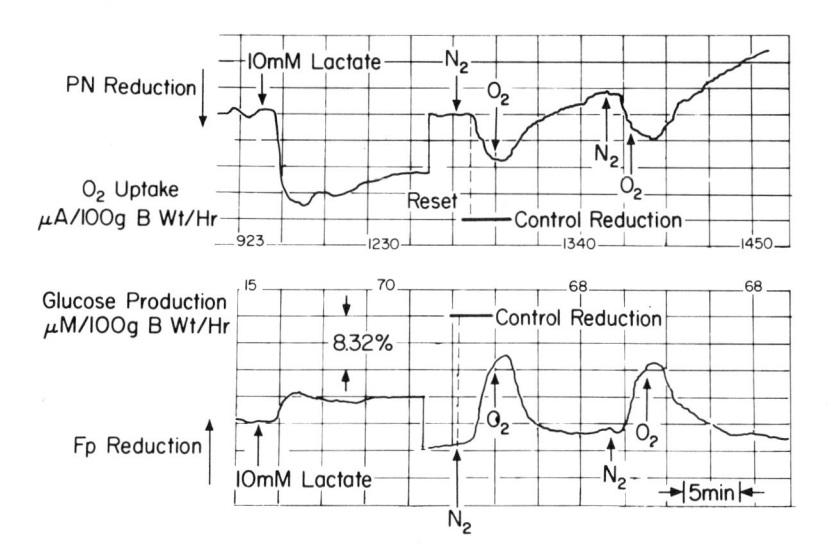

Fig. 5.: Oxidation-reduction changes of flavin and pyridine nucleotides in perfused rat liver. The value of the anoxyc control cycles is also indicated ─── on the figure. The rat was in early shock state (one hour after reinfusion of the shed blood) at the time liver perfusion started.

the ATP is not resynthetized in states when oxygen consumption is still maintained.

The protective action of phenoxybenzamine against the alterations in the perfused shock liver may be due to the metabolic (Kovách et al. 1971) as well as circulatory effects of the drug.

Discussing the mechanism of the metabolic defects in the liver the role of microcirculatory alteration leading to tissue hypoxia in the terminal stage of shock can not be excluded. The fact that the perfused late shock liver still uses oxygen at a normal rate does not mean that before the perfusion started in vivo it was also well supplied with oxygen. How far these results are causal producing the irreversibility or secondary terminal consequences has to be further investigated.

The appearance of these alterations in other forms of injury could give the answer to its general importance in the development of irreversible shock.

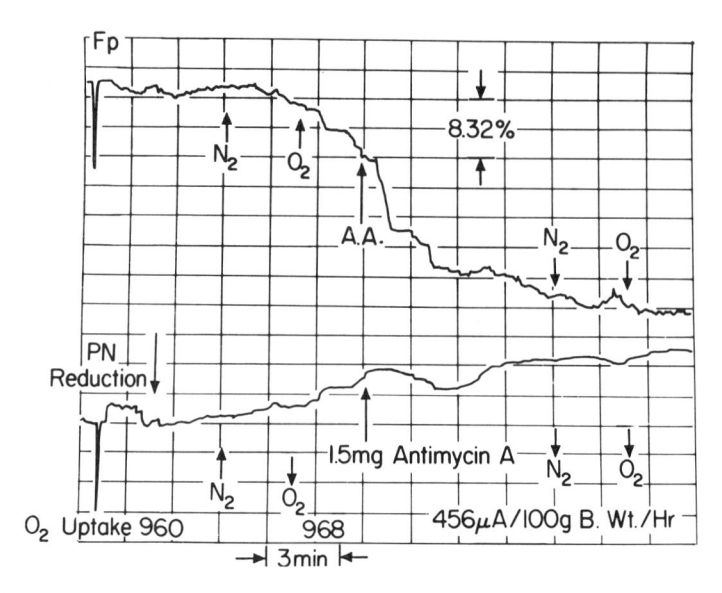

Fig 6: **Oxidation reduction changes of flavin and pyridine nucleotides in perfused rat liver.** The rat was in late shock (terminal phase) at the time starting liver perfusion. N_2 perfusion is not producing further reduction in flavin and pyridine nucleotides. The oxygen uptake is still 960 µA/100 g body wt./hr.

REFERENCES

Alldridge, W.N. & Stoner, H.B. (1960) Biochem J. 74:148.

Baue, A.E.& Sayeed, M.M. (1970) Surgery 68:40.

Baue, A.E., Sayeed, M.M., Schieber, R., Plauer, I. & Wurth, A.A. (1971) Abstract. Amer. Physiol. Soc. Fed. Meeting Chicago. Fed. Proc. 30:261.

Bergmeyer, H.V. ed. (1965) Methods of Enzymatic Analysis 2nd ed. Acad. Press. New York. p.253.

Chance, B., Cohen, P., Jobsis, F., Schoener, B. (1962) Science, 137:499.

Chance, B., Graham, N. & Mayer, D. (1971) Rev. Sci. Instr. 42:951.

De Palma, R.G., Levey, S. & Holden, W.D. (1970) J. Trauma, 10:122.

Fonnesu, A. (1960) In: The Biochemical Response to Injury. Springfield, Charles C. Thomas. p. 85.

Hift, H. & Strawitz, I.G. (1961) Am. J. Physiol. 200:264.

Holden, W.D., De Palma, R.G., Drucker, W.R. & McKalen, A. (1965). Ann. Surg. 162:517.

Hugget, A. St.G. & Nixon, D.A. (1957) Lancet, 273:368.

Kovách, A.G.B., Bagdy, D., Balázs, R., Antoni, F., Gergely, J., Menyhárt, J., Irányi, M. & Kovách, E. (1952) Acta physiol. Acad. Sci. hung. 3:331.

Kovách, A.G.B., Chance, B. & Salkovitz, I. (1971) Amer. Physiol. Soc. Fed. Meeting Chicago. Fed. Proc. 30:267. Abstr. 412.

Kovách, A.G.B., Koltay, E., Fonyó, A. & Kovách, E. (1971) Biochem. Med. 5:384.

Kovách, A.G.B., Kovách, E. (1969) Unpublished results.

Levenson, S.M., Einheber, A. & Malm, O.I. (1961) Fed. Proc, 20. Suppl. 9. 99.

Mela, L., Bacalzo, L.V. Jr., White, R.R. & Miller, L.D. (1970) Surg. Forum. 21:6.

Mela, L., Bacalzo, L.V. Jr. & Miller, L.D. (1971) Am. J. Physiol. 220:511.

Scholz, R. & Bücher, T. (1965) In: Control of Energy Metabolism. Ed. B. Chance, R.W. Estabrook and J.R. Williamson. New York, Acad. Press. p.393.

Strawitz, I.G. & Hift, H. (1965) In: Shock and Hypotension. Ed. L.C. Mills and H.J. Mayer. New York, Grune and Stratton Inc. p. 637.

Williamson, J.R., Refino, C. & La Noue, K. (1970) In: Energy Metabolism in Trauma. Ciba Foundation Symposium, Ed. R. Porter and J. Knight, London, Churchill. p. 145.

POTENTIAL RELATIONSHIPS OF CHANGES IN CELL

TRANSPORT AND METABOLISM IN SHOCK

A. E. Baue, M. M. Sayeed, and M. A. Wurth

Department of Surgery, Washington University School
of Medicine and The Jewish Hospital of St. Louis,
St. Louis, Missouri

INTRODUCTION

Circulatory failure or shock is known to produce metabolic changes, including anaerobiosis with increased lactate production, a reduction in ATP stores, and an increase in inorganic phosphate. These changes indicate only a lack of availability of oxygen with depression of the Kreb's cycle and electron transport system which should be reversible if blood flow or oxygen supply is restored. Ultimately, however, there must be not only a depression of cell function with shock but alterations and aberrations which may limit or destroy the cell's capabilities even if its environment is improved. Early studies of mitochondrial metabolism in shock were reviewed by Levinson et al. in 1961,[8] and they concluded that no significant changes in these cell systems had been found. Hift and Strawitz[6] found normal P/O ratios in liver mitochondria of dogs subjected to severe hemorrhagic shock, although oxygen uptake was found to be higher than normal. DePalma et al.[4] also demonstrated a lack of alteration of the P/O ratio of liver mitochondria with severe hemorrhagic shock using succinate as a substrate. Exceptions to this have been the demonstration by Fonnesu[5] of partial inhibition of phosphorylation in livers of rats given F-typhimurum toxin, by Moss et al.[10] of a depression of respiration in rat liver mitochondria exposed to E. coli endotoxin, and by Strawitz and Hift[15] of lowered oxidative phosphorylation in heart mitochondria from dogs in hemorrhagic shock. In addition, profound ultrastructural changes in cells and in mitochondria have been demonstrated after prolonged hemorrhagic shock by Holden and associates.[7] We have studied respiration and ion contents of mitochondria and the Na-K transport enzyme system to determine if abnormalities in such cell function would offer insight into the pathophysiology of circulatory failure and better approaches to treatment.[1,2,12,13,14]

METHODS

A standard hemorrhagic shock model was developed in albino rats of the Holtzman strain which has been used for all studies. The femoral arteries of ether anesthetized rats were cannulated and the animals were allowed to awaken. The cannula in one artery was allowed to bleed into a reservoir; arterial pressure was measured with the other cannula. Animals were bled to a mean arterial blood pressure of 40 mm Hg which was maintained at that level by removing or giving blood from the reservoir as needed. The time when blood had to be returned to the animal from the reservoir to maintain this pressure was called "early shock." The time when 70 per cent of the blood was returned to maintain this pressure was called "late shock." All studies were carried out with paired animals, one animal being bled and the other a control which was prepared in the same way but not bled. At appropriate times the animals were sacrificed by decapitation. Organs were removed and homogenized in a chilled medium containing 225 mM sucrose, 10 mM tris(hydroxymethyl)aminomethane (Tris), pH 7.3, and 1 mM ethylenediaminetetraacetic acid (EDTA). Mitochondria and microsomal cell fraction were prepared from these homogenates. Oxygen uptake by mitochondria were measured by the polarographic method in a sucrose-Tris-EDTA medium with substrate [3.3 mM succinate, β-hydroxybutyrate (β-Hb) or α-ketoglutarate (α-Kg)] and with or without added 1 mg/ml bovine serum albumin (BSA). The oxygen measurement was first made in the presence of substrate alone and was referred to as R_1, and then in the presence of substrate plus 0.16 mM adenosine diphosphate, ADP, (R_2). R_2 was measured until it returned to a near-R_1 level. The respiratory control ratio (RCR) was determined by dividing R_2 into R_1. Paired determinations of respiratory activity of liver mitochondria with and without added BSA (1 mg/ml), or diphosphopyridine nucleotide (DPN) (0.066 mM) or Mg (6 mM) were made from control and late shock animals. Cation contents including Na, K, Mg and Ca were analyzed in ashed liver mitochondria by using atomic absorption spectroscopy. Respiratory activity of approximately 0.5 mm thick liver slices were studied also. Slices were incubated at 28°C in an air-saturated 250 mM sucrose and 10 mM Tris medium with 10 mM succinate or α-Kg and oxygen uptake measured by the polarographic method.

Adenosine triphosphatase (ATPase) activities were measured in the microsomal fractions of organ homogenates. The enzyme activity was assayed by quantitating hydrogen ion (H^+) release from ATP hydrolysis in a radiometer auto-titrator operated in the pH-stat mode. The reaction mixture for the determination of total ATPase activity consisted of 5 mM MgCl, 20 mM KCl, 120 mM NaCl, and 3 mM adenosine triphosphate (ATP), pH 7.45 and enzyme preparation containing approximately 1 mg protein. The Mg-ATPase activity was determined in the reaction mixture from which KCl and NaCl were omitted. (Na + K)-ATPase activity was determined by subtracting the Mg-ATPase activity from the total ATPase activity.

PROCEDURE AND RESULTS

Initial studies were carried out in animals some time after death. Mitochondrial metabolism with succinate was shown to continue at a fairly normal level for a prolonged period after death but with α-Kg there was early depression of respiration after death. The α-Kg system was much more sensitive and susceptible to injury than was the succinate system. A comparison was then made of mitochondrial metabolism with succinate, α-Kg, and β-Hb in hepatic mitochondria from animals in late shock. The results are shown in Table 1 and indicate a profound decrease in the capability of mitochondria to metabolize α-Kg in late shock. There was also a large decrease in capability to utilize β-Hb. The decrease with succinate was significant but was small. This indicates again the stability of the succinate system. The α-Kg system and β-Hb are linked to the electron transport system through DPN and the former is a more complex enzyme system[11] which seems to be more susceptible to this type of injury. To determine if mitochondrial changes were progressive as shock continued, animals were sacrificed and mitochondria studied at various periods beginning with early shock to late shock.

TABLE 1. Respiration in rat liver mitochondria in prolonged shock

	SUCCINATE CONTROL	SUCCINATE SHOCK	β-Hb CONTROL	β-Hb SHOCK	α-Kg CONTROL	α-Kg SHOCK
R_1	30.30 ±2.04	32.98 ±1.54	9.42 ±0.72	9.49 ±0.65	8.99 ±0.57	4.33* ±0.67
R_2	75.23 ±4.74	67.99 ±3.71	30.99 ±2.40	21.61* ±1.49	39.51 ±2.49	7.69* ±1.34
RCR	2.36 ±0.07	2.07 ±0.08	3.25 ±0.10	2.27* ±0.05	4.18 ±0.16	1.74* ±0.09

All values are mean (±SEM) of 10 or more animals except those of β-Hb control group which are means of 7 animals.
*Values were significantly different from control ($p < .01$).

TABLE 2. α-Kg respiration in liver mitochondria at various stages of shock and effect of infusions on respiration in early shock

	CONTROL (N=39)	EARLY SHOCK I (N=6)	EARLY SHOCK II (N=6)	EARLY SHOCK III (N=6)	LATE SHOCK (N=38)
R_1	11.13 ±0.43	12.06 ±1.89	14.72 ±0.66	14.56 ±1.68	5.79 ±0.51
R_2	42.88 ±1.36	29.84 ±6.80	38.23 ±2.95	52.02 ±3.92	13.48 ±1.70
RCR	3.88 ±0.09	2.41 ±0.21	2.59 ±0.11	3.65 ±0.16	2.15 ±0.09

Values are means (±SEM); N - Number of animals. I. Before infusion; II. After infusion of blood; III. After infusion of blood plus Ringer's lactate.

Table 2 shows that these changes were, indeed, progressive as shock continued. Animals in early shock with the defect in the α-Kg metabolism were treated at this stage by return of the shed blood and infusion of an additional volume of Ringer's lactate solution. It was found that the previously altered function returned toward a normal 30 to 60 minutes after such a treatment program. (Table 2)

The decrease in metabolic capability of mitochondria could be due totally or in part to the release within the cell of free fatty acids which impair mitochondrial metabolism.[3] To investigate this, mitochondria were studied with and without addition of BSA which is known to bind free fatty acids. Mitochondria from both control and shock animals showed better capabilities with albumin present, but the defect produced by shock was not totally corrected.[12] Free fatty acids may play some role in metabolic deterioration, but did not seem to be a very important factor. We questioned whether some of the changes found in these studies may have been produced in vitro in cells from shocked animals during the preparation of mitochondria rather than in vivo during shock. A study was carried out in which oxygen consumption was measured with succinate or α-Kg in liver slices taken from control animals and from animals in late shock. The results shown in Fig. 1 support the concept that the changes that we were measuring were taking place in vivo during shock.

Possible alterations in the ion transport properties of cells with shock were studied by quantitating the (Na + K)-activated ATPase activity of the microsomal fractions of cells of various organs as this enzyme is thought to be an important part of the

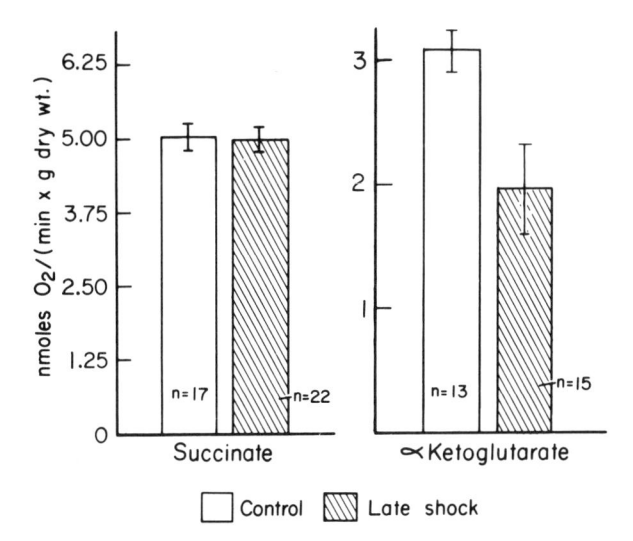

Fig. 1. Respiration by liver slices.
N = number of animals. Vertical lines
at the top of bars represent SEM.

mechanism for the transport of Na and K across cell membrane. The (Na + K)-ATPase was found to increase from 10.6 ± 1.0 (SEM) nmoles/(mg protein x min) in control liver to 19.4 ± 2.3 in early shock animal liver and to 41.0 ± 2.0 in late shock. The progressive increase in the transport enzyme activity occurred concomitantly with the metabolic deterioration of mitochondria. An increase in (Na + K)-ATPase activity can be associated with an increase in intracellular Na which in turn might alter intramitochondrial Na. The possibility of mitochondrial ion changes in shock was investigated by analyzing Na, K, Mg, and Ca contents in this organelle in early and late shock. It can be seen that when mitochondrial capability was beginning to decrease, but was reversible, Na had already increased significantly in mitochondria (Table 3). In late shock, Na in mitochondria was even higher, K had decreased and Ca had increased. There were no changes in Mg when measurements were made without added EDTA. Thus cation shifts in mitochondria evident in early shock and the (Na + K)-ATPase increase indicate early and progressive changes in membrane permeability in shock.

To determine if cell function was altered in other organs with late shock, mitochondrial capability and (Na + K)-ATPase were studied in the lung and brain of rats in shock and the results are shown in Table 4. The previous changes found in the liver were not found in the lung and in the brain. This suggests that nutrient

TABLE 3. Cation contents of liver mitochondria at various stages of shock

	CONTROL (N=15)	EARLY SHOCK (N=7)	LATE SHOCK (N-11)
Na	20.18 ±1.84	42.01 ±4.09	48.22 ±2.12
K	109.10 ±2.92	137.22 ±8.54	74.15 ±5.63
Mg	30.91 ±1.76	47.17 ±4.23	31.77 ±1.53
Ca	5.89 ±0.27	6.79 ±0.62	9.92 ±0.77

Mean values (±SEM) have units of nmoles/mg protein.

flow to these organ systems was better maintained even in late shock than in the splanchnic bed.

Because of the potential relationship of DPN to the abnormalities found in mitochondria in shock, studies were carried out to determine the effect of DPN addition on β-Hb respiration in rat liver mitochondria. Very little change was found with the addition of DPN.[12] The effect of the addition of Mg on β-Hb respiration was also studied in liver mitochondria from late shock animals. There was some improvement of activity with Mg additions but the improvement of substrate-plus-ADP respiration was not found (Table 5).

TABLE 4. Mitochondrial respiration and microsomal (Na + K)-ATPase activities in lung and brain.

			LUNG CONTROL	LUNG LATE SHOCK	BRAIN CONTROL	BRAIN LATE SHOCK
Mitochondrial	R_1-	Kg	10.31 ±0.82	9.83 ±0.61	5.43 ±0.49	5.69 ±0.47
	R_2-	Kg	29.75 ±3.63	24.22 ±2.34	13.27 ±1.66	13.64 ±1.43
Microsomal	(Na+K)-ATPase		159.00 ±18.00	152.00 ±19.00	898.00 ±24.00	865.00 ±73.00

Values (±SEM) are means of 5 or more animals.

TABLE 5. Effect of added magnesium on β-hydroxybutyrate respiration in liver mitochondria

	CONTROL		LATE SHOCK	
	- Mg	+ Mg	- Mg	+ Mg
R_1	6.52 ±0.39	5.58 ±0.44	5.41 ±1.58	3.82 ±0.32
R_2	23.57 ±1.24	29.42 ±1.87	16.55 ±0.98	18.53 ±1.34
RCR	3.60 ±0.21	4.90 ±0.21	2.99 ±0.08	4.74 ±0.30

Values are means (±SEM) of 10 or more animals.

By electronmicroscopy isolated mitochnodria from the liver of animals in late shock were found to be swollen when compared with similar preparations of mitocondria from control animals (Fig. 2). Mitochondria from control animals exhibited the condensed structure which is typical for active mitochondria.

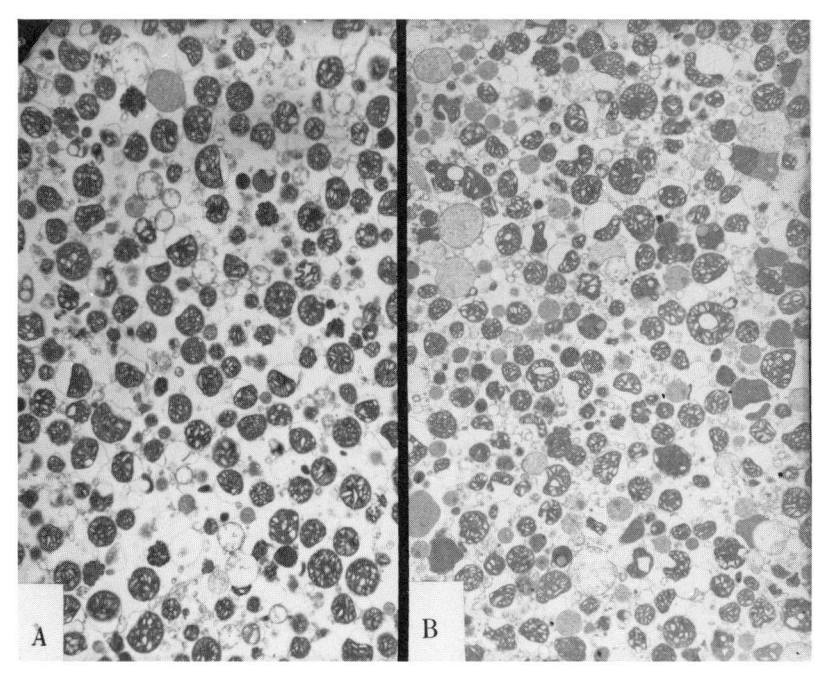

Fig. 2. Electron micrographs of liver mitochondrial pellets from a control animal (a) and a late shock animal (b).

Mitochondria isolated from shock livers were generally less dense and a small proportion of those mitochondria appeared to have undergone a disruption of intromitochondrial structures (cristae). These observations in late shock are in agreement with those of DePalma et al.[4]

DISCUSSION

These studies have indicated clearly that the metabolism of liver mitochondria was adversely and progressively affected by hemorrhagic shock. Such changes were not found in the lung or brain. A more pronounced inhibition of respiratory activity was obtained with the DPN-linked substrates than with succinate. This finding is compatible with present knowledge of the differential stabilities of the succinate dehydrogenase enzyme systems within mitochondria. The former enzyme is known to be membrane-bound and thus more resistant and less likely to be damaged under the altered conditions prevailing in tissue during shock. Since more than 80 per cent of ATP is synthesized by normal mitochondria from the metabolism of DPN-linked substrates, a block in the metabolism of these substrates would produce a significant impairment of the capability of liver mitochondria to synthesize ATP. The exact magnitude of this mitochondrial defect cannot be ascertained at the present time. The failure of α-Kg metabolism of liver mitochondria of shocked animals could also produce a block in amino acid metabolism.

Studies by other investigators have indicated that oxidized DPN in liver mitochondria was converted almost completely to the reduced from in late shages of shock.[9] This could explain the respiratory change in shock with the DPN-linked substrates. However, in our studies added DPN did not restore the respiratory activities to normal. There was also no significant effect of added albumin on the respiratory activities of liver mitochondria in late shock, suggesting that free fatty acids were not a major problem. Mg is another substance which if deficient produces derangements in functional capability of mitochondria. It is known that Mg is required for the synthesis of mitochondrial cofactors and for maximal activity of oxidative phosphorylation, particularly for α-Kg. The results presented here show that the addition of exogenous Mg to shock liver mitochondria improved function but did not restore the mitochondrial activities completely to normal. An alteration of mitochondrial membrane permeability is indicated from the data on Ca, K, and Na contents of these organelles in shock. The mitochondrial ion changes in all probability reflect alterations in the intracellular ionic millieu. The increase in mitochondrial Na, for example, suggests an increase in cell Na, and the decrease in K seems related to a loss of cell K. The increase in Mg content in early shock may well be related to the increased respiratory activity with succinate observed in this stage of shock. The increase in Ca in late shock may be associated with decreased mitochondrial metabolism. The increase in (Na + K)-ATPase activity with shock indicates an in vivo activation of this transport enzyme to above normal levels. Such activation may result from an increased intracellular availability of one or both of the reactant substances, ATP and Na. Increased ATP availability seems improbable with the hypoxic conditions which are found in shock. An increase in intracellular Na consequent to an initially depressed activity of the Na pump mechanism or to an increased passive entry of Na into cells, appears to be a more plausible explanation of the increased (Na + K)-ATPase activity.

From this series of studies we have concluded that a sequence of events occurs with the hypoxic hypoperfusion of shock which may be initiated by changes in cell membrane. This could be by a reduction in cell membrane potential with sodium entering cells which activates the (Na + K)-activated ATPase system. This system, may not be able to keep up with the increased entry of Na. Such alterations in cationic concentrations in the cell would also involve mitochondria with progressive alterations in Na, K and Ca contents. This could then lead to depression of the metabolic capability of mitochondria. Other changes within the cell could be involved in this process, including lysosomal disruption and the effects of lysosomal enzymes on the mitochondria and cell membranes. Whether or not studies such as these will eventually lead to more specific approaches to the treatment of shock remains to be determined.

SUMMARY

Progressive alterations in cellular and subcellular function in the liver were found to be produced in vivo by hemorrhagic shock. These changes were reversible at least in early shock. There was limitation of the metabolic capability of mitochondria which was most marked in the DPN-linked reactions of the electron transport system. There were marked alterations in the enzyme system thought to be involved in Na and K transport and this seemed to be related to alterations in the cation content of mitochondria with a decrease in K, and increase in Na. From this work an hypothesis has been developed of the events that occur in hepatic cells in shock. Shock may produce an initial change in membrane permeability both of the cell and mitochondria. This results in cation alterations and swelling which lead to a compensatory increase in ATPase activity and results in depression of mitochondrial metabolic capability.

(This study was supported by NIH Grant HE-12278 and U. S. Army Contract DADA-17-69-9165.)

REFERENCES

1. Baue, A. E. and Sayeed, M. M. Surgery 68: 40-47, 1970.
2. Baue, A. E., Sayeed, M. M., Schieber, R., Planer, J., and Wurth, M. A. (Abstract) Fed. Proc. 30: 261, 1971.
3. Boime, I., Smith, E. E., and Hunter, F. E., Jr. Arh. Biochem. Biophys. 128: 704-715, 1968.
4. DePalma, R. G., Levey, S., and Holden, W. D. J. Trauma 10: 122-134, 1970.
5. Fonnesu, A., Changes in energy transformation as an early response to cell injury. Springfield, Charles C. Thomas, 1960, p. 85.
6. Hift, H. and Strawitz, J. G. Am. J. Physiol. 200: 264-267, 1961.
7. Holden, W. D., DePalma, R. G., Drucker, W. R., and McKalen, A. Ann. Surg. 162: 517, 1965.
8. Levenson, S. M., Einheber, A., and Malm, O. J. Fed. Proc. 20, Suppl. 9: 99-119, 1961.
9. Loiselle, J. M. and Denstedt, O. F. Can. J. Biochem. 42: 21-34, 1964.
10. Moss, G. S., Erve, P. P., and Schumer, W. Surg. Forum XX: 24-25, 1969.
11. Reed, L. J. and Cox, D. M. Ann. Rev. Biochem. 35: 57-85, 1966.
12. Sayeed, M. M. and Baue, A. E. Am. J. Physiol. 220: 1275-81, 1971.
13. Sayeed, M. M., Baue, A. E., and Planer, J. (Abstract) Proceedings of the 25th International Congress of Physiological Sciences, 9: 495, 1971.
14. Sayeed, M. M., Wurth, M., Planer, J., and Baue, A. E. (Abstract) Fed. Proc. 29: 712, 1970.
15. Strawitz, J. G. and Hift, H. Shock and Hypotension, edited by L. C. Mills and H. J. Moyer, New York, Grune & Stratton, Inc., 1965, p. 637.

DEXAMETHASONE (DXM) EFFECT ON THE HISTAMINE-RELEASING ACTIVITY OF ENDOTOXIN

William Schumer, R.P. Obernolte and P.R. Erve

Department of Surgery, University of Illinois College of Medicine at Veterans Administration West Side Hospital (Supported by a Veterans Administration Research Grant)

It has been proposed that bacterial endotoxins possess little or no intrinsic toxic activity, but exert their myriad pathophysiological effects by an antigenic action (12). Weil and Spink reported a marked resemblance between the effects of anaphylactic and endotoxic shock (14). Lichtenstein and associates incubated endotoxin from Veillonella alcalescens or Serratia marcescens with guinea pig complement and were able to produce an anaphylatoxin which had a contractile effect on guinea pig ileum (8). This action was significantly inhibited by antihistamines, indicating a histamine release. In this connection, Hinshaw and co-workers found that plasma histamine increased following induction of endotoxic shock (5). The role of histamine in anaphylactic shock has been well documented (7), however its role in endotoxic shock has received scant attention. The research efforts of our laboratory have been directed toward the elucidation of the mechanism and treatment of septic shock. Therefore, we decided to study the conditions required for Escherichia coli endotoxin-induced histamine release and then, the effect of the potent anti-endotoxic agent, DXM, on this releasing activity (4).

In the present investigation we used the rat peritoneal cavity as a test system in the following manner: white male rats were adrenalectomized and 48 hr later were injected intraperitoneally with modified Tyrode's solution using a canula. In control animals, the lateral tail vein was injected with saline. Experimental animals similarly received E. coli endotoxin solution either alone or containing DXM. A zero time sample of peritoneal exudate was withdrawn using the canula.

At one hr post-challenge, plasma was analyzed for histamine by Shore's spectrofluorometric method (11). Concomitantly, the peritoneal cavity was opened and a sample removed. Both peritoneal samples were analyzed for histamine according to Norn's method (10).

Table 1 shows that after a one-hour endotoxin challenge there was an increase in plasma histamine concentration, supporting Hinshaw and co-workers (5). DXM given simultaneously with endotoxin completely suppressed plasma histamine release and significantly decreased the plasma histamine content below the control. During the same period, there was a twofold increase in peritoneal exudate histamine over the controls; DXM reduced the peritoneal exudate histamine by 40% (Table 2). It has been reported that glucocorticoids inhibit the hypersensitivity phenomenon at the level of antibody production (1, 9); however, once immunity is established, they suppress the cytopathogenic response partially or not at all (2, 3). Recently Jennings showed that glucocorticoids can interfere with the lytic response of sensitized cells by forming an unstable association with immune reactants. He proposed that steroids block the combination of complement with the antigen-antibody complex (6).

An *in vitro* study using peritoneal exudate cells from intact rats was performed to establish the requirements for histamine release by endotoxin, and to evaluate the effect of DXM on this activity. Preliminary results revealed that endotoxin alone was ineffective in releasing histamine. However, upon addition of reconstituted lyophilized guinea pig serum and magnesium chloride, a consistent increment in free histamine occurred. Guinea pig serum alone did not cause release of histamine. After incubation of endotoxin with guinea pig serum, it was found that the formation of the histamine-releasing factor could be inhibited by either decreasing the temperature to 4°C, by adding ethylene diamine tetraacetic acid, or by heating at 100°C for 5 min. This histamine-releasing factor could be separated from endotoxin and serum by filtering with a CF50A membrane ultra-filter (Amicon Corp., Lexington, Mass.).

Treatment	N	One** Hour	Increment (%)	P Value*
Saline	5	158	–	–
Endotoxin (12 mg/kg)	5	216	+58 (27%)	<0.01
Endotoxin and Dexamethasone (25 μg/kg)	5	110	-48 (44%)	<0.05

* Student's t test. **values expressed in ng/ml.

Table 1. Plasma histamine concentration (ng/ml) in endotoxin-challenged rats. Sprague-Dawley adrenalectomized white male rats weighing 165-225 g received 0.5 ml challenges of saline or 12 mg/kg endotoxin solution intravenously (E. coli O111:B4 lipopolysaccharide, Difco Laboratories, Detroit, Michigan) or endotoxin solution containing 25 μg/kg dexamethasone sodium-phosphate (Merck, Sharpe and Dohme, Rahway, New Jersey). The blood was collected after 60 min.

Treatment	N	Initial**	One** Hour	Increment	P Value*
Saline	8	34.0	42.2	+8.2	–
Endotoxin (12 mg/kg)	8	34.5	59.1	+24.6	<0.01
Endotoxin and Dexamethasone (25 μg/kg)	9	34.1	52.0	+17.9	<0.05

* Student's t test comparing endotoxin and dexamethasone to saline-treated rats. ** values expressed in ng/ml.

Table 2. Histamine content of rat peritoneal exudate after endotoxin-challenge. Mean values (ng/ml of exudate). Experimental animals were prepared as described in Table 1 and were injected intraperitoneally with 25 ml Tyrode's solution using a 16 gauge canula. Zero and 60 min time samples of 4 ml peritoneal exudate were removed. These were centrifuged and analyzed for histamine content.

The filtrate, containing molecules below 50,000 molecular weight, revealed the histamine-releasing activity. This supports the contention that endotoxin induces the histamine release through the formation of anaphylatoxin (3).

The effect of DXM on the histamine-releasing activity was determined by the incubation of guinea pig serum with saline, endotoxin and DXM. Evidently, the histamine-releasing activity generated by incubation of serum with endotoxin is only minimally depressed when DXM is added after incubation. However, if it is added before incubation, histamine release is significantly diminished (Fig. 1).

Further support for the DXM effect was provided by determining the amount of intracellular histamine. This was calculated according to the following scheme: the cells were sedimented and histamine concentration in the supernatant measured. Subtracting this value from the total histamine yielded the amount of intracellular histamine. Figure 2 reveals that the generation of anaphylatoxin results in a three- to four-fold decrease in intracellular histamine levels. Experiments in which anaphylatoxin generation was inhibited by DXM showed a decrease in intracellular histamine content one-half the value of the control. These data would support a proposal that DXM prevents the generation of anaphylatoxin from complement, thus reducing histamine release. A second but not exclusive alternative, that DXM may lyse immature cells releasing acid-mucopolysaccharides which bind histamine and result in a lower percentage of free histamine, was proposed by Surján and Csaba (13). The lytic effect hypothesis was supported by our experiments in which DXM alone was incubated with serum and the resulting mixture used to challenge the peritoneal exudate. The results revealed free histamine levels below saline controls (Fig. 1). This effect was also noted in the in vivo plasma histamine determinations (Table 1).

In determining intracellular and free histamine, we found that their sum never equalled the total histamine, the residual being denoted as "bound" histamine. The nature of this extracellular "bound" histamine is unknown, but it supports the possibility that a combination of histamine with heparin or other acid-mucopolysaccharide occurs, since the proportion of this fraction increases significantly over saline controls when either anaphylatoxin or DXM is present (Fig. 3).

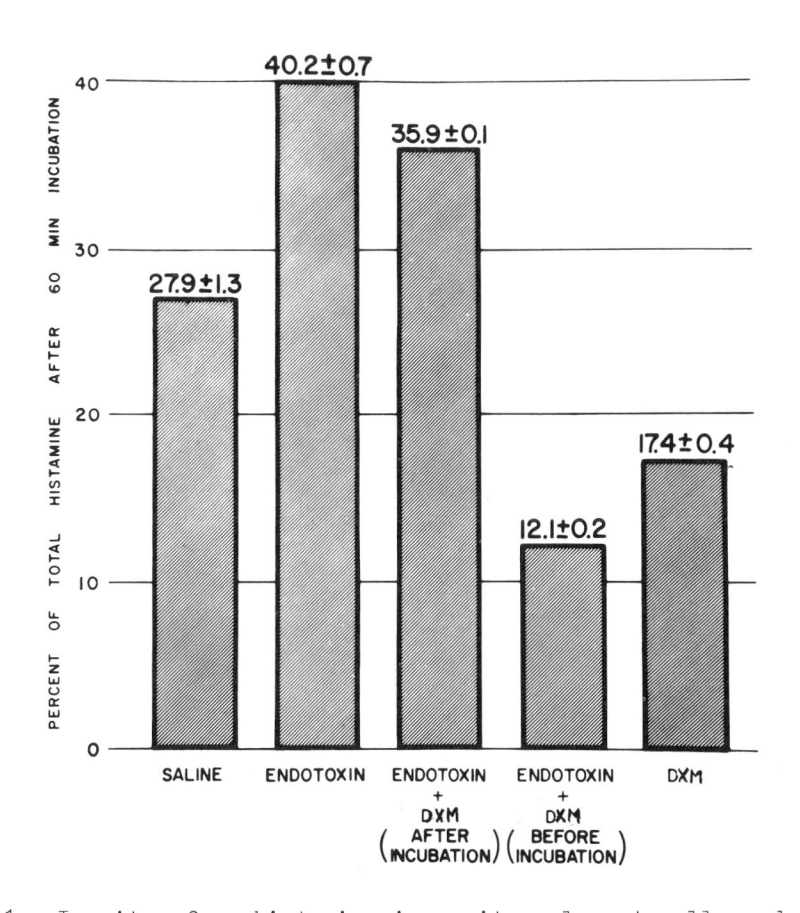

Fig. 1. In vitro free histamine in peritoneal mast cell exudate. Control tubes containing 6.3 ml of reconstituted guinea-pig serum (Hyland, Costa Mesa, California) were incubated for 60 min at 37°C with 0.5 ml saline. Experimental tubes contained serum with 0.5 ml (2.6 mg) endotoxin solution, 0.5 ml endotoxin solution with 4 μg DXM added after incubation but before challenge 0.5 ml endotoxin solution with DXM added before incubation and challenge and 0.5 ml DXM solution. Twenty nanomoles of magnesium chloride was added to each tube before incubation. After incubation, 20 ml freshly pooled peritoneal exudate was added and tubes were incubated for 60 min at 37°C. Free histamine concentration in the exudate after challenge is expressed as a percentage of total histamine after a correction for guinea-pig serum histamine.

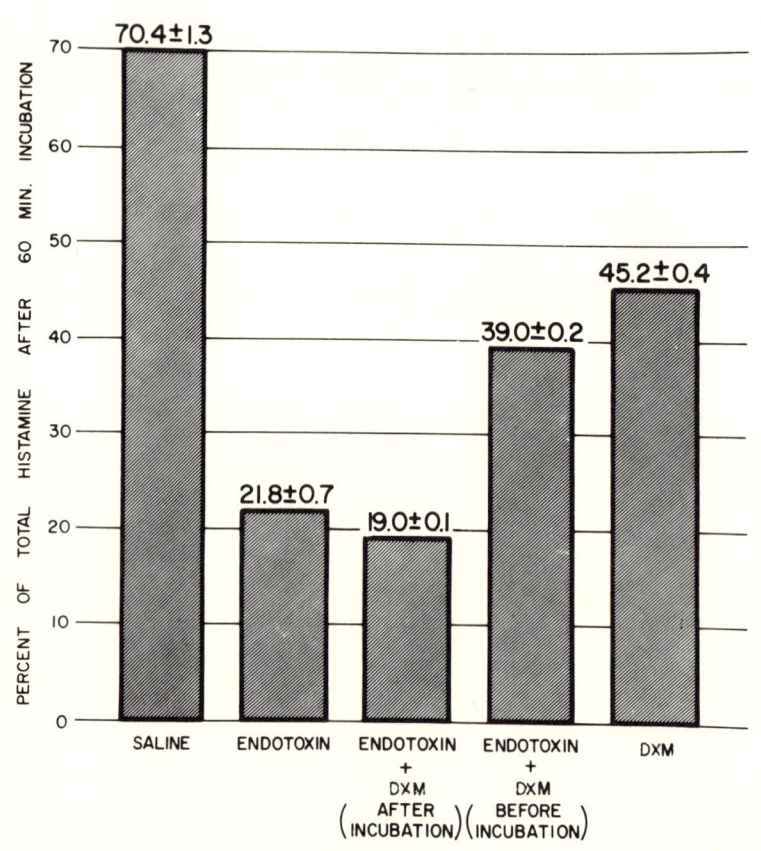

Fig. 2. Percentage of calculated intracellular histamine in the exudate.

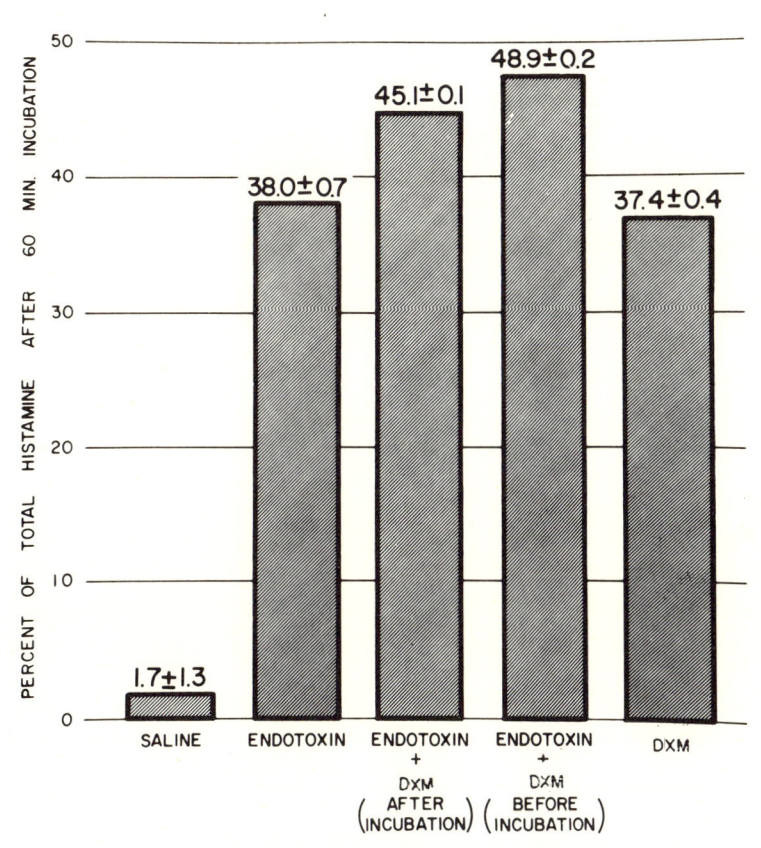

Fig. 3. Percentage of calculated histamine appearing as "bound".

From these observations we propose that DXM, when administered in close temporal proximity to endotoxin, interferes with histamine release by preventing the generation of anaphylatoxin, and that these findings may help explain the anti-endotoxic effects ascribed to this drug. If we can apply these findings clinically, they would support the contention that DXM must be administered in the early stages of septic shock in order to be effective.

REFERENCES

1. Bjørneboe, M., Fischel, E.M. & Stoerk, H.C. (1951) J. Exp. Med. 93, 39.
2. Blumer, H., Richter, M. Cua Lim, F. & Rose, B. (1962) J. Immunology 88, 669.
3. Fischel, E.E., Vaughan, J.H. & Photopoulos, C. (1952) Proc. Soc. Exp. Biol. Med. 81, 344.
4. Fukuda, T. & Hata, N. (1969) Japan J. Physiol. 19, 509.
5. Hinshaw, L.B., Vick, J.A., Carlson, C.H. & Fan, Y. (1960) Proc. Soc. Exp. Biol. Med. 104, 379.
6. Jennings, J.F. (1966) J. Immunology 96, 409.
7. Lecomte, J. CIBA Foundation Symposium on Histamine, edited by G.E.W. Wolsteinholme and C.M. O'Connor. Boston: Little, Brown and Company, 1956 p.173.
8. Lichtenstein, L.M., Gewurz, H., Adkinson, N.F.Jr., Shin, H.S. & Mergenhagen, S.E. (1969) Immunology 16, 327.
9. Malkiel, S. & Hargis, B.J. (1952) J. Immunology 69, 217.
10. Norn, S. (1966) Acta Pharmacol. Toxicol. 25, 281.
11. Shore, P.A., Burkhalter, A. & Cohn, V.H.Jr. (1959) J.Pharmacol 127, 182.
12. Stetson, C.A., Jr. (1955) J. Exp. Med. 101, 421.
13. Surján, L., Jr. & Csaba, C. (1969) Z. Mikro. Anat. Forsch. 80, 321.
14. Weil, M.H. & Spink, W.W. (1957) J. Lab. Clin. Med. 50, 501.

LIVER METABOLISM AFTER INJURY

D. F. Heath

Experimental Pathology of Trauma Section, MRC Toxicology Unit, Medical Research Council Laboratories Woodmansterne Road, Carshalton, Surrey, England

Most of my talk is on liver metabolism in the post-absorptive rat during the ebb phase after non-haemorrhagic injuries: bilateral hind-limb ischaemia and 20% full-thickness scald.

Before describing the abnormalities in liver metabolism I will give some of the reasons for studying these particular systems. Firstly, Stoner (1969; 1970) has shown that non-shivering thermogenesis is greatly affected in the rat during the ebb phase. Both in the rat (R.W. Brauer, personal communication) and in man (Gump, Price & Kinney, 1970) there is evidence that the liver consumes a substantial proportion of the total O_2 consumption, probably in the range 10-20%. It therefore makes an important contribution to non-shivering thermogenesis, and anything found out about heat-production in it and how it is altered by injury is likely to be of interest. Secondly, glucose metabolism is grossly disturbed, and this involves the liver.

The ebb phase is better studied in animals than in man. It precedes the crisis the outcome of which decides whether the animal will live or die, and it is reasonable to suppose that the processes taking place during it decide what that outcome will be. But the sort of experiments most likely to throw light on what is happening are not ones to carry out on critically ill patients.

After haemorrhagic injury changes in liver metabolism may be dominated by hypoxia (see, e.g. Wilhelmi, 1948), although this may not always be the case, because the arterial blood supply can compensate to a considerable extent for the deficiencies in the portal supply brought about by injury (see, e.g. Ankeney, Coffin & Littell, 1967; Greenway & Starke, 1971). There are, however, significant changes after non-haemorrhagic injuries which cannot be attributed to hypoxia in the liver (although they may, of course, be influenced by hypoxia elsewhere), and which may be of clinical interest as one major type of injury in man, burn injury, is essentially non-haemorrhagic.

The oxygenation of the liver has been studied using biochemical measures of the redox potential. These show the conditions within the cell, and there the oxidation state in the liver is entirely normal in the ebb phase of the injured post-absorptive rat (Table 1).

Table 1. Biochemical measures of oxygenation in rat liver after 4 hr bilateral hind-limb ischaemia.

State of rat	[Lactate]/[Pyruvate]	[HO-Butyrate]/[Aceto-Ac]	Energy Charge
post-absorbtive control	6.2	1.2 ± 0.1	0.82 ± 0.01
post-absorptive injured	11.8	1.5 ± 0.1	0.78 ± 0.01
starved control	30	1.6 ± 0.1	0.76 ± 0.02
necrobiotic injured	~140	11.0 ± 2	~0.5

[Lactate]/[Pyruvate] ratios and energy charges by Threlfall (1970 and personal communication.) corrected for compound in blood and extravascular space. (The correction raised only the value for starved controls.)

[HO-Butyrate]/[acetoacetate] ratios by Barton (1971).

The [lactate]/[pyruvate] ratio measures the redox potential in the liver cytosol; the [hydroxybutyrate]/[acetoacetate] ratio (Barton, 1971) that in the liver mitochondria; and the energy charge (Threlfall, 1970) the phosphorylating potential. In all three ways the oxidation state in the liver of the injured post-absorptive rat lies in the normal range, between the states in the livers of control post-absorptive and starved rats. The terminal necrobiotic stage, however, brings about very big changes to values which are unmistakably different both from those in controls and from those in injured rats in the ebb phase.

Liver metabolism is obviously abnormal in relation to the supply of substrate as shown by plasma concentrations. Free fatty acid concentrations are essentially normal in post-absorptive injured rats (Stoner, 1962; Barton, 1971). Glucose concentrations are,

LIVER METABOLISM AFTER INJURY

however, greatly elevated within about 1.5 hr of the injury and remain steady for several hours in surviving animals if the injury is severe (Table 2).

Table 2. Plasma glucose concentrations (mg/ml).

Injury	Time of day	1000–1200	1200–1400	1400–1600
None		1.52 ± 0.03 (38)	–	1.49 ± 0.04 (16)
4 hr tourniquet released at 1330		–	–	1.83 – 6.61 mean = 3.76
20% scald at 930–1000; 20°		2.00 ± 0.10 (5)	2.68 ± 0.16 (6)	2.0 ± 0.08 (3)
20% scald at 1115–1230; 28°		–	–	2.74 ± 0.14 (15)

Means ± S.E.M.'s given except for tourniquetted rats (v.variable).

Viewed in the light of these substrate concentrations liver metabolism in the post-absorptive injured rat shows four surprising features.
(1) Despite the very high blood sugar glycogenolysis is more rapid than in the controls, rapid both after the release of the tourniquets in ischaemic injury (Stoner, 1958) and after scald (Fig. 1).

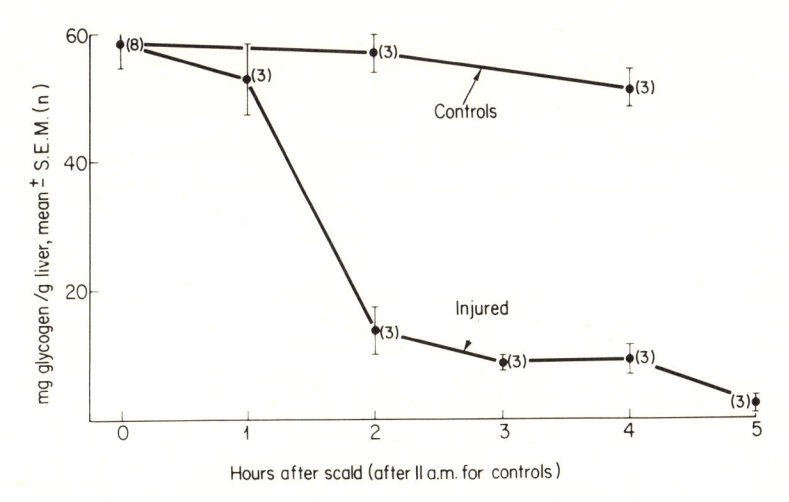

Fig.1. Glycogen concentrations in livers of normal rats and in rats given a 20% full thickness scald. All rats were placed in a room at 29°C at zero time.

Two explanations of this effect can be excluded. As after injury glycogenolysis proceeds as rapidly in adrenal-medullectomized rats catecholamine release is not an explanation (Stoner, 1958). Resistance to insulin likewise cannot be the sole explanation. Considerable resistance to insulin after injury (very well demonstrated in man after burns by Hinton, Allison, Littlejohn & Lloyd, 1971) is required to explain how the high glucose concentrations can be maintained in blood, as in normal animals such levels would soon fall owing to deposition of muscle glycogen. Liver, however, is permeable to glucose whether insulin is present or not, and a glucose load increases glycogen deposition even after anti-insulin serum has been given (see e.g. Hers, de Wulf & Stalmans, 1970). The effect is consistent with the action of glucagon in an animal made resistant to insulin by injury, and Williamson (1967) has produced a similar effect in the rat, i.e. glucose release from the liver against a high glucose concentration, by the combined action of anti-insulin serum and glucagon. Glucagon both stimulates glycogenolysis and reverses the activation of glycogen synthetase by glucose (de Wulf & Hers, 1968). Why the pancreas should release glucagon at these concentrations is another problem.

(2) Labelled pyruvates confer a proportion of their label onto glucose (Ashby et al., 1965). The fraction of injected label which becomes incorporated is much higher in injured rats than in post-absorptive controls, and in fact is nearly as high as in starved controls. A detailed interpretation of these and similar redistributions of label is complex. One possible approach has been made by Heath (1968) and Heath & Threlfall (1968), and still stands up in essentials. This detailed treatment indicates that in the liver of the injured rat at 20°C the citrate cycle is only rotating at about one third of its normal rate. This cannot be regarded as established. It is not, however, inconsistent with any existing data and fits in well with observations on the reduction of non-shivering thermogenesis in rats at 20°C (Stoner, 1969). The basic finding - heavy labelling of glucose from pyruvate against a high glucose concentration - may, like the preceding observation, be explicable in terms of glucagon action (Williamson, 1967).

(3) Ketone body concentrations were raised, from about 0.26 to about 0.46 μmol/g, in line with the increase in acetylCoA from 17.6 to 37.3 nmol/g (Threlfall, 1970) but systematically more than one would expect from the negligible increase in plasma FFA concentration (Barton, 1971), with which ketone body concentrations are closely correlated in normal rats. Glucagon, however, can stimulate lipolysis within liver cells (Bewsher & Ashmore, 1966). A specific release of FFA within liver cells could give higher intracellular FFA concentrations than are expected from the plasma FFA concentrations.

In the three ways already described the liver is apparently behaving as though it were in a starved rat although it is in one

LIVER METABOLISM AFTER INJURY 275

with substrate levels closer to those found in a very recently fed one. In the following way it actually behaves as though it were in one more recently fed than the controls.

(4) More of the acetylCoA utilized by the citrate cycle comes from pyruvate than in the controls. AcCoA is formed from FFA and pyruvate:

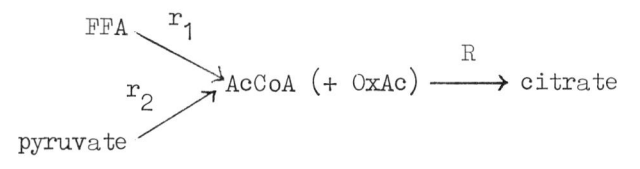

$$\text{FFA} \xrightarrow{r_1} \text{AcCoA (+ OxAc)} \xrightarrow{R} \text{citrate}$$
$$\text{pyruvate} \xrightarrow{r_2}$$

$R = r_1 + r_2$ = rate of citrate cycle. The fraction from pyruvate is r_2/R; and it can be roughly measured by the ratio:

[C5-label in glutamate]/[total label in glutamate] after injection of $2\text{-}^{14}C$-pyruvate or $2\text{-}^{14}C$-alanine (Freedman & Graff, 1958; Koeppe, Mourkides & Hill, 1959; Heath, 1968; Heath & Threlfall, 1968). Dr. J. Phillips in these laboratories has carried out some recent determinations of this ratio (personal communication). The values (Table 3) basically confirm some earlier ones carried out on rats under less standardised conditions (Heath & Threlfall, 1968).

Table 3. Labelling in C5-glutamate and mitochondrial acetyl-CoA formation in rat liver.

State of rat	$\dfrac{\text{C5-label}}{\text{total label}}$	% of AcCoA from Pyr
Injured post-absorptive	0.28 ± 0.04 (4)	25 - 40
Control post-absorptive	0.10 ± 0.01 (4)	5 - 10
Starved*	~ 0.02	~ 1

*Koeppe et al., (1959), Freedman et al., (1960)

The ratio in the injured is very big, and corresponds to some estimations made in more recently fed rats by Mrs. P. Corney of these laboratories. The difference from starved rats is very marked.

At several stages I have put forward possible explanations. I must emphasize that there is no evidence that there is a sufficient glucagon release in the injured rat to cause the first three effects, although there is a good case for thinking that glucagon is implicated, and we hope to investigate this possibility soon. It does not, however, seem likely that glucagon action can explain the fourth effect or the apparent slowing of the citrate cycle. These effects call out for quantitative experimental studies.

REFERENCES

Ankeney, J.L., Coffin, L.H., Jr., & Littell, A.S. (1967). Ann. Surg. 166, 365.
Ashby, M.M., Heath, D.F. & Stoner, H.B. (1965). J. Physiol. 179, 193.
Barton, R.N. (1971) Clin. Sci. 40, 463.
Bewsher, P.D. & Ashmore, J. (1966). Biochem. Biophys. Res. Comm. 24, 431.
Freedman, A.D. & Graff, S. (1958). J. biol. Chem. 233, 292.
Freedman, A.D., Rumsey, P. & Graff, S. (1960). J. biol. Chem. 235, 1854.
Greenway, C.V. & Stark, R.D. (1971). Physiol. Rev. 51, 23.
Gump, F.E., Price, J.B. & Kinney, J.M. (1970). Ann. Surg. 171, 321.
Heath, D.F. (1968). Biochem.J. 110, 313.
Heath, D.F. & Threlfall, C.J. (1968) Biochem.J. 110, 337.
Hers, H.G., de Wulf, H. & Stalmans, W. (1970). Febs Letters, 12, 73.
Hinton, P., Allison, S.P., Littlejohn, S. & Lloyd, I. (1971). The Lancet, Vol i, April, 767.
Koeppe, R.E., Mourkides, G.A. & Hill, R.J. (1959). J. biol. Chem. 234, 2219.
Stoner, H.B. (1958). Brit. J. exp. Path. 39, 635.
Stoner, H.B. (1962). Brit. J. exp. Path. 43, 556.
Stoner, H.B. (1969). Brit. J. exp. Path. 50, 125.
Stoner, H.B. (1970). Energy Metabolism in Trauma (CIBA Foundation Symposium) Ed. Ruth Porter & Julie Knight, Churchill, London, p.1.
Stoner, H.B. & Matthews, J. (1967). Brit. J. exp. Path. 48, 58.
Threlfall, C.J. (1970). Energy Metabolism in Trauma (CIBA Foundation Symposium proceedings). Ed. Ruth Porter & Julie Knight, Churchill, London, p.127.
Wilhelmi, A.E. (1948). Ann. Rev. Physiol. 10, 259.
Williamson, J.R. (1967). Advances in Enzyme Regulation. Pergamon - Oxford and New York, p. 229.
de Wulf, H. & Hers, H.G. (1968). Europ. J. Biochem. 6, 558.

AEROBIC AND ANAEROBIC GLYCOLYSIS IN THE LIVER AFTER HEMORRHAGE

M. Kessler, H. Lang, H. Starlinger, J. Höper and
M. Thermann. With the technical assistance of
K. Fehlau, K. Joachimsmeier, and D. Schmeling

Max-Planck-Institut für Arbeitsphysiologie, Dortmund,
W. Germany

The local O_2 supply of liver tissue during the development of hemorrhagic shock was investigated with O_2 electrodes. It was found that local anoxia and disturbances in the microcirculation could be detected even in the early phase of hemorrhagic shock, when systolic blood pressure was still at or above 100 mm Hg. Fig 1 shows a typical example of early tissue anoxia in dog liver.

The changes in the microcirculation which occur in the rat liver in hemorrhagic shock are shown in Fig. 2. The microcirculation in the liver tissue was measured by H_2 washout curves (Clark, 1956; Aukland, 1965; Lübbers, 1968; Thermann & Kessler, 1972). Bleeding of the animal was started about 60 min after beginning the experiment. It can be seen that the local microcirculation breaks down before the arterial blood pressure is affected.

The tissue O_2 supply to different organs and the changes in lactate/pyruvate ratio were correlated in a series of shock experiments in dogs. It became obvious that the lactate/pyruvate ratio did not change significantly unless local anoxia of tissue exceeded 15 - 20 % (Kessler et al., 1970; Görnandt et al., 1972).

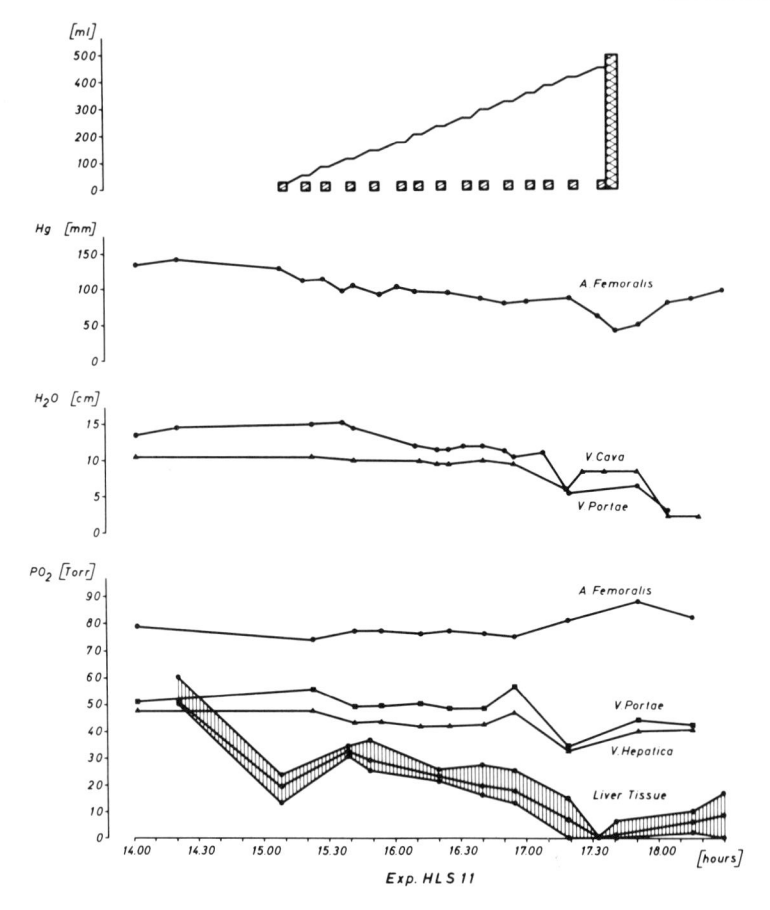

Fig 1 Hemorrhage in the dog.

To aid in interpreting these phenomena experiments were performed on an isolated and hemoglobin-free perfused rat liver at 22° C (Kessler, 1967; Kessler & Schubotz, 1968; Schubotz, 1968; Lang, 1971). The rat liver preparation was perfused in a closed circuit with a medium containing Krebs-Ringer solution with 35 g/l purest bovine albumin.

The surface fluorescence of the pyridine nucleotides (PN) was measured with a fluorometer (Chance & Jöbsis, 1959). Our instrument also enables us to do studies of vital microscopy (Lang et al., 1972).

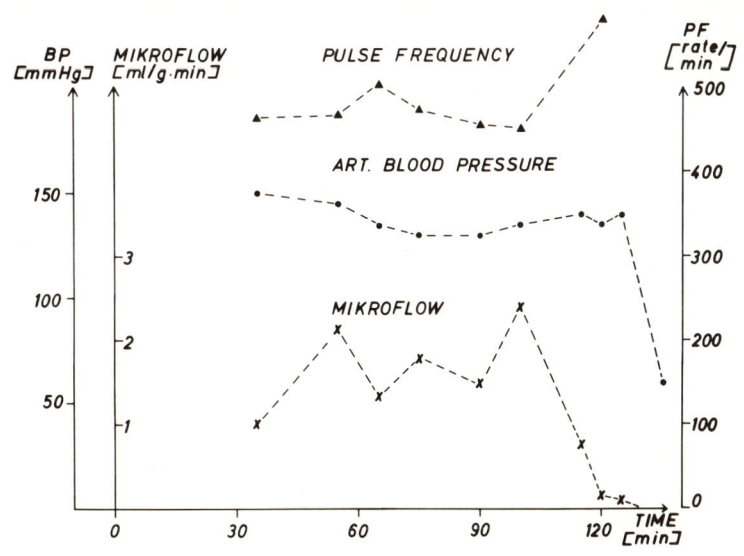

Fig 2 Microcirculation in rat liver during hemorrhage.

In our control experiments the changes in lactate, pyruvate and glucose were studied when the perfusate was equilibrated with carbogen (95% O_2, 5% CO_2) or with pure O_2. Partial anoxia of liver tissue was produced by equilibration of the perfusate with gas mixtures containing O_2, N_2 and 5% CO_2.

A liver preparation which is supplied with a perfusate equilibrated with carbogen (pH 7.4) shows characteristic reactions (Lang, 1971). We found a continuous decrease in lactate, a rapid decrease in pyruvate and a continuous increase in glucose (Fig. 3).

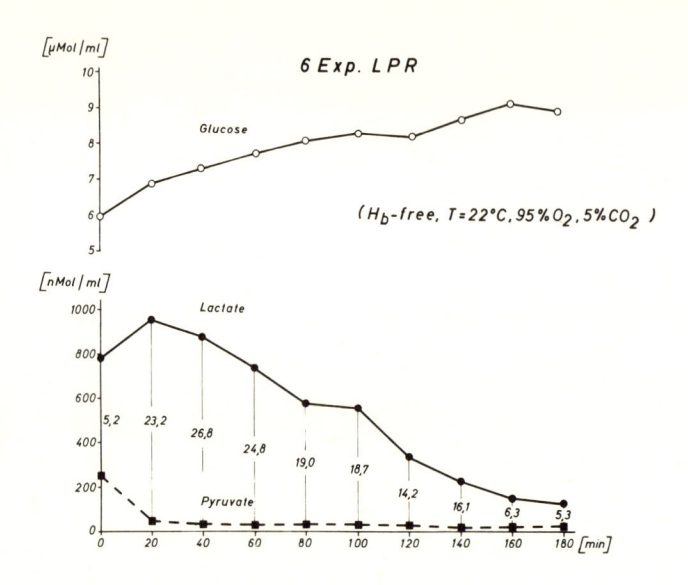

Fig 3 Changes in the lactate/pyruvate ratio and glucose in the perfusate of a perfused rat liver.

When the perfusate is equilibrated with pure O_2 we get a biphasic change in lactate and a continuous increase in pyruvate (Fig 4). The carbogen type reaction can be changed into the CO_2-free type reaction by adjusting the pH of the perfusate to 7.7 (Lang, 1971).

The O_2 supply of a liver which is perfused under conditions of normoxia was investigated with multiwire Po_2 electrodes (Kessler et al., 1965; Kessler & Grunewald, 1969). The Po_2-distribution curve of the liver tissue when the perfusate is equilibrated with carbogen is shown in Fig 5. Using a CO_2-free medium (100% O_2) for perfusion we found a rather high percentage of low Po_2 values which indicated disturbances in the microcirculation (Fig 6). In this context it should be mentioned that important effects of blood Pco_2 on blood flow and on the survival of hypothermic animals have been found by Miller (1971).

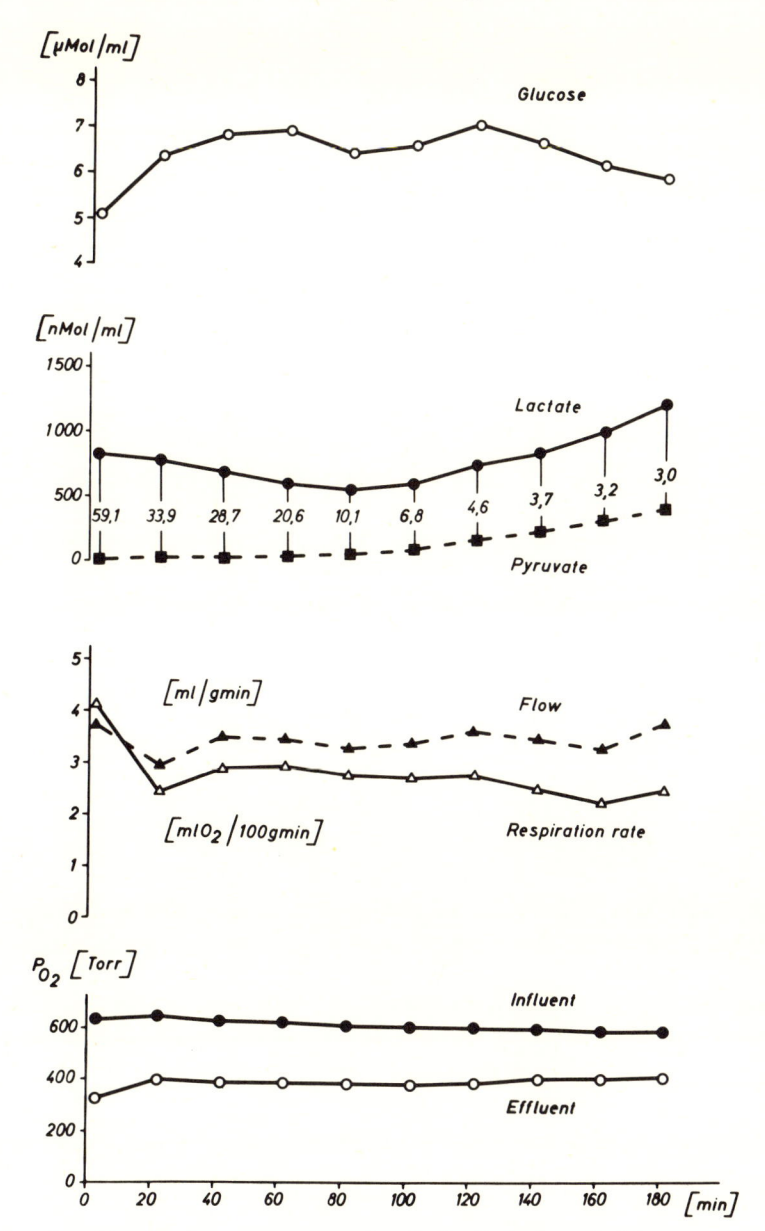

Fig 4: Perfusion experiment of isolated rat liver (T = 22° C, equilibration of the medium with 100% O_2, pH 7.4).

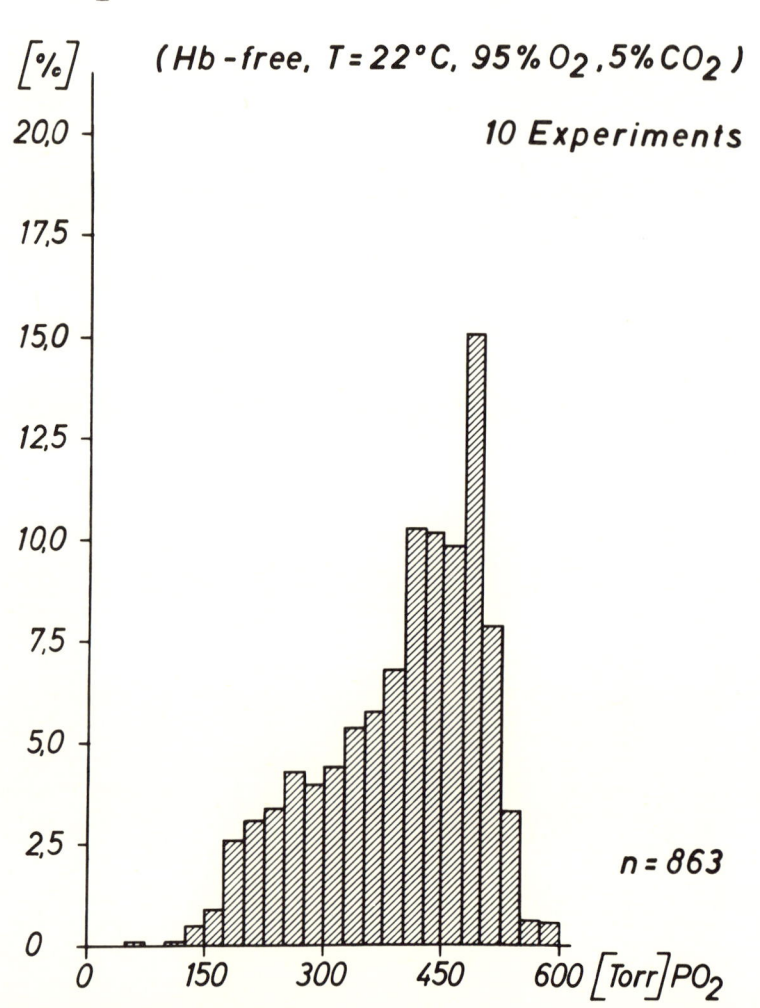

Fig 5 PO_2-distribution in the isolated and Hb-free perfused rat liver.

P_{O_2} - Distribution in the tissue

(Hb-free, T=22°C, 100% O_2, 0% CO_2)

5 Experiments

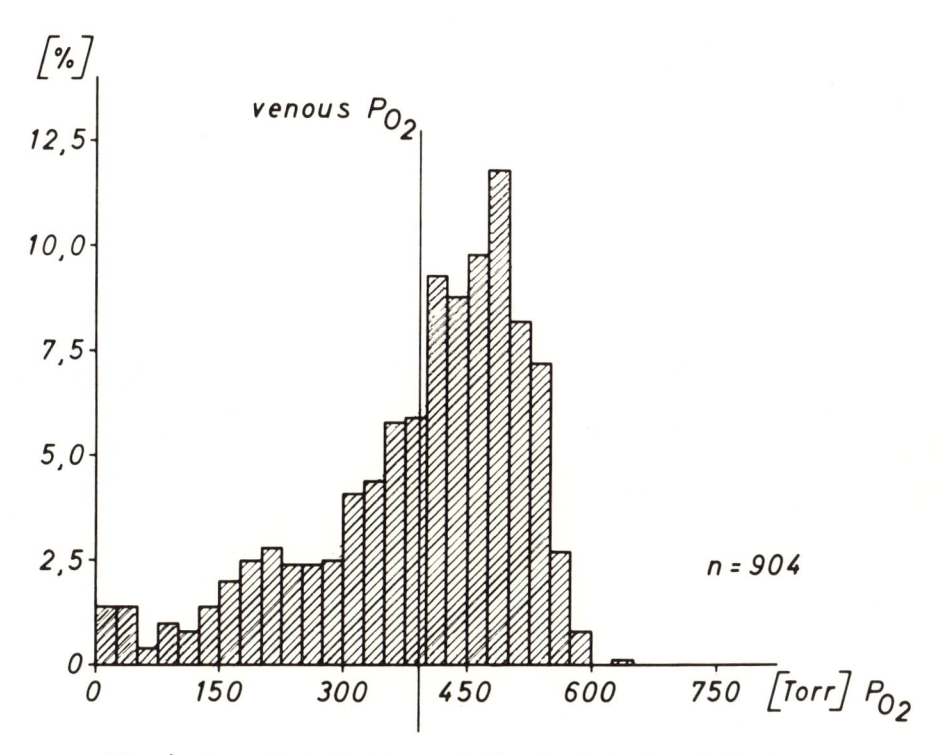

Fig 6 P_{O_2}-distribution of the isolated and Hb-free perfused rat liver (equilibration of the medium with pure O_2).

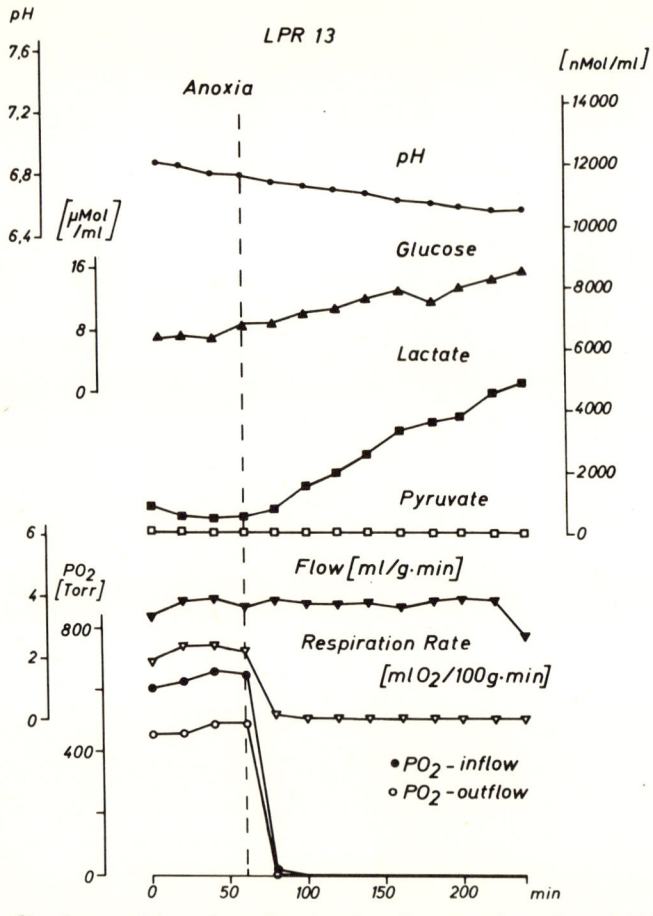

Fig 7 Anaerobic glycolysis in the perfused rat liver in acidosis.

The influence of the pH of the perfusate on anaerobic glycolysis in the liver was investigated. An example of anoxic perfusion with an initial pH of 6.8 is given in Fig 7. Compared to another experiment with initial pH of 7.4 it becomes very

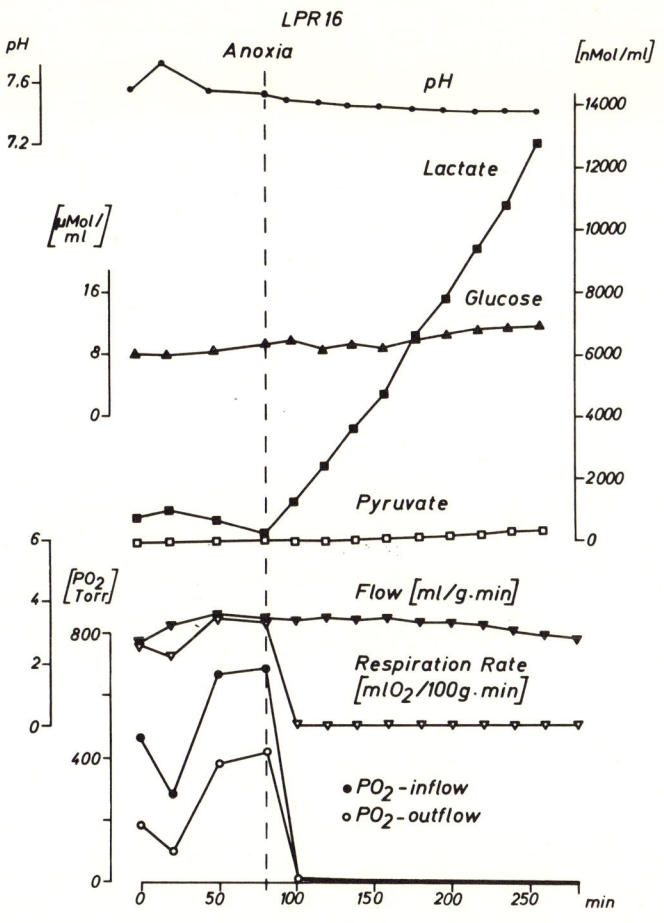

Fig 8 Anaerobic glycolysis in Hb-free perfused rat liver (pH 7.5).

obvious that anaerobic glycolysis and the release of glucose under conditions of anoxia are strongly influenced by the pH of the perfusate (Fig 8). Fig 9 summarizes the effects of pH on the release of lactate and glucose to the perfusate by anaerobic liver

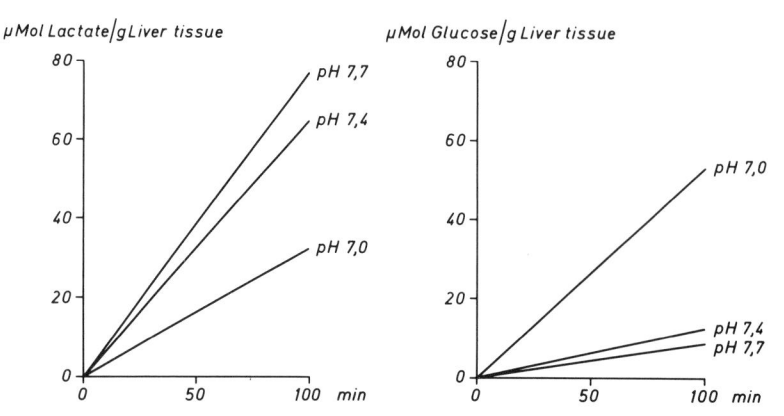

Fig 9 Anaerobic formation of lactate and glucose by Hb-free perfused rat liver (T = 22° C) under different pH conditions.

tissue. At pH 7.0 the formation of lactate is decreased to less than 50%, whereas a large release of glucose occurs. When we calculate the amount of energy formed by anaerobic glycolysis in the liver we get a value of 7-8%, compared to that of perfused organs (T = 22°C) with normal O_2 uptake.

There is some evidence that the fructose-6-phosphate/fructose-diphosphate ratio can give information on the pH dependance of anaerobic glycolysis. In Fig 10 the anaerobic formation of lactate is compared to the F6P/FDP ratio and it is seen that both parameters alter when pH of the perfusate is changed (see also Barwell & Hess, 1971).

In further experiments the respiratory rate of the perfused liver was measured continuously in different states of partial tissue anoxia and correlated with the following parameters:

1. fluorescence of pyridine nucleotides

2. Po_2 in liver tissue

3. formation of lactate

4. lactate/pyruvate ratio

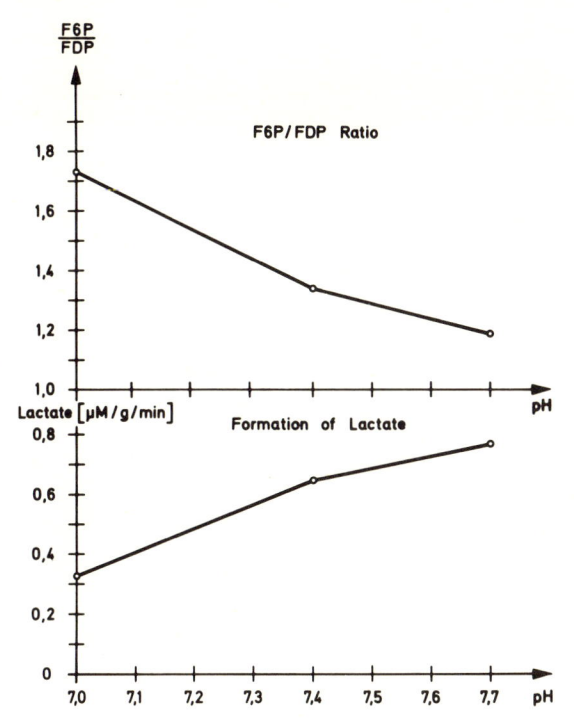

Fig 10 Relationship between the anaerobic formation of lactate by Hb-free perfused rat liver and the F6P/FDP ratio.

Partial anoxia of tissue was caused by decreasing the P_{O_2} of the medium while the flow was kept constant. Fig 11 shows that, with a stepwise decrease in the respiratory rate caused by P_{O_2} changes, PN fluorescence and respiratory rate were well correlated. The correlation was not linear however. The points which are plotted in Fig 11 are from 7 experiments. Great attention was given to the fact that before and after each measurement complete steady state conditions were present.

For this reason only 2-3 cycles of partial and total anoxia were made during a perfusion time of 2-3 hr. A similar relationship exists between respiratory rate and O_2 tension in tissue. The points in Fig 12 correspond to the % total anoxia determined by P_{O_2}-tissue electrodes. They represent the results of P_{O_2} measurements at 1096 different points in the liver.

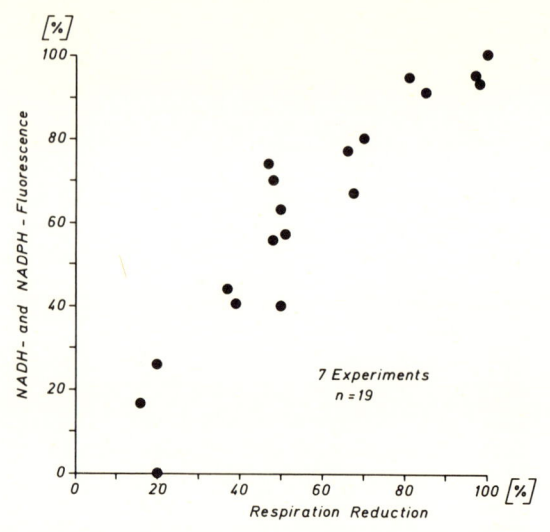

Fig 11 Relationship between the respiratory rate (changed by partial anoxia) and PN-fluorescence

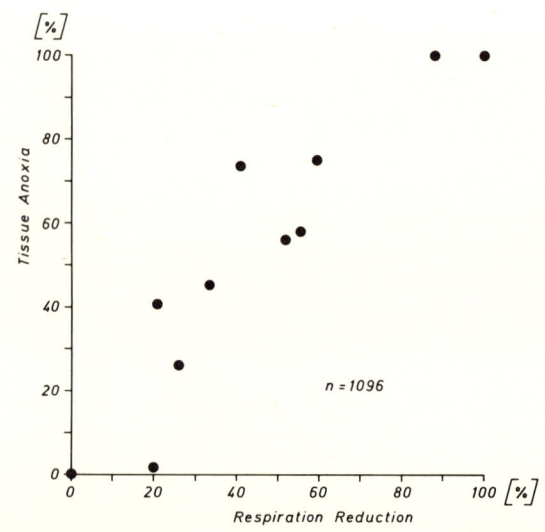

Fig 12 Relationship between the reduction in respiration and tissue O_2 supply

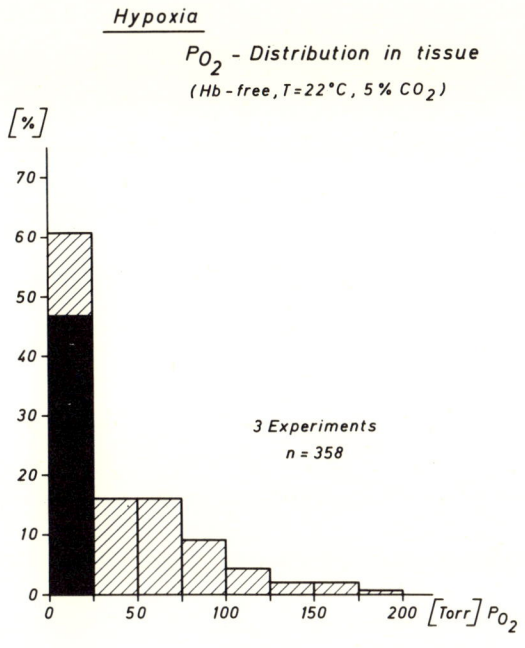

AEROBIC AND ANAEROBIC GLYCOLYSIS IN THE LIVER AFTER HEMORRHAGE

Fig 13 P_{O_2} histogram of liver tissue during partial anoxia.

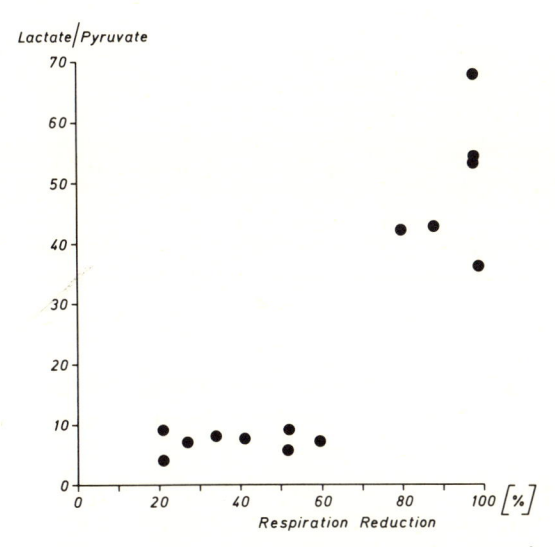

Fig 14 Relationship between the reduction in respiration and the lactate/pyruvate ratio.

Fig 13 shows a Po_2 histogram in partial anoxia. From 358 Po_2 measurements a partial anoxia of 47% was estimated.

The correlation between respiratory rate which is decreased by partial anoxia and the formation of lactate is very interesting. When the respiratory rate is reduced from 0% to 40% by O_2 deficiency we found a very small increase in lactate formation. Only when reduction exceeded 40% did lactate formation increase markedly. This was more clearly expressed by the L/P ratio which only rose when the reduction in respiration exceeded 60% (Fig 14).

How can we explain these unexpected results? We will try to do so by means of a theoretical model (Fig 15). In the Krogh cylinder model sketched in this figure, a partial anoxia of 50% is assumed. Lactate which is formed in the anaerobic part passes into the perfusate by diffusion. As we have a closed perfusion circuit this lactate is transported to the normoxic part of the Krogh cylinder by the recirculation where it rediffuses into the respiring cells. There it is oxidized to

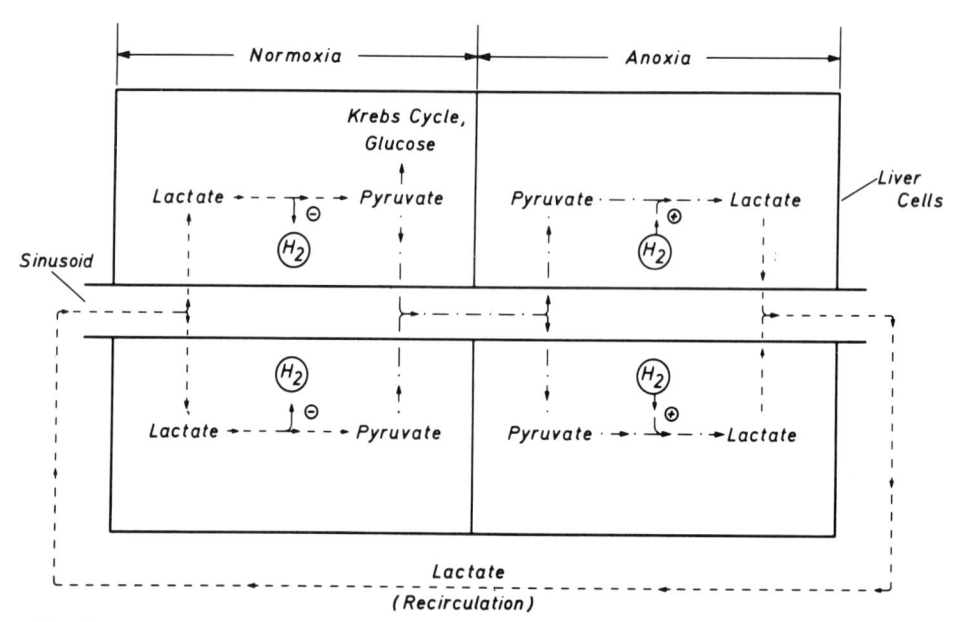

Fig 15 Model for illustration of intercellular hydrogen transfer shown on a Krogh-cylinder.

pyruvate. For the pyruvate formed in this way there are three main possibilities:

1. uptake by mitochondria and turnover in the Krebs cycle
2. formation of glucose by gluconeogenesis
3. return of pyruvate to the anoxic cells where it is again reduced.

The third pathway offers the possibility that there may be a cyclic, intercellular transfer of hydrogen.

REFERENCES

Aukland, A. (1965) Acta neurol. scand. Suppl. 14:26.

Barwell, C.J. & Hess, B. (1971) FEBS Letters 19:1.

Chance, B & Jöbsis, F. (1959) Nature 184:195.

Clark, L.C., Jr. (1959) Surgery 46:797.

Kessler, M. (1967) Normale und kritische Sauerstoffversorgung der Leber bei Normo-und Hypo-thermie. Habilitationsschrift, Marburg/Lahn.

Kessler, M., Görnandt, L., Thermann, M., Lang, H., Brand, K. & Wessel, W. (1972) In: Oxygen supply, theoretical and practical aspects of oxygen supply and microcirculation of tissue. Eds by: Kessler, M., Bruley, D.F., Clark, L.C., Lübbers, D.W., Silver, I.A. & Strauss, J. Urban & Schwarzenberg, München and Berlin, University Park Press, Baltimore.

Kessler, M., Grunewald, W. & Lübbers, D.W. (1965) Pflügers Arch. 283, R 37.

Kessler, M. & Grunewald, W. (1969) In: Int. Symposium on oxygen pressure recording, Nijmegen, 1968 Prog. Resp. Res. Vol 3, p. 136 (Karger, Basel - New York)

Kessler, M. & Schubotz, R. (1968) In: Stoffwechsel der isoliert perfundierten Leber / 3. Konf. d. Ges. f. Biol. Chemie (Springer, Berlin - Heidelberg - New York) p.12.

Kessler, M., Thermann, M., Lang, H. & Schneider, H. (1970) In: Schock, Stoffwechselveränderungen und Therapie. F.K. Schattauer Verlag, Stuttgart - New York.

Lang, H. (1971) Die Regulation der Lactat- und Pyruvat-
konzentrationen durch die hämoglobin-frei perfundierte Leber
bei unterschiedlichen CO_2- und O_2- Drucken Dissertation,
Bochum.

Lang, H., Kessler, M. & Starlinger, H. (1972) In: Oxygen supply
theoretical and practical aspects of oxygen supply and
microcirculation of tissue. Eds: Kessler, M., Bruley, D.F.,
Clark, L.C., Lübbers, D.W., Silver, I.A. & Strauss, J.
Urban & Schwarzenberg, München and Berlin, University Park
Press, Baltimore.

Lübbers, D.W. (1968) Regional blood flow and microcirculation
Proc. int. Conf., Glasgow, p. 162 (Livingstone, Edinburgh.)

Miller, J.A. (1971) Amer. J. Obstet. Gynec. 110:1125.

Schubotz, R. (1968) Die Sauerstoffversorgung der perfundierten
Leber Dissertation, Marburg/Lahn.

SEQUENTIAL CIRCULATORY AND METABOLIC CHANGES IN THE LIVER AND WHOLE BODY DURING HEMORRHAGIC SHOCK

W.C. Shoemaker, L.J. Stahr, S.I. Kim, and D.H. Elwyn

The Department of Surgery, The Mount Sinai School of Medicine of the City University of New York, New York

Since Claude Bernard first observed the presence of hyperglycemia in hemorrhagic shock, the metabolic effects of shock and trauma have been extensively studied. In the past, most investigations have been principally concerned with concentration changes of various plasma and tissue constituents in the anesthetized dog subjected to rapid hemorrhage and subsequent reinfusion of the shed blood (19). Usually, these changes represent late effects which may have particular relevance to terminal physiologic mechanisms.

Changes in carbohydrate metabolism after injury also have been described in the rat and dog subjected to tourniquet shock, hind-limb freezing, drum shock, etc. (5,17). Early hyperglycemia and increased glucose oxidation have been associated with the subsequent development of hyperglycemia and increased concentrations of lactic and pyruvic acids, as well as failure to clear exogenous intravenous loads of these organic acids (11,19). These and other data have been interpreted as due to incomplete oxidation of glucose resulting from reduced blood flow and anaerobiosis.

Progressively increasing rates of hepatic glucose and K^+ output were measured in the unanesthetized dog subject to gradual prolonged hemorrhage (14). Observations of the effects of epinephrine, norepinephrine, cortisol, glucagon and insulin on hepatic metabolism suggested that many of the hepatic hemodynamic and metabolic effects of shock could be explained on the basis of increased adrenal liberation of catecholamines and cortisol (9,10). Many of the metabolic changes in shock have been summarized recently (5,6,7,9,11,17,19,21).

In the present studies, the transport of glucose, electrolytes and plasma lipid fractions between liver, nonhepatic splanchnic area and peripheral tissues were measured in dogs subjected to hemorrhagic shock by a protocol designed to simulate clinical shock conditions (3). The purpose of the studies was to describe the sequence of hepatic blood flow and hepatic metabolic events in relation to each other and to the metabolism of nonhepatic splanchnic area as well as the whole body.

METHODS AND MATERIALS

Experimental Preparation

Experiments were performed on 4 mongrel dogs. The hepatic vessels were catheterized with plastic tubing, as previously described (15), 4 to 7 days beforehand. On the day of the experiment, a plastic catheter was inserted into the right ventricle via the anterior jugular vein and a second catheter was placed in the femoral artery. All dogs were studied in an unanesthetized state without sedation after fasting overnight.

Protocol (Fig 1)

6 to 8 control measurements, described below, were made while the animal was resting quietly in a Pavlov stand. Hemorrhage was then produced by slow withdrawal of blood over 2 to 3 hr until the mean arterial pressure reached 50 mm Hg (3). Pressure was maintained at this level by additional withdrawal or reinfusion of blood for 10 hr or until one fifth of the shed blood had been replaced. Then, all of the remaining shed blood was returned and measurements were continued in this normovolemic period until the animal died (3).

Both the hypovolemic period and the post-transfusion normovolemic shock periods were divided into early, middle and late stages to evaluate the sequential pattern more easily.

Analyses

Cardiac output was measured by the indocyanine green dilution technique (4). Arterial, portal venous, hepatic venous and right ventricular (mixed venous) blood samples were collected anaerobically in each sampling period at the time of the cardiac output determinations for measurement of PO_2, PCO_2, pH and O_2 saturation. Hepatic blood flow was measured by a modified BSP method (8).

In all instances, blood from a donor dog was used to replace quantitatively the blood losses from sampling. Arterial, portal and hepatic venous blood was withdrawn into heparinized, iced plastic centrifuge tubes by a constant withdrawal pump. 6 to 8 contemporaneous, time-integrated samples from each catheter were obtained over 10 to 20 min intervals in the control period; sampling was continued during the removal of blood and after reinfusion of the shed blood in the normovolemic period.

The blood samples were promptly centrifuged. Plasma glucose concentrations were measured by the glucose oxidase method. K^+ and Na^+ concentrations were measured with a flame photometer. Non-esterified fatty acids (FFA) determinations were made on 1 ml aliquots of plasma (1). A lipid extract was prepared from a 2.0 ml aliquot of plasma by the method of Folch et al. (2). Free and total cholesterol were measured in duplicate on the dried residue of the Folch extract (16). Triglycerides were measured on aliquots of the Folch extract, in duplicate with the appropriate reagent blank, using a modified van Handel and Zilversmit method (18,19). Phospholipid phosphorus was measured in duplicate on aliquots of the plasma extract which had been dried and digested with perchloric acid; the color was developed with ammonium molybdate followed by treatment with the reducing agent, ferrous sulfate. In general, the methods and procedures have been published (13).

Calculations

Total body O_2 consumption ($\dot{V}O_2$) was calculated as the product of the cardiac output and the simultaneously measured arterial-mixed venous O_2 content difference. Good agreement between this calculated value and direct measurements of O_2 consumption was observed (3).

The net output of the liver, nonhepatic splanchnic (gut) area and the total splanchnic area was calculated as the product of each of the concentration differences and corresponding blood flow rates (13). The nonhepatic splanchnic area represents the area drained by the portal vein; i.e. the gut, from the stomach to the colon, pancreas, fat depots of mesentery and omentum, etc. The total splanchnic area represents the liver and nonhepatic splanchnic area. The output of an organ was expressed as a positive number in m mole/min or mEq/min and the uptake as a negative number.

Four studies on the sequential changes in cardiac output, hepatic blood flow (HBF), total body $\dot{V}O_2$ and hepatic $\dot{V}O_2$ are presented. Because of the complexities of the inter-organ movements of each of the plasma lipid fractions as well as glucose and electrolytes and their dependence on variable physiologic conditions in the control state, only one illustrative metabolic experiment is described. Description of the inter-organ transport patterns of plasma constituents in this experiment is not made in order to draw firm conclusions, but rather as a prototype to illustrate the kinds of questions which may arise from this approach to the biologic alterations of trauma.

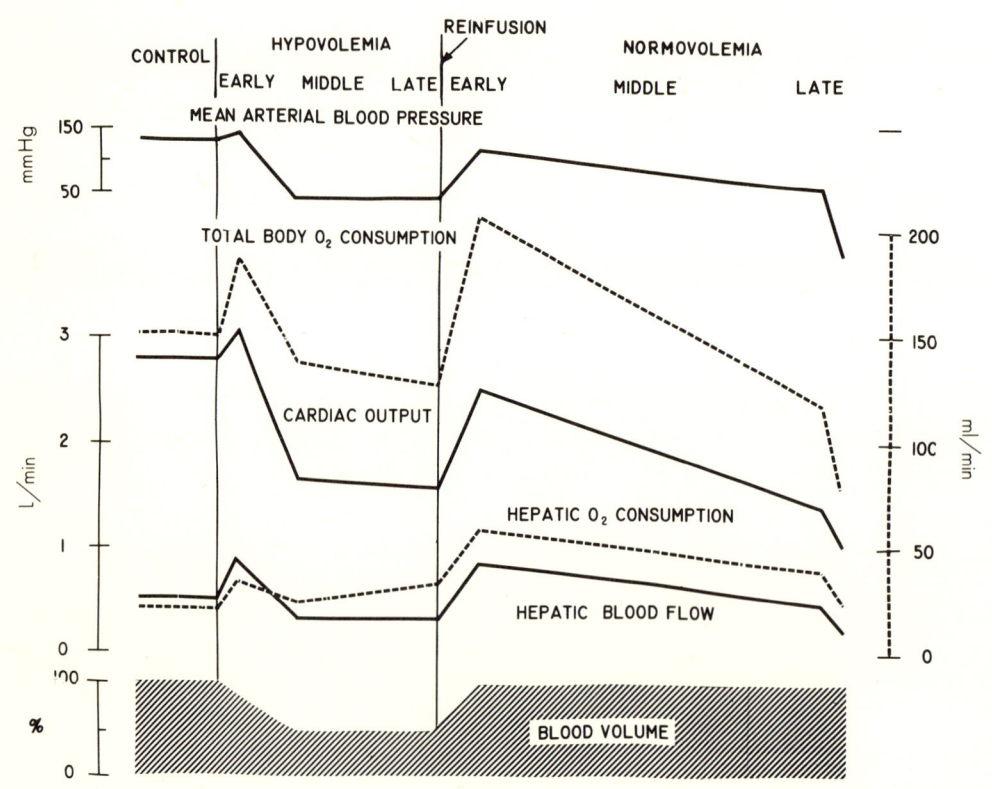

Fig 1. Illustration of the protocol showing (from above, downward): designated stages, mean arterial pressure, total body oxygen consumption, cardiac output, hepatic oxygen consumption, hepatic blood flow and the blood volume.

RESULTS

Sequential Hemodynamic and Oxygen Transport Changes

The measurements of cardiac output, hepatic blood flow (HBF), total body $\dot{V}O_2$ and liver $\dot{V}O_2$ are shown in Fig 1. Data illustrating a representative metabolic experiment is shown in figures 2-6 and tables 1-4. Immediately after onset of bleeding, cardiac output, HBF and total body $\dot{V}O_2$ increased slightly; but with continued bleeding, these values fell significantly and remained low in the middle and late hypovolemic stage. Hepatic $\dot{V}O_2$ initially increased and remained slightly elevated throughout the hypovolemic period.

In the early normovolemic period, immediately after reinfusion of the shed blood, these values increased from their hypovolemic levels; all but cardiac output reached and exceeded control values. The HBF and liver $\dot{V}O_2$ were relatively well maintained in middle normovolemic stage. In the late normovolemic stage, each of these values fell progressively to, or below, the corresponding values at the end of the hypovolemic period (Fig. 1).

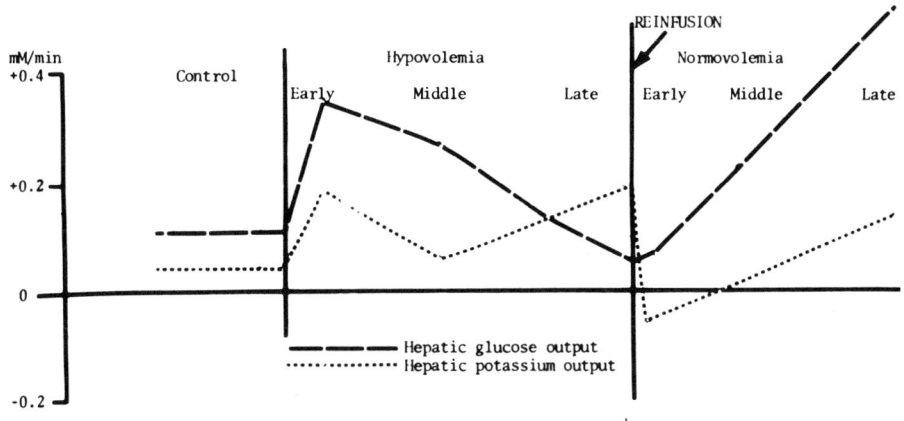

Fig 2. Mean values of hepatic glucose and K^+ movements at each stage in hypovolemic and normovolemic shock. Hepatic output rates are represented by positive values above the horizontal zero line; hepatic uptake values are shown by negative values below the zero line. Note the early increase in HGO with onset of blood withdrawal and the progressively increasing HGO in normovolemic shock; hepatic K^+ outputs roughly paralleled the glucose movements.

Glucose movements of the liver and non-hepatic splanchnic area

The control hepatic glucose output (HGO) of 0.1 m moles/min (20 mgm/min) increased initially after onset of hemorrhage and then returned slowly to control values by the late hypovolemic stage. After reinfusion of the shed blood, HGO rose progressively until the late normovolemic stage, where massive glucose outpouring occurred (Fig 2).

The nonhepatic splanchnic gut area removed glucose slowly in the post-absorptive control state. Immediately after onset of blood loss, the rate of uptake increased sharply, but during the middle and late hypovolemic stages, glucose uptake returned toward control values. After reinfusion of the shed blood, the gut transiently released glucose, but then returned toward the control range in the middle stage of the normovolemic period. In the late normovolemic stage, the gut again removed glucose rapidly (Table 3, fig 6).

The total splanchnic glucose output increased in the early and middle hypovolemic stage and fell in the late hypovolemic stage. Splanchnic glucose output increased two to three-fold in the ensuing normovolemic period.

Potassium movements of the liver and nonhepatic splanchnic area

Usually the net hepatic potassium movements were associated with corresponding glucose movements into or out of the liver and gut. The relatively trivial hepatic K^+ output in the control period (0.04 mEq/min) increased about four-fold with the onset of bleeding; the K^+ release was sustained throughout the middle and late hypovolemic stages. In the early stage after reinfusion of the shed blood, K^+ entered the liver; later it was released at progressively increasing rates, until terminally, when there was a massive hepatic K^+ release (Fig 2). In general, hepatic K^+ movements roughly paralleled glucose movements into and out of the liver.

K^+ was released by the nonhepatic splanchnic area at greater than control rates in the early and middle stages of both the hypovolemic and the normovolemic period.

K^+ was released from the total splanchnic area at greater than control rates in all stages except the early normovolemic period.

Hepatic lipid movements

In the control period, the liver released triglycerides but removed FFA, phospholipids, free cholesterol and cholesterol esters. After the onset of hemorrhage those lipids which were released in the control period were released at greater than control rates, while those which were removed by the liver in the control period were taken up at greater rates. This initial response was short lived; in the middle hypovolemic period the rates of hepatic metabolism returned to the control values and in the late hypovolemic period, the liver cleared cholesterol esters at less than control rates and released both phospholipids and FFA (Fig 3).

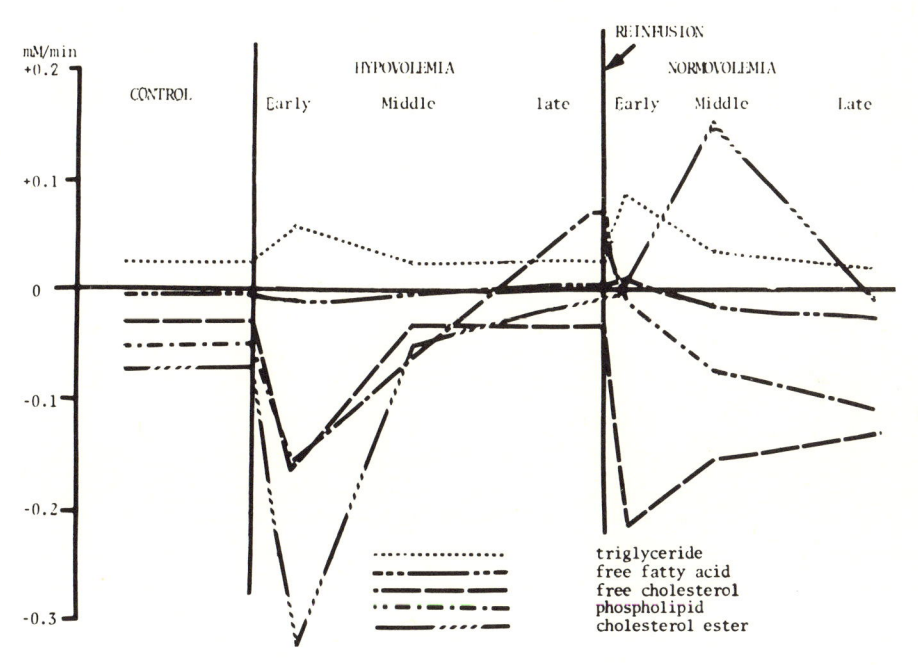

Fig 3. Hepatic lipid output rates at each of the various stages.

In general, after reinfusion of the shed blood, the response pattern was similar to that which occurred in hypovolemia, except that cholesterol esters were released, especially in the middle normovolemic stage and FFA were removed in the late stage. The hepatic uptake of phospholipids and free cholesterol was more prolonged in the normovolemic shock period than in the hypovolemic period.

Nonhepatic splanchnic lipid movements

The nonhepatic splanchnic area released the various lipid fractions, presumably, from mesenteric and omental fat stores.

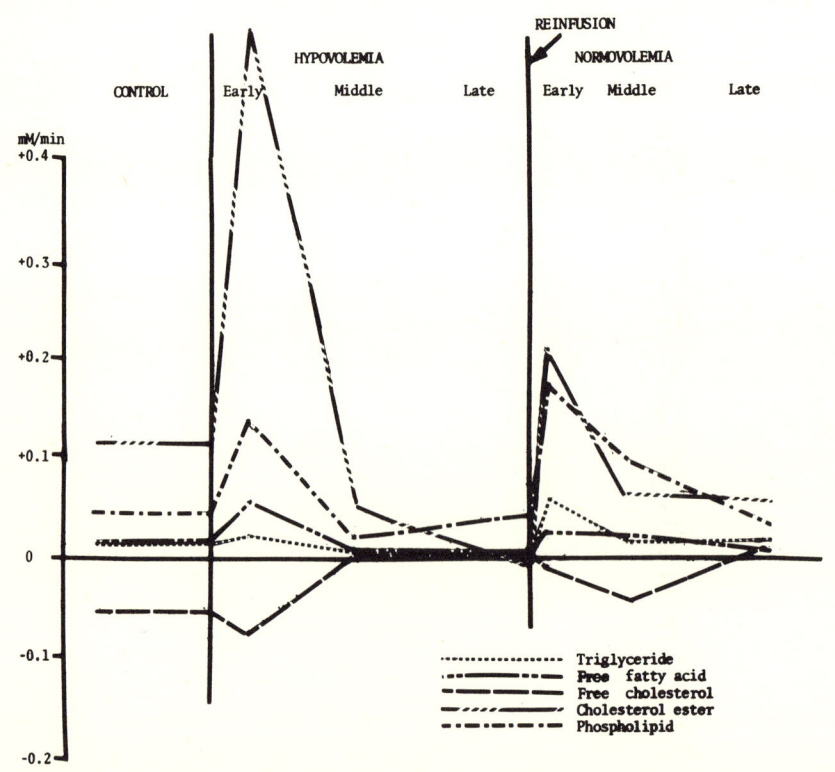

Fig 4. Nonhepatic splanchnic lipid output rates at each stage.

An exception was free cholesterol which was cleared by the gut area (Fig 4). The initial response to hemorrhage was an increase in the control rates of output or uptake. In the middle and late stages of hypovolemia, the rates returned to, or below, control values except for free cholesterol.

In the normovolemic period, after reinfusion of the shed blood, a similar pattern developed but to a somewhat lesser degree.

Total splanchnic lipid movements

The total splanchnic area i.e. both the liver and nonhepatic splanchnic area, released cholesterol esters, triglycerides, and FFA in the control period. As the plasma concentrations were stable in the control period, the net output from total splanchnic area reflected a comparable uptake by peripheral tissues. After the onset of hemorrhage, there was an initial outpouring of cholesterol esters, triglycerides and free fatty acids by the splanchnic area. The rates fell to, or below, control values in the middle and late hypovolemic stages.

After reinfusion, the total splanchnic area released cholesterol esters, triglycerides and phospholipids at increased rates; these rates subsequently returned to or below control values. The rate of FFA output increased appreciably in the early and late normovolemic stages.

DISCUSSION

Previous studies have shown close relationships between plasma glucose and K^+ concentrations in the Wiggers' type of experimental shock (19) as well as between hepatic release of glucose and K^+ during slow exsanguination (14). In the present studies we used a protocol which simulates more closely hemorrhagic shock in man (3) and confirmed the initially increased hepatic output of both glucose and K^+ which were approximately parallel. These hepatic hemodynamic and metabolic events have been explained on the basis of catecholamine action; presumably the hepatic response provides oxidative substrates for increased tissue requirements. An alternative explanation of these data is that peripheral glucose uptake represents a passive tissue response to hyperglycemia due to increased hepatic glycogenolysis which only incidentally has resulted from increased secretion of epinephrine.

TABLE 1

HEPATIC PLASMA FLOW AND ARTERIAL CONCENTRATIONS OF LIPID FRACTIONS, GLUCOSE AND ELECTROLYTES

	Control Period	Hypovolemic Period			Normovolemic Period		
		Early stage	Middle stage	Late stage	Early stage	Middle stage	Late stage
Hepatic plasma flow	186 ± 12*	621 ± 161	219 ± 34	184 ± 23	528 ± 338	444 ± 53	370 ± 30
Cholesterol esters	2.44 ± .14	2.15 ± .07	2.09 ± .09	2.36 ± .17	2.28 ± .26	2.30 ± .04	2.35 ± .06
Free cholesterols	1.04 ± .07	.96 ± .02	.755 ± .139	.571 ± .080	.712 ± .090	.532 ± .029	.469 ± .018
Phospholipids	3.39 ± .083	3.51 ± .129	2.97 ± .113	2.32 ± .039	2.86 ± .005	2.99 ± .046	3.13 ± .083
Triglycerides	.280 ± .014	.314 ± .012	.286 ± .014	.281 ± .011	.215 ± .030	.232 ± .010	.210 ± .01
Free fatty acids	.596 ± .009	.584 ± .007	.623 ± .025	.491 ± .011	.459 ± .002	.679 ± .043	.672 ± .018
Total fatty acids	9.62 ± .19	9.64 ± .23	8.62 ± .18	7.65 ± .19	8.25 ± .18	8.78 ± .06	8.98 ± .16
Glucose	5.44 ± .14	5.67 ± .07	7.95 ± .59	6.86 ± .121	2.53 ± .03	3.32 ± .28	3.66 ± .52
K^+	3.5 ± .06	3.8 ± .09	3.6 ± .07	5.1 ± .60	5.4 ± .10	5.0 ± .10	4.9 ± .10
Na^+	138 ± .20	138 ± .20	138 ± .50	140 ± .80	140 ± .80	139 ± .90	139 ± .50

* Mean and standard error of the means for hepatic plasma flow, expressed in ml/min, and the arterial concentrations expressed in mM/liter or mEq/liter.

TABLE 2
HEPATIC OUTPUT OF LIPID FRACTIONS, GLUCOSE AND ELECTROLYTES

	Control period	Hypovolemic Period Early Stage	Hypovolemic Period Middle stage	Hypovolemic Period Late stage	Normovolemic Period Early Stage	Normovolemic Period Middle stage	Normovolemic Period Late stage
Cholesterol esters	-.073 ± .022*	-.314 ± .117	-.041 ± .019	-.012 ± .053	-.004 ± .064	+.160 ± .015	-.003 ± .025
Free cholesterol	-.031 ± .007	-.169 ± .061	-.031 ± .010	-.034 ± .015	-.211 ± .154	-.153 ± .027	-.128 ± .004
Total cholesterol	-.104 ± .025	-.484 ± .138	-.071 ± .018	-.046 ± .041	-.215 ± .218	+.008 ± .032	-.126 ± .022
Phospholipids	-.052 ± .014	-.153 ± .031	-.051 ± .029	+.074 ± .031	+.025 ± .015	-.071 ± .018	-.105 ± .124
Triglycerides	+.022 ± .007	+.053 ± .019	+.024 ± .006	+.025 ± .007	+.086 ± .039	+.037 ± .008	+.022 ± .008
Free fatty acids	-.005 ± .003	-.011 ± .010	-.002 ± .003	+.003 ± .004	+.013 ± .011	-.015 ± .010	-.029 ± .006
Total fatty acids	-.099 ± .025	-.424 ± .181	-.056 ± .055	+.192 ± .125	+.309 ± .089	+.137 ± .055	-.080 ± .071
Glucose	+.111 ± .021	+.330 ± .016	+.236 ± .044	+.060 ± .087	+.081 ± .011	+.220 ± .124	+.463 ± .106
K⁺	+.042 ± .020	+.181 ± .070	+.062 ± .010	+.193 ± .080	-.065 ± .010	+.017 ± .050	+.136 ± .050
Na⁺	+.870 ± .280	+.360 ± .220	+.031 ± .090	-.039 ± .190	-.025 ± .350	-.756 ± .440	-.304 ± .710

* Mean lipid and glucose values ± standard error of the mean expressed as mM/min, electrolytes as mEq/min; positive values represent hepatic output rates, negative values represent hepatic uptake rates.

TABLE 3

NONHEPATIC SPLANCHNIC OUTPUT OF LIPID FRACTIONS, GLUCOSE AND ELECTROLYTES

	Control period	Hypovolemic Period			Normovolemic Period		
		Early stage	Middle stage	Late stage	Early stage	Middle stage	Late stage
Cholesterol esters	+.117 ± .025	+.506 ± .145	+.043 ± .016	-.003 ± .010	+.209 ± .125	+.057 ± .034	+.049 ± .031
Free cholesterol	-.053 ± .013	-.073 ± .020	+.003 ± .013	+.001 ± .007	-.003 ± .004	-.040 ± .013	+.011 ± .009
Total cholesterol	+.064 ± .034	+.433 ± .134	+.040 ± .009	-.002 ± .010	+.212 ± .129	+.098 ± .028	+.060 ± .027
Phospholipids	+.047 ± .018	+.124 ± .085	+.021 ± .011	+.040 ± .006	+.135 ± .084	+.083 ± .026	+.028 ± .024
Triglycerides	+.016 ± .003	+.020 ± .008	+.003 ± .002	+.001 ± .001	+.046 ± .042	+.016 ± .004	+.013 ± .003
Free fatty acids	+.015 ± .003	+.051 ± .015	+.005 ± .002	+.003 ± .002	+.020 ± .010	+.017 ± .006	+.008 ± .007
Total fatty acids	+.258 ± .043	+.829 ± .210	+.093 ± .027	+.069 ± .023	+.597 ± .404	+.263 ± .041	+.145 ± .018
Glucose	-.037 ± .015	-.230 ± .058	-.062 ± .038	-.047 ± .013	+.047 ± .080	-.011 ± .080	-.259 ± .010
K⁺	+.017 ± .010	+.087 ± .060	+.028 ± .020	+.003 ± .005	+.027 ± .030	+.054 ± .020	+.016 ± .010
Na⁺	-.484 ± .240	-.113 ± .250	+.183 ± .146	+.051 ± .054	+.362 ± .180	+.990 ± .399	+.143 ± .220

* Mean lipid and glucose values expressed in mM/Min, electrolytes in mEq/min, positive values represent nonhepatic splanchnic output rates, negative values represent nonhepatic splanchnic uptake rates.

TABLE 4
TOTAL SPLANCHNIC OUTPUTS OF LIPID FRACTIONS, GLUCOSE AND ELECTROLYTES

	Control period	Hypovolemic Period			Normovolemic Period		
		Early stage	Middle stage	Late stage	Early stage	Middle stage	Late stage
Cholesterol esters	+.044 ± .030	+.192 ± .124	+.002 ± .030	-.015 ± .057	+.205 ± .061	+.218 ± .040	+.052 ± .04
Free cholesterol	-.085 ± .015	-.244 ± .071	-.034 ± .021	-.033 ± .020	-.208 ± .149	-.112 ± .026	-.118 ± .00
Total cholesterol	-.040 ± .028	-.051 ± .073	-.032 ± .014	-.048 ± .039	+.029 ± .089	+.106 ± .055	-.065 ± .04
Phospholipids	-.005 ± .020	-.029 ± .110	-.030 ± .033	+.114 ± .034	+.160 ± .100	+.013 ± .018	-.077 ± .04
Triglycerides	+.038 ± .006	+.074 ± .025	+.028 ± .006	+.026 ± .007	+.132 ± .081	+.053 ± .009	+.036 ± .00
Free fatty acids	+.010 ± .001	+.040 ± .011	+.003 ± .003	+.005 ± .004	+.033 ± .021	+.014 ± .011	+.036 ± .00
Glucose	+.074 ± .026	+.099 ± .120	+.174 ± .047	+.013 ± .088	+.121 ± .069	+.210 ± .124	+.205 ± .03
K^+	+.059 ± .030	+.268 ± .090	+.091 ± .030	+.195 ± .080	-.038 ± .040	+.070 ± .050	+.152 ± .03
Na^+	+.393 ± .130	+.247 ± .13	+.214 ± .080	+.012 ± .220	+.338 ± .530	+.234 ± .320	-.160 ± .74

Mean lipid and glucose values ± standard error of the mean expressed as mM/Min, electrolytes as mEq/min; positive values represent total splanchnic output rates, negative values represent uptake rates.

In early hypovolemia, the increased HGO was associated with an increased uptake of free cholesterol, cholesterol esters, phospholipids, FFA and total fatty acids. This pattern of hepatic responses suggests several possible interactions between carbohydrate and lipids (Fig 5). The energy requirements of the liver itself may be readjusted to accommodate to the available oxidative substrates; that is, the liver may metabolize a higher percentage of lipids when glucose is released at greater rates. There may be metabolic conversions of various lipid fractions. For example, the increased hepatic output of triglycerides associated with the simultaneous uptake of FFA, phospholipids and cholesterol esters suggests that the fatty acids of the latter compounds are being used in the synthesis of triglycerides.

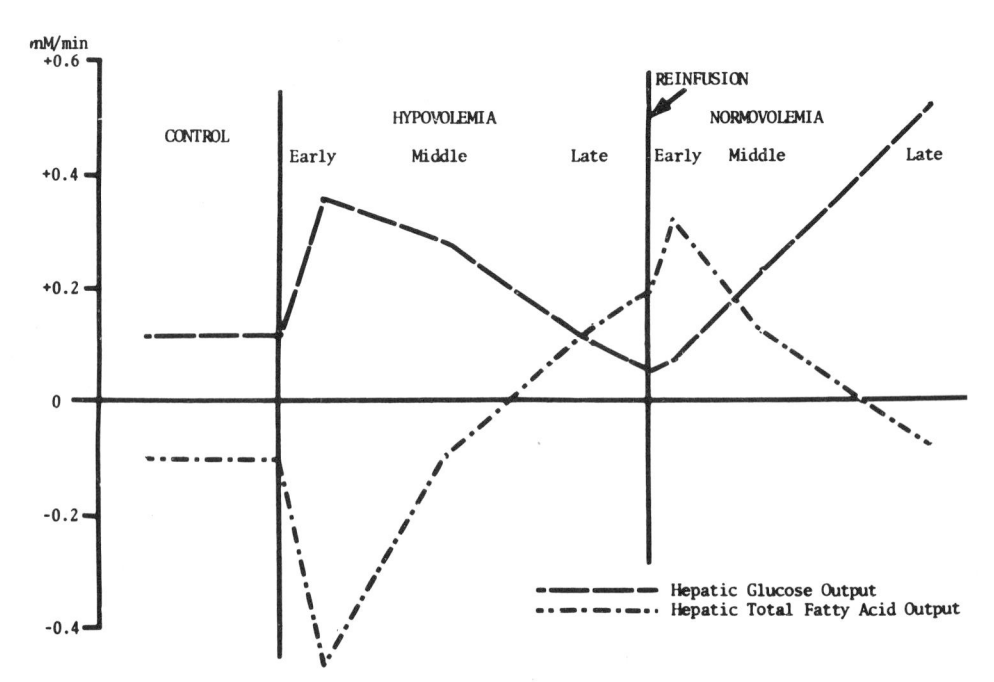

Fig 5 Hepatic glucose and total fatty acid output rates at each stage.

The early responses to hypovolemia by the nonhepatic splanchnic area were generally the opposite to those of the liver. The gut area removed glucose at markedly increased rates and released most lipid fractions at comparably increased rates (Fig 6). The roughly opposite responses of glucose and lipids in this area also suggest possible interrelationships. Presumably, the lipids are initially released by the fat depots of the mesentery and omentum by neural and neurohumeral mechanisms and the liver may respond to the increased portal venous lipid concentrations by increased lipid uptake rates.

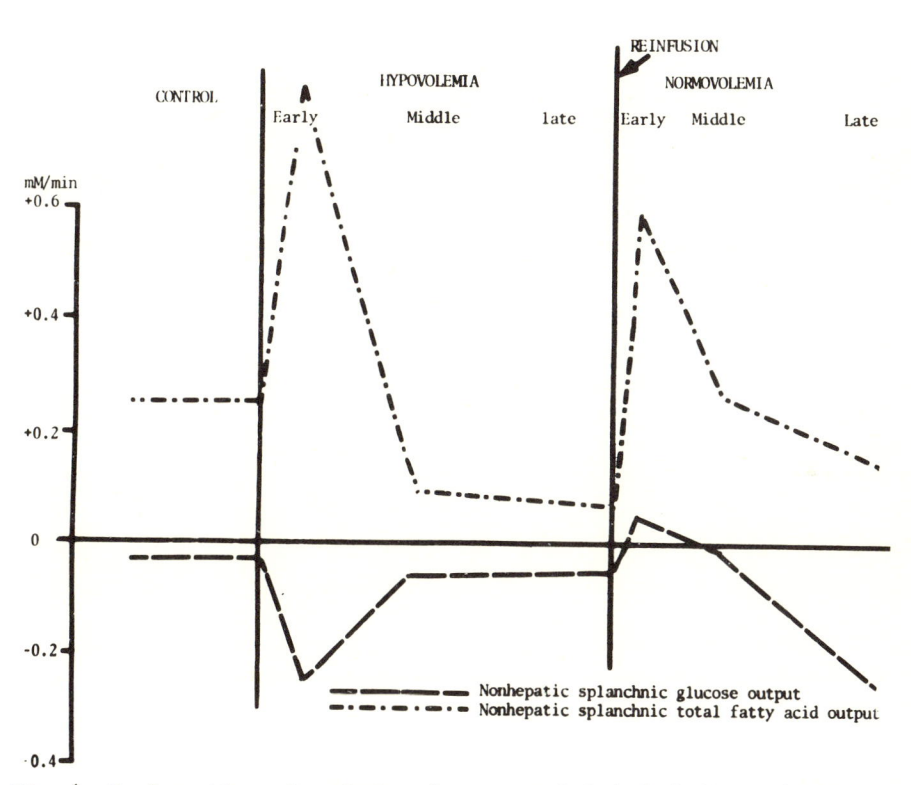

Fig 6 Nonhepatic splanchnic glucose and total fatty acid output rates at each stage.

In the total splanchnic area in the early hypovolemic period there was an increased output rate of glucose, cholesterol esters, triglycerides and FFA. These responses may reflect peripheral tissue needs, since an output of a given plasma constituent by the total splanchnic area approximately equals the peripheral uptake when the circulating concentration changes are relatively small compared with the net movements. These changes may be interpreted as: (a) increased metabolic demands of peripheral tissues accompanying the onset of stress, (b) the metabolic consequences of neurohormonal regulatory mechanisms, (c) uncoupling of oxidative phosphorylation, and (d) inefficient metabolism from incomplete, i.e. anaerobic, glucose oxidation.

The increased total body $\dot{V}O_2$ in the initial phase suggests that anaerobic glycolysis is not the major factor in the initial response. Because of the relatively normal mixed venous PO_2 values, it is unlikely that there were sufficient degrees of tissue hypoxia to explain the increased HGO. However, with redistribution of blood flow, it is possible that some tissues were anaerobic even though the whole body consumed O_2 at greater than normal rates.

In the late hypovolemic stage, as the high rate of HGO diminished, the liver released phospholipids, FFA and total fatty acids at greater than control rates. The fall in the HGO was probably not due to exhaustion of hepatic glycogen stores, since the liver was able to release substantial amounts of glucose in the terminal stage although the rate limitations of reduced HBF may have played a role.

In the nonhepatic splanchnic area the initial responses were also not sustained during the middle and late hypovolemic stages. The outputs of most of these lipid fractions fell below the control rates. This change was most marked with the cholesterol esters. The failure to sustain lipid release from the fat depots may reflect reduced blood flow, but it can hardly be explained by exhaustion of lipid stores.

The total splanchnic glucose output continued at a high rate through the middle hypovolemic stage, but fell below control values at the late stage. At this time availability of glucose to the peripheral tissues was severely limited. The markedly elevated output of cholesterol esters in the early hypovolemic stages fell below control values in the middle stage and became an uptake in the late stage. The opposite occurred with free cholesterol, so that relatively little change occurred in the total cholesterol movement. However, in the late hypovolemic stage, the total splanchnic area released phosphlipids and total fatty acids at greater than control rates.

After reinfusion of the shed blood and the reestablishment of a relatively normovolemic state, the patterns of carbohydrate and lipid transport changed remarkably. The HGO progressively increased in the late stage of this normovolemic period to values four-fold greater than control. Thus, hepatic glycogen stores were present up to the final agonal stage, indicating the animal was in a satisfactory stage of nutrition at the time of the experiment. Previously, we reported no significant glycogenolysis with maximum stimulation with glucagon in the depleted animal (12).

In contrast to the pattern of HGO, there was an initially increased hepatic output of total fatty acid which progressively fell to an uptake; this diminishing hepatic lipid release was primarily due to the fall of cholesterol ester, phospholipid and triglyceride outputs. The patterns of change in hepatic glucose and total fatty acid were approximately opposite to each other in these circumstances.

The pattern of changes in the nonhepatic splanchnic area during the normovolemic period were only roughly opposite to those of the liver (Fig 3,4). Glucose uptakes increased progressively until the terminal stage. At the same time, total fatty acid outputs progressively decreased; the latter reflected the composite fall in cholesterol esters, phospholipids, triglycerides and FFA. As in the hypovolemic period, the reduced output of lipids may be due to reduced blood flow or reduced O_2 availability to this area.

The total splanchnic area in the normovolemic period had a slowly but progressively increasing glucose output and a rapidly decreasing output of total fatty acids; the latter largely reflected cholesterol esters, phospholipids and triglycerides. The progressive diminution of lipids supplied to the peripheral tissues was associated with reduced blood flow, low venous PO_2 and diminished $\overset{\circ}{V}O_2$. This may be interpreted as: (a) a failure of O_2 delivery to the tissues, (b) an unavailability of oxidative substrates, (c) alteration of the cell membrane integrity, (d) failure to take up and utilize the available oxidative substrates, and (e) altered cellular metabolism and cell death in the terminal stage. In the presence of tissue hypoxia, glucose is incompletely and inefficiently burned and in the absence of adequate oxalacetate, the fatty acids may not be oxidized at satisfactory rates.

In essence, reduced transport of, or inability to utilize, O_2 or available calories at one stage or another may compromise cellular metabolism. This occurs as the agonal stage is ushered in.

SUMMARY

Hepatic blood flow, cardiac output, hepatic $\dot{V}O_2$ and total body $\dot{V}O_2$ were measured together with the inter-organ movements of glucose, K^+ and various lipid fractions in the unanesthetized dog. Measurements were made under control conditions and throughout early, middle and late stages of both hypovolemic and normovolemic shock. Initially in hypovolemia, there was a pronounced outpouring of glucose from the liver associated with hepatic uptake of lipid fractions; these metabolic responses were attributed to neurohormonal mechanisms. Later, as the initial HGO response diminished in hypovolemia as well as in the middle and late stage of normovolemia, the liver released the lipid fractions. The responses in the nonhepatic splanchnic area were qualitatively in the opposite direction to those of the hepatic area. The metabolic needs of the peripheral tissues, as reflected by peripheral uptake rates, increased in both hypovolemic and normovolemic shock. The final terminal mechanisms may result from an enormous number and variety of interacting components. Nonetheless, one of the limiting factors may be a lack of availability of oxidative substrates.

REFERENCES

1. Dole, V.P. (1956) J. Clin. Invest. 35:150.

2. Folch, J., Lees, M. & Sloan-Stanley, G.H. (1957) J. Biol. Chem. 226:497.

3. Kim, S.I., Desai, J.M. & Shoemaker, W.C. (1969) Am. J. Physiol. 216:1044.

4. Kinsman, J.M., Moore, J.W. & Hamilton, W.F. (1929) Am. J. Physiol. 89:322.

5. Kovach, A.G.B. & Fonyo, A. (1960) In: The biochemical response to injury, Stoner, H.B. & Threlfall, C.J., Eds. Oxford, Blackwell.

6. Moore, F.D. The metabolic care of the surgical patient. Philadelphia, W.B. Saunders, 1959.

7. Seeley, S.F. & Weisinger, J.R. Eds. Recent progress and present problems in the field of shock. Fed. Proc. 20, Suppl 9, 1961.

8. Shoemaker, W.C. (1960) J. Appl. Physiol. 15:473.

9. Shoemaker, W.C. Shock: Chemistry, physiology and therapy. Springfield, Charles C. Thomas, 1966.

10. Shoemaker, W.C. (1968) Rev. Surg. $\underline{25}$:9.

11. Shoemaker, W.C. & Elwyn, D.H. (1969) Ann. Rev. Physiol. $\underline{31}$:227.

12. Shoemaker, W.C. & Van Itallie, T.B. (1960) Endocrinol. $\underline{66}$:260.

13. Shoemaker, W.C., Carruthers, P.J., Elwyn, D.H. & Ashmore, J. (1962) Am. J. Physiol. $\underline{203}$:919.

14. Shoemaker, W.C., Walker, W.F. & Turk, L.N. (1961) Surg. Gynec. & Obst. $\underline{112}$:327.

15. Shoemaker, W.C., Walker, W.F., Van Itallie, T.B. & Moore, F.D. (1959) Am. J. Physiol. $\underline{196}$:311.

16. Sperry, W.M. (1938) Am. J. Clin Path., Tech. Suppl. $\underline{2}$:91.

17. Stoner, H.B. & Threlfall, C.J. Editors. Biochemical responses to injury. Oxford, Blackwell. 1960.

18. van Handel, E. & Zilversmit, D.B. (1957) J. Lab. Clin. Med. $\underline{50}$:152.

19. Wiggers, C.J. Physiology of Shock. New York, The Commonwealth Fund, 1950.

20. Zilversmit, D.B. Personal communication.

21. Zweifach, B.W. Annotated bibliography of shock, 1950-1962. Washington, D.C., National Academy of Sciences, National Research Council, 1963.

Supported by U.S. Army contract C9089 and N.I.H. grants GM 1706 and AM 13323.

EFFECT OF PORTO-CAVAL SHUNT ON SPLANCHNIC BLOOD FLOW IN HEMORRHAGE

J. Hamar, M. Gergely, I. Nyáry & A.G.B. Kovách

Experimental Research Department, Semmelweis Medical

University, Budapest, Hungary

Hypovolemic shock of the dogs is regularly accompanied by hemorrhagic necrosis of the intestinal epithelium. According to Lillehei et al., (1962), Selkurt (1970) and Lefler (1970) hemorrhagic enteritis appears to play an etiologic role in the shock state. Although the cause of the widespread intestinal damage is not yet clear, hemodynamic changes undoubtedly play a relevant role in its development. It was repeatedly shown (Fine, 1967; Lillehei et al., 1967; Bashour & McClelland, 1967) that blood flow through organs of the splanchnic area decreases considerably both in hemorrhagic and other forms of shock. Among possible etiologic factors, the long-lasting hypotension, the increased sympathetic activity (Chien & Hitzig, 1964; Fedina et al., 1965; Thämer et al. 1969) and, in dogs only, the constriction of the hepatic sphincters (Cohn & Parsons, 1950; MacLean et al., 1956; Knisely et al., 1957) should be considered. However the precise role of the hepatic sphincters in initiating necrosis of the intestinal epithelium has not yet been established. In order to clarify their role we studied the effect of porto-caval shunt on the development of shock in anesthetized dogs.

Splanchnic blood flow decreases considerably in shock and tissue damage is also seen at the same time. No data are available, however, on whether the extent of damage is uniform along the entire length of the intestine, or, alternatively, the various intestinal sections behave differently.

METHODS

Experiments with porto-caval shunts were performed on 50 mongrel dogs weighing 10-18 kg. The animals were anesthetized with pentobarbital sodium (30 mg/kg) and were divided into 2 groups. In the first group porto-caval shunt was established (p-c group) and shock was induced 3-5 weeks after the operation. The other group served as controls. Sham operation was performed in 8 of them, and the others were kept under identical circumstances for 4 weeks without sham operation. There was no hemodynamic difference between the two control groups. All the animals received the same diet.

Arterial and central venous pressures were measured with electro-manometers and heart rate from the ECG. Cardiac output and its subfractions were estimated by means of the thermodilution method (Kovách & Mitsányi, 1964). Polyethylene tubes for injection were placed at the root of the aorta, at the level of the diaphragm and below the renal arteries. A thermistor was inserted at the bifurcation of the abdominal aorta to measure the blood temperature changes due to the injection of saline at room temperature. The fractions of the cardiac output were calculated with a Fisher cardiac output computer from the differences in the flow values at the 3 injection sites.

After making control measurements the dogs were bled into a Lamson-bottle to keep their blood pressure first at 50-60 mm Hg for 90 min (first bleeding period BI) and later at 30-40 mm Hg for a 90 min period (BII). Following this all the blood in the reservoir-bottle was reinfused (R) within 10 min. Measurements were continued in most of the animals for 2 hr after reinfusion. Survival time was only studied in 4 control and 6 p-c animals after shock. The hemodynamic observation in this group was continued for 3 hr after reinfusion.

Measurements of intestinal blood flow. Fifteen mongrel dogs anesthetized with 30 mg/kg pentobarbital sodium were used for this study. Intestinal loops, both of the ileum and jejunum, were isolated through paramedian laparotomy. They were placed on a wooden plate, covered with warm, wet gauze, and their veins cannulated by a polyethylene tube. Venous outflow was continuously measured by a drop-counter unit and recorded on a kymograph. Arterial blood pressure was recorded by a mercury manometer.

RESULTS

Mean arterial blood pressure (Fig. 1) was significantly lower in the p-c group (Control = 150 ± 4.0, p-c = 123 ± 5.0 mm Hg; $P < 0.01$). Following reinfusion the restoration of blood pressure was not complete but there was no significant difference between the two groups. The bleeding volume (Fig. 1) of the p-c group was significantly lower during BI than in the controls. In the control group during the second part in BII there was spontaneous re-uptake from the reservoir. In the p-c group spontaneous re-uptake was not observed and the maximal bleeding volume was reached at the end of BII. The maximal bleeding volume was lower in the control group (48.1 ± 2.5 ml/kg) than in the p-c group (55.4 ± 3.1 ml/kg). The difference is statistically significant (P < 0.05).

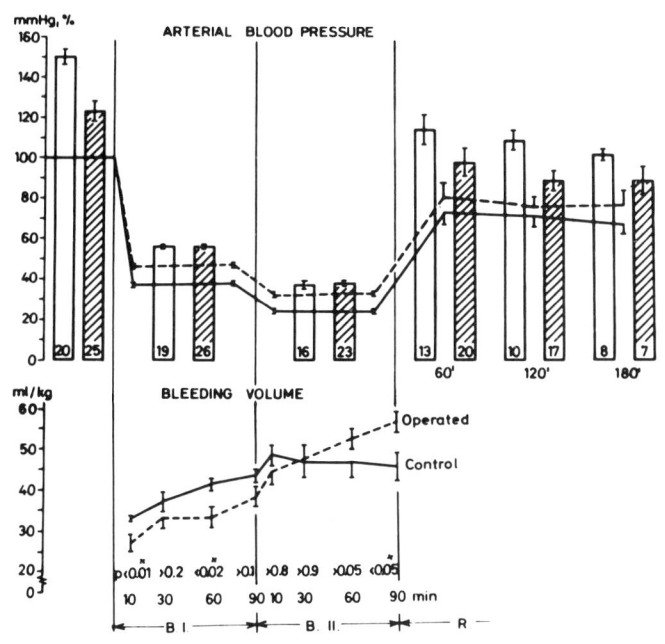

Fig. 1. Mean arterial blood pressure and bleeding volume in control and porto-caval (p-c) shunt dogs before (first 2 columns), during bleeding and after reinfusion. Arterial blood pressure (mm Hg) is indicated by columns and as a percentage of the initial blood pressure (lines). Open columns and solid lines = control shock, shaded columns and dotted lines = p-c shunt group. Numbers at the bottom of columns indicate the number of observations. First bleeding period = BI, second bleeding period = BII, reinfusion = R. The significant P values are marked by stars.

The central venous pressure (Fig. 2) decreased during bleeding in both groups. The decrease in the p-c group at the end of BI was significantly greater than in the control (P < 0.05). After reinfusion the central venous pressure only returned to the prebleeding level in the p-c group. The control group remained lower (significant difference 120 min after reinfusion (P = 0.05).

The cardiac output (Fig. 3) was 2.83 ± 0.08 l/min in the control and 2.63 ± 0.06 l/min in the p-c group before bleeding. Thirty min after starting bleeding the cardiac output of the control group was lower (P = 0.05). The difference between the 2 groups disappeared later. Reinfusion restored the cardiac output temporarily in the p-c group, the values of the controls remaining below the initial level. The difference in cardiac output between the 2 groups after reinfusion was statistically significant (P < 0.05).

The blood flow values before bleeding were as follows: head, forelimb, control: 1.04 ± 0.08 l/min, p-c: 1.15 ± 0.08 l/min; splanchnic, control: 1.34 ± 0.08 l/min, p-c group: 1.16 ± 0.07 l/min; pelvic hind-limb, control: 0.45 ± 0.08 l/min, p-c group: 0.36 ± 0.04 l/min.

The blood flow of both the splanchnic area and the hind-leg fraction in the p-c group was significantly higher than that of the control in the 30th minute of BI (Fig. 4). The splanchnic fraction of the cardiac output was higher in the p-c animals 60 and 120 min after the reinfusion. The blood flow of the hind-leg fraction of the p-c dogs differed from the controls only 180 min after reinfusion.

At autopsy there was severe hemorrhagic necrosis in the entire small intestine of the control shock dogs, while this phenomenon was not observed in the p-c shunt group. The survival rate in the porto-caval group was not different from that in the controls.

Blood flow studies in different segments of the intestine (Fig.5). The blood flow through the jejunum was 44 ± 3.5 ml/min/100g and through the ileum 36 ± 2.9 ml/min/100g before bleeding. During bleeding the reduction in the flow was the same, it decreased in the jejunum in BI to 39 ± 7.1% and in BII to 11.5 ± 9%, the corresponding values in the ileum were 37 ± 5.9% and 17.8 ± 5.5%.

There was a considerable difference in recovery of blood flow after reinfusion between the two regions. In the jejunum the recovery was 59 ± 9% of the initial flow level, in the ileum 90 ± 15%. The increase in resistance was the same in both segments during bleeding.

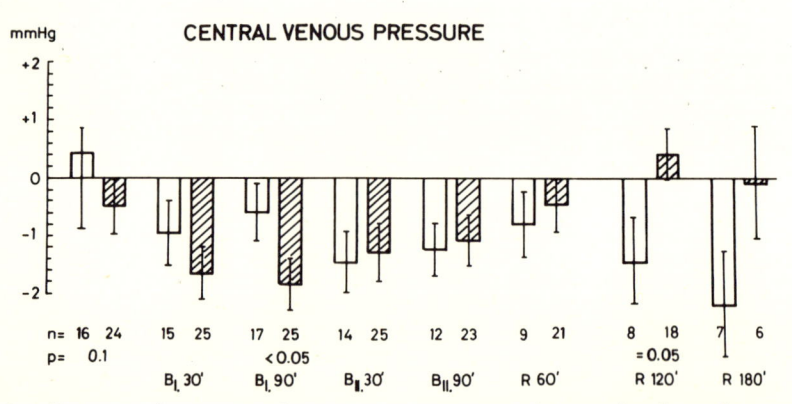

Fig.2. Changes in central venous pressure in control and p-c shunt group before (first two columns), during bleeding and after reinfusion. Open columns = control shock, shaded columns = p-c shock group, "n" denotes the number of experiments. Results of statistical evaluation are indicated by P values. First bleeding = BI, second bleeding = BII, reinfusion = R.

CARDIAC OUTPUT

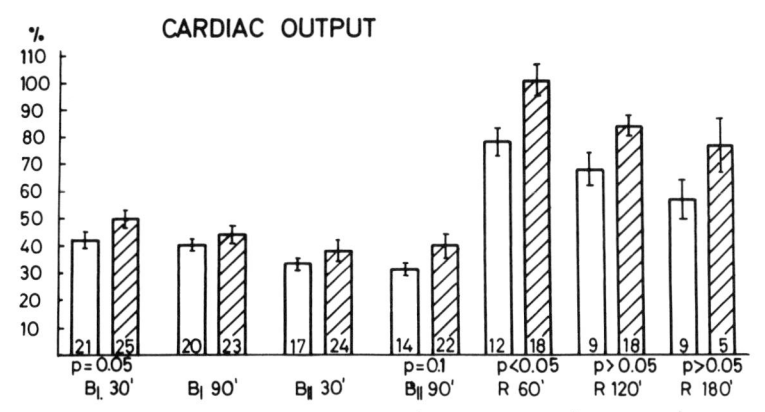

Fig. 3. Changes in cardiac output (relative, % values) in control and p-c shunt group, before, during bleeding and after reinfusion. Open columns = control shock, shaded columns = p-c shock group. Numbers of observations are indicated by numbers at the bottom the columns. Results of statistical evaluation are indicated by the P values. First bleeding = BI, second bleeding = BII, Reinfusion = R.

It reached a level of +50% in BI and +200% in the BII. After reinfusion the resistance of the ileum returned to the initial value which remained high in the jejunum (Fig. 6)

DISCUSSION AND CONCLUSIONS

According to our present studies the porto-caval (p-c) shunt eliminated the consequences of the constriction of hepatic sphincters and abolished the intestinal epithelial necrosis. This suggests that the hepatic sphincter has a role in the development of necrosis. Survival of the animals was not influenced by the p-c shunt, so we might conclude that there was no correlation between intestinal necrosis and the survival of dogs in shock. Because of the small number of animals in the survival studies we have to be careful about the above conclusion.

The p-c shunt operation and its later consequences can change the sensitivity of the dogs to hemorrhage. In p-c animals the peripheral vascular resistance was lower than in the control dogs. The bleeding volume was not equal in the 2 groups which can be explained by possible different degrees of stress after bleeding to the same level of blood pressure. This phenomenon might explain the higher values of cardiac output and of its splanchnic fraction at the same time. It is interesting to note that the vascular resistance of the p-c dogs was similar to the control animals after the first 30 min of the bleeding. This result suggests that the role of the hepatic sphincters in the increase of vascular resistance in the splanchnic region is subordinated during long lasting bleeding.

The spontaneous reuptake of blood from the reservoir in the control shock dogs at the end of the prolonged hypotension could be due to decreased sympathetic nervous activity (Fedina et al. 1965; Thämer et al., 1969).

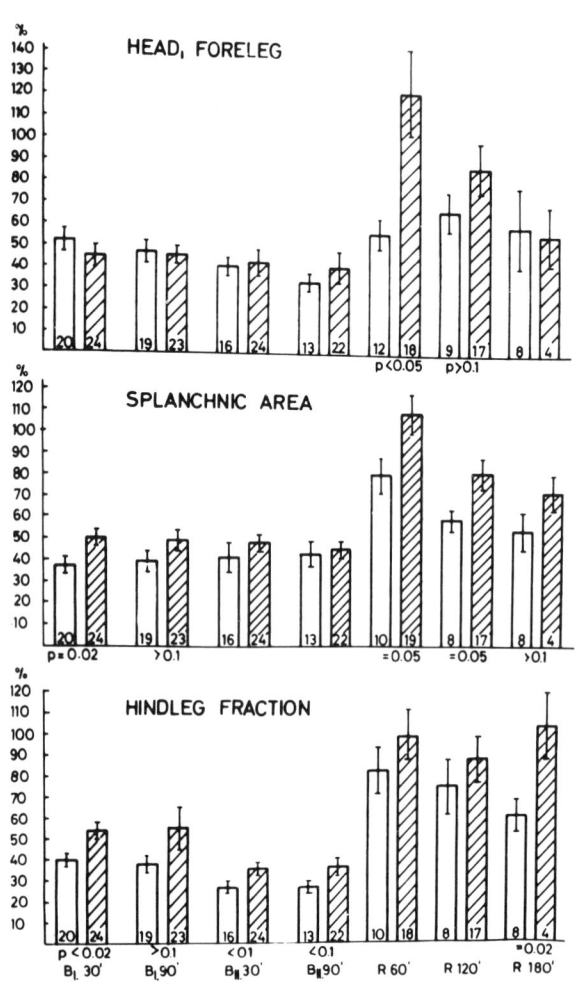

Fig. 4. Changes in head-forelimb, splanchnic and pelvic-hindlimb fractions of the cardiac output (relative % values) in control and p-c shunt group, before, during hemorrhage and after reinfusion. Open columns = control shock, shaded columns p-c shock group. Numbers of experiments are indicated at the bottom of the columns. Results of statistical evaluation are indicated by the P values. First bleeding = BI, second bleeding = BII, reinfusion period = R.

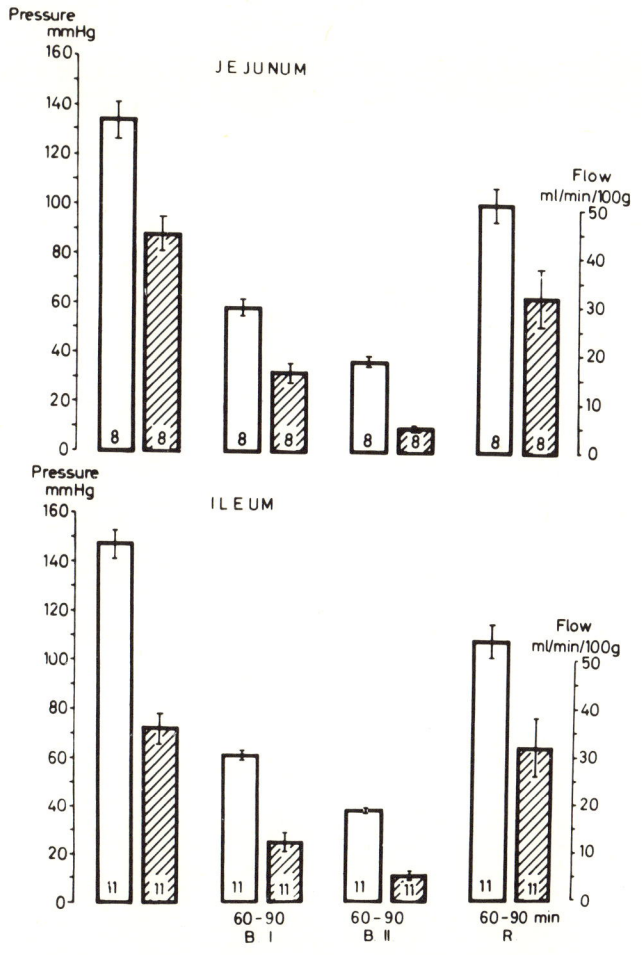

Fig. 5. Changes in intestinal (jejunum and ileum) blood flow in dogs, before (first two columns), during bleeding and after reinfusion. Blood flow was measured by the venous outflow method. Open columns mean arterial blood pressure values (left ordinate), shaded columns = blood flow values (right ordinate). Number of experiments are indicated at the bottom of the columns. First bleeding = BI, second bleeding = BII, reinfusion = R.

According to the blood flow studies we can conclude that the small intestine does not react uniformly to heamorrhagic shock. The blood flow through the jejunum and ileum was equally decreased and vascular resistance was similarly increased during hypovolaemia. However, after reinfusion the flow values approached the initial level only in the ileum. Because of the considerable difference in blood flow in the different layers of the intestinal wall the possibility of uneven vascular damage (Gouriz & Nickerson, 1965) has to be studied in more detail.

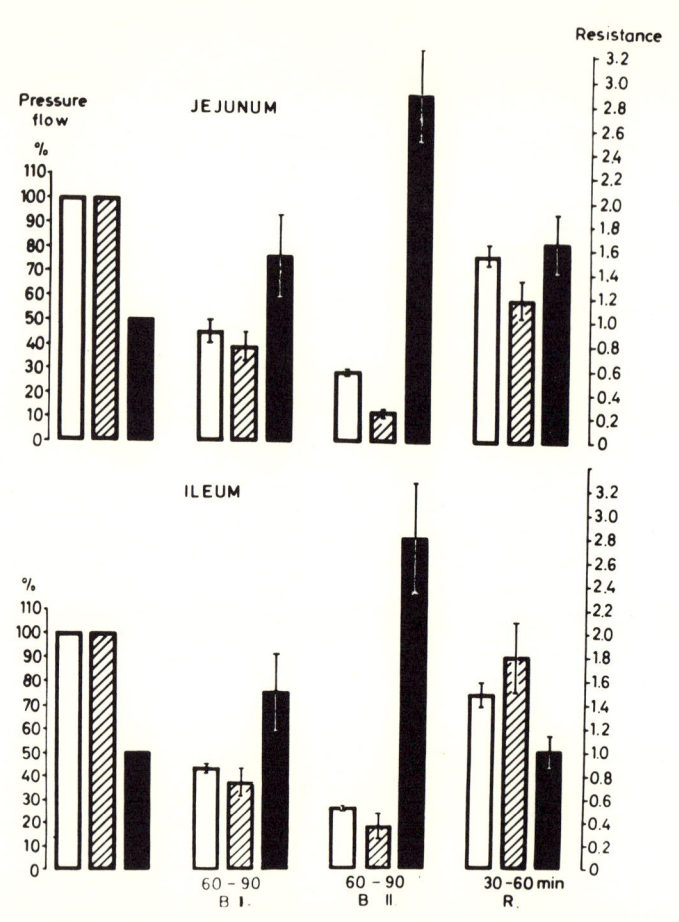

Fig. 6. Changes in intestinal (jejunum & ileum) vascular resistance (relative, % values) before, (first 3 columns), during bleeding and after reinfusion. Open columns = mean arterial pressure, shaded columns = blood flow, black columns = vascular resistance values.

REFERENCES

Bashour, F.A. & McClelland, R. (1967) Gastroenterology. 52, 461.
Chien, S. & Hitzig, B. (1964) Am.J.Phys. 206, 21.
Cohn, R. & Parsons, H. (1950) Am. J. Phys. 160, 347.
Fedina, L., Kovach, A.G.B. & Vanik, M. (1965) Acta physiol.hung. Suppl. 26, 35.
Fine, J. (1967) Gastroenterology 52, 454.
Gouriz, J.T. & Nickerson, M. (1965) In: Shock and Hypotension. Ed. L.C.Mils and J.H. Moyer. Grune & Stratton, New York-London.
Knisely, M., Harding, F. & Debacker, H. (1957) Science, 125, 602.

Kovách, A.G.B. & Mitsányi, A. (1964) Wien Med Wschr.114,401.
Leffer, A. (1970) Fed. Proc. 29, 1836.
Lillehei, R.C., Dietzman, R. & Movsas, S. (1967) Gastroenterology 52, 468.
Lillehei, R.C., Longbeam, J.K.& Rosenberg, J.C. (1962) In: Shock, Pathogenesis and Therapy. Ed. K.D. Bock, Berlin, Göttingen, Heidelberg, Springer Verl. p. 106.
MacLean, L.D., Weil, M.H., Spink, W.W. & Visscher, M.B. (1956) Proc.Soc.Exp.Biol.Med. 92, 602.
Selkurt, E. (1970) Fed. Proc. 29, 1832.
Thämer, V., Weidinger, H. & Kirchner, F. (1969). Z. Kreislaufforsch 58, 472.

ADIPOSE TISSUE AND HEMORRHAGIC SHOCK

S. Rosell, P. Sándor and A.G.B. Kovách

Dept. of Pharmacology, Karolinska Institute, Stockholm, Sweden and Semmelweis Medical University, Budapest, Hungary

The average man has around 15 kg of adipose tissue and as much as 90% of this may be triglycerides. As a consequence, adipose tissue may yield close to the theoretical 9.4 cal/g of pure lipid. The adipose tissue is then by far the largest energy source of the body. It is therefore obvious that it must be of great interest to find out how the adipose tissue reacts during different forms of trauma. There have been a limited number of studies on lipid metabolism, including changes in the plasma FFA level, during trauma. These studies have indicated that there is an increased mobilization of FFA from adipose tissue (Wadström, 1959, Stoner, 1962). Therefore the question has been raised whether high plasma FFA levels produced by elevated mobilization from adipose tissue may be harmful to the individual. It has been argued that heart failure may partly be due to oxidation of fat, since this requires more oxygen than carbohydrate oxidation. Furthermore, high plasma levels of FFA may be the cause of lung embolism.

However, there is also another side of the coin. If, by any means, the mobilization of fat from adipose tissue is inhibited during trauma then the organism may be in a serious situation as far as energy requirements are concerned. The carbohydrate stores of the body are very limited and will not last so long. Consequently, a risk of internal starvation must be considered.

This paper reviews our data on the blood flow and lipid mobilization during hemorrhagic shock which indicate that adipose tissue is very sensitive to hypotension and that the mobilization of fat from adipose tissue may be seriously impaired during trauma.

The adrenergic neuro-humoral system and adipose tissue

Before considering the situation in shock some of our relevant data concerning the sympathetic control of blood circulation and metabolism will be discussed.

The work of Dole (1956) and Gordon and Cherkes (1956) has shown that the mobilization of FFA is due to the hydrolysis of triglyceride stores in the adipocytes. Lipolysis leads to an outflow of FFA and glycerol from adipose tissue. Of the two products glycerol is generally used as a measure of the rate of lipolysis, since it is utilized in white adipose tissue to only a minor extent, or not at all, because adipose tissue has a very limited capacity to phosphorylate glycerol. FFA, on the other hand, is re-esterified to a certain extent as indicated by the finding that less than three moles of FFA per mole of glycerol are released in vitro (Steinberg & Vaughan, 1965) and in vivo (Fredholm & Rosell, 1968).

Electrical stimulation of the appropriate sympathetic nerves to subcutaneous adipose tissue produces increased outflow of FFA and glycerol, as well as vascular responses. These effects are seen with stimulation frequencies which are supposed to be within the physiological range (1-3 Hz) (Folkow, 1952)

Interestingly enough, adrenergic α-receptor blocking agents potentiate the venous outflow of FFA and glycerol following stimulation of sympathetic nerves (Fredholm & Rosell, 1968). One factor of importance for the potentiation seems to be the vascular reactions to sympathetic nerve activity in canine subcutaneous adipose tissue. As is indicated in Fig. 2 the vasoconstriction following sympathetic nerve stimulation is converted into vasodilatation following stimulation after administration of α-receptor blocking agents. It is conceivable that this promotes a more rapid and effective venous outflow of the products of lipolysis and may thus partly explain the potentiation.

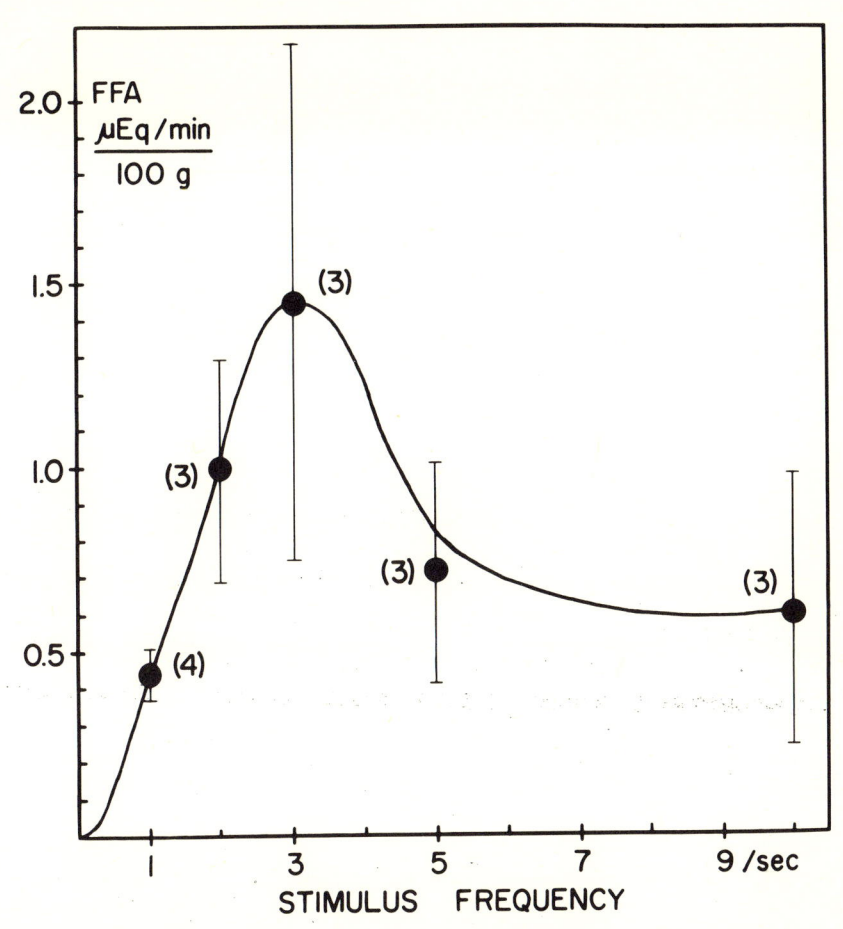

Fig. 1. Relation between stimulation frequency and the rate of net release of FFA in canine subcutaneous adipose tissue. Electrical stimulation of the nerve to the tissue. The net release rate is the value obtained during the stimulation period of 24 to 30 min. The figures within parentheses indicate the number of stimulations. The vertical bars represent ± S.E.M. (From Rosell, 1966, By permission).

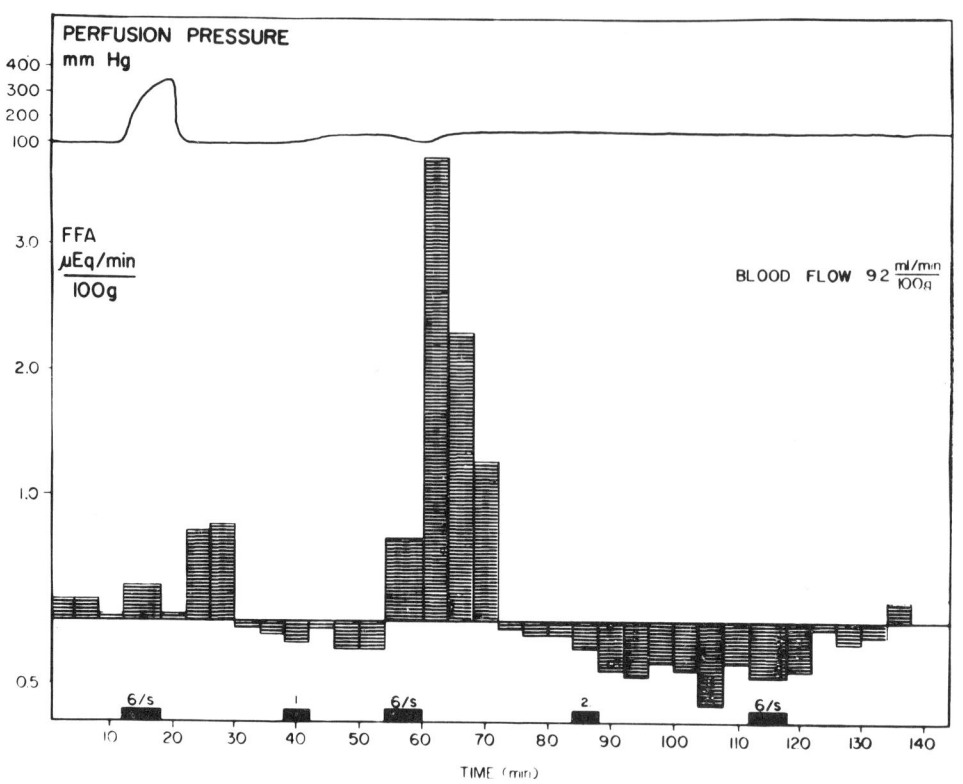

Fig. 2. Canine subcutaneous adipose tissue perfused with a constant blood flow of 9.2 ml/min/100 g. Electrical stimulation of the nerve to the tissue (6 Hz). Samples for FFA determinations collected during 3-min intervals. At 1, 60 μg of dihydroergotamine i.a. At 2, 100 μg of propanolol i.a. (Fredholm & Rosell, 1968. By permission).

There are important qualitative as well as quantitative regional differences in the responses to sympathetic nerve activity. Thus, in the adipose tissue of the mesentery factors other than sympathetic activity must be of greater importance for the control of circulation and metabolism as reported by Ballard (Ballard & Rosell, 1969). Although catecholamines are potent lipolytic agents, our studies indicate that in dogs blood-borne norepinephrine or epinephrine are not as effective in producing lipolysis as the activity in sympathetic nerves to adipose tissue (Ballard, Cobb & Rosell, 1971, Rosell, 1966). To increase the lipolytic rate in subcutaneous adipose tissue the concentrations of added norepinephrine or epinephrine in the arterial blood had to exceed 0.01 μg/ml and such levels have been reported for norepinephrine only at supramaximal work loads in man and during stress conditions, including bleeding, in dogs (Millar & Benfey, 1958). This should be compared with the fact that both the omental and subcutaneous adipose tissue react with elevated fat mobilization already during stimulation of the adrenergic sympathetic nerves with a frequency of 1/sec.

As is well known, there are several other lipolytic factors than the adrenergic neuro-humoral system, including growth hormone and glycocorticoids. These may induce lipolysis after a latency of more than 1 hr (Engel et al., 1959; Zahnd et al., 1960; Pearson et al., 1960; Fain et al., 1965). These so-called "slow acting hormones" may be of importance for the lipid metabolism during trauma but their importance in this respect does not seem to have been elucidated. The same is true for the "fast acting hormones" from the pituitary gland (ACTH) and (TSH) and from the pancreas (glucagon).

Blood flow in subcutaneous adipose tissue during hypotension

Studies on the adipose tissue during heomorrhagic shock have been performed in Budapest in recent years in collaboration with Dr. Kovách and his associates (Kovach et al., 1970, 1971). A standardized shock procedure has been used (Kovách, 1967). Chloralose anaesthetized dogs were bled to 55 mm Hg for 90 min and for a further 90 min period to 35 mm Hg. The shed blood was then reinfused. In all of the experiments there was a severe reduction of the blood flow in subcutaneous adipose tissue (Fig. 3). On the average, during bleeding to 55 mm Hg for 90 min the blood flow was reduced to about 10% of the resting blood flow (Fig. 4). This decrease is much more pronounced than that which occurs in most other organs under similar conditions. Thus in skeletal muscle, liver, myocardium and hypothalamus the blood flow fell to about 60% of the resting flow and the renal cortical blood flow to about 40%. Moreover, after 180 min of hypotension reinfusion of the shed blood failed to restore a normal blood flow in the subcutaneous adipose tissue, indicating vascular damage. This suggests that subcutaneous adipose tissue may be one of the organs where irreversible shock is manifested at an early stage. In contrast, dogs pretreated with phenoxybenzamine, an α-receptor blocking drug, did not develop signs of vascular damage during the bleeding periods. The blood flow remained at about 70% of the resting blood flow and reinfusion restored blood flow to normal. It is conceivable that the protective effect of phenoxybenzamine was largely a consequence of the comparatively high blood flow during the hypotensive period. Phenoxybenzamine treatment maintained tissue blood flow for two reasons. It gave protection from vasoconstriction by α-receptor blockade and caused vasodilation due to β-receptor stimulation (Fredholm & Rosell, 1968).

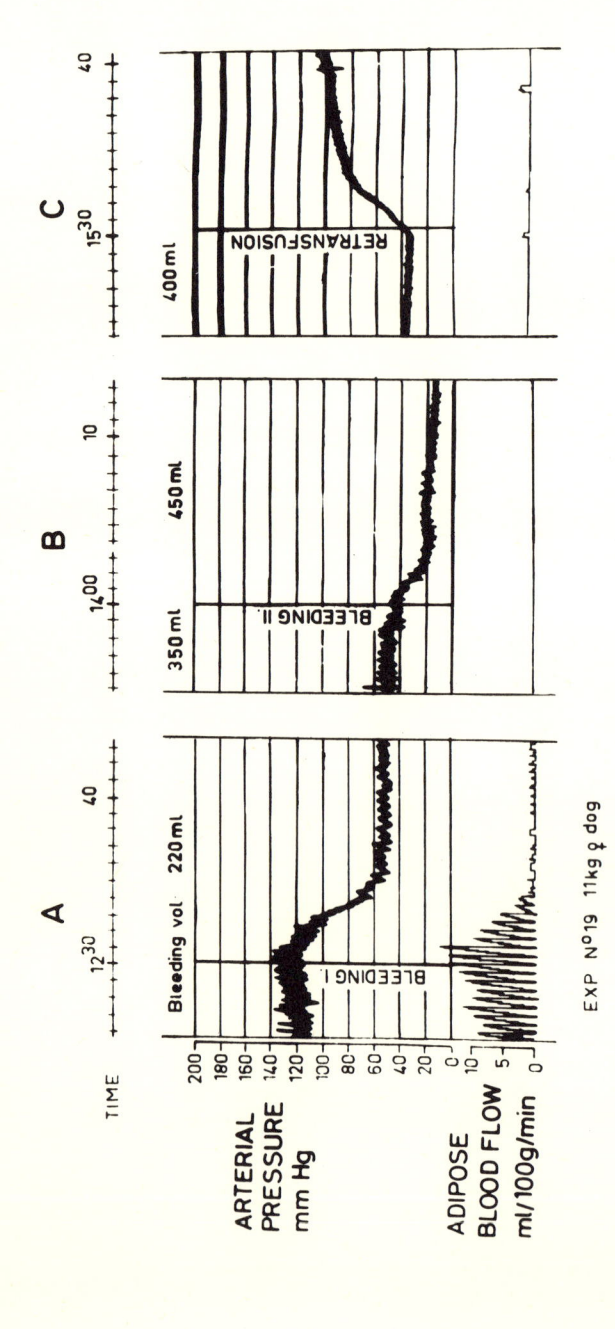

Fig. 3. Sections of an experimental record during standardized hemorrhagic shock. Note that the blood flow in the canine subcutaneous adipose tissue stopped completely during Bleeding I and Bleeding II and could not be restored during the retransfusion period (Kovách et al., 1970, by permission).

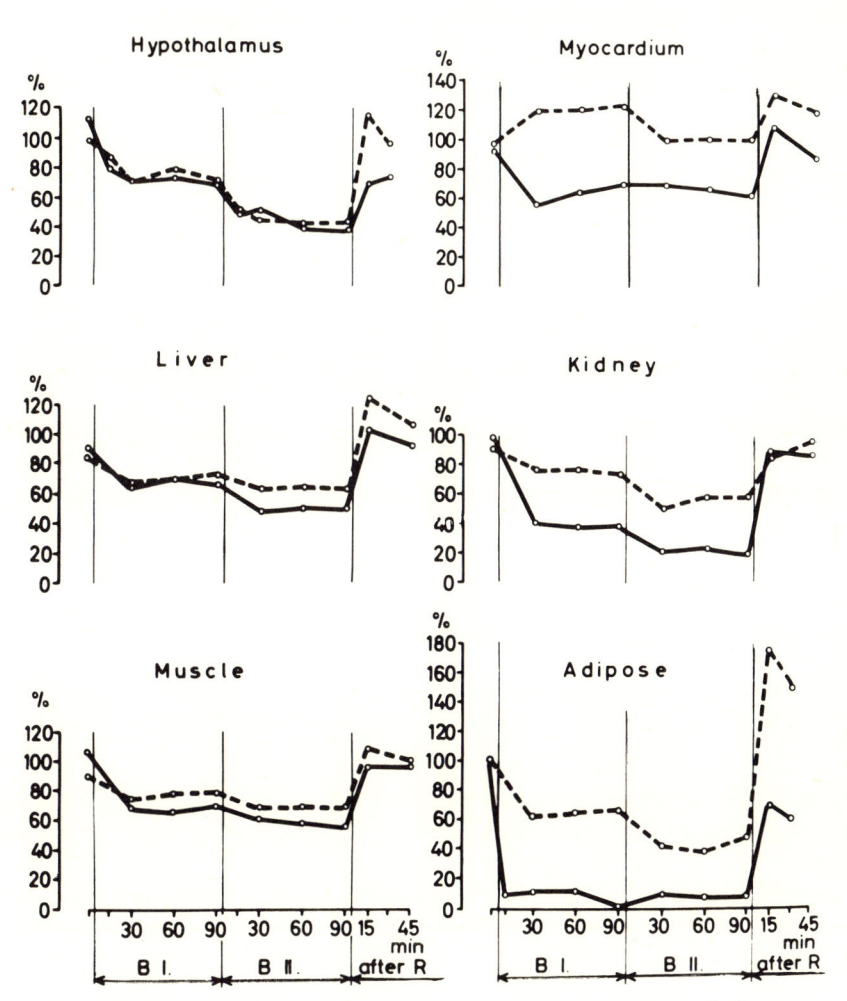

Fig. 4. Changes in blood flow in different organs of dogs during bleeding to 55 mm Hg for 90 min (B I) and to 35 mm Hg for a further 90 min period (B II). R = reinfusion of the shed blood.

Metabolic consequences of restricted blood flow

The impairment of blood flow in the subcutaneous adipose tissue had metabolic consequencies (Fig. 5) as well. Thus, despite a presumably high sympatho-adrenal activity during bleeding, the release rate of FFA did not rise. There was also a severe restriction in oxygen uptake in the subcutaneous adipose tissue. These local changes in metabolism of adipose tissue evidently had consequences for lipid metabolism in the whole animal, as indicated by a tendency for the plasma FFA concentration to fall. The uptake of FFA from blood into organs like skeletal muscle is related to the inflow of FFA (plasma flow x arterial FFA concentration) (Hagenfeldt & Wahren, 1971). Consequently, the diminished plasma FFA concentration during bleeding, as well as impaired blood flow in peripheral organs, may restrict the supply of FFA.

Rise in the re-esterification rate due to high lactate

The diminished blood flow in adipose tissue is presumably not the only reason why FFA release is severely restricted during bleeding. Other factors which may have a role in this connection are the elevated lactate concentration, as well as acidosis, both of which are known to inhibit the release of FFA (Issekutz & Miller, 1962; Issekutz et al., 1965; Nahas & Poyart, 1967; Triner & Nahas, 1965). Furthermore, Fredholm (1970; 1971) in our laboratory has demonstrated that Na-L(+) lactate at blood concentrations above 5 mM inhibits the release of FFA caused by electric stimulation of the sympathetic nerves to the subcutaneous adipose tissue (Fig. 6). Interestingly enough, the release of glycerol was not significantly affected by lactate infusion. The decreased release of FFA concomitantly with an unchanged glycerol release suggested an enhanced re-esterification of fatty acids in the adipose tissue. These findings thus indicate that not only the rate of hydrolysis of triglycerides but also the rate of re-esterification may be of importance in regulating the outflow of FFA from adipose tissue. These findings by Fredholm have a direct bearing upon our study of the metabolic events in adipose tissue during hemorrhagic shock since we found lactate concentrations of the same order of magnitude as those which elevated the re-esterification rate. As can be seen in Fig. 7, the plasma lactate concentration during the second bleeding period was around 10 mM which may explain why the arterial glycerol concentration increased but not that of FFA.

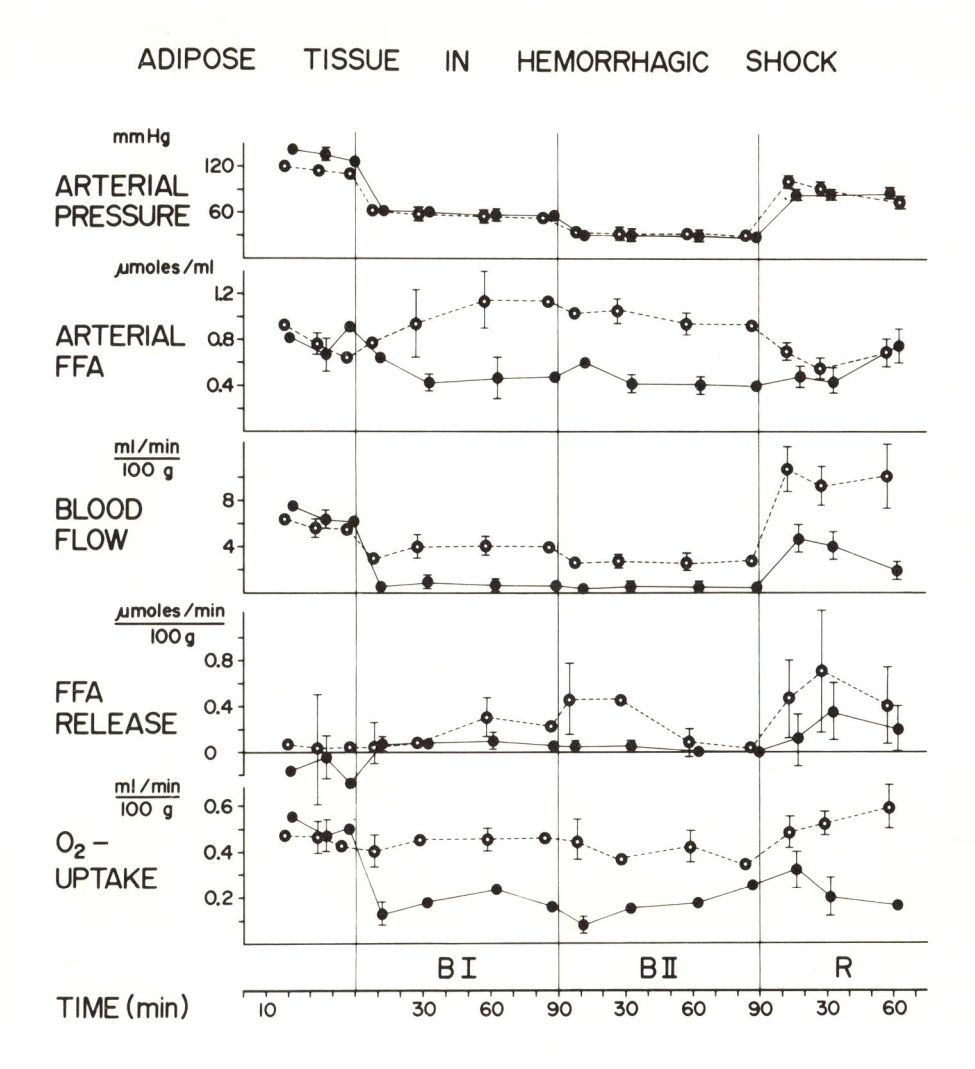

Fig. 5. Mean values (± S.E.M.) during control period, bleeding periods (B I and B II) and after reinfusion of the shed blood (R). Solid circles show experiments without phenoxybenzamine and open circles show experiments in phenoxybenzamine pretreated dogs (Kovách et al., 1970. By permission).

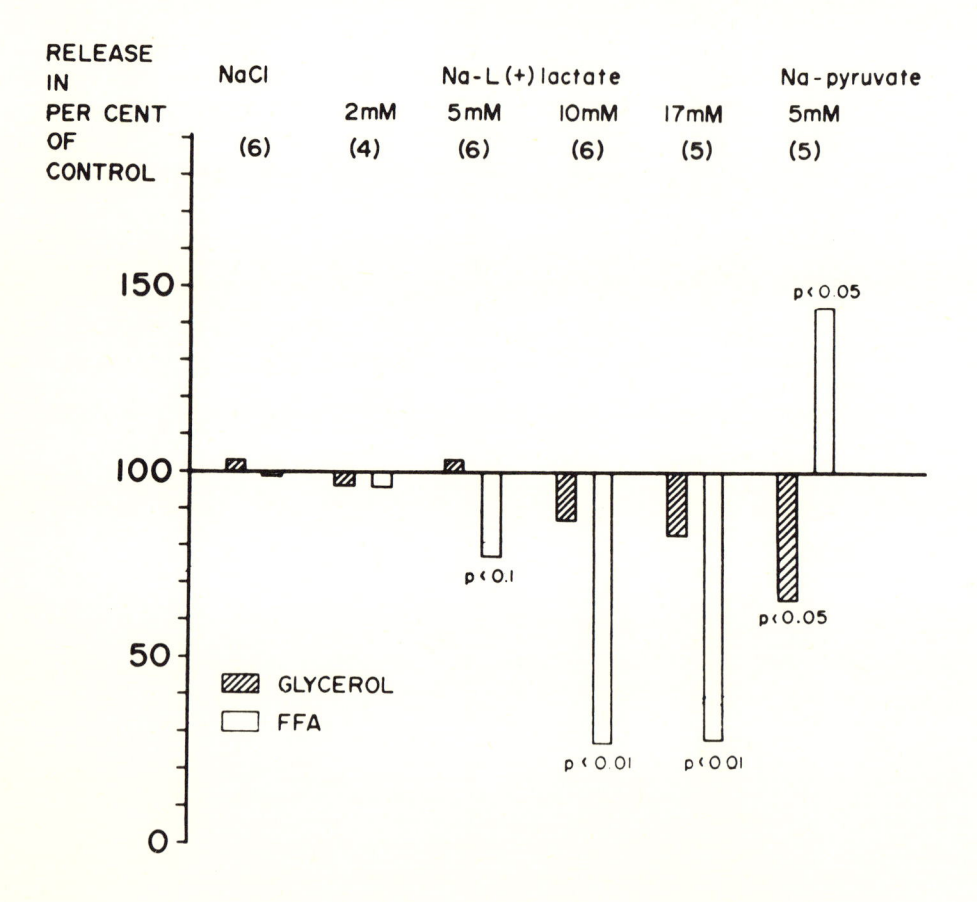

Fig. 6. The release of FFA and glycerol by sympathetic nerve stimulation during the infusion of lactate and pyruvate in per cent of the release caused by a similar nerve stimulation during saline infusion (canine subcutaneous adipose tissue) Figures within parentheses denotes the number of experiments. (Fredholm, 1971., By permission).

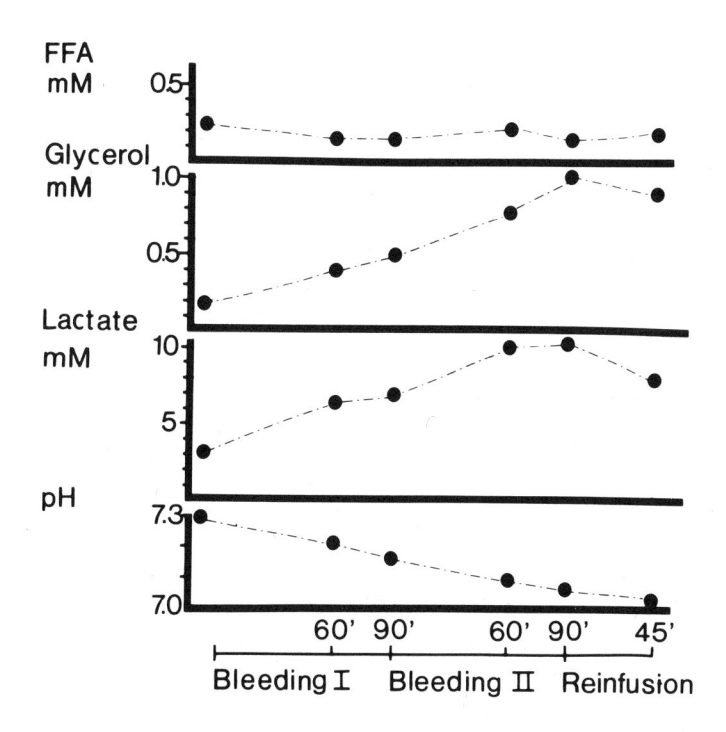

Fig. 7. Arterial concentrations of FFA, glycerol lactate and pH during hemorrhage in dogs. For further information see legend to Fig. 4.

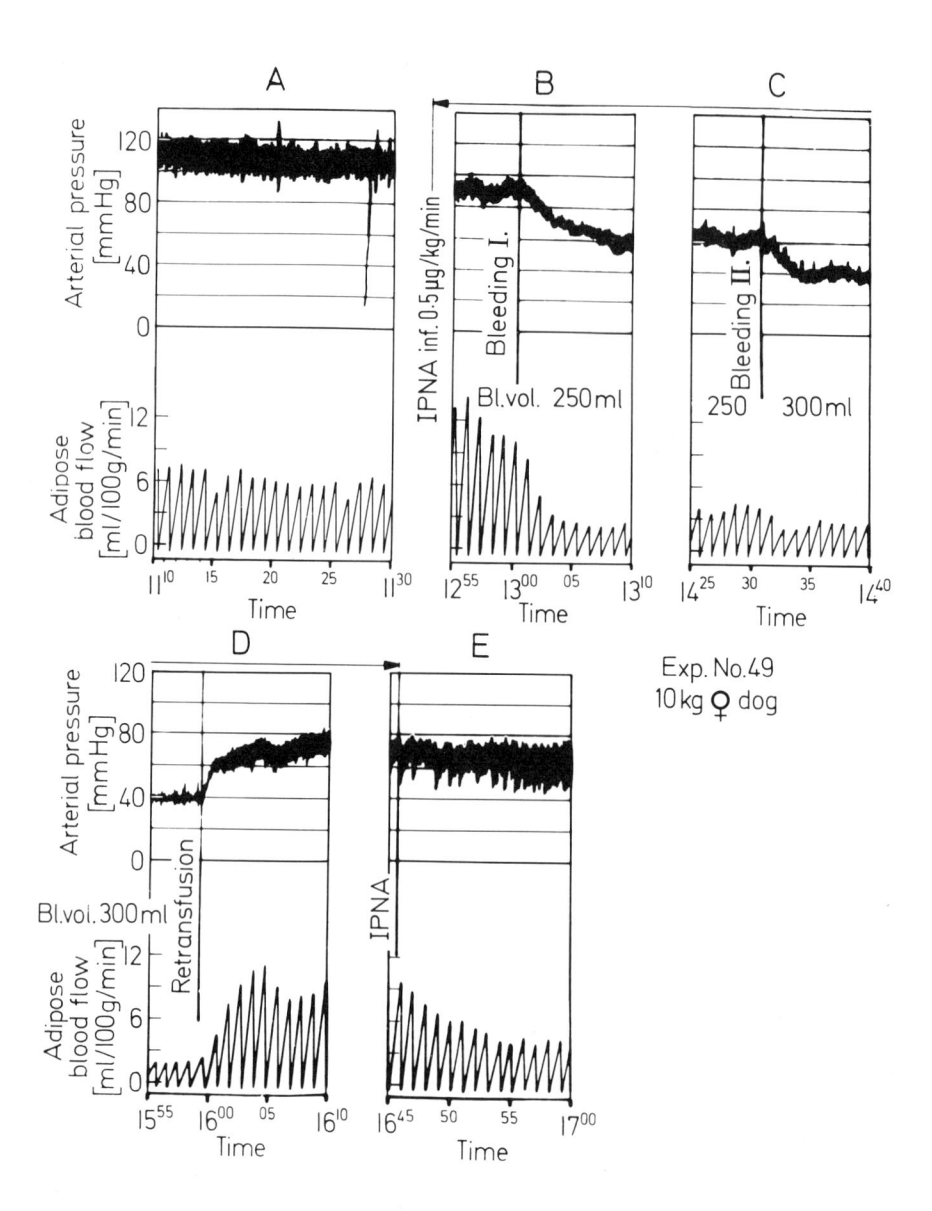

Fig. 8. Sections of an experimental record during standardized hemorrhagic shock. Isoprenaline 0.5 µg/kg/min, was infused i.v. IPNA = Isoprenaline; Bl.vol. = bleeding volume (Kovach et al., 1971, By permission).

Conclusions

Our experiments have shown that the outflow of FFA from subcutaneous adipose tissue is severely restricted during standardized hemorrhagic shock, despite a presumably high adrenergic neurohumoral activity. Some of the mechanisms for this have also been elucidated as being partly due to severe limitation of blood flow in adipose tissue and partly to an increased re-esterification rate. There are regional differences in the adrenergic neuro-humoral control of adipose tissue (Ballard & Rosell, 1969, 1971) and there also seem to be quantitative regional differences in the response to hemorrhagic shock (Kovách et al, to be published). In any event, our experiments show that trauma such as severe bleeding may profoundly limit the supply of energy to the peripheral tissues because of the fact that the largest energy store is not working properly. The next question is how to prevent this. Our experiments indicate that adrenergic α-receptor blockade is one way and, as is indicated in Fig. 8, adrenergic β-receptor stimulation may be another. Both treatments keep the blood flow in the subcutaneous adipose tissue going and one of the results is that there is no indication of irreversible damage. Furthermore, adipose tissue can release FFA into the circulation even during the hypotensive period.

REFERENCES

Ballard, K. & Rosell, S., (1969) Acta physiol. scand. 77, 442.
Ballard, K., Cobb, C.A. & Rosell, S. (1971) Acta physiol.scand. 81, 246.
Dole, V.P. (1956) J. Clin. Invest. 35, 150.
Engel, H.R., Bergenstal, D.M., Nixon, W.E. & Patten, J.M. (1959) Proc. Soc. Exptl. Biol. Med. 100, 699.
Fain, J.N., Kovacev, V.P. & Scow, R.O. (1956) J. Biol. Chem. 240, 3522.
Folkow, B. (1952) Acta physiol. scand. 25, 49.
Fredholm, B. & Rosell, S. (1968) J. Pharmacol. exp. Therap. 159, 1.
Fredholm, B. (1970) Acta physiol. scand. suppl. 354.
Fredholm, B. (1971) Acta physiol. scand. 81, 110.
Gordon, R.S., Jr. & Cherkes, A. (1956) J. Clin. Invest. 35, 206.
Hagenfeldt, L. & Wahren, J. (1968) Scand. J. Clin. Lab.Invest. 21, 314.
Issekutz, B., Jr., & Miller, H.J. (1962) Proc.Soc.exp.Biol. Med. 110, 237.
Issekutz, B., Jr., Miller, H., Paul, P. & Rodahl, K. (1965) Am. J. Physiol. 209, 1137.
Kovách, A.G.B. (1967) Fed. Proc. 20, 122.
Kovách, A.G.B., Rosell, S., Sándor, P., Koltay, E., Kovách, E. & Tomka, N. (1970) Circ. Res. 26, 733.

Kovách, A.G.B., Rosell, S., Sándor, P., Koltay, E., Hámori, M. & Kovách, E. (1971) Naunyn-Schiedebergs Arch. Pharmak. 268, 140.

Kovách, A.G.B., Rosell, S., Sándor, P., Hamar, J., Ikrenyi, N. & Kovách, E. to be published.

Millar, R.A. & Benfey, B.G. (1958) Brit. J. Anaesth. 30, 158.

Nahas, G.C. & Poyart, C. (1967) Amer. J. Physiol. 212, 765.

Pearson, O.H., Dominiquez, J.M., Greenbert, E., Pazianos, E. & Ray, B.S. (1960) Trans. Assoc. Am. Physicians. 73, 217.

Rosell, S. (1966) Acta physiol. scand. 67, 343.

Steinberg, D. & Vaughan, M. (1965) In Handbook of Physiology Section 5: Adipose Tissue. Eds. A.E.Renold & G.F.Cahill, American Physiological Society, Washington, D.C. p. 239.

Stoner, H.B. (1962) Brit. J. Exp. Path. 43, 556.

Triner, L. & Nahas, G.G. (1965) Science 150, 1725.

Wadström, L.B. (1959) Acta Chir. Scand. suppl. 238.

Zahnd, G.R., Steinke, J. & Renold, A.E. (1960) Proc. Soc. Exptl. Biol. Med. 105, 455.

INFLUENCE OF ENDOTOXIN ON ADIPOSE TISSUE METABOLISM

J.A.Spitzer, A.G.B.Kovach[1], S. Rosell, P. Sandor[2],
J.J. Spitzer and R. Storck
Departmentof Physiology and Biophysics,Hahnemann
Medical College and Hospital, Philadelphia,
Pennsylvania, 19102, U.S.A.

INTRODUCTION

Adipose tissue representa a relatively large tissue mass, about 15% of body weight, receiving a substantial portion of the cardiac output. Regional vasomotor adjustments in this vascular bed, therefore, are very likely to influence the "overall" hemodynamics of the animal. In addition, the mobilization of free fatty acids (FFA) from the energy rich triglyceride stores of adipose tissue is also dependent on an adequate blood supply. Since in endotoxic shock vascular responses are drastically altered, it became of interest to study the effect of endotoxin administration on adipose tissue metabolism.

It has been previously suggested that adipose tissue from different sites in the same animal shows quantitative

1.-Senior Foreign Scientist Fellow of the National Science Foundation.

2.-Research Fellow of the Heart Association of South-eastern Pennsylvania.

Abbreviations used: MABP = mean arterial blood pressure, CO = cardiac output, HR = heart rate, BF = blood flow, NE = norepinephrine, FFA = free fatty acids, GLY = glycerol, L = lactate, GLU = glucose, SC = subcutaneous, O = omental, M = mesenteric, TG = triglyceride, cAMP = cyclic 3',5'-adenosine monophosphate.

differences in responding to norepinephrine stimulation (1,2,3). Therefore, the present experiments were planned to study the metabolic changes occurring in subcutaneous and omental adipose tissue in response to endotoxin administration in dogs.

MATERIALS AND METHODS

Female, mongrel dogs, 16 to 23 kg were used after an 18 hr fast. They were anesthetized with Na pentobarbital given intravenously initially in a dose of 30 mg/kg and throughout the experiment at a constant infusion rate of 4 mg/kg/hr. Following spenectomy, each animal received 700 U.S.P. units heparin/kg body weight. Blood pressure was measured through the left common carotid artery, systemic arterial blood samples were taken from the left femoral artery. Cardiac output was determined by the dye dilution technique(4).

The subcutaneous fat pad in the suprapubic region of the dogs was completely isolated from the abdominal muscles. It remained connected with the dog's circulation and nervous system only by one supplying artery, one vein and a sympathetic nerve (5). The vein was cannulated and the outflowing blood collected in order to determine the subcutaneous tissue blood flow. Venous blood was continuously returned to the animal by transfusion through the right femoral or saphenous vein. Tissue samples for isolated fat cell studies were taken from the contralateral subcutaneous fat pad.

The omental adipose tissue was prepared according to Ballard's technique (3). The mesenteric adipose tissue was prepared in the following manner. An approximately 10 inch long segment of the small intestine - 15-25 inches below the duodenum - was isolated with a number of fine ligatures from the mesenteric adipose tissue. This segment was removed while the two lateral parts of the bowel with intact circulation were connected to each other by end-to-end anastomoses. The main mesenteric adipose tissue vein was then cannulated.

Blood and adipose tissue were sampled before, < 60 min after, and > 60 min after the injection of Escherichia coli endotoxin (Difco) in a dose that corresponded to approximately LD_{90}. Each blood sample for FFA, glucose (GLU), glycerol (GLY), and lactate (L) analysis was anticoagulated in vitro with heparin and kept in an ice bath. Blood GLY and L concentrations were determined enzymatically (6,7). Plasma FFA were measured by the Dole-Meinertz method (8), plasma GLU by the autoanalyzer.

Fat cells were obtained by enzymatic digestion of subcutaneous and omental adipose tissue, following the procedure of Rodbell (9) with minor modifications. The washed cells were incubated in plastic, capped culture tubes, in Krebs-Ringer bicarbonate buffer containing 4% bovine serum albumin in the presence of GLU (5.5 μMole/ml), and norepinephrine (NE, 0.2 μg/ml). The gas phase for incubation was 95% O_2 - 5% CO_2. Incubation was carried out at 37°C in a metabolic shaker for 2 hr. At the end of the incubation period aliquots of the medium were taken for subsequent FFA and for GLY analyses, the latter by an enzymatic fluorometric micro method (10).

Release of FFA and GLY was expressed in units per mmole triglyceride (TG) content of the cells (11), calculated on the basis of total fatty acid content after saponification.

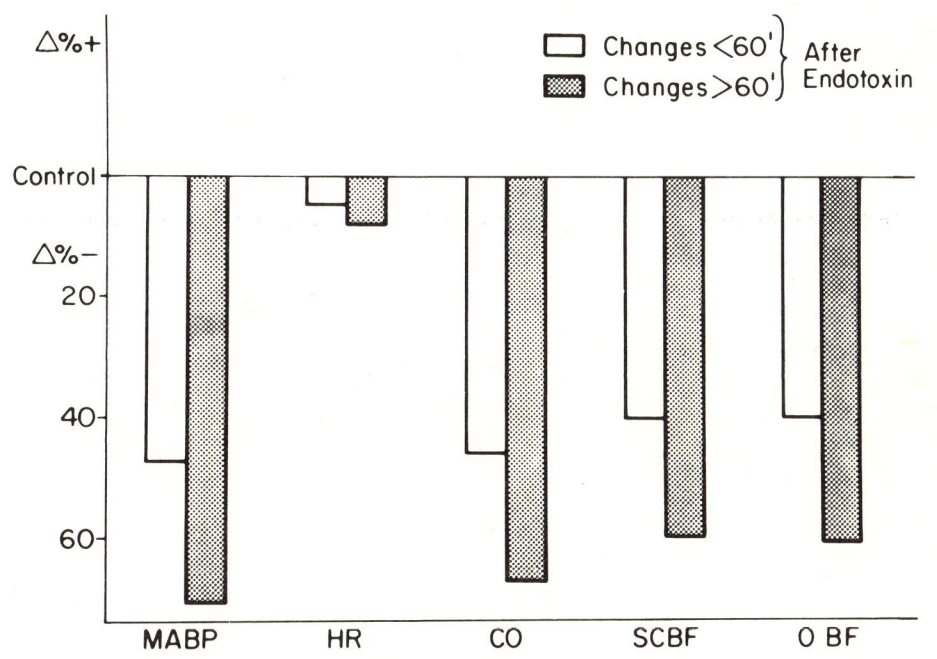

Figure 1; Hemodynamic changes following endotoxic shock (n=13).

Abbreviations used: MABP=mean arterial blood pressure, HR=heart rate, CO=cardiac output, SCBF=subcutaneous blood flow, OBF=omental blood flow.

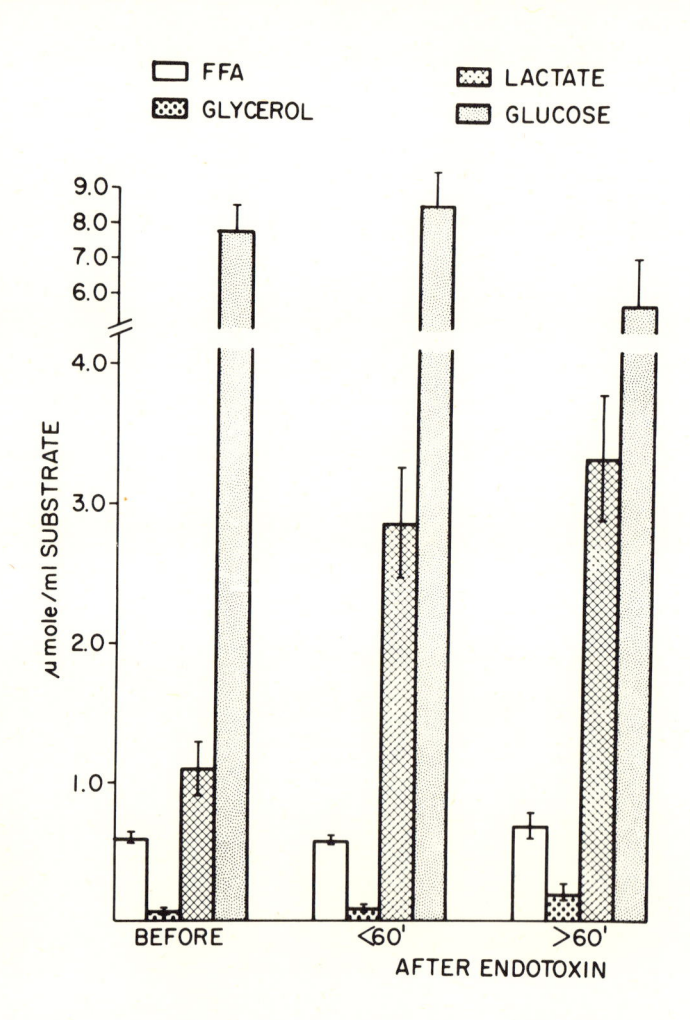

Figure 2: Arterial levels of various substrates before and after endotoxin administration.

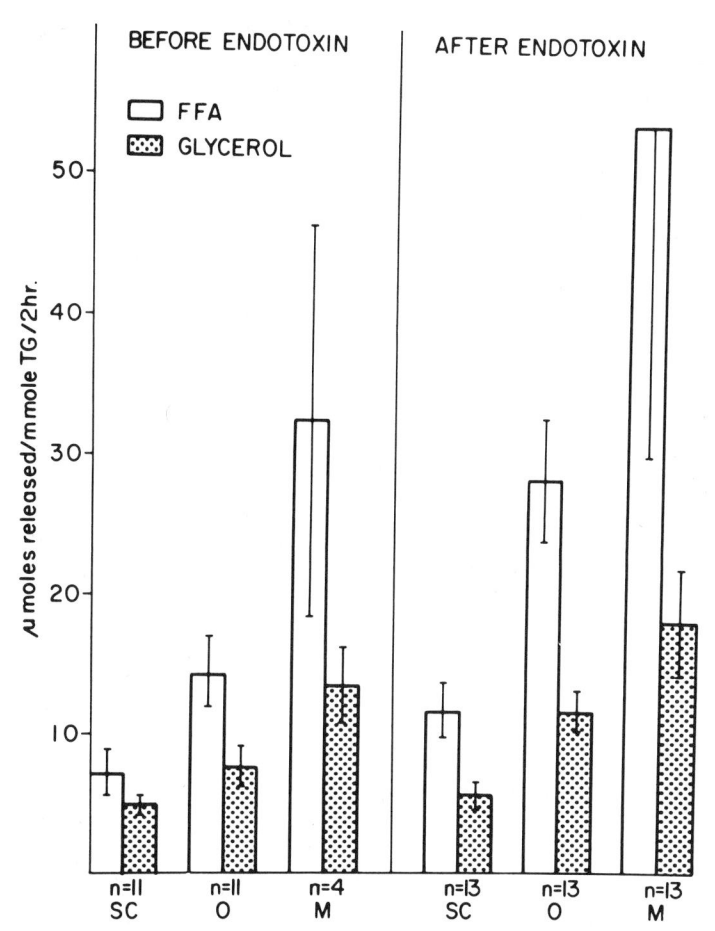

Figure 3: Norepinephrine stimulated FFA and glycerol release by isolated canine fat cells from various sites. Abbreviations used: SC=subcutaneous, O=omental, M=mesenteric adipose tissue, TG=triglyceride

RESULTS

MABP, CO, and BF decreased significantly, HR remained unchanged following E. coli endotoxin injection. As Fig. 1 illustrates, the percentage decrease of CO was quite similar to decrements of BF in the two adipose tissue regions at both time periods.

Mean values of the arterial concentrations of the metabolites studied are presented in Fig. 2. FFA levels remained unchanged throughout the experiments, GLY levels increased significantly >60 min following endotoxin administration. L levels rose progressively and significantly at both sampling times, GLU levels remained unchanged at <60 min, and decreased at >60 min after endotoxin administration.

NE-stimulated release of FFA and GLY was higher in O and mesenteric (M), than in SC fat cells both under control conditions and at both sampling times after endotoxin administration.

Following endotoxin administration there is increased NE invoked lipolysis at all three sites studied as shown in Fig. 3.

Under control conditions, the NE stimulated release of FFA by O cells was significantly higher ($P<.05$) than by SC cells. In 3 out of 4 experiments performed with M cells, they also released more FFA than SC fat cells. GLY release under the same experimental conditions was significantly higher by M cells ($P<.05$) than by SC cells. After endotoxin administration, the NE invoked release of both FFA and GLY was significantly higher by O ($P<.01$) and M adipocytes ($P<.05$ for FFA and $P<.01$ for GLY) than by SC fat cells.

Blood L levels became elevated, whereas FFA levels remained unchanged during endotoxic shock. It has been known for some time that an inverse relationship exists between plasma L and FFA levels (12). L might also serve as a feedback regulator of FFA mobilization (13). In a different experimental shock model, in hemorrhagic shock, we noted a negative correlation between blood L and plasma FFA (14).

GLY levels increased progressively after endotoxin administration. Since high circulating catecholamine levels are prevalent in endotoxic shock, this finding agrees with the observation of Ballard and Rosell (15) that the outflow of GLY from canine O adipose tissue was increased during sympathetic nerve stimulation or NE infusion.

In the course of our _in vitro_ fat cell studies we found that adipocytes of O adipose tissue consistently released more FFA and GLY upon NE stimulation, than adipocytes from SC adipose tissue, both before and after endotoxin administration.

Regional differences in enzyme activities and flavo-protein content in canine adipose tissue have already been reported (16). Response to epinephrine and NE varies with adipose tissue location in the rat (17) and in the human (18). Canine M fat was found to be relatively unresponsive to sympathetic nerve stimulation _in vivo_, in contrast to the strong lipolytic effect brought about in the SC adipose tissue (3). However, when isolated M fat cells were stimulated by NE *in vitro*, a more powerful lipolytic response than that of SC fat cells was observed before endotoxin (in 3 animals out of 4), as well as after endotoxin administration (in each of 4 animals).

The same dose of NE evoked a greater lipolytic response - as demonstrated by increased FFA and GLY release - in SC, O, and M fat cells after endotoxin, than before. In searching for an explanation for this endotoxin induced hypersensitivity, we turned our attention to the severely reduced blood flow in the adipose tissue regions studied, with their probably attendant hypoxia. We found (unpublished observation) that fat cells prepared from SC canine adipose tissue following experimentally induced hypoxia - achieved by clamping the artery supplying the tissue mass - also exhibited greater FFA and GLY release upon NE stimulation, than under normal conditions.

At the present time we can only speculate as to the exact mechanism of the accentuated lipolytic response to NE brought about by endotoxin. Circumstantial evidence indicates that tissue hypoxia may well be a contributing factor. Endotoxin may exert its effect on adipose tissue by promoting the accumulation of cAMP directly or indirectly, by releasing or unmasking unknown endogenous materials, or by some combination of these actions. Elucidation of some of these mechanisms will be the subject of further investigations.

SUMMARY

Adipose tissue metabolism was studied in the isolated SC and O regions in female mongrel dogs before and after *E. coli* endotoxin administration. MABP, CO, and BF decreased significantly, HR remained unchanged after endotoxin injection. Arterial concentration of FFA remained unchanged, GLY and L increased progressively, and GLU decreased only >60 min after endotoxin administration.

Both basal and NE stimulated lipolysis by isolated fat cells were studied. Adipocytes of O adipose tissue consistently released more FFA and GLY upon NE stimulation, than adipocytes from SC adipose tissue, both before and after endotoxin administration. M adipocytes behaved like those of O origin in 3 out of 4 instances under control conditions, and in each instance after endotoxin. Endotoxic shock increased the NE induced lipolytic response in SC, O and M fat cells, as illustrated by increased FFA and GLY release.

REFERENCES

1. Aronovsky, E., Levari, R., Vornblueth, W., & Wertheimer, E., (1963) Invest.Ophthal. 2: 259.

2. Carlson, S.A. & Hallberg, D., (1968) J. Lab. Clin. Med 71: 368.

3. Ballard, K. & Rosell, S., (1969) Acta Physiol. Scand. 77: 442.

4. Kinsman, J.M., Moore, J.W., & Hamilton, W.F., (1929) Am. J. Physiol. 89: 322.

5. Rosell, S., (1966) Acta Physiol. Scand. 67: 343.

6. Wieland, O., (1963) Glycerol, In:Methods of Enzymatic Analysis (H.V. Bergmeyer, ed.), New York-Academic Press.

7. Hohorst, H.J., (1963) L-(+)-lactate determination with lactic dehydrogenase and DPN. In: Methods of Enzymatic Analysis (H.V.Bergmeyer, ed.), New York-Academic Press.

8. Dole, V.P. & Meinertz, H., (1960) J. Biol. Chem. 235: 2595.

9. Rodbell, M., (1964) J. Biol. Chem. 239: 375.

10. Laurell, S. & Tibbling, G., (1966) Clin. Chim. Acta 13: 317.

11. Fain, J.N. & Reed, N. (1970) Lipids 5: 210.

12. Miller, H.I., Issekutz, B., Jr., Paul P., & Rodahl, K., (1964) Am. J. Physiol. 207: 1226.

13. Weil, R., Hovand, P., Altszuler, N., (1965) Am. J. Physiol. 208: 887.

14. Spitzer, J.A., & Spitzer, J.J., J. Trauma, in press.

15. Ballard, K. & Rosell, S., (1971) Circul. Res. 28: 389.

16. Spitzer, J.A., (1967) Proc. Soc. Exptl. Biol. Med. 124: 640.

17. Wertheimer, E., Hamosh, M., & Shafrir, E., (1960) Am. J. Nutr. 8: 705.

18. Carlson, L. & Hallberg, D., (1965) J. Lab. Clin. Med. 71: 368.

Aided by grants from the John A. Hartford Foundation, Inc., and from the National Institutes of Health (HE 03130).

CHARACTERIZATION OF RAT BRAIN MITOCHONDRIA, THE EFFECT OF INJURY

J. Somogyi[+], Jill E. Cremer and Kornélia Ikrényi

Experimental Research Dept., Semmelweis University
Medical School, Budapest, Hungary and Biochemical
Mechanism Section, Medical Research Council Toxicology
Unit, Carshalton, Surrey

Data concerning metabolism of brain mitochondria are contradictory in more than one respect. It is beyond doubt that most of the problems are associated with the fact that crude mitochondrial fractions of the brain homogenate always contain other cellular elements, e.g. cell membrane fragments, synaptosomes, etc. (Klingenberg et al. 1959; Somogyi & Vincze, 1961; Gray & Whittaker, 1962; Jobsis, 1963; Abood, 1970; Cunningham & Bridgers, 1970).

For extensive study of the metabolism of normal and, in particular, of pathologically altered brain mitochondria two requirements must be fulfilled: selection of a suitable, standardisable method for the preparation of pure mitochondria, and description of the basic properties of the isolated mitochondria.

Very pure mitochondria can be prepared from brain tissue by using density gradient- or zonal-centrifugation. Electron microscopic examination of these preparations shows a very homogeneous population. However, their biochemical study is quite problematic because of the small amount of material in a pure mitochondrial fraction and because the original properties of the mitochondria change almost in all cases after a hypertonic treatment which can hardly be avoided during the gradient separation (see in Abood's Review, 1970).

In this paper some data will be discussed concerning the study of oxygen uptake of rat brain mitochondria prepared with two different methods. According to the first procedure (Somogyi & Vincze, 1961), the contaminating impurities of mitochondria were

---

[+] Supported by a research fellowship from the Wellcome Trust, London

removed by several washings of the mitochondrial pellet. Using the other method as described by Clark & Nicklas (1970), the mitochondria were purified by means of a Ficol gradient. The two procedures yielded essentially the same results. By using our method, 3 times more mitochondrial protein could be isolated than with the other procedure. On the other hand, oxygen uptake of the mitochondria, prepared according to Clark & Nicklas (1970), was 20-30 per cent higher than that found in our preparation indicating a higher purity.

The various dehydrogenase activities were measured spectrophotometrically. Oxygen consumption was recorded using a Clark oxygen electrode. All the experiments were carried out at $37^{\circ}C$.

Table 1 presents some dehydrogenase activities. As it can be seen, dehydrogenases acting in the citrate cycle show similar activities, ranging between 200 to 500 nmoles/mg protein/min. Malate dehydrogenase, being the sole exception where the specific activity was about 23,000. It is very characteristic that all brain mitochondrial preparations contain a lactate dehydrogenase activity indicating a synaptosomal contamination. The brain mitochondria exert a beta-OH-butyrate dehydrogenase activity which is comparable with the other dehydrogenase activities functioning in the citrate cycle. It should be mentioned briefly that the activity of this dehydrogenase is extremely dependent on the age of the rats. Maximal activity can be measured when the rats are 3-4 weeks old. The composition of food may perhaps influence the activity but other conditions, such as starvation are surely without effect.

Table 1

Dehydrogenase Activities of Rat Brain Mitochondria

Dehydrogenase	No. of mitochondrial preparations	Activity nmoles/mg protein/min
malate	13	23,300 ± 2,100**
isocitrate (NAD)	7	510 ± 40
isocitrate (NADP)	7	260 ± 35
glutamate*	7	195 ± 45
succinate	6	495 ± 42
β-hydroxybutyrate	12	145 ±. 32
lactate	14	590 ± 74

*glutamate dehydrogenase activity was measured in direction
glutamate ⟶ α ketoglutarate

** ± S.D.

Table 2 included the values of oxygen uptake of brain mitochondria using different substrate pairs of citrate cycle intermediates. The highest oxygen consumption was measured with malate + pyruvate. The average value of state 3 respiration was near 500 n atoms oxygen/mg protein/min. The highest respiratory rate was 720 but the lowest one was never under 380. Respiration in the presence of DNP was always less than that measured in the presence of ADP in all cases tested. It is very important that oxygen uptake with beta-OH-butyrate and glutamate were similar. The lowest oxygen consumption was measured with succinate. However, respiration was more pronounced when succinate oxidation was recorded in the presence of rotenone.

Table 2

Oxygen Uptake of Rat Brain Mitochondria

No. of mitochondrial preparations	Substrate	State 4	With ADP	With DNP	Resp. control ratio
16	mal + pyr	108 ± 8.4	545 ± 62.0	515 ± 55.4	5.05
12	OOA + pyr	88 ± 7.2	394 ± 37.6	382 ± 36.6	4.47
8	mal + glut	100 ± 9.6	398 ± 41.2	378 ± 40.4	3.98
8	succ	124 ± 10.9	324 ± 36.4	313 ± 34.6	2.62
6	succ with rotenone	148 ± 16.2	440 ± 31.6	432 ± 33.1	2.70
8	mal + β-OH-but	112 ± 14.1	322 ± 39.4	303 ± 35.0	2.88

nAtom O/mg protein/min

The incubation medium contained 10 mM tris-phosphate (pH 7.4) 6 mM $MgCl_2$, 58 mM KCl, and 26 mM NaCl, total volume 1.6 ml; substrate concentrations: malate (mal), oxalacetate (OAA) 1 mM; pyruvate (pyr), succinate (succ), glutamate (glut) 5 mM; dl-β-hydroxybutyrate (β-OH-but) 10 mM; 1 μmole ADP and 40 nmoles DNP were added respectively.

Fig. 1 shows the action of inorganic phosphate on respiration with two substrate pairs, i.e. malate + pyruvate and oxaloacetate + pyruvate in the presence of ADP or DNP. Up to a limited concentration of P_i oxygen consumption increased. Further rise in P_i concentration caused a definite decrease in respiration. It was very characteristic that the inhibitory effect of P_i was always more pronounced in the presence of DNP than in the presence of ADP. This biphasic effect of P_i can be explained as follows: according to Chappel & Haarhoff's (1967) investigations the penetration of substrate dicarboxylic acid requires the presence of some phosphate. Therefore low P_i concentration stimulates respiration also in the presence of DNP by increasing the penetration of substrate anions. However, it seemed that at

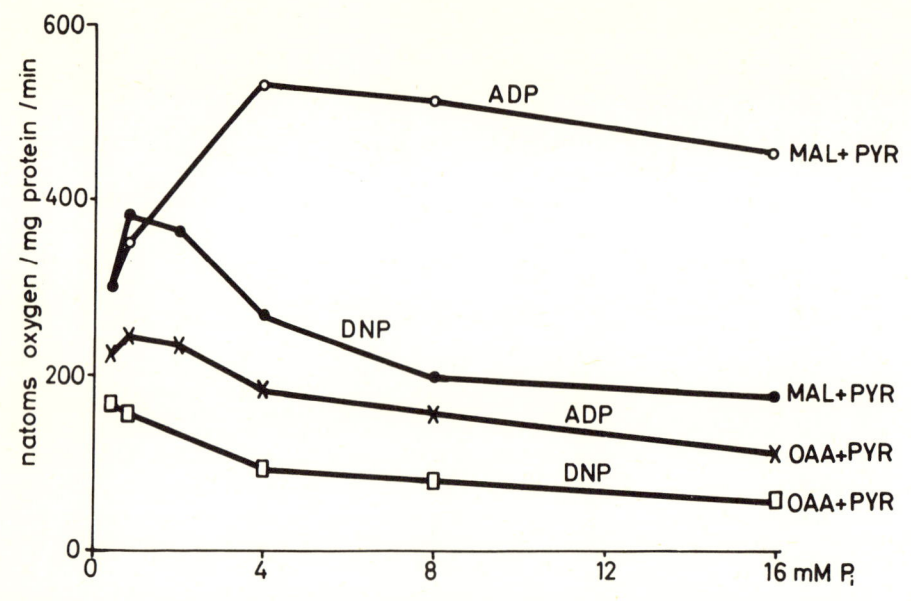

Fig. 1. Effect of P_i on stimulation of oxygen uptake by ADP and DNP of rat brain mitochondria. Inorganic phosphate was added as K salt, for other experimental conditions see Table 2.

a higher P_i concentration there was a competition between substrate anions and P_i for the entry capacity.

Depending on the experimental conditions the penetration of either substrates or P_i could be restricted. In the presence of adenine nucleotide, the entry of substrates was less inhibited by higher P_i concentration than it was in the presence of DNP. In a few experiments the substrate and P_i content of mitochondria were estimated after equilibration with different concentrations of P_i in the presence of both substrates and either ADP or DNP. When the media contained adenine nucleotides, P_i entry was restricted in comparison with the samples containing DNP instead of ADP.

Figure 2 includes a few records of oxygen uptake. The applied substrate pair was oxaloacetate + pyruvate, since this combination was particularly sensitive to higher concentrations of P_i in the presence of DNP. The first record demonstrates the control measurement, the trace B shows the effect of added P_i. Trace C indicates that ATP can counteract the inhibitory effect of P_i. This effect of adenine nucleotide seems to be very sensitive, because ADP or ATP in a 200 times less concentration than that of the applied P_i could prevent the inhibitory effect of P_i. Not only adenine mononucleotide but adenine dinucleotide were also effective. NAD or NADH showed effects similar to ATP

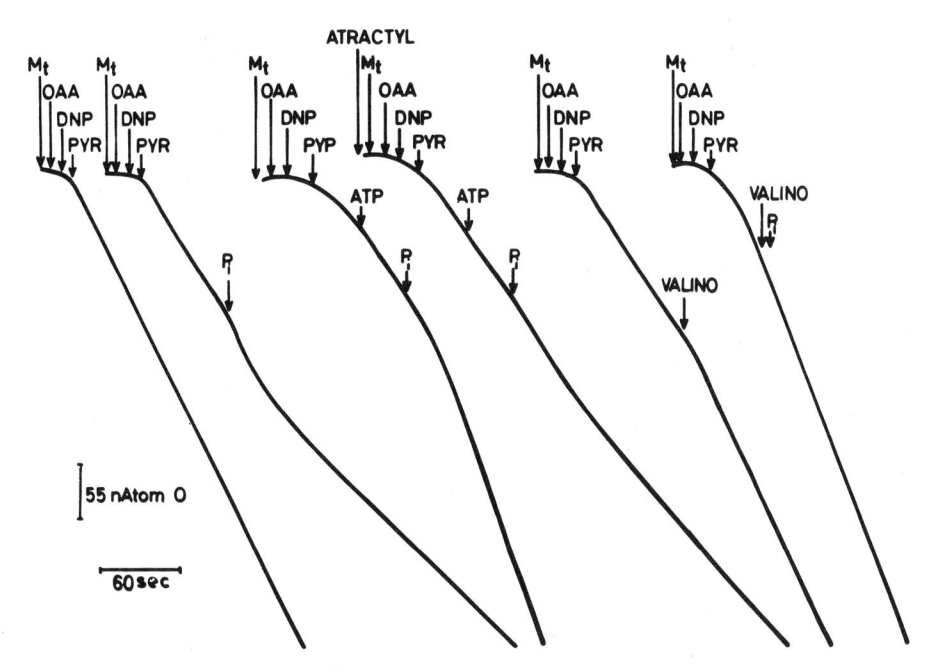

Fig. 2. Effect of P_i on the DNP stimulated oxygen uptake; modification of P_i effect by ATP, atractyloside and valinomycin. The media contained tris-Cl, substrates and DNP as indicated in Table 2. Total volume 1.2ml; at the arrows the following substances were added: K-phosphate, 20 µmoles; ATP, 50 nmoles, atractyloside, 336 µg; valinomycin, 16 µg.

Atractyloside prevents the interaction between adenine nucleotides and mitochondrial membranes (Vignais et al., 1961). If brain mitochondria were pretreated with atractyloside and subsequently substates, ADP, DNP and finally P_i were added, the inhibition of respiration remained unchanged indicating that atractyloside abolished the preventive effect of ATP (trace D). Furthermore this observation suggests that the preventive effect of adenine nucleotides is somehow connected with their binding to the mitochondrial membrane. It is not indicated in the figure, but atractyloside was able to abolish the preventive effect of NAD or NADH too.

Trace E demonstrates the effect of valinomycin. Valinomycin alone increases DNP-stimulated respiration to an irrelevant extent only. If valinomycin was added together with 20 µmoles P_i, no inhibition of respiration by P_i could be observed. Valinomycin increases K^+ entry into mitochondria; together with the raised penetration of cations, the anion entry is also increased.

Fonyo and Bessman (1968) and Tyler (1968) demonstrated that P_i entry into mitochondria can be blocked selectively by mercurials. It is not shown in the figure, but after mersalyl treatment of the mitochondria the inhibition of respiration by P_i was almost totally abolished. When the blocked SH groups in mitochondrial membrane were reactivated by cystein or mercaptoethanol, the inhibition of respiration reappeared.

It seems very probable that there is a competition between the entry of substrate anions and that of P_i into the mitochondria. It is likely that their ratio is different under energized and de-energized conditions.

In a few experiments we investigated the respiration capacity of rat brain mitochondria following injury. The animals were narcotized and their legs were frozen with liquid nitrogen. After the development of the typical shock condition, the rats were decapitated and the brains were placed in an ice-cold homogenizing medium. For each experiment 8 brains were used. In order to be able to select 8 brains from animals in an identical state of shock from among the 20-25 shocked rats about 30 min were required.

Table 3 shows a diminished rate of respiration of the shocked rat brain mitochondria. Data refer to state 4, the ADP and DNP stimulated respiration measured with different substrate pairs. All data were expressed in percent of the normal control values. The decrease in oxygen uptake was especially significant when β-OH-butyrate was used as substrate. State 4 respiration was about 30 per cent of the control value and ADP as well as DNP stimulated respiration did not exceed 12 per cent of the corresponding control value. β-OH-Butyrate dehydrogenase and succinate/dehydrogenase are located in the mitochondrial membrane. However, as it can be seen respiration with succinate did not decrease under identical conditions. Both membrane-bound dehydrogenases require some lipid for their full activities (Cerletti et al., 1967, Jurkshut et al., 1961). It can be presumed that the lipoprotein structure of β-OH-butyrate dehydrogenase is selectively damaged following the hypoxia due to shock. Furthermore, the fatty acids which might be liberated during hypoxia can themselves cause a restricted respiration.

Table 3

Rate of Oxygen Uptake of Brain Mitochondria following Freezing of the Legs

No. of mitochondrial preparations	Substrate	Oxygen uptake (per cent of control values)		
		State 4	With ADP	With DNP
5	mal+pyr	80 ± 9.4	58 ± 7.8	69 ± 9.1
5	mal+glut	72 ± 8.6	42 ± 6.9	58 ± 7.6
5	mal+ β-OH-but	34 ± 5.8	12 ± 3.4	12 ± 3.7
5	succ	80 ± 9.9	80 ± 16.2	88 ± 18.2

Table 4 demonstrates an *in vitro* effect on respiration of brain mitochondria. Rats were anaesthetized and the removed brains were kept in ice-cold homogenizing medium. After standing for 60 min the brains were homogenized and the mitochondria prepared. The results indicate that under *in vitro* conditions, when the brains were kept for 60 min in the homogenizing medium, damage can occur similar to that demonstrated under *in vivo* conditions.

Table 4

Rate of Oxygen Uptake of Rat Brain Mitochondria following Keeping Whole Brains at 0° for 60 minutes

No. of mitochondrial preparations	Substrate	Oxygen uptake (per cent of control values)		
		State 4	with ADP	with DNP
5	mal+pyr	58 ± 7.2	49 ± 6.2	49 ± 6.2
5	mal+glut	53 ± 6.4	42 ± 5.2	53 ± 6.7
5	mal+ β-OH-but	42 ± 4.5	17 ± 3.1	17 ± 3.2
5	succ	80 ± 9.9	82 ± 10.2	84 ± 8.9

Addition of albumin to the assay system caused a restoration of oxygen uptake in both *in vitro* and *in vivo* damaged mitochondria, except for the oxidation of β-OH-butyrate. This indicated that the liberated free fatty acids played a significant role in the injury of mitochondria.

REFERENCES

Abood, L.G. (1970) Brain Mitochondria. In: Handbook of Neurochemistry. Vol. 2 p.303. Edited by Lajtha, A. New York, Plenum Publishing Co.

Chappel, J.B. & Haarhoff, K. (1967). The penetration of the Mitochondrial Membrane by Anions and Cations. In: The Biochemistry of Mitochondria. p.75. Edited by Slater, E.C., Kaniuga, Z. & Wojtczak, L. New York, Academic Press.

Clark, J.B. & Nicklas, W.J. (1970). J. Biol. Chem. 245:4724.

Cunningham, R.D. & Bridgers, W.F. (1970). Biochem. Biophys. Res. Commun. 38:99.

Fonyo, A & Bessman, S.P. (1968). Biochem. Med. 2:145.

Gray, E.G. & Whittaker, V.P. (1962). J. Anat. 96:79.

Jobsis, F.F. (1963) Biochem. Biophys. Acta, 74:60.

Klingenberg, M., Slenczka, W. & Ritt, E. (1959) Biochem. Z. 332:47.

Somogyi, J. & Vincze, S. (1961) Acta physiol. Acad. Sci. hung. 20:325.

Tyler, D.D. (1968) Biochem. J. 107:121.

Vignais, P.V., Vignais, P.M. & Stanislas, E. (1961) Biochem. Biophys. Acta. 51:394.

BRAIN AND LIVER INTRACELLULAR COMPARTMENTAL REDOX STATES IN HYPOXIA, HYPOCAPNIA AND HYPERCAPNIA

A. T. Miller, Jr. and Francis M. H. Lai

Department of Physiology, University of North Carolina

School of Medicine, Chapel Hill, N. C.

This paper gives an account of our experiences with the substrate ratio technique for calculating cytoplasmic and mitochondrial $NAD^+/NADH_2$ changes in brain and liver in several metabolic states. The NAD redox system is especially suitable for studies on oxygen availability since it occurs in both the cytoplasm and the mitochondria; this makes it possible to distinguish changes in the $NAD^+/NADH_2$ ratio in the two compartments. NAD exists in cells in both free and bound forms. Krebs (1967) and Krebs and Veech (1969) recommend that, on thermodynamic grounds, only the free nucleotide concentrations should be used in calculating redox ratios.

Studies on the redox state of NAD have generally employed one of two methods: the microfluorometric procedure of Chance et al. (1964) or the substrate ratio technique introduced by Holzer, Schultz and Lynen (1956) and elaborated by Bücher and Klingenberg (1958), Krebs (1967), and Krebs and Veech (1969). The fluorometric method permits repeated measurements of rapid changes in exposed organs in living animals, but it measures predominantly bound rather than free nucleotides, and the distinction between cytoplasmic and mitochondrial changes must be made indirectly, e.g., by the use of iodoacetate to block cytoplasmic changes.

The substrate ratio method is based on the following reaction involving an NAD-linked dehydrogenase:

$$\text{Oxidized substrate} + NADH_2 \underset{k_2}{\overset{k_1}{\rightleftharpoons}} \text{Reduced substrate} + NAD^+$$

$$\frac{\text{Oxidized substrate}}{\text{Reduced substrate}} = k \cdot \frac{NAD^+}{NADH \cdot H^+}$$

The $NAD^+/NADH_2$ ratio in a particular cell compartment may be calculated from measured values of the oxidized and reduced substrate concentration, provided: (1) the dehydrogenase involved in the reaction is confined to that compartment, (2) the activity of the dehydrogenase is high enough to assure virtual equilibrium of the reaction, and (3) the distribution of the oxidized and reduced substrates is uniform throughout the cell. In addition, the value of intracellular H^+ must be either measured or assumed. Lactic dehydrogenase, β-hydroxybutyrate dehydrogenase, and glutamate dehydrogenase have been shown to be suitable enzymes for measuring the $NAD^+/NADH_2$ ratios in the cytoplasmic, mitochondrial cristae and mitochondrial matrix compartments, respectively, in rat liver cells (Krebs, 1967). We have confirmed this claim with respect to rat liver, and turtle liver and brain, but not with respect to rat brain.

EXPERIMENTAL

Male albino rats and fresh-water turtles (Pseudemys species) were used in order to compare results in higher and lower vertebrates. Rats were anesthetized with pentobarbital and their pulmonary ventilation was controlled with a pump. Turtles breathed spontaneously during exposure to nitrogen, but were pump-ventilated in the hypercapnia and hypocapnia experiments. At the conclusion of an experiment a sample of liver was obtained by freeze-clamping; the head was then frozen by immersion in liquid Freon-12 at -140°C and the brain was removed under liquid nitrogen. The frozen tissues were homogenized at liquid nitrogen temperature and used for measurements of substrate concentrations by enzymatic methods, and of intracellular pH by the CO_2 method (Siesjö and Ponten, 1966).

RESULTS

If a glutamate dehydrogenase and β-hydroxybutyrate dehydrogenase share a common nucleotide pool, as reported for rat liver by Krebs (1967), the two systems should be in equilibrium and their combined substrate ratio should be constant under varying conditions. Data for rat liver and brain are presented in Table 1. It is apparent that the two systems are approximately in equilibrium in rat liver, under a variety of conditions, but they are not in equilibrium in rat brain under the same conditions. This means that either of the two substrate ratios can be used for the estimation of mitochondrial $NAD^+/NADH_2$ in rat liver, while further study of these (and perhaps other) systems is needed in rat brain. In the turtle, the two systems are in equilibrium in both liver and brain, perhaps indicating a less complex metabolic compartmentation in the more primitive brain of the turtle.

Table 1 Substrate systems ratios* in rat liver and brain under various conditions.

Condition	Liver	Brain
Well-fed	7.3×10^{-2}**	
Starved	6.6×10^{-2}**	
Alloxan diabetes	8.4×10^{-2}**	
Breathing air	8.2×10^{-2}	0.3×10^{-2}
Breathing 10% O_2	7.0×10^{-2}	0.4×10^{-2}
Breathing 7% O_2	7.9×10^{-2}	0.4×10^{-2}
Hypervent. ($PaCO_2$ = 20 mm Hg)	7.6×10^{-2}	0.5×10^{-2}
Hypervent. ($PaCO_2$ = 10 mm Hg)	6.2×10^{-2}	1.1×10^{-2}
Breathing 5% CO_2 in air	6.9×10^{-2}	0.1×10^{-2}
Breathing 5% CO_2 + 7% O_2 in N_2	7.1×10^{-2}	0.6×10^{-2}

* $\dfrac{\beta\text{-hydroxybutyrate} \cdot \alpha\text{-ketoglutarate} \cdot NH_4^+}{\text{acetoacetate} \cdot \text{glutamate}}$

**Krebs (1967)

The effects of a 30-minute administration of 10 or 7% oxygen on the cytoplasmic and mitochondrial $NAD^+/NADH_2$ ratios in rat liver and brain are shown in Figure 1. In the case of liver, exposure to 10 percent oxygen produced little if any change in the mitochondrial $NAD^+/NADH_2$ ratios, but a substantial decrease in the cytoplasmic ratio. This illustrates the advantage of the substrate ratio method which detects separately the changes in the two compartments. The administration of 7% oxygen resulted in a marked decrease in both mitochondrial $NAD^+/NADH_2$ ratios, and a slight further decrease in the cytoplasmic ratio. The close agreement of the two mitochondrial compartment redox changes should be noted.

The rat brain results are difficult to interpret. The calculated cytoplasmic $NAD^+/NADH_2$ ratio was reduced as expected. The calculated mitochondrial $NAD^+/NADH_2$ ratios were, as in liver, unaffected by 10% oxygen, but were increased when 7% oxygen was breathed. It should be noted that the mitochondrial substrate ratios were unchanged by hypoxia (Figure 2) so that the increase in the calculated $NAD^+/NADH_2$ ratios was the mathematical consequence of an increase in intracellular acidity. A similar result, based on the glutamate dehydrogenase system only, has recently been reported by Siesjö and Nilsson (1971).

When turtles were made hypoxic by exposure to an atmosphere of nitrogen for periods of 3 or 6 hours, the calculated cytoplasmic and mitochondrial $NAD^+/NADH_2$ ratios were reduced in both brain and liver (Figure 3). The fact that the mitochondrial substrate ratios were altered in the expected direction in the turtle brain, but were unchanged in rat brain may provide a clue to the unexpected results in rat brain; this will be considered in a later section of this report.

It has been suggested from time to time that hyperventilation might produce cerebral hypoxia due to the hypocapnic

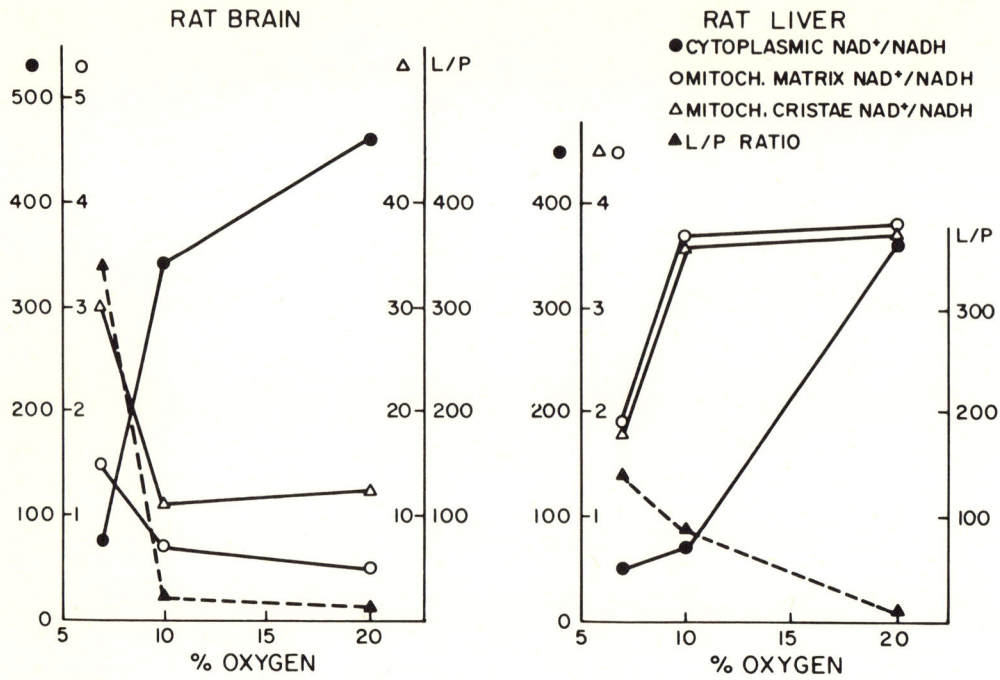

Figure 1 Changes in cytoplasmic and mitochondrial $NAD^+/NADH_2$ ratios in the liver and brain of rats breathing 10 and 7% oxygen.

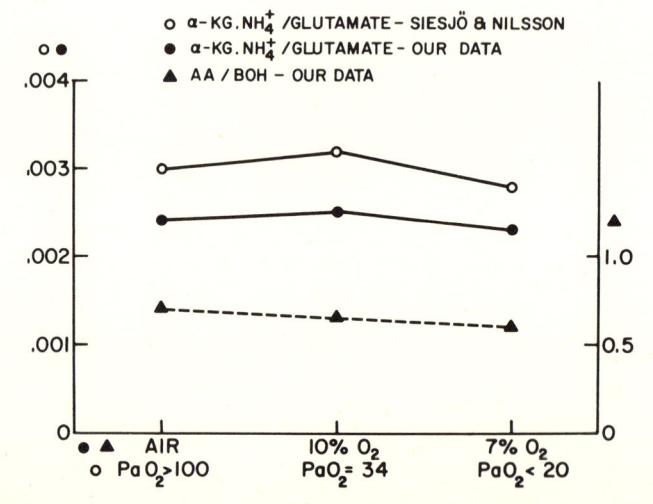

Figure 2 Rat brain mitochondrial substrate ratios in hypoxia.

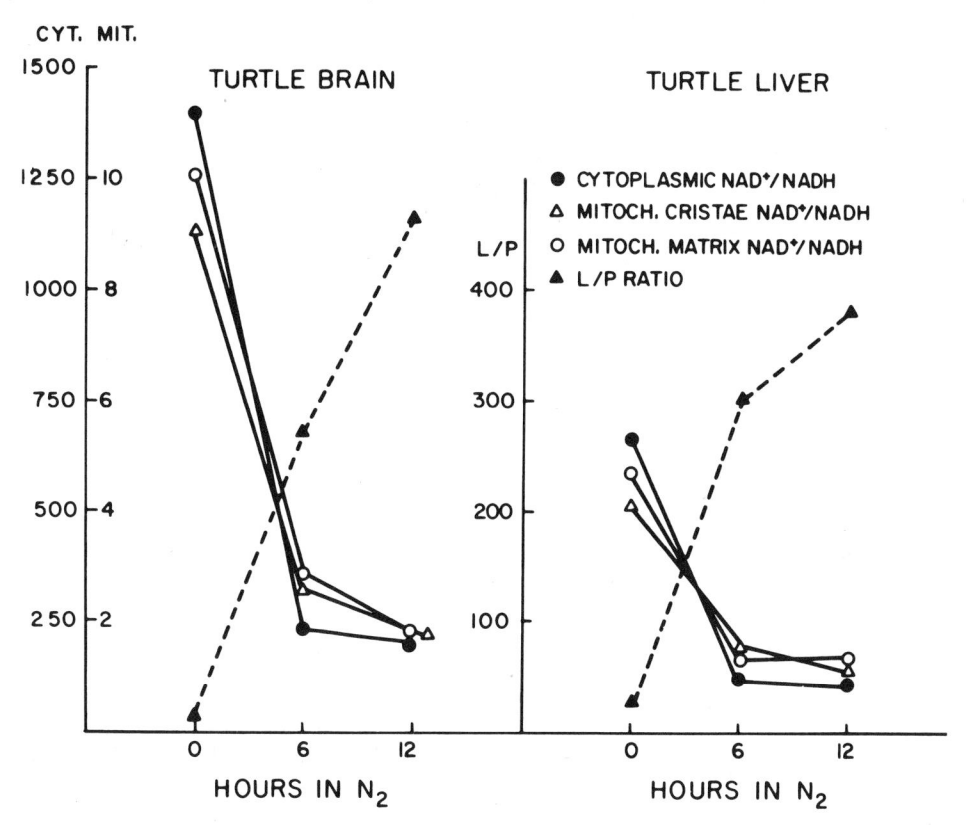

Figure 3 Changes in cytoplasmic and mitochondrial $NAD^+/NADH_2$ ratios in the liver and brain of turtles exposed to a nitrogen atmosphere.

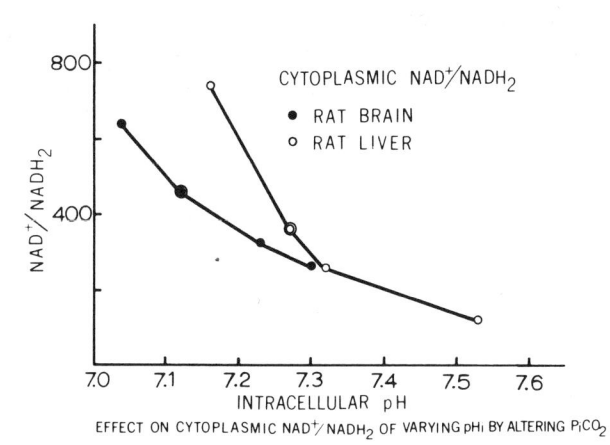

Figure 4 Effect of changes in intracellular pH on cytoplasmic $NAD^+/NADH_2$ ratios in rat liver and brain.

Figure 5 Effect of changes in intracellular pH on mitochondrial $NAD^+/NADH_2$ ratios in rat liver and brain.

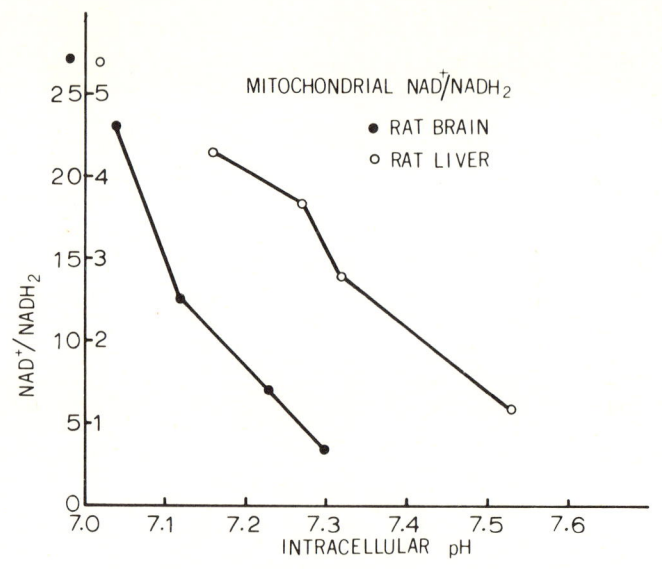

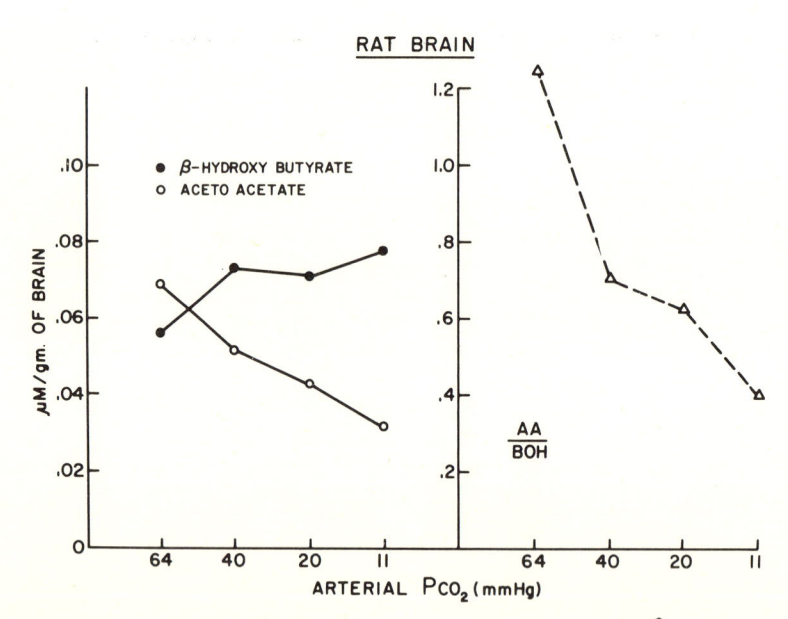

Figure 6 Changes in rat brain acetoacetate and β-hydroxybutyrate in response to alterations in intracellular acidity (produced by changes in arterial P_{CO_2}).

reduction in cerebral blood flow and release of oxygen from hemoglobin (Bohr effect). We (Miller et al., 1970) hyperventilated rats to a PaCO$_2$ of about 10 mm Hg and observed no change in brain lactate/pyruvate ratio or creatine phosphate concentration. Granholm et al. (1968), using the microfluorometric technique of Chance, reported marked increase in NADH$_2$ fluorescence in the exposed cerebral cortex of rats hyperventilated to a PaO$_2$ of 11.5-12.5 mm Hg. Figures 4 and 5 show the effects of alterations in intracellular pH, produced by hyperventilation and by the administration of CO$_2$, on the calculated cytoplasmic and mitochondrial NAD$^+$/NADH$_2$ ratios in rat brain and liver. Similar results were obtained in turtle brain and liver. Figure 6 shows that the acetoacetate/β-hydroxybutyrate ratio in rat brain responded to alterations in intracellular acidity, even though it did not change in response to hypoxia. This may have significance in relation to the argument (see Discussion) that the lack of change in this substrate ratio in hypoxia is due to lack of access of the dehydrogenase to its substrate in rat brain.

Some additional experiments were performed with the object of separating the effects of hypoxia and the associated increase in intracellular acidity on brain mitochondrial redox ratios. Rats were hyperventilated with 7% oxygen, to various arterial PCO$_2$ values. It was then observed that the calculated brain mitochondrial NAD$^+$/NADH$_2$ ratios were reduced just as they were in rats hyperventilated with air. The redox ratios corresponding to a particular intracellular pH were, in fact, about the same whether the rats were breathing air or 7% oxygen.

DISCUSSION

Our experiments have probably raised more questions than they have answered. The most difficult problem concerns the interpretation of NAD$^+$/NADH$_2$ ratios calculated from substrate ratios and intracellular pH. Do the NAD$^+$/NADH$_2$ ratios calculated in this manner provide more information than the substrate ratios alone? If so, the additional information must come from the intracellular pH measurements. This might, however, lead to confusion, since the calculated effects of increased intracellular acidity on the NAD$^+$/NADH$_2$ ratio and on the oxidized/reduced substrate ratio are opposite in direction as shown by the following relation:
$$\frac{\text{reduced substrate}}{\text{oxidized substrate}} = k \cdot \frac{\text{NADH}}{\text{NAD}^+} \cdot \text{H}^+$$
In the response of rat brain to hypoxia the mitochondrial substrate ratio remains unchanged while H$^+$ increases. The mathematical result is a decrease in the calculated NADH/NAD$^+$ ratio, which is opposite in direction to the alteration in the brain cytoplasmic NADH/NAD$^+$ ratio. In the latter case, the measured increase in H$^+$ is, of course, the same as that used in calculating the mitochondrial NAD$^+$/NADH$_2$ ratios (since separate measurements

of mitochondrial and cytoplasmic pH are not feasible), but now the large increase in the ratio of reduced to oxidized substrate (lactate/pyruvate ratio) more than balances the increase in H^+ concentration, with the result that the calculated cytoplasmic NADH/NAD$^+$ ratio is increased. Since it is highly improbable that hypoxia would actually cause the mitochondrial redox ratios to go more oxidized and the cytoplasmic redox ratio to go more reduced, it must be concluded that the substrates ratio method is not suitable for the study of the redox responses of brain mitochondrial to hypoxia in the adult rat, at least with the substrate systems that we have used. It is tempting to attribute the non-responsiveness of the rat brain mitochondrial substrate ratios in hypoxia to the complex metabolic compartmentation in rat brain (Van den Berg et al., 1969; Patel and Balazs, 1970), which might make the substrate pools inaccessible to the NAD-linked dehydrogenases. In this case, one would have to assume that the acetoacetate/β-hydroxybutyrate changes in rat brain in response to alterations in intracellular pH (by hypercapnia and hypocapnia) do not involve the β-hydroxybutyrate dehydrogenase. In experiments in progress, we are studying the response to hypoxia in very young rats (in which brain maturation has not yet resulted in complex metabolic compartmentation, multiple citric acid cycles, etc.) and in birds (whose phylogenetic development from the reptiles differs from that of mammals). The bird experiments were suggested by the difference in response to hypoxia of rat and turtle brain mitochondrial redox ratios.

At the present time, we can only state that the substrate ratio approach appears to give valid information about hypoxic changes in the cytoplasmic and mitochondrial $NAD^+/NADH_2$ ratios in rat liver and in turtle liver and brain; and also about the cytoplasmic but not the mitochondrial ratios in rat brain. Whether the calculated $NAD^+/NADH_2$ ratios provide more information than the substrate ratios alone, and whether the use of measured rather than assumed values for intracellular pH increases the accuracy of the procedure, remain to be determined.

REFERENCES

Bücher, Th. & Klingenberg, M. (1958) Angew. Chem. 70, 552.
Chance, B., Schoener, B. & Schindler, F. The intracellular oxidation-reduction state. In: Oxygen in the Animal Organism, edited by F. Dickens and E. Neil. New York, MacMillan, 1964, p. 367.
Granholm, L., Lukjanova, L. & Siesjö, B.K. (1968) Acta physiol. scand. 72, 533.
Holzer, H., Schultz, G. & Lynen, F. (1956) Biochem.Z. 328, 252.
Krebs, H.A. (1967) Adv. Enzyme Reg. 5, 409.
Krebs, H.A. & Veech, R.L. (1969) FEBS Symposium, 17, 101.

Krebs, H.A., & Veech, R.L. Pyridine nucleotide interrelations. In: The Energy Level and Metabolic Control in Mitochondria, edited by S. Papa, J.M. Tager, E. Quagliariello and E.C. Slater. Bari, Adriatica Editrice, 1969, p. 329.

Miller, A.T., Jr., Curtin, K.E., Shen, A.L. & Suiter, C.K. (1970) Am. J. Physiol. 219, 798.

Patel, A.J. & Balazs, R. (1970) J. Neurochem. 17, 955.

Siesjö, B.K. & Ponten, U. (1966) Ann. N.Y. Acad.Sci. 133, 78.

Siesjö, B.K. & Nilsson, L. (1971) Scand. J. Clin. Lab. Invest. 27, 83.

Van den Berg, C.J., Krzalic, L.J., Mela, P. & Waelsch, H. (1969) Biochem.J. 113, 281.

PERIPHERAL AND CARDIAC FACTORS IN EXPERIMENTAL SEPTIC SHOCK

Lerner B. Hinshaw

Veterans Administration Hospital, and Departments of
Physiology and Biophysics and Surgery, University of
Oklahoma Medical Center, 921 Northeast 13th Street,
Oklahoma City, Oklahoma 73104, U.S.A.

The insidious and sometimes precipitous development of
septic shock in man involves mechanisms not clearly understood.
Cardiovascular changes have been reported to occur within
seconds following an intravenous injection of endotoxin in dogs
(5,7,19,22), and at variable times after injection of endotoxin
in monkeys (8,10,13) or live E. coli organisms in the canine and
primate species (3,11,14,15). Although the animal shock model
has been studied extensively in recent years, the precise
mechanisms involved in the early development of shock are not
clearly understood. Recent emphasis has been focused on the
causes of inadequate tissue perfusion in this form of shock.
Of particular importance have been studies in which cardiac
output decreased significantly in the early phase of shock.
Several explanations have been proposed to account for the early
decrease in flow and these will be the subject of this presentation.

Peripheral pooling mechanisms

In the mammalian species the decrease in venous return
after administration of endotoxin (13,22) may readily account for
the reported decrease in cardiac output. Peripheral mechanisms
resulting in the early sequestration of blood in various organs
or anatomical regions have been described (4,6,9,12,13,16,19,22)
and seem to vary with the species (12). There is evidence for
hepatosplanchnic pooling in the dog (6,9,19,22), extra-
hepatosplanchnic pooling in both dog and monkey (4,12,13), and
pulmonary pooling in the latter species (13). There is no
evidence for early pooling in skin and muscle in any species
(7,16). Early sequestration of perfusate in the monkey is

primarily intravascular and is apparently quite generalized according to a recent study carried out in this laboratory (8,13). Trapping of blood in the subhuman primate appears to occur not only in the pulmonary bed, but in the capacitance vessels of the systemic circulation and in dilated precapillary vessels (13).

Cardiac Mechanisms

Although the evidence is strong for a significant degree of peripheral pooling of blood in the endotoxin animal model, the possibility exists that cardiac mechanisms may also be concerned in the diminished cardiac output in the early phase of shock. Endotoxin could conceivably poison the heart directly, or indirectly bring into play detrimental circulating factors which themselves might depress myocardial contractility. Evidence for heart failure in endotoxin shock in animals (1,21) and septic shock in man (2,20) has been published.

This report is focused on the role of the heart in the early phase of shock before long standing effects of poor perfusion have probably damaged the myocardium. In the latter situation, all organs would eventually fail to function. However, the present discussion is focused on the earlier period of shock which might conceivably provide insight into the role of the heart. If the heart is found to fail early, that is, within 2-4 hr after endotoxin, cardiac failure could be established as a precipitating factor in the development of irreversible shock.

Research from this laboratory has failed to provide evidence for an early detrimental action of endotoxin on the myocardium (17,18), and even the effects of systemic hypotension and acidosis after endotoxin have provided no evidence for early heart failure. Death is seen to occur in animals receiving endotoxin, but a normal heart continually exchanging blood with the shocked animal performs normally (17,18). Three hr of systemic hypotension and depressed cardiac output after endotoxin fails to elicit myocardial damage, while left ventricular end diastolic pressure, myocardial contractility, cardiac power (work/sec), dP/dT, O_2 uptake and CO_2 production are unchanged from control values 200 min after endotoxin injection (17). Cardiac performance is also found to be unimpaired after endotoxin in the presence of β-adrenergic blockade (18). Normal cardiac contractility is observed in the presence of acidosis, and myocardial O_2 uptake is found to be independent of pH values. Coronary blood flow is diminished in the hypotensive state but elevated markedly above control values when mean arterial pressure (afterload) is returned to the control (pre-shock) level. Oxygen delivery to the heart appears adequate since O_2 uptake is normal and coronary venous O_2 content is elevated above control values when afterload is restored to pre-shock levels.

These studies clearly demonstrate that endotoxin exerts no early direct toxic action on the myocardium but that indirect factors most probably intervene during the intermediate or later stages of shock. The effects of prolonged systemic hypotension and progressive peripheral pooling would be expected to result in the eventual depression of cardiac function at later stages of shock on the basis of diminished coronary perfusion pressure and insufficient myocardial blood flow.

Summary

Peripheral pooling mechanisms apparently predominate in the early phase of endotoxin shock in bringing about a decrease in cardiac output. Left ventricular function appears to be entirely normal and only a longer sustained period of hypotension would be expected to bring about impairment of myocardial performance. A variety of indirect mechanisms resulting from poor systemic hemodynamic status may ultimately contribute to myocardial failure in the later stages of shock.

REFERENCES

1. Alican, F., Dalton, M.L. & Hardy, J.D. (1962) Am. J. Surg. 103:702.

2. Bell, H., & Thal, A. (1970) 48:106.

3. Guenter, C.A., Fiorica, V. & Hinshaw, L.B. (1969) J. Appl. Physiol. 26:780.

4. Hinshaw, L.B., Gilbert, R.P., Kuida, H. & Visscher M.B. (1958) Am. J. Physiol. 195:631.

5. Hinshaw, L.B., Vick, J.A., Jordan, M.M. & Wittmers, L.E. (1962) Am. J. Physiol. 202:103.

6. Hinshaw, L.B. & Nelson, D.L. (1962) Am. J. Physiol. 203:870.

7. Hinshaw, L.B., Brake, C.M., Emerson, T.E., Jr., Jordan, M.M. & Masucci, F.D. (1964) Am. J. Physiol. 207:925.

8. Hinshaw, L.B., Emerson, T.E., Jr., & Reins, D.A. (1966) Am. J. Physiol. 210:335.

9. Hinshaw, L.B., Reins, D.A. & Hill, R.J. (1966) Canad. J. Physiol. Pharm. 44:529.

10. Hinshaw, L.B., Jordan, M.M. & Vick, J.A. (1967) J. Clin. Invest. 40:1631.

11. Hinshaw, L.B., Solomon, L.A., Holmes D.D. & Greenfield, L.J. (1968) Surg. Gynec. Obstet. 127:981.

12. Hinshaw, L.B. (1968) J. Surg. Res. 8:535.

13. Hinshaw, L.B., Shanbour, L.L., Greenfield, L.J. & Coalson, J.J. (1970) Arch. Surg. 100:600.

14. Hinshaw, L.B., Mathis, M.C. Nanaeto, J.A. & Holmes, D.D. (1970) J. Trauma 10:787.

15. Hinshaw, L.B. (1970) Internat. Symposium, Resistance to Infectious Disease. Saskatoon Mod. Press. Saskatchewan, Saskatoon, Canada, pp. 85-91.

16. Hinshaw, L.B. & Owen, S.E. (1971) J. Appl. Physiol. 30:331.

17. Hinshaw, L.B., Archer, L.T. Greenfield, L.J. & Guenter, C.A. Am. J. Physiol. (in press).

18. Hinshaw, L.B., Greenfield, L.J., Archer, L.T. & Guenter, C.A. Proc. Soc. Exptl. Biol. Med. (in press).

19. MacLean, L.D., & Weil, M.H. (1956) Circulation Res. 4:546.

20. Siegel, J.H., Greenspan, M. & Del Guercio, L.R.M. (1967) Anns. Surg. 165:504.

21. Solis, R.T. & Downing, S.E. (1966) Am. J. Physiol. 211:307.

22. Weil, M.H., MacLean, L.D., Visscher, M.B. & Spink, W.W. (1965) J. Clin, Invest. 35:1191.

Research supported by Veterans Administration Hospital and U.S. Navy.

THE LYSOSOMAL PROTEASE – MYOCARDIAL DEPRESSANT FACTOR SYSTEM IN CIRCULATORY SHOCK

Allan M. Lefer and Thomas M. Glenn

Department of Physiology, University of Virginia

School of Medicine, Charlottesville, Virginia 22901 USA

Circulatory shock is characterized by severe perturbations in the homeostatic regulation of cardiovascular function leading to conditions which are incompatible with life. The nature of these perturbations is not clearly understood, nor are the origins of the positive feedback loops generated by these disturbances. Despite the obvious gaps in our knowledge, certain findings have been uncovered in recent years which, in the light of new information, point to a final common pathway in the pathogenesis of circulatory shock. These findings are:
1. a marked reduction in blood flow to the splanchnic viscera very early in circulatory shock of varying etiologies (1)
2. a marked alteration in lysosomal membranes with the attendant release of lysosomal hydrolases into the plasma during many forms of shock (2)
3. a profound impairment of myocardial performance late in circulatory shock (3,4)

These findings superficially appear to be unrelated events in circulatory shock. However, they interrelate in a special way to play a key role in the derangement of circulatory homeostasis in circulatory shock. The thread connecting these events is a toxic factor which accumulates in the plasma during circulatory shock. This substance, termed myocardial depressant factor (MDF), has as its primary action a profound negative inotropic effect on the heart. Figure 1 illustrates the steps in the formation of MDF and thus connects the links among the previously cited findings.

Thus, the decrease in splanchnic blood flow occurs as a compensatory response to the hypotension and/or trauma, and although it is a beneficial effect in the short run, prolongation of

Events Leading to the Formation of MDF in Circulatory Shock

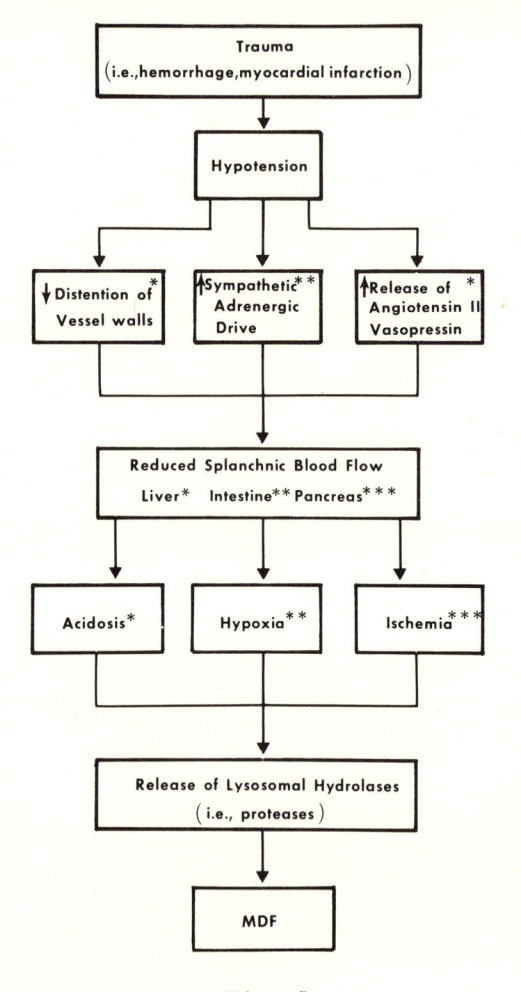

```
                    ┌─────────────────────────────────┐
                    │            Trauma               │
                    │(i.e.,hemorrhage,myocardial       │
                    │         infarction)             │
                    └─────────────────────────────────┘
                                   │
                                   ▼
                         ┌──────────────────┐
                         │   Hypotension    │
                         └──────────────────┘
                          │      │       │
              ┌───────────┘      │       └────────────┐
              ▼                  ▼                    ▼
    ┌──────────────────┐ ┌──────────────────┐ ┌──────────────────┐
    │ ↓Distention of * │ │ ↑Sympathetic* *  │ │ ↑Release of   *  │
    │   Vessel walls   │ │    Adrenergic    │ │   Angiotensin II │
    │                  │ │      Drive       │ │   Vasopressin    │
    └──────────────────┘ └──────────────────┘ └──────────────────┘
                                   │
                                   ▼
              ┌────────────────────────────────────────┐
              │     Reduced Splanchnic Blood Flow      │
              │   Liver*   Intestine** Pancreas***     │
              └────────────────────────────────────────┘
                   │            │            │
                   ▼            ▼            ▼
            ┌──────────┐ ┌──────────┐ ┌────────────┐
            │ Acidosis*│ │ Hypoxia**│ │ Ischemia***│
            └──────────┘ └──────────┘ └────────────┘
                                   │
                                   ▼
                    ┌─────────────────────────────┐
                    │ Release of Lysosomal         │
                    │      Hydrolases              │
                    │      (i.e., proteases)       │
                    └─────────────────────────────┘
                                   │
                                   ▼
                              ┌────────┐
                              │  MDF   │
                              └────────┘
```

Fig. 1

splanchnic ischemia elicits an enormous price in the long haul. The payment for conservation of blood flow for the brain and the heart is autolysis of the splanchnic viscera, particularly the pancreas. The currency for this is in the form of toxic substances, first in the release of lysosomal hydrolases, and secondly in the formation of MDF.

MDF has been identified as a peptide or glycopeptide with a molecular weight of 800 to 1000 (5). It is dialyzable, heat stable, soluble in aqueous media, is a charged species and is relatively stable in the plasma of shocked animals. MDF has been found in a variety of shock states including hemorrhagic, endotoxic, cardiogenic, splanchnic ischemic, pancreatitis and burn shock (6). Furthermore, it is found in cats, dogs, baboons and man. Thus, MDF appears to be common to a variety of forms of shock as well as to a multitude of mammalian species. It is tempting to think of this toxic agent as a common denominator of circulatory shock, since the MDF found in all of these forms of shock and in all these species is both qualitatively and quantitatively indistinguishable.

MDF fulfills the basic criteria of a toxic factor in shock (6) since:
- (a) it is produced in several forms of circulatory shock
- (b) it is not present to any significant extent in non-shocked animals
- (c) it can be isolated from the plasma
- (d) it is capable of inducing shock when injected into a normal animal
- (e) it exerts a serious pathophysiologic action (i.e., myocardial depression)
- (f) it is present in man as well as other mammals

Considerable evidence has been obtained which indicates that MDF is produced almost exclusively by the ischemic pancreas (7). These points are as follows:
- (a) Occlusion of just the pancreaticoduodenal arteries for two hours followed by release of the occlusion results in high plasma MDF activity. This is specific, since occlusion of the carotid or iliac arteries does not produce MDF.
- (b) Pancreatectomy prior to induction of hemorrhagic shock prevents the accumulation of MDF in the plasma.
- (c) Selective irritation of the pancreas (i.e., by injecting trypsin and bile salts into the pancreatic ducts) produces MDF, whereas the same stimuli given intravenously are ineffective in forming MDF.
- (d) Perfusion of the isolated pancreas under ischemic or hypoxic conditions results in amounts of MDF comparable to those formed in shock.
- (e) MDF can be recovered from the pancreas as well as from the plasma during circulatory shock.
- (f) Chronic ligation of the pancreatic ducts prevents MDF formation upon induction of shock in these animals.
- (g) Incubation of homogenates of pancreatic tissue results in a large amount of MDF production, whereas homogenates of liver, spleen or intestine do not produce MDF under identical conditions. Unincubated homogenates of pancreas do not produce MDF, indicating that MDF is not just a storage product of the normal pancreas.

Thus, it appears that all the components necessary for the formation of MDF (i.e., substrates, enzymes, cofactors, etc.) are present in sufficient amounts within the pancreas to form lethal amounts of MDF. Also, no other organ thus far studied has the necessary ingredients or conditions to form MDF. The pancreas, which contains many enzymes and hormones that are regulatory in nature to a wide variety of metabolic reactions and pathways, is capable of generating a lethal reaction when its organization reaches a sufficient state of disarray such as that which occurs during shock.

One of the important sequelae of splanchnic hypoperfusion is the disruption of the normal stability of the lysosomal membrane within splanchnic organs. Splanchnic lysosomes are very sensitive to ischemia and/or hypoxia and contain a variety of hydrolytic enzymes which are capable of inducing great intracellular and extracellular damage. Some of the cogent facts implicating lysosomes in the pathogenesis of circulatory shock are (8):
  (a) The plasma activity of lysosomal enzymes increases 3 to 4 fold early in the course of many forms of shock and remains elevated throughout the shock state.
  (b) Splanchnic organs, particularly the pancreas and liver, release large amounts of lysosomal enzyme activity since their lysosomal content is markedly decreased, and one can recover much of this "lost" enzyme in the plasma and lymph draining these organs.
  (c) There is a marked increase in the fragility of splanchnic lysosomes during shock. This is manifested by a doubling of the "free" or non-bound form of lysosomal enzymes as well as by the elevation of the rates of release of lysosomal enzymes by suspensions of lysosomes isolated from pancreatic tissue during shock.
  (d) The activity of virtually all of the known lysosomal hydrolases are optimal at acid pH, a condition which occurs during shock in plasma as well as in splanchnic tissues.
  (e) The morphology of lysosomes in the liver and pancreas is greatly altered during shock. Hepatocytes and pancreatic acinar cells exhibit great numbers of large vacuoles which appear to be associated with disruption of the lysosomal membrane and the subsequent release of intralysosomal hydrolases.
  (f) The lysosomal hydrolases remain in the circulation for a much longer period of time in circulatory shock since the kidneys and the reticuloendothelial system are not functioning adequately during shock.
  (g) Lysosomal enzymes (i.e., cathepsins) appear to play a key role in the formation of MDF in circulatory shock by acting as hydrolases of intracellular proteins in the pancreas.

(h) Agents which prevent the release of lysosomal enzymes, such as glucocorticoids, prevent MDF formation and markedly prolong survival in several forms of shock.

Thus, all available evidence points to the lysosomal hydrolases, particularly the cathepsins, as being important agents in the pathogenesis of shock. These enzymes are released in massive amounts under conditions which are optimal for their action during shock. These enzymes remain in circulation for a long time during shock and originate largely from the sensitive splanchnic organs, particularly the liver and pancreas. The pathophysiologic actions of these lysosomal enzymes are gradually becoming known (9). At present there are three major actions of these hydrolases that appear to play a role in the pathogenesis of circulatory shock.

Firstly, the lysosomal hydrolases play a role in the formation of MDF, probably by the action of cathepsins on intrapancreatic proteins in cleaving peptides from the larger intracellular proteins. One of these peptides appears to be MDF. MDF is then transported via the lymph and the plasma to the heart where it exerts a direct negative inotropic effect.

Secondly, the lysosomal hydrolases directly constrict the resistance vessels of the splanchnic vasculature both in the whole animal and in the isolated perfused pancreas. This, of course, represents a positive feedback loop, since it renders the splanchnic area more ischemic which, in turn, effects the release of more lysosomal enzymes. In this regard, the lysosomal enzymes may also act on the microcirculation, enhancing the loss of intravascular fluid and increasing the resistance of the vessels to flow.

Thirdly, the lysosomal hydrolases exert a small negative inotropic effect and sensitize the myocardium to the later action of MDF, thus consuming valuable reserve capacity of the heart early in shock and rendering the heart more vulnerable to the depression observed in later stages.

By this combination of events, the lysosomal enzymes act in concert with MDF in a two phase temporal sequence to activate both simple and complex mechanisms which have a negative survival value. Figure 2 illustrates the present conceptualization of the scope of these effects which thus suggests the name "The Lysosomal Protease - MDF System". The basic effects of MDF are outlined (5) as follows:

(1) The direct cardiac depressant effect of MDF has been observed in isolated cardiac tissue as well as in intact animals (both control and shocked animals). This negative inotropic effect is independent of changes in heart rate, rhythm, or coronary flow. The depression of cardiac contractility induced by plasma extracts containing MDF is approximately 45 to 65% in all of these systems.

The Lysosomal Protease - MDF System in Circulatory Shock

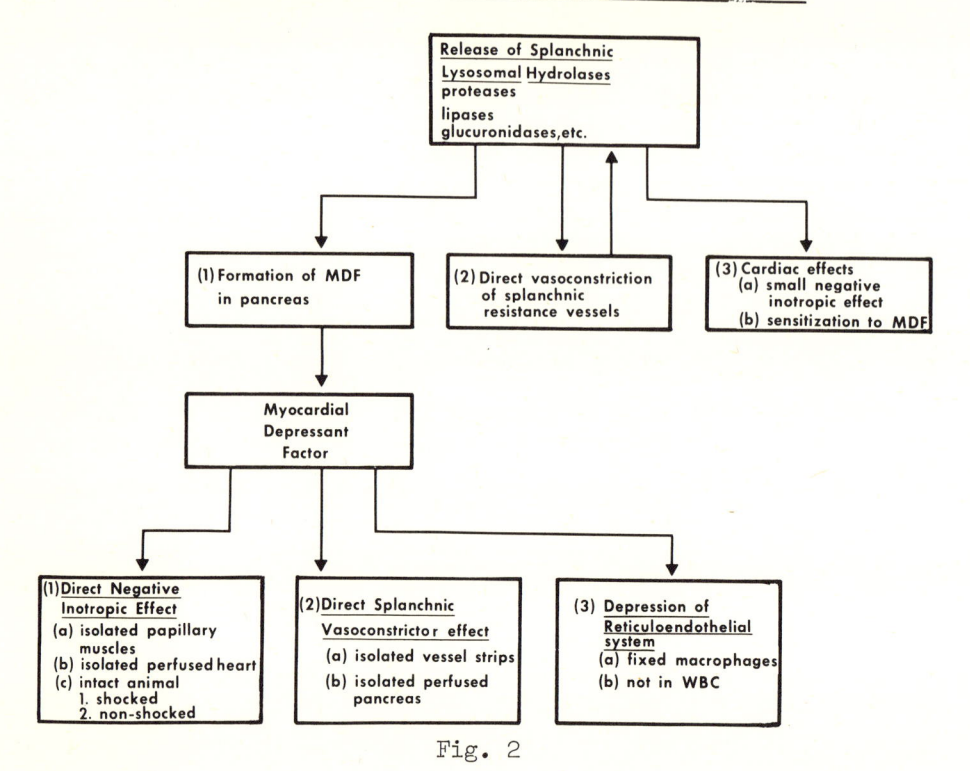

```
                    ┌─────────────────────────┐
                    │ Release of Splanchnic   │
                    │ Lysosomal Hydrolases    │
                    │  proteases              │
                    │  lipases                │
                    │  glucuronidases, etc.   │
                    └─────────────────────────┘
          ↓                    ↓                       ↓
┌──────────────────┐  ┌──────────────────────┐  ┌──────────────────────────┐
│(1) Formation of  │  │(2) Direct            │  │(3) Cardiac effects       │
│    MDF in        │  │    vasoconstriction  │  │   (a) small negative     │
│    pancreas      │  │    of splanchnic     │  │       inotropic effect   │
│                  │  │    resistance vessels│  │   (b) sensitization to MDF│
└──────────────────┘  └──────────────────────┘  └──────────────────────────┘
          ↓
┌──────────────────┐
│   Myocardial     │
│   Depressant     │
│     Factor       │
└──────────────────┘
     ↓                       ↓                           ↓
┌──────────────────┐  ┌──────────────────────┐  ┌──────────────────────────┐
│(1) Direct Negative│  │(2) Direct Splanchnic │  │(3) Depression of         │
│   Inotropic Effect│  │  Vasoconstrictor     │  │    Reticuloendothelial   │
│  (a) isolated     │  │  effect              │  │    system                │
│     papillary     │  │  (a) isolated vessel │  │  (a) fixed macrophages   │
│     muscles       │  │      strips          │  │  (b) not in WBC          │
│  (b) isolated     │  │  (b) isolated        │  │                          │
│    perfused heart │  │      perfused        │  │                          │
│  (c) intact animal│  │      pancreas        │  │                          │
│    1. shocked     │  │                      │  │                          │
│    2. non-shocked │  │                      │  │                          │
└──────────────────┘  └──────────────────────┘  └──────────────────────────┘
```

Fig. 2

When very pure MDF is applied, the depression is about 90%. The nature of the negative inotropic effect of MDF is such that it is characterized by a reduction in developed tension rather than a change in time to peak tension. Furthermore, no notable electrophysiological changes occur in the cardiac muscle cell resting or action potentials. These findings suggest that the cardiodepressant effect is manifested by inhibition of the active state of the contractile elements of the heart.

(2) MDF exerts a splanchnic vasoconstrictor effect which occurs both in isolated vessel strips and in the isolated perfused pancreas. This effect is relatively specific and does not occur in aortic strips or on the total peripheral resistance. The mechanism of the effect of MDF is largely unknown, although it appears to be direct and is not dependent upon the release of other vasoactive agents, since it occurs in isolated vascular strips bathed in a simple physiological salt solution. This splanchnic vasoconstrictor effect constitutes a positive feedback loop with MDF production, since splanchnic vasoconstriction contributes to splanchnic hypoperfusion resulting in MDF production, which, in turn, causes further splanchnic vascular constriction.

(3) MDF appears to contribute to a depression of reticuloendothelial function (i.e., clearance of carbon particles from the circulation), although this may be a function of another toxic factor in shock, a reticuloendothelial depressant substance (RDS) which may be similar or even identical to MDF (10). This effect is manifested on the fixed macrophages (i.e., Kupfer cells) rather than on the leucocytes. In any event, this RE depression constitutes another positive feedback loop, since the RE depression favors the accumulation of MDF in the plasma, and thus MDF is free to exert its toxic effects without being cleared from the plasma.

Thus, MDF is a peptide which is produced in the ischemic pancreas as a result of cellular disruption and the release of lysosomal proteases which catalyze the cleavage of peptides from proteins. MDF accumulates in the plasma where it exerts a prominent negative inotropic effect. Two positive feedback loops (e.g. splanchnic vasoconstriction and RE depression) magnify its toxic effect. MDF itself can induce a state of shock when injected into non-shocked control animals. MDF fulfills all the basic criteria of a toxic factor in shock and thus appears to be a crucial factor in the induction of lethality in the pathogenesis of circulatory shock.

REFERENCES

1. Lillehei, R.C., Longerbeam, J.K., Block, J. & Manax, W.G. Ann. Surg. 160, 682.
2. Weissman, G. & Thomas, L. (1962) J. Exp. Med. 116, 433.
3. Crowell, J. & Guyton, A.C. (1962) Am. J. Physiol. 203, 248.
4. Gomez, O.A. & Hamilton, W.F. (1964) Circ. Res. 14, 327.
5. Lefer, A.M. (1970) Fed. Proc. 29, 1836.
6. Lefer, A.M. & Glenn, T.M. The role of the pancreas in the pathogenesis of circulatory shock. In press. Oklahoma Shock Symposium, 1971, Oklahoma City, Oklahoma.
7. Lefer, A.M. & Glenn, T.M. Interaction of lysosomal hydrolases and a myocardial depressant factor in the pathogenesis of circulatory shock. In: Shock in High and Low Flow States, In press, Exerpta Medica, 1972.
8. Glenn, T.M. & Lefer, A.M. (1971) Circ. Res. 29, 338.
9. Glenn, T.M., Morris, J.R.III, Lefer, A.M. Lopez-Rasi, A.M., Serati, R.S., Ferguson, W.W. and Wangensteen, S.L. (1972) Ann. Surg. In press.
10. Lefer, A.M. & Blattberg, B. (1968) J. Reticuloendothel.Soc. 5, 54.

MYOCARDIAL METABOLISM DURING ENDOTOXIC SHOCK

John C. Scott, Jen Tsoh Weng, and John J. Spitzer

Department of Physiology and Biophysics Hahnemann
Medical College and Hospital Philadelphia, Pennsylvania,
19102

There is little evidence of myocardial depression during the initial fall in arterial pressure following the injection of endotoxin in the dog[1]. It appears more likely that the initial changes are due to an inadequate venous return[2]. However, if this condition persists for a longer period of time, myocardial changes may also be observed.

Since subtle alterations of myocardial metabolism may precede overt functional changes of the heart, the present study was undertaken to assess some of the alterations of substrate utilization by the myocardium during the early phases of endotoxic shock and to relate these changes to the hemodynamic responses during this period.

MATERIALS AND METHODS

Experiments were performed on 9 fasted mongrel dogs weighing 20 to 27 kg. They were anesthetized with sodium pentobarbital (30 mg/kg) given intravenously. An additional dose of 30-70 mg was given during the experiment when needed. Catheters were introduced through the external jugular vein into the coronary sinus vein and the pulmonary artery under fluoroscopic guidance. Heparinized saline (10 units/ml) drip was maintained through the catheters at a rate of 6 to 12 drops per min.

Mean arterial blood pressure, heart rate, ECG, and total O_2 consumption were recorded intermittently before and during a

2 hr period, following the intravenous injection of E. coli endotoxin, using conventional techniques. Blood samples from a systemic artery, and the coronary sinus vein were drawn to determine the coronary blood flow and some parameters of myocardial metabolism. Blood from the pulmonary artery was analyzed to calculate cardiac output by the Fick principle. Left ventricular work and efficiency were calculated as described previously (3). The blood samples were taken in a control period before, at 60 min and 120 min after the injection of endotoxin. Coronary blood flow was measured by the I^{131}-iodoantipyrine saturation method (4).

A continuous infusion of albumin-bound $1-C^{14}$-palmitate was given in tracer doses through a catheter inserted into a brachial vein (5). A period of at least 30 min was allowed for the labeled FFA infusion before the first run, which consisted of control determinations. This was followed by a dose of E. coli endotoxin of $\sim$ LD 90 through a venous catheter. Sixty and 120 min after endotoxin injection, the second and third runs were performed. In each run, samples for the measurement of coronary blood flow were taken first, followed by blood samples for O_2 and CO_2 content determination (by the Van Slyke method) and metabolite assays.

Each blood sample was heparinized in vitro and kept in an ice bath. Two aliquots of 0.5 ml were immediately transferred into chilled tubes containing 3.6 ml of 6.0% perchloric acid for determination of blood lactate by enzymatic analysis (6). The remainder of each sample was centrifuged in a refrigerated centrifuge and 2 aliquots of plasma, 1 ml each, were added to 5 ml of Dole's solution to be used for FFA estimation by the method of Dole and Meinertz (7). The lower phase was then acidified and radioactive FFA extracted and counted as previously described (8).

The blood samples for $C^{14}O_2$ determination were drawn in duplicate aliquots and estimated by the method of Passman et al (9). The arterial pH was measured anaerobically at 38°C using a Radiometer Model 27 pH meter.

All calculations were based on the assumption that labeled and unlabeled fatty acids are metabolized in the same way by the tissues and that the different individual fatty acids comprise a single pool with respect to uptake and release. The following equations were used for the information presented in the Tables (5).

(1) total body FFA turnover (μmole/min) =

$$\frac{\text{rate of palmitate-}^{14}\text{C infusion (dpm/min)}}{\text{FFA}_a \text{ SA (dpm/μmole)}}$$

(2) myocardial FFA uptake (μmole/min) =

$$\frac{\text{labeled FFA}_a - \text{labeled FFA}_v}{\text{labeled FFA}_a \text{ SA}} \times \text{plasma flow}$$

(3) rate of FFA oxidation to CO_2 (μmole/min) =

$$\frac{^{14}CO_{2\,v} - {}^{14}CO_{2\,v}}{^{14}C\text{-FFA}_a \text{ SA}} \times \text{blood flow}$$

Radioactivity is expressed in dpm/ml. Subscripts a and v denote arterial and coronary sinus blood.

RESULTS

Table 1 lists the hemodynamic changes before and during a 2 hr period following the intravenous injection of endotoxin.

TABLE 1. Hemodynamic changes during endotoxic shock (N=8)

	Control	1 hr after endotoxin	(%) Change	2 hr after endotoxin	(%) Change
MABP (mm Hg)	134±4	66±5	-51†	74±5	-45†
HR (per min)	180±12	175±9	-3	173±10	-4
CO (ml/min)	34.80±5.50	1820±230 (N=7)	-39†	1930±230	-45‡
CBF (ml/min/100 g)	125.3±14.0	85.0±10.5	-32‡	90.5±9.8	-29‡

MABP = mean arterial blood pressure
HR = heart rate
CO = cardiac output
CBF = coronary blood flow
± = SEM
†P < 0.01
‡P < 0.05

As expected, arterial blood pressure, cardiac output and coronary blood flow decreased significantly. The heart rate remained unchanged.

Table 2 lists the changes of O_2 content of the arterial and mixed venous blood during the course of shock. (Total body O_2 consumption and the accumulated O_2 deficit are expressed as O_2 ml/min/kg in this table.)

TABLE 2. Changes in O_2 consumption and blood content during endotoxic shock

	Control	1 hr after endotoxin	2 hr after endotoxin
Arterial O_2 (vol %)	14.9±0.67	14.6±0.64	12.9±1.1
Pulmonary artery O_2 (vol. %)	9.8±0.77	7.0±0.58‡	6.2±1.0‡
A-V O_2 (arteriovenous difference)	5.1±0.53	8.0±0.65†	6.7±0.53
Total O_2 extraction (%)	34±3.7	53±3.7‡	54±4.7‡
Total body O_2 consump. (ml/kg/min)	6.9±0.41	5.8±0.41	5.6±0.50
Accumulated O_2 deficit (ml/kg)	---	73	136

± = SEM
†P <0.01
‡P <0.05

In addition to the 60 and 120 min observations, more frequent measurements were made of mean arterial blood pressure, heart rate and total body O_2 consumption. These are plotted as per cent changes from control values in Fig. 1. It will be noted that there is a sharp dip in each of these parameters which reaches a low value in 3-5 min and then gradually returns towards normal over a period of about 30 min. During the next 90 min, they level off somewhat below the control values.

TABLE 3. Changes in O_2 consumption, CO_2 production, left ventricular work and efficiency during endotoxic shock (N=8)

	Control	1 hr after endotoxin	(%) Change	2 hr after endotoxin	(%) Change
Left ventric. O_2 consump. (μmole/min/100g)	701±94	449±39 (N=7)	-41	385±51	-45‡
Left ventric. CO_2 produc. (μmole/min/100g)	559±122	413±68 (N=7)	-36	319±53	-43
Total body O_2 consumption (μmole/M^2/min)	8634±716	7544±620	-13	7231±740	-16
Arterial O_2 conc. μmole/ml)	6.69±0.33	6.66±0.29	-2	6.14±0.40	-8
Left ventric. work (kg m/min)	6.13±0.91	1.74±0.31 (N=7)	68†	1.97±0.32	64†
Left ventric. efficiency (%)	27.9±6.0	10.9±1.9	48†	13.9±0.9	32‡

± = SEM
†P <0.01
‡P <0.05

The changes in O_2 consumption and CO_2 production by the myocardium, and in total body O_2 consumption are presented in Table 3. The left ventricular O_2 consumption decreased by 41 and 45% 1-2 hr after endotoxin injection. The left ventricular CO_2 production diminished also, but there were two out of eight dogs which showed an increase instead of a decrease during endotoxic shock. Total body O_2 consumption decreased slightly and the change was not statistically significant. Left ventricular work decreased by more than 60%. Left ventricular efficiency also diminished significantly.

TABLE 4. Metabolic changes during endotoxic shock (N=8)

	Control	1 hr after endotoxin	(%) Change	2 hr after endotoxin	(%) Change
Arterial FFA (μmole/ml)	0.732±0.198	0.653±0.065	-11	0.848±0.202	+15
FFA flux (μmole/min)	311±76	291±67	-7	296±52	-5
Arterial glucose (μmole/ml)	5.56±0.34	5.04±0.73	-9	3.24±0.53	-42†
Arterial lactate (μmole/ml)	1.53±0.18	3.62±0.46	+137†	3.87±0.53	+154†
Arterial pH	7.45±0.03	7.32±0.04	-1.7†	7.22±0.04	-3.1†
Arterial hematocrit (%)	41.4±2.2	48.0±2.5	+14‡	45.9±2.5	+11‡

± = SEM
†P <0.01
‡P <0.05

Arterial concentration of various metabolites and arterial pH following endotoxin injection are presented on Table 4. Arterial FFA concentration and FFA flux did not change significantly from control during one and two hr after endotoxin injection. Arterial glucose concentration did not significantly change in 1 hr, but did decrease by 42% in 2 hr. The expected increase in arterial lactate and decrease in arterial pH are also shown in Table 4.

The changes in myocardial metabolism of FFA during endotoxic shock are shown in Table 5. Although the myocardium extracted the same fraction of the arterial FFA, uptake decreased by 35 and 22%, oxidation decreased to a greater extent at both 1 and 2 hr following endotoxin. Thus, the ratio of FFA oxidation to FFA uptake decreased during shock. The calculated contribution of FFA to myocardial CO_2 production decreased from 60% in the control period to 32 and 44% following endotoxin. Although the mean values of lactate uptake increased following endotoxin, the changes were not statistically significant.

TABLE 5. Changes in myocardial metabolism of FFA and lactate during endotoxic shock (N=8)

	Control	1 hr after endotoxin	(%) Change	2 hr after endotoxin	(%) Change
FFA extraction (%)	41.5±5.1	43.9±6.6		36.2±5.4	
FFA uptake (μmole/min/100 g)	17.96±3.05	11.64±1.45	-35†	14.09±3.83	-22
FFA oxidation (μmole/min/100 g)	16.10±4.20	7.59±1.86	-53†	7.88±1.71	-51†
FFA oxidation/ FFA uptake	90%	65%		56%	
Lactate uptake (μmole/min/100 g)	76.3±29.7	101.1±33.4	+33	93.4±21.6	+22

± = SEM
†P <0.05

Calculated contribution of FFA and of lactate to myocardial CO_2 production are shown in Table 6. FFA oxidation accounted for a greater portion of CO_2 production under control conditions than following endotoxin. The opposite held true for lactate.

TABLE 6. Calculated contribution of FFA and lactate to myocardial CO_2 production during endotoxic shock (N=8)

	Control*	1 hr after endotoxin*	2 hr after endotoxin*
†Calculated contribution of FFA to myocardial CO_2 production (%)	60.1±20.6	32.3±5.9	43.8±10.7
‡Calculated contribution of lactate to myocardial CO_2 production (%)	41.0±12.6	68.4±29.3	82.8±16.6

* † ‡ (See over)

(See Table 6)

* Mean ± SEM
† Calculated as [(FFA oxid. X 17)/Left ventric. CO_2 prod.] X 100
‡ Calculated as [(lact. upt. X 3)/Left ventric. CO_2 prod.] X 100

DISCUSSION

The blood pressure changes in response to endotoxin (Fig.1) are almost identical to those shown in a report by Brobman et al (2). On the basis of earlier reports, these authors explained

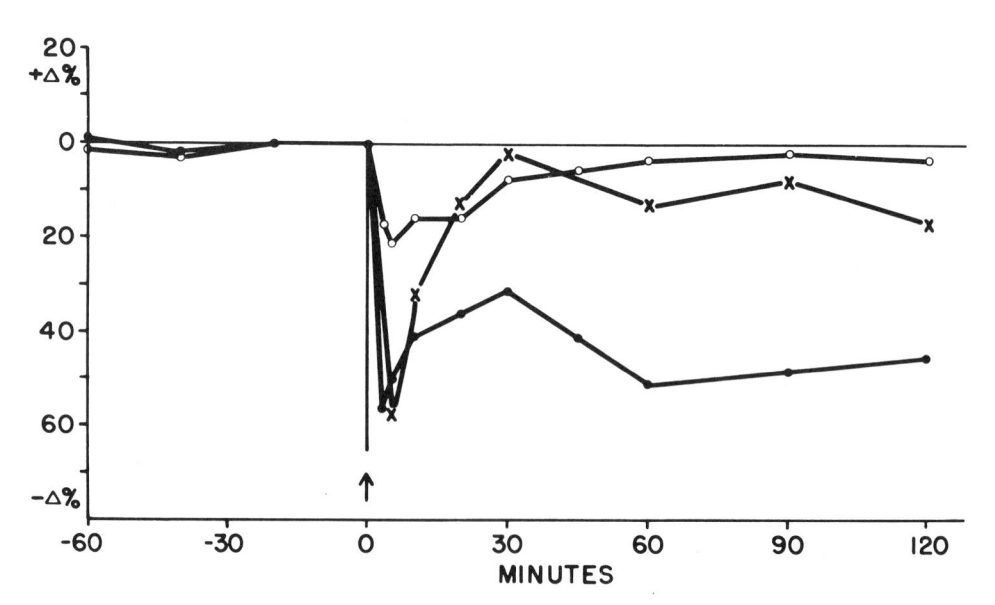

Fig. 1. Per cent changes in heart rate, mean arterial blood pressure and total O_2 consumption before and during endotoxic shock.

  o = heart rate
  x = total body O_2 consumption
  ● = mean arterial blood pressure
  ↑ = time of endotoxin injection

the fall in blood pressure as being due to a pooling of blood in the splanchnic viscera which reduced the venous return and hence the cardiac output. However, equal volumes of blood may be pooled elsewhere according to a recent report (1). Fig. 1 also shows parallel changes in heart rate and total body O_2 consumption in the very early period following endotoxin. The fall in heart rate which is contrary to the familiar Marey's Law has not been explained satisfactorily but may be due to the release of histamine which is commonly associated with the symptoms of endotoxic shock (11).

The O_2 consumption curve follows quite closely the changes in blood pressure. Both of these parameters may in fact be the result of corresponding changes in cardiac output. This would mean that the reduction in cardiac output and hence in O_2 delivery to the tissues would be indicated by a fall in mixed venous blood O_2 content and a rise in the total body O_2 extraction. Oxygen changes of this type at 60 and 120 min following the injection of endotoxin are shown in Table 2*. Thus, it would appear that inadequate perfusion of the tissues accounts for the major hemodynamic changes during endotoxic shock. This table also shows the accumulation of an oxygen deficit during the two hours following the injection of endotoxin. According to studies of hemorrhagic shock reported by Crowell (12) an accumulation of 120 ml O_2/kg deficit produced a 50% mortality, due to irreversible shock. If this interpretation is applicable to endotoxin, the metabolic data obtained at 120 min may reflect irreversible changes.

The main objective of the present study was to assess some of the alterations of myocardial substrate metabolism during endotoxic shock whether or not they are primary cardiac changes or secondary reflections of peripheral tissue changes. It is well known that FFA and lactate are the two major metabolic fuels for the myocardium (13, 14). In the present study FFA accounted for 60% and lactate for about 40% of the myocardial CO_2 production under control conditions, assuming that all of the removed lactate was oxidized. Following endotoxin injection, the myocardial CO_2 production derived from FFA was decreased, and that derived from lactate was increased.

---

* If these assumptions are correct, the cardiac output should show the same precipitous drop in the first 3 min as that which occurred with blood pressure and total body oxygen consumption. Preliminary experiments with a catheter-tip flowmeter placed in the ascending aorta have indeed shown a rapid fall in cardiac output during the first 3 min after the injection of endotoxin.

The metabolic alterations described in this communication are strikingly similar to those found in a group of dogs subjected to severe hemorrhage (15). Myocardial uptake and oxidation of FFA decreased markedly, lactate uptake did not change significantly. The hemodynamic changes accompanying hemorrhagic shock were also comparable to those found in the present group of dogs.

In the present study, myocardial glucose uptake amounted to 29.4 ± 13.6 µmole/min/100 g under control conditions and 14.3 ± 4.9 and 10.4 ± 5.3 during runs 2 and 3. The large standard errors were due to the fact that there were 2 out of 8 dogs which failed to show any A-V differences in all 3 runs.

From Table 3, it is evident that the percentage reduction of left ventricular O_2 consumption is far greater than that of total body O_2 consumption during endotoxic shock. The important question arises whether these changes indicate myocardial hypoxia. It can be observed that left ventricular O_2 consumption decreased more than did the coronary blood flow, while arterial O_2 concentration did not change significantly. Thus, the reduction of ventricular O_2 consumption was due to the decreased O_2 requirement of the ventricle, accompanying the diminished cardiac output and arterial blood pressure, rather than to inadequate supply of O_2. The above considerations assume that no direct toxic action of endotoxin on myocardial tissue takes place (1).

Left ventricular O_2 consumption decreased less percentage-wise than did left ventricular work. Thus, the efficiency of the heart diminished following endotoxin. This is presumably due to the decreased stroke volume, while the heart rate remained unchanged.

It has been noted (16) that diminished myocardial lactate utilization may be a more sensitive indicator of myocardial ischemia than electrocardiographic or hemodynamic parameters. The present finding of no significant decrease of lactate uptake during endotoxic shock appears to be additional evidence against the presence of myocardial hypoxia.

It is not clear whether the cause of the increased myocardial reliance on lactate for oxidizable substrate is due to a primary change in myocardial tissue metabolism or is the consequence of increased availability of lactate because of peripheral release. It may be inferred from the experiments of Hirche and Langohr (17) that the latter may be operational, at least under certain experimental conditions.

In addition to the diminished myocardial FFA uptake, a smaller fraction of the removed FFA was oxidized which may indicate a diversion of the removed FFA to triglyceride and phospholipid synthesis. These results are again in accordance with those reported to occur with hemorrhagic shock (15). It may be argued that the increased lactate availability might inhibit some steps in the pathway of FFA oxidation, and hence cause the accumulation of FFA in the myocardium. This, in turn, would inhibit FFA uptake and divert FFA to triglyceride and phospholipid synthesis.

The present studies do not define the exact mechanism of the observed shift in myocardial metabolism. Assuming that there is no specific myocardial response to endotoxin, the mechanism may be due to decreased metabolic requirements imposed upon the heart, secondary effects of peripheral hypoxia, or the increased availability of blood lactate. Since all three alternatives also hold for the metabolic changes following hemorrhage and because the metabolic changes following hemorrhage and endotoxin administration are quite similar, it is suggested that the inadequate perfusion of peripheral tissues is indirectly responsible for the major myocardial metabolic changes in both types of shock.

SUMMARY

Endotoxic shock was produced in 9 anesthetized dogs in order to compare hemodynamic and metabolic changes of the myocardium with those resulting from severe hemorrhage. Myocardial metabolism of FFA was studied by a continuous infusion of albumin-bound 1-C^{14}-palmitate. Left ventricular O_2 consumption, work and efficiency decreased following endotoxin. Marked decrease of myocardial FFA uptake and oxidation was also noted, although arterial FFA concentration and total body FFA flux did not change significantly. At the same time, lactate uptake by the myocardium showed a tendency to increase. Thus, a larger fraction of myocardial energy was supplied by lactate than by FFA after endotoxin, while the opposite was true under control conditions. These changes bore marked similarity to those observed following hemorrhage. Therefore, it seems that the changes may be due to one or more of the following: decreased metabolic requirements imposed on the heart, secondary effects of peripheral hypoxia, or increased availability of blood lactate.

REFERENCES

1. Hinshaw, L.B., Archer, L.T., Greenfield, L.J. & Guenter, C.H. (1971) Am. J. Physiol. 221:504.

2. Brobman, G.F., Ulano, H.B., Hinshaw, L.B. & Jacobson, E.D. (1970) Am. J. Physiol. 219:1464.

3. Scott, J.C. & Balourdas, T.A. (1959) T.A., Circul. Res. 7:162.

4. Krasnow, N., Levine, H.J., Wagman, R.J. & Gorlin, R. (1963) Circul. Res. 12:58.

5. Little, J.R., Goto, M. & Spitzer, J.J. (1970) Am. J. Physiol. 219:1458.

6. Hohorst, H.J. In: Methods of Enzymatic Analysis (H.V. Bergmeyer, ed.), New York-Academic Press, 1963.

7. Dole, V.P. & Meinertz, H. (1960) J. Biol. Chem. 235:2595.

8. Gold, M. & Spitzer, J.J. (1964) Am. J. Physiol. 206:153.

9. Passman, J.M., Radin, N.S. & Cooper, J.A.D. (1956) Anal. Chem. 28:484.

10. Blattberg, B. & Levy, M.N. (1971) Am. J. Physiol. 220:1267.

11. Aviado, D.M. 12th Hahnemann Symposium P.192, L.C. Mills and J.H. Moyer, eds., Grune & Stratton, New York, 1965.

12. Crowell, J.W. & Smith, E.E. (1964) Am. J. Physiol. 206:313.

13. Scott, J.C., Gold, M., Bechtel, A.A. & Spitzer, J.J. (1968) Metab. 17:370.

14. Opie, L.H. (1969) Am. Heart J. 77:100.

15. Spitzer, J.J. & Spitzer, J.A. (1971) Am. J. Physiol. 222, 101.

16. Scheuer, J. & Brachfeld, N., (1966) Circul. Res. 18:178.

17. Hirche, H. & Langohr, H.D. (1967) Pflugers Archiv. 293:208.

These studies were supported by grants HE 04916 and HE 03130 from the National Heart Institutes and by the U.S. Navy Themis Project.
We acknowledge with appreciation the valuable contribution of Mr. A.A. Bechtel to the success of this project.

THE ACTIVITY OF LIPOPROTEIN LIPASE IN RAT HEART AFTER

TOURNIQUET STRESS

J. Borbola, A. Gecse and S. Karady

Institute of Pathophysiology, University of Szeged
School of Medicine Szeged, Hungary

Previously (1971) we found a considerable hyperlipaemia
in rats subjected to sublethal tourniquet stress and 48-72 hr
after stress the values returned to normal. Robinson and
Jennings (1965 proved that perfusion of rat heart with
heparinized McEwen-albumin solution by the Langendorf technique
resulted in the release of clearing factor lipase into the
perfusate. Later Gartner and Vahouny (1966) demonstrated the
presence and heparin activation of lipoprotein lipase in the
subcellular fractions of rat heart homogenate. To get a
detailed answer to the changes in the lipoprotein lipase
activity in rat heart we adopted the Gartner and Vahouny method
(1966) as preferable to the acetone extracted powder method
proposed by Korn (1955a, 1955b). The presence of the
lipoprotein lipase in the cardiac muscle enables the heart to
utilise circulating triglyceride fatty acids effectively as
source of energy. The main purpose of the present study was
to determine the pathophysiological significance of the heart
lipoprotein lipase in sublethal tourniquet stress of rats.

The experiments were carried out on male rats of
R-Amsterdam strain weighing 180-200 g, kept on standard diet.
The rats were fasted overnight before the experiment.

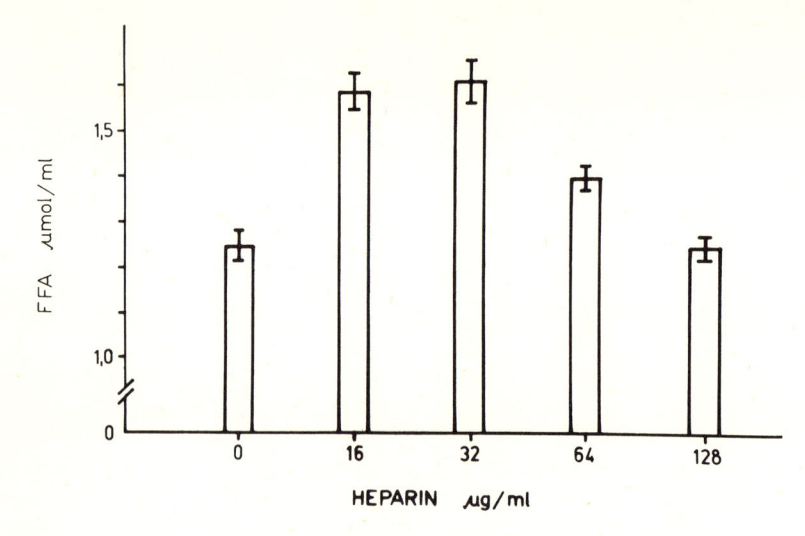

Fig. 1. The lipoprotein lipase activity of rat heart at different heparin concentrations. On the ordinate the quantity of FFA (μmol/ml) released by the enzyme from Ediol at 37°C, pH 7.4 in 60 min.

Sublethal tourniquet stress was induced by the method of Stoner (1958). The hindlimbs of rats were excluded from the circulation for 3 hr. Extraction of the heart lipoprotein lipase was carried out according to the method of Gartner and Vahouny (1966). Ediol (Riker Lab. USA) was used as triglyceride substrate. Five percent fatty acid-poor bovine albumin (Calbiochem) was used as fatty acid acceptor. The albumin pH was adjusted to 8.5 with ammonium hydroxide. The substrate mixture contained the following components: 0.2 ml Ediol (4% emulsion), 0.2 ml rat serum (cofactor), 1.0 ml bovine albumin. Equal volumes of enzyme and substrate were incubated at 37°C and aliquots were taken every 30 min and their FFA content was determined according to the method of Mosinger (1965).

Mosinger's method was modified as follows: 0.05% sulfuric acid extraction was carried out to remove the organic acids from

ACTIVITY OF LIPOPROTEIN LIPASE IN RAT HEART AFTER TOURNIQUET STRESS

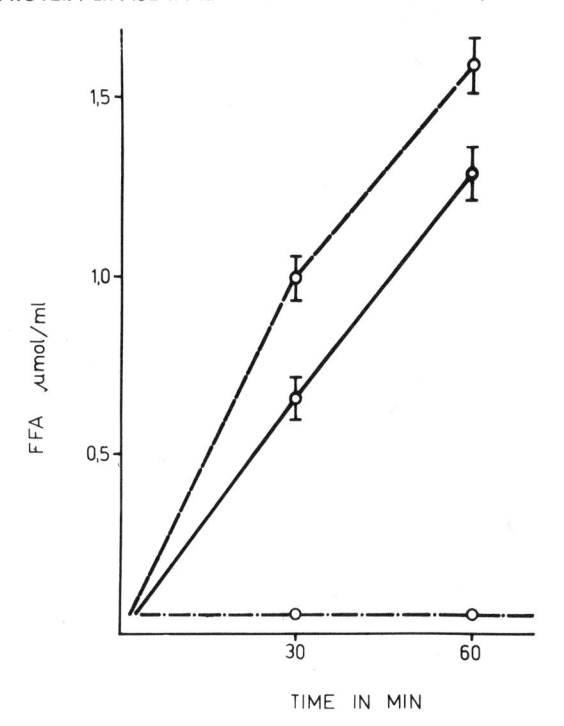

Fig. 2. Activation and inhibition of lipoprotein lipase obtained from rat heart.
On the ordinate the amount of FFA released by the enzyme.
o———o without activation, o— — —o enzyme activated with 16 µg/ml heparin, o—•—•—o effect of 1 M NaCl.

the mixture. In each experimental group both normal rat serum and the serum of stressed animal was used as cofactor. Enzyme activity was determined both before and after activation with heparin. For statistical evaluation Student "t" test was used.

The effect of different heparin concentrations on the rat heart lipoprotein lipase activity is illustrated on Fig.1. In the presence of 16-32 µg/ml heparin the activity of enzyme considerably increased. The further elevation of heparin concentration failed to elicit enzyme activation.

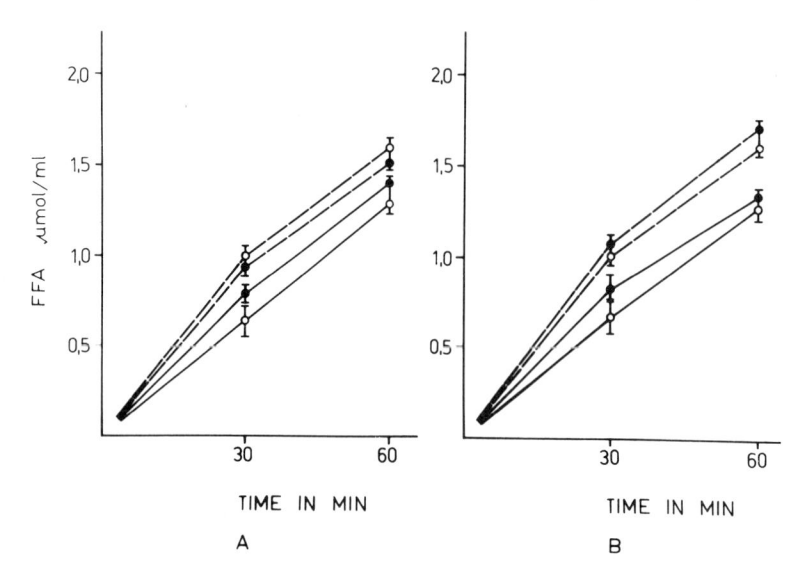

Fig. 3. Alterations of rat heart lipoprotein lipase activity immediately after immobilisation.
A. Cofactor - normal rat serum, B: cofactor - serum of immobilised rat. o———o control enzyme without activation, o- - -o heparin activated control enzyme, •———• enzyme action after immobilisation (without activation) •- - -• heparin activated enzyme in immobilised rat.

The lipolytic effect of rat heart was observed without heparin activation. Activating the enzyme at room temperature with 16µg/ml heparin for 15 min significantly increased the activity of heart lipase (Fig. 2). In one series of experiments the enzyme containing heart homogenate was incubated at 100°C for 15 min, it caused total disappearance of the lipolytic activity. Adding 1 M NaCl solution inhibited the activity of the enzyme in each case. This proved that the enzyme isolated by us was lipoprotein lipase since it was activated by heparin and inhibited by 1 M NaCl solution.

The effect of immobilisation of rats on the lipolytic activity of their hearts is illustrated in Fig. 3.

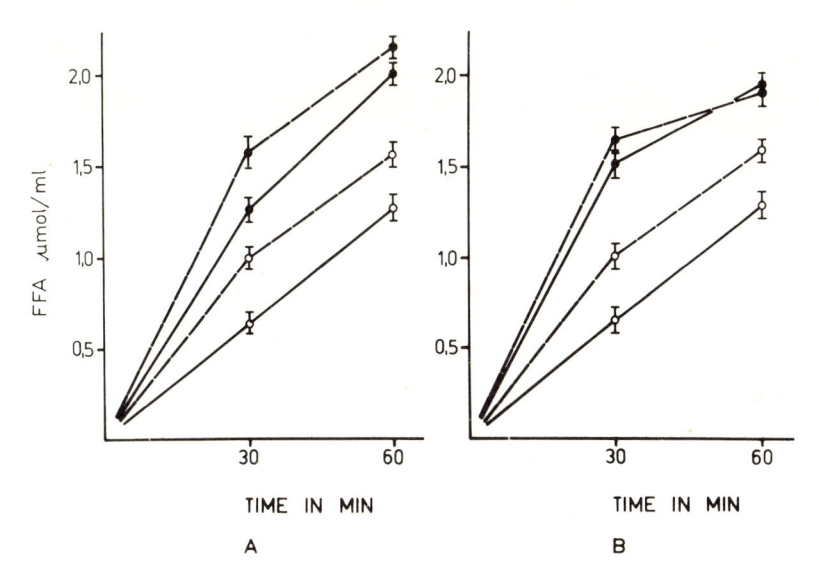

Fig. 4. The activity of heart lipoprotein lipase in rats subjected to sublethal tourniquet stress.
The determinations were carried out 6 hr after releasing the tourniquets. A: cofactor - normal rat serum, B: cofactor - serum of rats subjected to stress 6 hr earlier. o———o control enzyme without activation, o- - -o control enzyme activated by heparin, •———• enzyme activity 6 hr after stress (without activation), •- - -• heparin activated enzyme 6 hr after stress.

The activity of clearing factor in the rat heart increased slightly immediately after releasing the tourniquets, when either normal rat serum or immobilised rat serum was used as cofactor in the incubation mixture.

A considerable increase in heart lipolytic activity was observed in the 6th hr after releasing the tourniquets when either in vitro heparin activated serum or unactivated serum was used. Increased clearing activity was also found when the applied serum-cofactor was taken in the 6th hr after the sublethal tourniquet stress. In this case there was no significant difference between the activity of heparin activated or unactivated enzyme (Fig.4). The heart lipolytic activity showed a gradual decrease from the 6th till the 48th hr after limb ischaemia when it approached the control value.

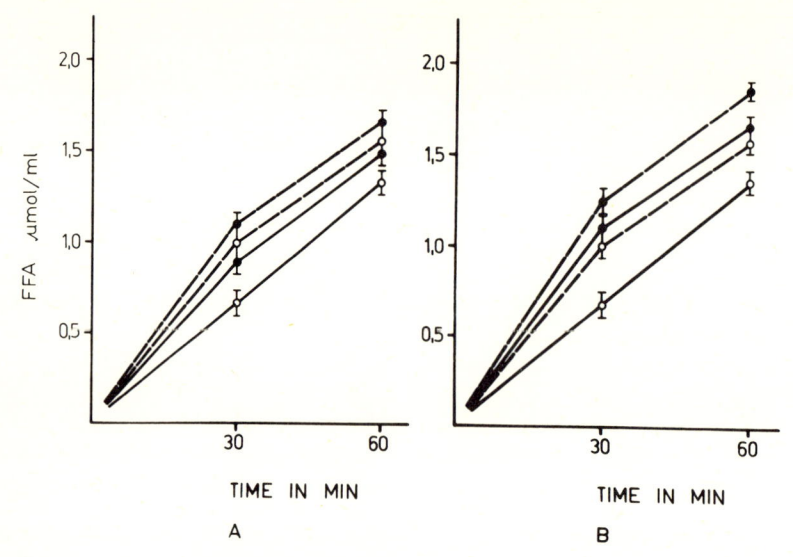

Fig. 5. The activity of heart lipoprotein lipase of rats subjected to sublethal tourniquet stress.
The determinations were carried out 48 hr after releasing the tourniquets. A: cofactor — normal rat serum, B: cofactor — serum of rats subjected to stress 48 hr earlier. o———o control enzyme without activation, o- - -o control enzyme activated with heparin, •———• enzyme action 48 hr after stress (without activation), •- - -• heparin activated enzyme 48 hr after stress.

Using serum from animals 48 hr after they had been exposed to sublethal tourniquet stress greater lipoprotein lipase activity was observed in the heart of rats than after the application of normal serum. The lipoprotein lipase was extracted from the hearts of rats, 48 hr earlier subjected to tourniquet stress (Fig. 5).

Summarizing our results we may say that a sublethal tourniquet stress induces a considerable increase in the activity of rat heart lipoprotein lipase. The highest lipase activity was observed in the 6th hr after the stress stimulus. Immediately after the 3 hr immobilisation the lipoprotein lipase activity of rat heart increased slightly. The isolated enzyme was found to be activated by 16 µg/ml heparin and inhibited by 1 M NaCl as well as incubation at 100°C. It is most likely that the endogenous heparin released from the mast cells is responsible for the presence of the lipoprotein lipase in great quantity found after sublethal tourniquet stress.

REFERENCES

Borbola, J., Gecse A. & Karady I. (1971). Kiserl, Orvostud. 23: 256.

Gartner, S.L. & Vahouny GV. (1966) Am. J. Physiol. 211:1063.

Korn, E.D. (1955a), (1955b) J. Biol.Chem. 215:1. J. Biol.Chem. 215:15.

Mosinger, F. (1965) J.Lip.Res. 6:157

Robinson, D.S. & Jennings M.A. (1965) J.Lip.Res. 6:222.

Stoner, H.B. (1958) Brit.J.exp. Path. 39:251.

TIME COURSE OF METABOLITE AND ENZYME CHANGES IN HYPOXIC CONDITIONS

G. Horpácsy, T. Barankay, K. Tarnoky, S. Nagy and
G. Petri
Institute of Experimental Surgery, University Medical
School of Szeged, Szeged, Hungary

Tissue hypoxia of shock is often characterized by
metabolites of anaerobic glycolysis released into the blood
plasma such as lactic and pyruvic acids and by the so-called
"excess lactate" of Huckabee. (Huckabee 1958; Rosenberg et al.,
1961; Peretz et al., 1965). Release of lysosomal and
cytoplasmic enzymes into the circulation also occurs in
hemorrhagic shock. We have shown earlier that their plasma
level correlates well with the severity of shock (Gergely et al.,
1970; Barankay and Petri 1969). According to the investigations
of Schmidt during hypoxic damage of isolated, perfused organs
a significant release of enzymes and metabolites takes place
(Schimdt et al., 1966). These substances appear in the
perfusate after the lapse of a certain period of time depending
on their intracellular localization, binding, etc.

This study aimed at the recognition of the time sequence
of appearance in the circulation or perfusate of lysosomal and
other intracellular enzymes and metabolites and their relationship
under hypoxic conditions.

Methods

The experiments were performed on 11 mongrel dogs of
10 kg mean body weight, and on 30 isolated dog kidneys.
The following methods were used: hemorrhagic shock was produced
by bleeding the animals into an elevated reservoir until a mean
arterial blood pressure between 30 and 40 mm Hg was reached.
This hypotension was maintained for 90 min and followed by
reinfusion of the blood. Animals were observed for an additional

90 min after reinfusion. Blood samples were taken every 30 min during the experiment.

Perfusion of isolated dog kidneys was done in an apparatus assembled in our Institute (Petri and Horpacsy 1971). The experiments were carried out at 3 temperatures, (a) room temperature, (b) 8-10°C and (c) 2-4°C. The mean duration of perfusion was 10 hr. Fresenius's solution was used for perfusion and samples were taken hourly from the perfusate (Vahlensiek, 1969).

The following enzymes and metabolites were determined in plasma and perfusate by fluorometric methods:

    acid phosphatase (Barankay 1969);
    β-glucuronidase (Greenberg 1966);
    leucine-aminopeptidase (Rockerbie and Rasmussen 1967);
    lactic dehydrogenase (Tarnoky 1970);
    lactic acid (Loomis 1961);
    pyruvic acid (Tarnoky 1969);
    excess lactate was calculated according to Huckabee (1958)

Measurements were made with a Carl Zeiss spectrofluorometer.

Results

Metabolite and enzyme changes during hemorrhagic shock are demonstrated in Fig. 1.

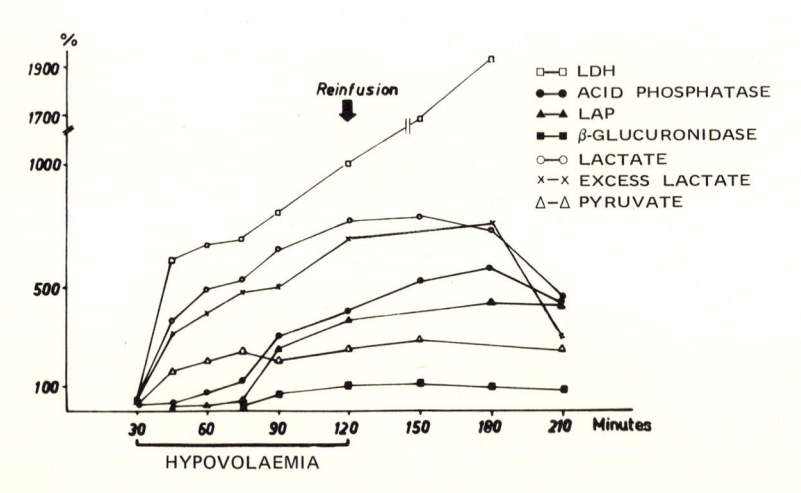

Fig. 1

TIME COURSE OF METABOLITE AND ENZYME CHANGES IN HYPOXIC CONDITIONS

Fig. 2

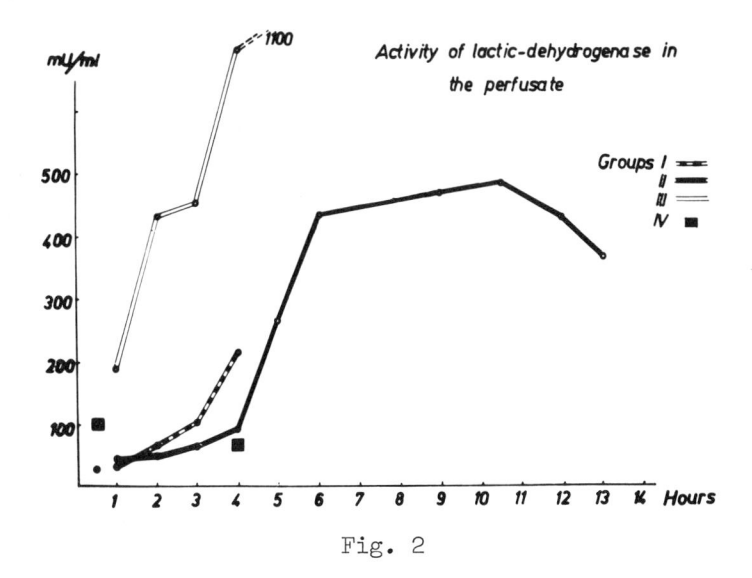

It can be seen that due to hypoxia the levels of LDH, lactic acid and excess lactate rose rapidly and the rise in pyruvic acid was also significantly greater than that of acid phosphatase, β-glucuronidase and LAP, the last three of which showed a significant rise only after the 45th min of hypotension. After the reinfusion there was a further elevation of the level of enzymes and metabolites which by the end of the experiment

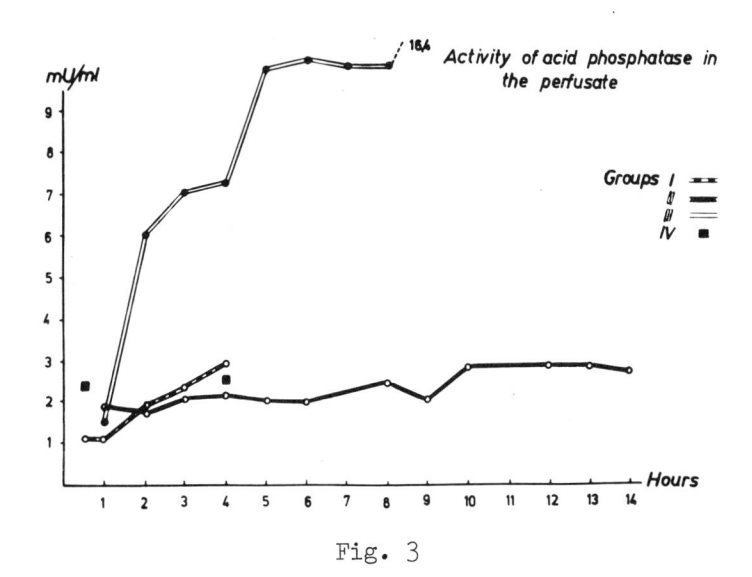

Fig. 3

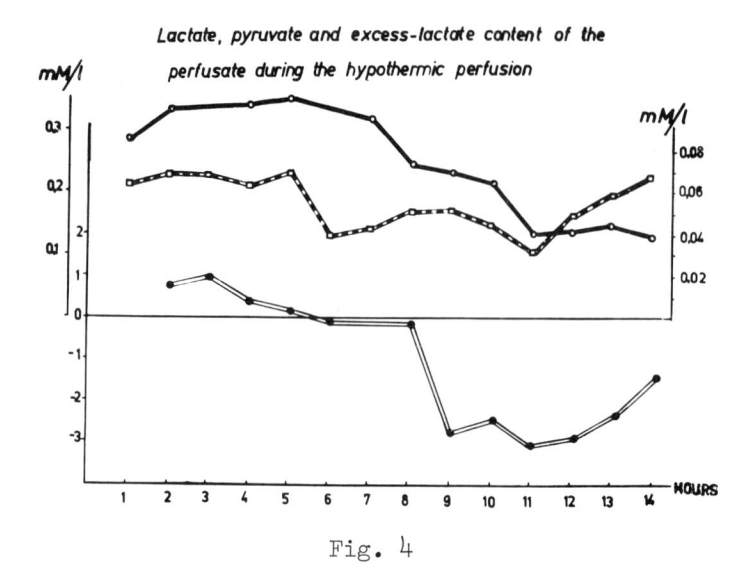

Lactate, pyruvate and excess-lactate content of the perfusate during the hypothermic perfusion

Fig. 4

showed a tendency to fall with the exception of LDH which remained persistently high. A similar elevation of LDH after reinfusion was observed by Vesell et al. (1959).

The next three figures demonstrate enzyme and metabolite changes in the perfusate during perfusion of isolated dog kidneys.

Figure 2 shows that with perfusion at room temperature LDH activity increased markedly, while during hypothermic perfusion following a rise, it began to decrease by the end of the perfusion.

The extensive hypoxic damage of room temperature perfusion resulted in a rapid elevation and very high levels of acid phosphatase activity. Under hypothermia damage to the kidney could be prevented and there is no significant change from the initial value (Fig. 3).

In Fig. 4 lactic and pyruvic acid and excess lactate levels in the hypothermic experiments are shown. It can be seen that at 2–4°C the increase of anaerobic metabolites in the perfusate can be prevented. This suggests that the danger of hypoxic damage can be greatly reduced at this temperature.

Discussion

Our experiments show that elevation of lysosomal enzyme activity in the plasma occurs later in the course of shock than lactic acid, lactic dehydrogenase and excess lactate.

This would suggest that the release of lysosomal enzymes is caused by a greater degree of hypoxia and/or after a longer period of time than that of the anaerobic metabolites. The latter possibility is supported by the fact that lysosomal enzymes have to cross first the lysosome membrane and the cell membrane in order to get into the circulation while the metabolites must get only through the cell membrane.

Another factor of importance here is the difference in molecular weights: the enzymes are large molecules while the metabolites are small.

We would comment on the results of our experiments on isolated perfused dog kidneys by saying that the extent of hypoxic damage of isolated organs can be characterized by the increase of enzymes and metabolites in the perfusate (Belzer et al., 1968; Beisang et al., 1969).

The efficacy of hypothermic or other kind of organ preservation can well be judged from the determination of enzymes and metabolites examined by us in the present experiments.

REFERENCES

Barankay, T., (1969) Kisérl.Orvostud. 21:646.

Barankay, T. & Petri, G. (1969) Kisérl. Orvostud. 21:18.

Beisang, A.A., Graham, E.F., Lillehei, R.C., Dietzman, R.H. & Carter, J.E., (1969) Transplantations Proc. 1:862.

Belzer, F.O., Ashby, S.B. & Donnes, G.L. (1968) Surg. Forum 19:205.

Gergely, M., Horpácsy, G., Barankay, T. & Hézsai, K. (1970) Kisérl. Orvostud. 22:488.

Greenberg, L.J. (1966) Ann. Biochem. 14:265.

Huckabee, W.E. (1958) J. clin. Invest. 37:244.

Loomis, M.E. (1961) J. lab. clin. Med. 57:966.

Peretz, D.I., Scott, H.M., Duff, J., Dossetor, J.B., McLean, L.D. & McGregor, M. (1965) Ann. N.Y. Acad.Sci. 119:1133.

Petri, G. & Horpácsy, G. (1971) Orvos és Techn. 9:19.

Rockerbie, R.A. & Rasmussen, K.L. (1967) Clin. chim. Acta 18:183.

Rosenberg, J.C., Lillehei, R.C., Longerbeam, J. & Zimmerman, B. (1961) Ann. Surg. 154:611.

Schmidt, E., Schmidt, F.W., Herfarth, C., Opitz, K. & Vogell, W. (1966) Enzymol. biol. Clin. 7:185.

Tárnoky, K. (1970) Kisérl. Orvostud. 22:197.

Tárnoky, K. (1969) Kisérl. Orvostud. 21:650.

Vahlensiek, W. (1969) Ther.Berichte (Bayer) 19, 650.

Vesell, E.S., Feldmann, M.P. & Frank, E. (1959) Proc.Soc.exp. Biol.Med. 101, 644.

ENERGY AND TISSUE FUEL IN HUMAN INJURY AND SEPSIS

John M. Kinney, Frank E. Gump & Calvin L. Long

Department of Surgery College of Physicians and
Surgeons of Columbia University

The loss of body weight has been a prominent feature of the response to injury or sepsis since the earliest clinical descriptions of these conditions. It was evident that the extent of weight loss tended to parallel the magnitude of the injury or sepsis. Subsequent observations revealed that the degree of weight loss is largest in males, younger adults, well-muscled individuals, and those with good previous nutrition. The weight loss is less in the female, the elderly, the poorly muscled, the nutritionally depleted. However, the underlying alterations in energy requirements and tissue fuel associated with injury and sepsis remain incompletely understood and are the subject of this discussion.

The weight loss of the injured or septic patient must include a consideration of calorie balance and nitrogen balance. Conventional calorie and protein intake is sharply diminished or even abolished during the acute response to injury or sepsis. Therefore the presence of a negative balance of calories and nitrogen was historically considered to be a form of starvation.

Some workers have reported that the weight loss and negative nitrogen balance after uncomplicated elective operation could be prevented by adequate feeding. Others have considered the nitrogen excretion following injury and sepsis to be increased because of obligatory changes in intermediary metabolism which were over and above any effects of starvation. It seems probable that with lesser degrees of injury and sepsis the factor of decreased intake is of primary importance in any negative balance of calories or nitrogen. However, with larger magnitudes of injury or sepsis the caloric expenditure and the nitrogen excretion may each be increased as a result of altered intermediary metabolism (1).

The pioneering observations of Cuthbertson (2) established that the negative nitrogen balance after injury was in part related to an increase in nitrogen excretion. During the subsequent half-century, many laboratories have confirmed the fact that a hallmark of injury and sepsis is not only an increase in total nitrogen excretion, but particularly urea excretion. Since urea is the major excretory product in man of the deamination of amino acids, and the blood level of urea does not significantly rise in these conditions, the increased urea excretion is accepted as increased urea synthesis from the breakdown of tissue protein.

An increase in caloric expenditure following injury or sepsis has been suggested in the literature for many decades. However, some of the estimates have suggested increases of two to three times or more the normal caloric expenditure for a given individual. Measurements in our laboratories (3), utilizing indirect calorimetry, suggest that the increases in resting caloric expenditure have been over-estimated in previous reports. Uncomplicated elective operation produces no significant increase in postoperative resting caloric expenditure. Multiple skeletal injury may result in increases of 10-25% over the first two weeks following injury. Major sepsis may provide sustained increases of up to 40 or 50% above normal. The only conditions we have studied with sustained increases exceeding 50% above normal have been major third degree burns. Thermal injury is capable of producing sustained increases of up to twice normal in certain individuals.

Increased urea synthesis has sometimes been explained on the assumption that a large extra need for calories due to injury or sepsis requires the breakdown of tissue protein to provide additional fuel. This seems unlikely for the following reasons. The predominant fuel for the operation of the Krebs cycle is the two carbon fragment. The major source for two carbon fragments in the starving human body is via the mobilization of fatty acids from adipose tissue stores. There is no evidence that mobilization of fatty acids is inadequate as a result of injury or sepsis. In fact Carlson (4) has suggested that under some circumstances increased catecholamine levels might result in excessive mobilization of free fatty acids, and excessive fat deposition in the liver. Body protein is not a major reserve of two carbon fragments for tissue fuel and therefore excessive protein breakdown must be related to some other cause when it occurs in association with injury or sepsis.

The increase in caloric expenditure following injury has been suggested by Cairnie and co-workers (5) to be associated with an increased nitrogen loss with excess non-usable heat being produced as a result of the specific dynamic action of protein. In this case the nitrogen loss would be from the breakdown of body tissues rather than food protein. The experimental basis for this suggestion arose from the examination of the increased caloric expenditure of rats sustaining a long bone fracture who were studied in a gradient layer calorimeter. The timing and magnitude of the increase in caloric expenditure closely followed the increase in protein metabolism as judged by nitrogen loss. Our data in man suggest that while this may be a factor in explaining the increased energy expenditure after skeletal injury, it is not enough to explain the increase seen in sepsis and burns. Both the acute and delayed responses to experimental injury are affected by changes in environmental temperature (6). Cuthbertson and co-workers (7) have suggested that raising the environmental temperature by 8-10°C, may reduce the increased nitrogen loss following skeletal injury.

While there is evidence that increased caloric expenditure and increased urea synthesis may be associated in time, probably neither is the sole cause of the other. An increase in caloric expenditure has at least three other possible explanations besides that of an "internal specific dynamic action" which was proposed by Cairnie and co-workers. The possibility of increased body temperature increasing caloric expenditure by a "Q_{10}" effect may be important to consider in febrile patients. An evaluation of this factor several years ago indicated that for surgical conditions with fever and increased caloric expenditure up to perhaps 25% above normal, the relationship was similar to that found by DuBois in various medical conditions. The so-called 7% rule, or 7% increase in metabolism for each increase of 1°F in body temperature, was only a rough average. Very similar findings were found with a series of postoperative patients (8). However for patients in the acute phase of major sepsis, usually peritonitis, and major burns, the caloric expenditure was markably in excess of that predicted from body temperature (9). Therefore the concept that increased metabolism can be explained by temperature effects on the reaction rate of enzymes throughout the body, appears to be an inadequate explanation for the more severe forms of surgical infection.

The possibility that injury or sepsis may produce uncoupling of oxidation and phosphorylation, has been suggested as an attractive hypothesis but remains to be documented. This concept is based on the body needing a certain amount of available energy at any given time in the form of high energy phosphate bonds.

If these bonds are being produced at a lower efficiency than normal the amount of fuel being oxidized and the amount of oxygen consumed and CO_2 being produced would be increased correspondingly above the actual production of high energy bonds, or usable energy being realized. If injury and sepsis were to produce such an un-coupling effect, it could theoretically explain increases in caloric expenditure as judged by indirect calorimetry (from gas exchange) or by direct calorimetry (actual heat loss) without the requirement for utilization of phosphate bond energy being increased.

The increase in urea synthesis and excretion following injury and sepsis may be the result of protein breakdown to produce glucogenic precursors which the breakdown of fatty acids cannot provide. Current biochemical evidence indicates that fatty acids can readily yield two carbon fragments but cannot produce a net gain of glucose, glycogen or any of the intermediates of the Embden-Meyerhof pathways. Therefore, if the body has an actual or potential need for a continuous supply of glucose and/or carbohydrate intermediates for the synthesis of glucose and perhaps other critical materials, the only major source is the breakdown of protein and the deamination of the resulting amino acids. Therefore the subject of gluconeogenesis is assuming central importance in understanding the increased nitrogen excretion after injury. It is possible that some step in the control of gluconeogenesis is modified as part of the metabolic response to injury and sepsis. Evidence for this has been obtained in our laboratories with the use of C^{14}-labelled glucose which has confirmed that small amounts of exogenous carbohydrate (approximately 6 g per hr) is sufficient to nearly shut off hepatic production of new glucose (10). However the infusion of this same amount of carbohydrate is not able to shut off the hepatic production of glucose during the acute phase following serious infection. Therefore the factors controlling gluconeogenesis have clearly been altered. It is also possible that there is an increased need for glucogenic precursors as part of the response to the injury and sepsis. This has lead Cahill (11) to suggest that in conditions where there is an increased amount of primitive tissue such as granulation tissue which relies on anaerobic metabolism for its energy needs, the recycling of glucose to lactate and back may be increased. The use of a mathematical model by Long, Spencer and others (12) with C^{14}-glucose given to critically ill surgical patients, suggests that sepsis is accompanied by an increased turnover rate of glucose in the bloodstream as well as some increase in glucose oxidation.

The consideration of increased urea synthesis following injury and sepsis as an external reflection of increased gluconeogenesis, turns attention to the liver. Thus previous knowledge of hepatic physiology and biochemistry may be applicable to an understanding of the metabolic response to injury and sepsis.

Cahill (13) has explored the mechanisms of human starvation, with special reference to protein breakdown in skeletal muscle which is presumed to be the major source of amino acids which are deaminated for the synthesis of urea in the liver. His work has suggested that perhaps not only protein breakdown occurs within the muscle cell but perhaps even the process of transamination takes place there. Then specific amino acids, particularly alanine and perhaps glutamine, appear to serve as transport mechanisms between peripheral muscle tissue and the liver. This concept may explain why the past efforts to demonstrate increased circulating levels of total amino acids or alpha amino nitrogen were not successful following clinical injury and sepsis.

Studies by Myers and others (14) have emphasized that the liver behaves in similar fashion to the whole body when considering the relationship of cardiac output to arterial mixed venous oxygen difference. The normal resting cardiac output is associated with an average A.V. oxygen difference of approximately four volumes percent. There is a curvilinear relationship such that when cardiac output increases in anemia the A.V. oxygen difference may decrease to approximately two volumes percent, while decreases in cardiac output cause a rise in A.V. oxygen difference up to 10 volume percent. The liver behaves in similar fashion to the whole body, in manipulating O_2 extraction to compensate for blood flow. Myers separated clinical conditions into those with low, normal and high hepatic metabolism. He further divided these three groups into those where changes in O_2 uptake were the result of an alteration in blood flow vs. changes associated with an alteration in A.V. extraction. Myers sought a means of increasing the metabolism of the liver in a brief reproducible manner and selected the rapid intravenous infusion of 50 g of amino acids over a 30 min interval. For the present discussion it is of particular interest that infusion of amino acids had a similar effect to that observed in hyperthyroidism: an increased A.V. oxygen difference with a normal hepatic blood flow. This was in contrast to the response after giving an intravenous pyrogen where there was a normal A.V. oxygen difference but an increase in hepatic blood flow which paralleled an increased whole body flow. He also observed that the infusion of glucose resulted in a low A.V. oxygen difference but a normal hepatic blood flow. Dis the glucose infusion result in less hepatic oxygen extraction because the glucose infusion was providing a material which was partially oxidized, or was it in addition tending to shut off the processes of gluconeogenesis which stimulates hepatic oxygen uptake as amino nitrogen was delivered to hepatic cells?

If it is true that urea synthesis and gluconeogenesis are centered in the liver, and the liver plays a central role in resting thermogenesis, any rise of resting caloric expenditure following injury and sepsis may likewise be centred in the liver. Gump and co-workers (15) studied patients who were febrile because of intraperitoneal sepsis. They compared whole body blood flow (cardiac output) and oxygen consumption with the corresponding measurements being made across the splanchnic bed utilizing hepatic vein catheterization. These studies revealed what had been demonstrated earlier, that a certain proportion of surgical patients may have moderate fever without any elevation in resting metabolism. However all patients who had fever and increased oxygen consumption, also had an increase in cardiac output. It was of particular interest that the septic surgical patients had increases in both blood flow and oxygen consumption across the splanchnic bed which accounted for 40-50% of the increases observed in the whole body. These findings are consistent with the concept that the processes of hepatic gluconeogenesis and urea synthesis are associated with an increased oxygen consumption and heat production. However they emphasize that the increases in caloric expenditure are only partly related to gluconeogenesis and approximately half of the increase in caloric expenditure in such patients must be sought in other viscera or peripheral tissues. The lack of correlation between urea excretion and caloric expenditure is particularly noticable in the burn patients where the total nitrogen excretion per day averaged an amount which is in the same order of magnitude as that seen following multiple fractures and certain other kinds of major sepsis (16). However the caloric expenditure of the burn patients have increases of twice or more that seen in the non-burn septic patients. Therefore it is of importance to examine not only the mechanisms for the production of new glucose but also the mechanisms for mobilizing and oxidizing fatty acids since they represent the major source of calories for all surgical patients not receiving hypertonic glucose solutions by central venous feeding or elemental diet into the G.I. tract.

REFERENCES

1. Kinney, J.M. In Symposium on Protein Metabolism: Influence of Growth Hormone, Anabolic Steroids and Nutrition in Health and Disease (F. Gross, ed.) p.275. Springer, Berlin, 1962.

2. Cuthbertson, D.P. (1932) Biochem.J. $\underline{25}$, 233.

3. Kinney, J.M., Duke, J.H.Jr., Long, C.L. & Gump, F.E. (1970) J. Clin. Path, $\underline{23}$, (Suppl 4) 65.

4. Carlson, L.A. Mobilization and utilization of lipids after trauma: relation to caloric homeostasis. Ciba Foundation Symposium. In Energy Metabolism in Trauma (R. Porter & J. Knight eds) p.155. J. & A. Churchill, London 1970.

5. Cairnie, A.B., Campbell, R.M., Cuthbertson, D.P. & Pullar, J.D. (1957) Brit.J. exp. Path. $\underline{38}$, 504.

6. Stoner, H.B. (1969) Postgrad Med.J. $\underline{45}$, 555.

7. Campbell, R.M. & Cuthbertson, D.P. (1967) Quart. J. exp. Physiol. $\underline{52}$, 114.

8. Kinney, J.M. & Roe, C.F. (1962) Ann. Surg. $\underline{156}$, 610.

9. Roe, C.F. & Kinney, J.M. (1965) Ann. Surg. $\underline{161}$, 140.

10. Long, C.L., Spencer, J.L., Kinney, J.M. & Geiger, J.W. (1971) J. Appl. Physiol. $\underline{31}$, 102.

11. Cahill, G.F. Jr. Personal communication.

12. Long, C.L., Spencer, J.L., Kinney, J.M. & Geiger, J.W. (1971) J. Appl. Physiol. $\underline{31}$, 110.

13. Cahill, G.F. Jr. (1970) New Eng. J. Med. $\underline{282}$, 668.

14. Myers, J.D. (1954) In Shock and Circulatory Homeostasis (H.D. Green, ed.) p. 121. Madison Printing Co. Inc., Madison, New Jersey.

15. Gump, F.E., Price, J.B.Jr. & Kinney, J.M. (1970) Ann. Surg. $\underline{171}$, 321.

16. Gump, F.E., Price, J.B.Jr. & Kinney, J.M. (1970) Surg. Gynec. Obstet. $\underline{130}$, 23.

ENVIRONMENTAL TEMPERATURE AND METABOLIC RESPONSE TO INJURY PROTEIN, MINERAL AND ENERGY METABOLISM

D.P. Cuthbertson, G.S. Fell, A.G. Rahimi and W.J. Tilstone

University Department of Pathological Biochemistry

Royal Infirmary, Glasgow C.4.

The post-shock generalized metabolic response to moderate to severe physical injury first described by Cuthbertson (1930, 1932, 1936, 1942) is a complex series of reactions with increased net protein catabolism and parallel increases in oxygen consumption as the main events. The changes in protein catabolism are exhibited as increased urinary losses of nitrogen (mainly as urea), sulphur (mainly as sulphate), phosphorus as phosphate, potassium, zinc and creatine but not creatinine (Cuthbertson, McGirr and Robertson, 1939; Cuthbertson, Fell, Smith and Tilstone 1972). This was termed the 'flow' **phase of metabolism** following the 'ebb' phase of diminished metabolism and heat production of the shock period (Cuthbertson, 1942). Muscle was implicated as the major source of nitrogen loss (Cuthbertson, 1942; Munro, 1964). The fractional catabolic rate of plasma albumin is increased and there is also a characteristic change in levels of plasma proteins with a fall in albumin and rises in the acute phase reactants (see Owen, 1967; Cuthbertson and Tilstone, 1969). The slight degree of fever which is generally present for a few days and which is apparently non microbial in origin was termed 'traumatic fever' and the whole reaction was considered as the generalized component of the inflammatory response.

The extent of this metabolic response depends on the nature and severity of the injury, being slight for uncomplicated surgery, greater for major skeletal trauma and considerable in sepsis and major burns (Kinney, Long and Duke, 1970). Of these injuries those associated with sepsis exhibit a greater increase in body temperature and burns present a unique situation because of increased extrarenal losses.

The usefulness of the rat as an experimental animal in this work was early recognised (Cuthbertson et al., 1939) and the permissive role of corticosteroid in the 'flow' phase was established (Ingle, Ward & Kuinzenga, 1947; Campbell, Sharp, Boyne and Cuthbertson, 1954). Munro and Cuthbertson (1943) and Munro and Chalmers (1945) reported that rats fed on a diet deficient in protein do not respond to trauma with an increased excretion in urinary nitrogen and that the response is directly related to the percentage of protein in the diet.

But the pathways responsible for the protein catabolic response are not at all clear. All endocrine systems so far studied are stimulated at some stage in the body's response to trauma (see review by Cuthbertson and Tilstone, 1969) but apart from the adrenal cortex playing a major permissive role in the protein metabolic response no causal relationship has been unequivocally defined.

Observations on energy metabolism in the 'ebb' phase in man are lacking, but Stoner (1970) has shown that in the rat following standardised injuries of either burning or hind limb ischaemia there is, first, a depressed oxygen consumption due to lowering of the critical temperature. The slope of the regression line of oxygen consumption on environmental temperature was the same in the injured animals as it was in the controls but displaced to the left.

In the post-shock period the increased oxygen consumption or heat production following fracture which was earlier noted in man (Cuthbertson, 1932) was also found to occur in the rat and in both it is possible to account for bulk of the increment from oxidation of endogenous protein equivalent to the extra urinary nitrogen (Cairnie, Campbell, Cuthbertson and Pullar (1957). These observations were made at normal temperatures 20-22°C. and indicate a rise in what Kinney (1967) calls 'resting metabolic energy expenditure' (RME) of about 9 per cent and of about 10 per cent in BMR. Further evidence for increased energy production, as observed by gradient layer calorimeter, due to additional protein catabolism was obtained by Cairnie et al.(1957) when they found no heat increment after fracture in the rat when fed a protein-free diet for a week beforehand and for ten days after fracture of the femur. On the other hand, Caldwell (1970) has studied heat production after bilateral fracture of the femora in the rat but can only account for 35 per cent of the heat increment of 10 to 20 per cent after fracture, as arising from increased protein oxidation. A further 40 to 60 per cent could be attributed to an increase in average body temperature assuming a Q10 of 2.3. Kinney et al. consider that protein provides only 20 per cent of total calories after major trauma in man and only 14 per cent in burns, with the major increase in heat production coming from the oxidation of fat.

Tracer studies of glucose metabolism show severely injured patients to have increased glucose oxidation and increased turnover of blood glucose, despite a slight hyperglycaemia. Infusion of glucose (6g/h) was shown by Long, Spencer, Kinney and Geiger (1971a, b) to shut off hepatic gluconeogenesis in normal subjects but did not shut off hepatic gluconeogenesis in the acute phase following serious infection. If there is an increased demand for gluconeogenesis, then oxidation of amino acids, but not of fatty acids, will produce the necessary fuel. Obviously further work needs to be done to clarify the situation. Erici (1954,1957) was probably the first to report an enhanced rate of wound healing in rabbits at a higher environmental temperature and Cuthbertson and Tilstone (1967) confirmed this in the rat with superficial wounds.

As the increase in energy expenditure in burns is generally greater than in other forms of trauma, probably because of the heat load of the evaporation of water from the burn it is of interest that Caldwell (1962) found that rats with burns did not show the characteristic hypermetabolism following such injury if they were housed at $30^\circ C$. These results have been confirmed on human patients with burns by Barr, Burke, Liljedahl and Plantin (1968) and Davies and Liljedahl (1970). They found that exposure to dry air at $32^\circ C$ improved the overall condition of their patients. Mortality and morbidity were improved in both rats and man. Increased protein catabolism was no longer so marked and energy metabolism was reduced compared with that at $20^\circ C$.

However, in neither the rats nor the patients was evaporative water loss changed. This is not surprising since examination of the relative humidity and temperature showed the water vapour pressure to be the same at both environmental temperatures. The absence of an obvious explanation for this effect of the higher environmental temperature had earlier been demonstrated by Campbell and Cuthbertson (1967) who found that at $30^\circ C$ the increased heat production and increased protein catabolism which normally follow fracture of a femur in the rat at $20^\circ C$ had been largely suppressed. This was followed by supporting observations

from Miksche and Caldwell (1968) in the rat and in man by
Cuthbertson, Smith and Tilstone (1968; 1971).

Present Findings

We now report on a more extensive series of observations on
male patients with injuries of the main long bones treated at
30°C for 10 to 12 days following injury and these are compared
with a group at 20°C. 29 patients were studied in the 30° group
and 28 in the 20° group. Results of all variables measured were
not available for every patient. Intake of energy and nitrogen
was calculated from food tables (McCance and Widdowson, 1960)
with occasional checks of nitrogen by Kjeldahl analysis. The
patients were offered diets according to preferences established
for each individual from dietary history. Urine nitrogen was
estimated by a partly mechanised Kjeldahl technique (Fleck, 1970),
creatine by a fluorescent method (Tilstone & Fell, 1971),
potassium and sodium emission by flame photometry; calcium,
magnesium and zinc by atomic absorption spectrophotometry.
Urinary chloride excretion was determined by coulometric titration.

The extent of net protein metabolism was estimated by
calculating net nitrogen excretion, that is, urine nitrogen minus
food nitrogen. The average net nitrogen excretion for the first
four days and subsequent 3 day periods were calculated. The urine
nitrogen values used were those for the day following food nitrogen
values. Statistical significance of differences in means between
the 30° group and the 20° group was calculated by Student's t-test
making no assumption about variance. Multiple linear regression
of urine nitrogen on food nitrogen and food energy was calculated
by the method of least squares and its significance by analysis of
variance. Single correlations were established for the urinary
data and significance checked by t-test. All statistical
calculations were performed on an IBM 1130 computer using standard
statistical packages.

Comparison of net nitrogen excretion for all periods showed
this to be significantly less at 30° than at 20° (P < .01).
Nitrogen intake and energy intake were not statistically different
when all periods were considered together, or in any of the
individual periods. Potassium, zinc and creatine levels in
urine likewise were not significantly different at 30° from those
at 20° when all periods were considered together. However,
creatine and net nitrogen excretion were significantly less at
30° in the first period (P < .05 in each case).

Urine nitrogen always correlated significantly with urine
zinc (P < than .005) with urine creatine (P < .001) and with urine

potassium (P < .001). The correlations are highly statistically significant and the values of t at 20° were greater than those at 30°. Correlation of creatine and zinc (P < .001) and of potassium and creatine (P < .01) were significant at 20° but not at 30°.

These results show that net nitrogen excretion consequent on trauma is reduced at an ambient temperature of 30°C, and that the increased urine nitrogen is probably of muscular origin (vide creatine and zinc excretion). This catabolism of muscle is not so evident at 30°.

Multiple linear correlation of urine nitrogen on food nitrogen and food energy shows that the expected linear relationship, with positive regression co-efficient for food nitrogen and negative regression co-efficient for food energy, is not found in any of the four periods in patients housed at 20°, and indeed the sign of the regression co-efficients is not as expected (Table I). At 30°, however, a significant regression is obtained in the first period and almost so in the fourth period. This is further evidence of the quantitative importance of loss of endogenous protein in the metabolic response to trauma at 20° ambient, and its reduction on exposure to 30° ambient. It also suggests that feeding high energy diets will not reduce the catabolism of endogenous protein after trauma, although such treatment will of course improve the deposition of dietary nitrogen.

The patients were of equivalent age in both groups (P > .05), and the degree of traumatic fever, according to number of days oral temperature was 37.2 to 37.7° or 37.8 to 38.2°, or 38.3 to 38.8°, was also equivalent in both groups (P > .05). Comparing the severity of trauma in two groups is more difficult. Assessment based on number of pints of blood transfused gave no significant difference in severity of those at 20° to those at 30° (P > .05), and this together with the traumatic fever result indicates equivalently injured groups at both temperatures.

TABLE I

Linear regression of urine nitrogen (UN) on food nitrogen (FN) and food energy (MJ, energy intake in megajoules). Statistical significance of regression as calculated by analysis of variance is also shown. NS = not significant.

Temperature °C	Period	P	Equation
20	1	NS	UN = -0.67FN + 0.87MJ + 16.3
	2	NS	UN = -0.30FN + 0.40MJ + 4.4
	3	NS	UN = -0.37FN + 0.95MJ + 12.5
	4	NS	UN = -0.40FN + 0.90MJ + 11.5

Temperature °C	Period	P	Equation
30°	1	<.05	UN = 0.45FN −1.24MJ + 7.8
	2	NS	UN = 0.88FN −0.59MJ + 10.7
	3	NS	UN = 0.48FN −0.40MJ + 13.8
	4	<0.1	UN = 0.80FN −0.81MJ + 13.5

Excretion of electrolytes was further examined in a separate series of 12 cases at 20° and 12 cases at 30°.

Sodium and chloride. As was to be expected there was very close parallellism between these two elements at each of the two temperatures. There was first a progressive fall lasting over 4 days followed by a slow rise to the early values. There was no significant difference between excretion of chloride at the two temperatures but in the case of sodium there was a lower excretion at 30° ($P < .05$).

Potassium. There was also a tendency for a fall over four days followed by a rise to above earlier values. This was most marked during the second period of six days and although when viewed by 3-day periods there was no significant difference between the excretion at the two temperatures, calculations on the whole of the data for the two temperatures revealed a significant overall difference ($P < .05$).

Calcium and Magnesium. The courses of excretion of calcium and magnesium were not very different at the two temperatures, and such fluctuations as occurred in these two elements were in parallel. There was a slight rise in each with time.

Summary

In male patients with moderate to severe fractures of the lower limbs net excretion of nitrogen, that is urine nitrogen minus food nitrogen, was significantly reduced in an environmental temperature of 30°C (29 patients) compared to a similarly injured group at normal ward temperatures of 20-22°C (28 patients). The urinary excretion of other products of tissue catabolism, namely creatine, potassium and zinc correlated well with urine nitrogen excretion, and excretion of creatine was significantly less in the first four days post injury for patients at 30°C.

Regression analysis of urine nitrogen on food nitrogen and food energy, together with the correlation of the other products of tissue breakdown excreted in urine with urine nitrogen confirmed that much of the post injury nitrogen excretion is due to catabolism of muscle protein. It is this which is reduced by exposure to a warm environment.

Excretion of electrolytes (Na, Cl, Ca, Mg) was not so affected by the environmental temperature other than for a slight reduction in sodium excretion.

Acknowledgements

We thank Professor W.A. Mackay and Mr. John White and their medical and nursing colleagues for clinical support. DPC thanks the Medical Research Council and AGR the World Health Organisation for financial support. Miss C.M. Smith gave excellent help with diets, and Miss C. Hall provided skilled technical assistance.

References

Barr, P.O., Burke, G. Liljedahl, S.O. and Plantin, L.O. (1968) Lancet 1, 164.
Cairnie, A.B., Campbell, R.M., Cuthbertson, D.P. and Pullar, J.D. (1967) Br. J. exp. Pathol. 38, 504
Caldwell, F.T. (1962) Ann. Surg. 155, 119
Caldwell, F.T. (1970). In Energy Metabolism in Trauma, Ciba Foundation Symposium, p.23 ed. Porter, R. and Knight, J.C. London, Churchill.
Campbell, R.M. and Cuthbertson, D.P. (1967). Q.Jl exp.Physiol.52, 114.
Cuthbertson, D.P. (1930). Biochem. J. 24, 1244.
Cuthbertson, D.P. (1932). Q.Jl Med. 25, 233.
Cuthbertson, D.P. (1936). Br. J. Surg., 23, 505.
Cuthbertson, D.P. (1942). Lancet, 1, 433.
Cuthbertson, D.P., Fell, G.S., Smith, C.M. and Tilstone, W.J. (1972) Nutrition and Metabolism, in press.
Cuthbertson, D.P., McGirr, J.L. and Robertson, J.S.M. (1939).Q.Jl exp. Physiol., 29, 18.
Cuthbertson, D.P., Smith, C.M. and Tilstone, W.J. (1968). Br. J. Surg. 55, 513.
Cuthbertson, D.P. and Tilstone, W.J. (1969). Adv. clin.Chem., 12,1.
Cuthbertson, D.P. Tilstone, W.J. and Green, J.A. (1969).Lancet,1, 987.
Davies, J.W.L. and Liljedahl, S.O. (1970). In Energy Metabolism in Trauma, Ciba Foundation Symposium, p.59.ed Porter, R. and Knight, J. London, Churchill.
Erici, I. (1954). Svensk, kir. Füren. Förh. 29th May. p.34. Also Abstr. in Nord.Med. (1955), 53, 128.
Erici, I. (1957). Acta Chir. Scand. 112, 345
Fleck, A. (1970). Proc.Nutr.Soc. 29, 81.
Henzel, J.H., de Weese, M.S. and Lichte, E.L. (1970). Arch.Surg. 100, 349.
Ingle, D.J., Ward, E.O. and Kuizenga, M.H. (1947).Am.J.Physiol. 149, 510.
Kinney, J.M., (1967). Br. J. Surg. Lister Cent. No.435.

Kinney, I.M., Long, C.L. and Duke J.H. (1970). In Energy Metabolism in Trauma. Ciba Foundation Symposium. p.103, ed. Porter, R. and Knight, J., London, Churchill.

Long, C.L., Spencer, J.L., Kinney, J.M. and Geiger, J.W. (1971a). J. Appl. Physiol, 31, 102.

Long, C.L., Spencer, J.L., Kinney, J.M. and Geiger, J.W. (1971b). J. Appl. Physiol. 31, 110.

McCance, R.A. and Widdowson, E.M. (1960). The Composition of Foods. M.R.C. Special Report Series, No.297. London, H.M.S.O.

Miksche, L.W. and Caldwell, F.T. (1967). Ann Surg. 54, 455.

Munro, H.N. (1964). In Mammalian Protein Metabolism, Vol.2. p.381; ed. Munro, H.N. New York and London, Academic Press.

Munro, H.N. and Chalmers, M.I. (1945). Br.J.exp.Pathol. 26, 396

Munro, H.N. and Cuthbertson, D.P. (1943). Biochem. J. 37, xii

Owen, J.A. (1967). Advan.Clin.Chem. 9, 1.

Stoner, H.B. (1970). In Energy Metabolism in Trauma. Ciba Foundation Symposium, p.1. ed. Porter, R. and Knight, J. London, Churchill.

Tilstone, W.J. and Fell, G.S. (1971). In: Automation in Analytical Chemistry, p.75. London Tech. Instrument Co. Ltd.

Tilstone, W.J. and Roach, R.J. (1969). Q.Jl. exp. Physiol. 54, 341.

ENVIRONMENTAL TEMPERATURE AND THE METABOLIC RESPONSE TO INJURY — PLASMA PROTEINS

A. Fleck, F.C. Ballantyne, W.J. Tilstone and D.P. Cuthbertson

Biochemistry Department, Glasgow Royal Infirmary
Glasgow, G4 OSF

Few studies of the now well-known changes in plasma proteins after injury take into account the environmental temperature of the subject; the work of Davies, Liljedahl & Birke (1969) on patients who had sustained burns is one exception. Changes in the plasma protein fractions after injury have been reviewed and summarised by Owen (1967):-- a_1-, and a_2-globulin increase and there is a transient decrease in plasma albumin concentration. With the introduction of quantitative immunodiffusion techniques it was found that numerous proteins change in a variety of ways after trauma. The group known as 'acute phase reactants' which include fibrinogen, C-reactive protein, a_1 acid glycoprotein, haptoglobin, and ceruloplasmin, all show an increased plasma concentration after injury (Werner & Cohnen, 1969). Each protein has a characteristic response which is not related to the nature of the trauma (Koj, 1970). In contrast, the plasma concentration of albumin, thyroxine-binding pre-albumin and transferrin decreases after injury (Gordon, 1970).

There is now an accumulation of evidence that environmental temperature influences protein metabolism after injury. (See Tilstone & Cuthbertson, 1970). The patients investigated by us had suffered fractures of the long bones, the basic findings in whom have been discussed in the previous paper (Cuthbertson et al. 1972).

Table 1. Serum Proteins after Injury (Man). Influence of environmental temperature on the concentration of serum proteins in man after injury (see text).

	Albumin		C-reactive protein		a_1- acid glycoprotein	
	20°	30°	20°	30°	20°	30°
No. of patients	12	11	6	6	9	9
Day 2	33.8 + 1.1	34.2 + 1.1	0.109 + 0.008	0.072 + 0.008	1.37 + 0.03	1.37 + 0.04
3	30.8 + 1.6	34.0 + 1.7	0.099 + 0.009	0.061 + 0.009	1.45 + 0.07	1.43 + 0.08
4	28.2 + 1.0	34.5 + 0.9	0.088 + 0.009	0.056 + 0.007	1.58 + 0.08	1.41 + 0.08
5	27.3 + 1.2	34.5 + 1.3	0.080 + 0.008	0.049 + 0.010	1.75 + 0.09	1.53 + 0.10
7	28.9 + 1.0	33.8 + 1.0	0.068 + 0.009	0.048 + 0.008	1.88 + 0.07	1.48 + 0.09
10	30.4 + 1.3	34.8 + 1.4	0.053 + 0.009	0.034 + 0.009	1.86 + 0.06	1.52 + 0.05
14	30.6 + 1.0	35.2 + 1.5	0.034 + 0.008	-	1.61 + 0.08	1.46 + 0.06
17	31.1 + 1.6	37 + 1.5	-	-	1.48 + 0.07	1.34 + 0.07

All values expressed as g/l; figures are mean + S.E.M. from the numbers of patients indicated.

Footnote to Table 1.

Albumin - At 20°C a significant fall in concentration occurred from the value on day 2 (day 2 v day 5 P < 0.001) but no significant change occurred at 30°C. The concentrations at the 2 temperatures were significantly different between days 3 and 17 after injury, e.g. day 5, P < 0.001, day 17, 0.025 > P > 0.010.

C-reactive protein - The fall in concentration was significant at both temperatures and the concentration was significantly greater at 20°C than at 30°C between days 2 and 7 after injury, e.g. day 3, 0.01 > P > 0.005, day 7, 0.05 > P > 0.025.

$α_1$- acid glycoprotein - A significant increase in concentration occurred at 20°C (P < 0.001) but not at 30°C. The concentration was significantly greater (0.005 > P > 0.001) at 20°C than at 30°C between days 7 and 10 after injury. (Comparisons by t-test).

In a group which could be regarded as 'control subjects' maintained at the usual ward temperature in Glasgow of approximately 20°C, the change in plasma albumin after injury was not related to haematocrit; the fall in albumin can be contrasted with the rise in C-reactive protein and in $α_1$-acid glycoprotein (Table 1). In contrast the changes found in patients maintained in the 30°C environment are much less than in those maintained in the open ward (Table 1). Both of the 'acute phase reactants' were determined by an immunodiffusion technique using 'Partigen'* plates. Albumin was determined after electrophoresis on cellulose acetate and elution (Webster, 1965) the technique being standardised against immunodiffusion (Becker, Rapp, Schwick & Störiko, 1968).

Table 2. Serum Albumin in Rats after Injury. Influence of environmental temperature on the serum albumin concentration of the rat after fracture of 1 femur.

Day	20°	30°
0	28 ± 0.4	28 ± 0.6
1	25.2 ± 0.9	25.7 ± 0.5
3	20.7 ± 0.9	*24.8 ± 0.8
5	24.9 ± 0.5	25.6 ± 0.6
7	25.7 ± 0.7	26.8 ± 0.6

Results are expressed as g/l and represent the mean and S.E.M. of concentration in 8 rats; 40 rats were studied at the environmental temperature of 20°C and 40 at 30°C. At both temperatures the fall in concentration was significant and the concentration was significantly lower in the rats at 20°C than in those at 30°C on day 3 (t- test). *P < 0.001.

*Hoechst Pharmaceuticals

Experimental studies on rats after injury confirmed the difference in response at 20° and 30° (table 2).

Studies of Albumin catabolism after injury were carried out using ^{125}I labelled human plasma albumin (commercially available from the Radiochemical Centre, Amersham, England). Preliminary studies with this prepartion were carried out on normal individuals, and on fasting obese patients. The results indicated that the preparation was satisfactory for albumin catabolism studies by the criteria of McFarlane (1965).

Results (table 3) were determined using the U/P radioactivity method (Campbell, Cuthbertson, Mathews & McFarlane, 1956). This was selected because injured patients are in a non-steady state when it is inadvisable to use Matthews' (1957) curve stripping technique, the method of Nosslin (see Andersen, 1964) or the method of Orr & Gillespie (1968) which is similar to Nosslin's.

There are clear differences in the fractional catabolic rates of albumin in patients maintained at 20° and 30° after injury (table 3). These results are essentially similar to those described by Davies et al (1970) in patients suffering from extensive burns.

Table 3. Fractional Catabolic Rate of Albumin in Man (d^{-1}) after injury.

	Normal	at 20° (11 patients)	at 30° (7 patients)
Mean	0.095	0.175	0.123
S.E.M.	0.003	0.013	0.010

• Values given are for day 5 after injury.

At 20° the value is significantly greater than that from the patients at 30° (t- test $0.010 > P > 0.005$)

It is important to note however that despite the difference in the fractional catabolic rates there was no demonstrable change in the absolute catabolic rate of albumin. Thus the alterations in the fractional catabolic rate can be assumed to be due mainly to the fall in plasma albumin concentration clearly shown at the fifth day after injury in the 20° group of patients. Thus plasma albumin apparently does not contribute to the loss of body protein after injury (see Fleck, 1971).

As has been reported elsewhere our preliminary investigations indicate that there is a reduction in albumin synthesis 18 hours after injury to experimental rabbits which is consonant with the increased breakdown pattern of liver polysomes in rats 18 hours after injury (Fleck, 1971). It would seem, in agreement with Mouridsen (1969) that the evidence indicates that the fall in plasma albumin concentration after injury is a result of a transient inhibition of albumin synthesis. At this stage no hypothesis can be proposed about the response of other proteins except that their response and the effects of environmental temperature indicate that losses due to haemorrhage or into the site of injury is not a major factor influencing the responses of the plasma proteins to trauma.

Acknowledgements

We wish to thank Professor W.A. Mackey and Mr. John White and other clinical and nursing colleagues for clinical support. Two of the authors (AF and DPC) wish to acknowledge financial support by the MRC which made this work possible.

References

Andersen, S.B. (1964). Metabolism of Human Gamma Globulin', appendices p. 103. Oxford: Blackwell.
Becker, W., Rapp, W., Schwick, H.G. & Storiko, K. (1968). Z. klin . Chem. klin. Biochem. 6, 113.
Caldwell, F.T. (1962). Ann. Surg. 155, 119.
Campbell, R.M. Cuthbertson, D.P., Matthews, C.M.E. & McFarlane, A.S. (1956). Internat. J. appl. Rad. Isot. 1, 66.
Cuthbertson, D.P., Fell, G.S. Rahimi. A.G. and Tilstone, W.J. (1972). Contribution to Symposium of Neurohumoral and Metabolic Aspects of Injury, August 3-7, 1971. Budapest. In press.
Davies, J.W.L. Liljedahl, S.O. & Birke, G. (1969). Injury 1, 43.
Davies, J.W.L. (1970). J. clin. Pathol. 23, suppl. (Roy. Coll. Pathol.) 4, 56.
Fleck, A. (1971). Proc. Nutr. Soc. In press.
McFarlane, A.S. (1965). Tech. Rep. Ser., Internat. Atom. Energy AG. (Vienna) 45, 3.
Matthews, C.M.E. (1957). Phys. Med. Biol. 2. 36.
Mouridsen, H.T. (1969). Clin. Sci. 37, 437.
Orr, J.S. & Gillespie, F.C. (1968). Science (N.Y.) 162, 138.
Owen. J.A. (1967). In: Advances in Clinical Chemistry, Vol. 9, p.1 ed Sobotka, H. and Stewart, C.P. New York, Academic Press.
Tilstone, W.J. & Cuthbertson, D.P. (1970). In: Energy Metabolism in Trauma, Ciba Foundation Symposium, p.43. ed Porter, R. and Knight, J. London, Churchill.
Webster, D. (1965). Clin. chim. acta. 11, 101.
Werner, M. & Cohnen G. (1969). Clin. Sci. 36, 173-184.

METABOLIC, HORMONAL AND ENZYMATIC FUNCTIONS OF RATS DURING RECOVERY FROM INJURY

S. Németh, M. Vigaš and A. Straková

Institute of Experimental Endocrinology
Slovak Academy of Sciences
Bratislava, Czechoslovakia

In our experiments on adult male rats Noble-Collip drum trauma (16) in doses not exceeding 400 revolutions in 6 min 40 sec was used. The general response of the animals may be divided into two parts. The first is observed during drumming and therefore lasts only a few minutes, whereas the second begins with the termination of the injury and lasts several hours or days (1,6). An important aspect of this first response, an increased rate of disappearance of blood glucose, becomes immediately apparent when rats starved for 18 or 24 hr are injured. In such animals during injury there are increases in blood pyruvate and lactate, and nearly total depletion of glycogen stores. These events should lead to hyperglycaemia; but exactly the reverse, hypoglycaemia, regularly occurs (12,18). Intravenous glucose tolerance tests devised for use during traumatization fully confirmed the increase in the disappearance rate of blood glucose (13). This increase does not only take place in starved animals, but is also observed in fed rats, in which a substantial increase in the arterial venous glucose difference takes place (Table 1).

Table 1. Blood sugar levels of rats in mg/100 ml (means of 8 ± S.E.M.) Immediately before injury 0.1 ml of 0.33% pentobarbital (SPOFA, Prague) was given intraperitoneally. Blood samples were taken immediately after injury.

	Fed Control	Fed Injured	P	Fasted Control	Fasted Injured	P
Aorta	126 ± 2	173 ± 8	< 0.001	113 ± 6	89 ± 10	N.S.
Posterior Vena cava	148 ± 6	117 ± 3	< 0.001	103 ± 4	82 ± 7	< 0.05

Unfortunately a study of the distribution and metabolic products of injected radioglucose seems impracticable because of the rapidly changing pools during injury. Even without one, however, experiments in our laboratory indicated that the glucose that disappeared was metabolized (14), so that net gluconeogenesis during recovery becomes quite understandable.

Ninety minutes after injury in fasted rat blood glucose, pyruvate and lactate levels are returning to their initial values, and in the liver a repletion of its glycogen stores takes place (5,19), very probably under the influence of glucocorticoids from the adrenals. In fact, plasma corticosterone levels reach very high values during this time (11) and after adrenalectomy even in fed animals liver glycogen continues decreasing and is very low 90 min after injury (20). It would be very tempting to account for the newly formed glycogen by the decrease in pyruvate and lactate levels. However, both are also readily metabolized in adrenalectomized rats without having any effect on liver glycogen formation (20). Furthermore, during this period, a great part of the lactate disappears even in rats with deranged liver function after an oral dose of 5 ml/kg of carbon tetrachloride in liquid paraffin given 4 hr before injury (10). We concluded from these results that during recovery the clearance from blood of most of its lactate may be performed by extrahepatic, very probably muscle, tissues, that removal by the liver is followed by oxidation, and that liver glycogen is more probably formed from amino acids (20).

There are no published data from which to confirm or reject these suppositions, so we carried out experiments which we would like to present here. The results are, of course, only preliminary ones.

Ninety min after injury of 18 hr fasted rats, when we supposed that liver glycogenosynthesis would be maximal, radioactive substrate was given intravenously. 1 µCi of each substrate was given in 0.5 ml 0.9% saline/200 g body wt, as follows: U-^{14}C-amino acids (mixed) in 10% bovine serum albumin; U-^{14}C-tyrosine; Na 2-^{14}C-pyruvate with about 7.5 µg of carrier; Na 2-^{14}C-lactate with about 100 µg of carrier; 6-^{14}C-glucose with about 220 µg of carrier; or U-^{14}C-palmitate in 5% bovine serum albumin. One hour later the animals were killed by decapitation and approximately equal parts of their livers (314-474 mg) were quickly removed for glycogen determination (9), half of which was used for the measurement of radioactivity by a Packard type scintillation counter, with Bray's solution as scintillation fluid. It must be emphasized that at the time of the injection the blood levels of the substrates were already in the normal range.

Under these circumstances the relative quantities of label transferred to glycogen in controls and injured measures the relative rates of synthesis of glycogen. Table 2 showed that there were no great differences; injury did not increase glycogen synthesis significantly from most of these substrates. These results are not conclusive, as it has not been established that the intracellular specific radioactivities of the substrate were the same in the controls and the injured.

Table 2. Total radioactivity in liver glycogen in rats, means ± per 100 mg liver.

	Controls n	Imp/10 min	Injured n	Imp/10 min	P
Amino acid m.	8	1695 ± 522	8	1773 ± 466	N.S.
Tyrosine	19	367 ± 26	20	462 ± 82	N.S.
Na pyruvate	14	180 ± 17	16	265 ± 23	< 0.01
Na lactate	16	1520 ± 228	15	1543 ± 339	N.S.
Glucose	15	1668 ± 255	15	1536 ± 232	N.S.
Palmitic acid	8	505 ± 41	8	783 ± 147	N.S.

Regarding the peripheral actions of glucocorticoid hormones, during recent years their stimulatory effect upon one of the amino acid transaminating enzymes, i.e. tyrosine-α-ketoglutarate transaminase (E.C. 2.6.1.5 TAT) was observed (8). This enzyme catalyzes the transformation of tyrosine to p-hydroxyphenylpyruvic acid. Its increased activity was found in several stress conditions: immobilization (4), septicaemia (17) and laparotomy (3). In our own laboratory increased TAT activity (for the method used see Ref.2) was found 90 min after 400 rotations in the Noble-Collip drum (15). This hyperactivity was of rather long duration in spite of the short-term nature of the injury and no substantial differences between the responses of fed and 18 hr fasted rats were found. In adrenalectomized animals decreased enzyme activity was observed before traumatization and the post-traumatic increase was substantially blunted as was that of rats pretreated with actinomycin D (100 μg in 0.2 ml saline/100 g body wt., i.p., 1 hr before injury), an agent inhibiting DNA-dependent messenger RNA synthesis. So the post-traumatic TAT hyperactivity seems to be dependent on the parallel corticosterone hypersecretion (11) leading to increased production of enzyme protein (15).

The crucial point regarding the eventual physiological significance of the post-traumatic TAT hyperactivity concerns the role of tyrosine as material for glycogenosynthesis. Tyrosine is still considered as a non-gluconeogenic amino acid. However, from one half of its molecule fumaric acid is produced, the use of which for gluconeogenic purposes cannot be excluded (7). No significantly increased labelling of glycogen after i.v. administration of ^{14}C-tyrosine was found by us during the recovery period (Table 2). It is possible that with more adequate techniques such a role of tyrosine could be shown. However, we are of the opinion that even then a more reasonable explanation for the post-traumatic TAT hyperactivity will have to be found.

SUMMARY

From previous experiments, gluconeogenesis from amino acids taking place as soon as 90 min after injury was considered as one of the main features of recovery from Noble-Collip drum trauma. The incorporation of label into liver glycogen from ^{14}C-amino acids injected intravenously was, however, not increased. Similarly there was no or little increase in labelling from $2-^{14}C$-lactate or $6-^{14}C$-glucose.

An increased activity of liver tyrosine-α-ketoglutarate transaminase was also found 90 min after injury. In adrenalectomized or actinomycin D treated animals the post-traumatic enzyme hyperactivity was substantially blunted. Its physiological significance for gluconeogenic purposes after injury remains uncertain. No statistically significant increase of activity in liver glycogen was found in experiments, in which instead of the radioactive amino acid mixture, tyrosine-U-^{14}C was injected.

The authors wish to state that a corrected form of this paper is presented here, after considering some critical remarks by Dr. D.F. Heath and Dr. J.M. Kinney concerning our experiments with radioactive substrates. Their friendly help in this regard is appreciated with many thanks.

REFERENCES

1. Chytil, F. & Hruza, Z. (1958) Can.J.Biochem.Physiol.36,457.
2. Diamondstone, T.I. (1966) Anal. Biochem. 16, 395.
3. Geller, E., Yuwiler, A. & Schapiro, S. (1969) Proc.Soc.exp. biol Med. 130, 458.
4. Hanninen, O. & Hartiala, K. (1967) Acta Endocr. (Kbh) 54, 85.
5. Hava, O., Mraz, M. & Triner, L. (1959) Naunyn-Schmiedebergs Arch. 236, 81.
6. Hruza, Z. & Chytil, F. (1959) Physiol. Bohemoslov. 8, 307.
7. Karlson, P. (1964) Biochemie. Stuttgart: Georg Thieme Verlag, p. 140.
8. Knox, W.E. (1963) Trans. N.Y.Acad.Sci. 25, 503.
9. Korec, R. (1967) Experimental Diabetes Mellitus in the Rat. Bratislava: Publishing House of the Slovak Academy of Sciences, p.16.
10. Magdolenova, A., Nemeth, S. & Vigas, M. (1970) Physiol. Bohemoslov. 19, 332.
11. Mikulaj, L. & Kvetnansky, R. (1966) Physiol. Bohemoslov, 15, 439.

12. Mraz, M., Triner, L. & Hava, O. (1959) Naunyn-Schmiedebergs Arch. 236, 83.
13. Nemeth, S. & Vigas, M. (1968) Endocrinologia Experimentalis (Bratislava) 2, 179.
14. Nemeth, S., Vigas, M. & Lichardus, B. J. Trauma (in press)
15. Nemeth, S., Strakova, A. & Vigas, M. (1971) Horm.Metab. Res. 3, 359.
16. Noble, R.L. & Collip, J.B. (1942) Quart. J. exp. Physiol. 31, 187.
17. Shambaugh, G.E. & Beisel, W.R. (1968) Endocrinology 83, 965.
18. Vigas, M. & Nemeth, S. (1966) Cs. Fysiol. (Prague) 15, 96.
19. Vigas, M. & Nemeth, S. (1968) Endocrinologia Experimentalia (Bratislava) 2, 91.
20. Vigas, M., Nemeth, S. & Strakova, A. Physiol. Bohemoslov. (In press)

DECREASED 2,3-DIPHOSPHOGLYCERATE AND REDUCED OXYGEN DELIVERY
FOLLOWING MASSIVE TRANSFUSION AND SEPTICEMIC SHOCK

Harvey J. Sugerman, Leonard D. Miller, Maria Delivoria-Papadopoulos and Frank A. Oski

Departments of Pediatrics and Surgery, University of Pennsylvania School of Medicine, Philadelphia, Pa.

Oxygen delivery has been demonstrated to be inadequate in patients in septicemic shock as manifested by an elevated mixed venous O_2 saturation (1). The subsequent narrow arterio-venous (A-V) O_2 content difference has been attributed to pre-capillary A-V shunting (2). Other possible causes could be decreased mitochondrial utilization of O_2 (3) and decreased release of O_2 from hemoglobin (4).

In 1967, Benesch and Benesch (5) and Chanutin and Curnish (6) reported that the affinity of a hemoglobin solution for O_2 is decreased by its interaction with organic phosphates. 2,3-Diphosphoglycerate (DPG), comprises approximately 70% of red cell organic phosphates and thus, quantitatively, DPG has the greatest influence on the oxygen-hemoglobin equilibrium curve. Therefore, the higher the DPG, the greater is the shift of the oxy-hemoglobin equilibrium curve to the right as shown by an increased P_{50}. The P_{50} is a convenient way of expressing the position of the curve and is the partial pressure of O_2 at which hemoglobin is 50% saturated. A normal P_{50} in our laboratory is 27 0 ± 1.1 (7). A fall in DPG levels produces a shift to the left, or lower P_{50}. This increases hemoglobin-oxygen affinity and decreases the efficiency of O_2 delivery to tissues.

In several clinical states with an imbalance between O_2 demand and supply, there is an elevation in DPG levels, with a proportionate shift of the curve to the right, facilitating O_2 unloading. These conditions include chronic lung disease (8), cyanotic heart disease, (9) anemia, (10) thyrotoxicosis, (11) liver disease (12) and adaptation to altitude (13). The

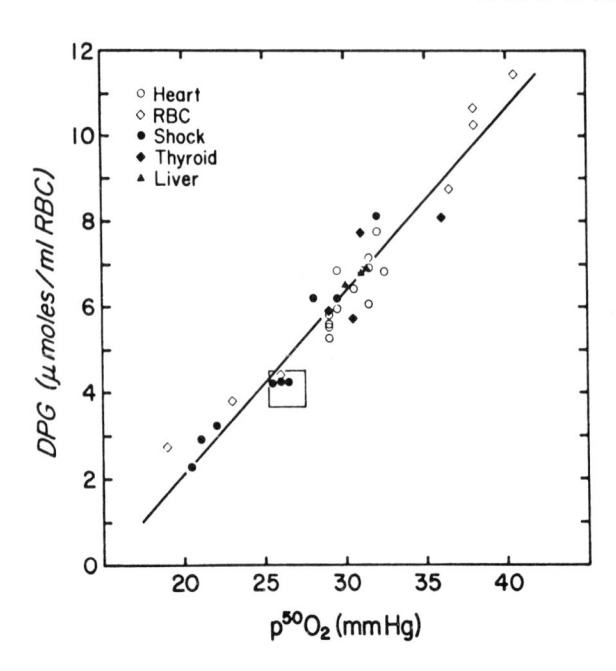

Fig. 1 Relationship between P_{50} of whole blood and red cell DPG in a variety of clinical disorders. These include cyanotic heart disease (o), red cell enzyme defects (◊), septic shock (●), hyperthyroidism (♦), and chronic liver disease (▲). Square block indicates normal range for P_{50} and red cell DPG. (reproduced with permission from Miller et al. 11)

relationship between DPG and oxygen-hemoglobin affinity in patients with these disorders is demonstrated by the strong correlation of 0.97 between the DPG and P_{50}, shown in Fig 1, such that a 1 mm Hg shift in the position of the curve in either direction equals a change of 420 nmoles DPG per ml red blood cells (7).

In septicemic shock and patients receiving large volumes of stored blood, despite the need for increased oxygen transport, there is a maladaptive shift to the left in the oxygen-hemoglobin equilibrium curve with low DPG levels (14). In septic shock, the DPG level appeared to mirror the patient's general clinical status; the poorer the clinical criteria, the lower the DPG and the lower the P_{50}; with significant clinical improvement, the DPG approached and often surpassed normality. The red cell DPG level and blood P_{50} was determined simultaneously on 12 occasions in 7 patients in septicemic shock. The correlation coefficient for this relationship was 0.87.

2, 3-DPG AND OXYGEN DELIVERY FOLLOWING TRANSFUSION AND SEPTICEMIA

With a curve moving steadily to the left, in an effort to maintain O_2 consumption, a patient is faced with two potentially detrimental possibilities either accepting a lower A-V O_2 difference and, therefore, extracting less O_2 with the necessity of attaining, and maintaining, a very high cardiac output (Fig 2),

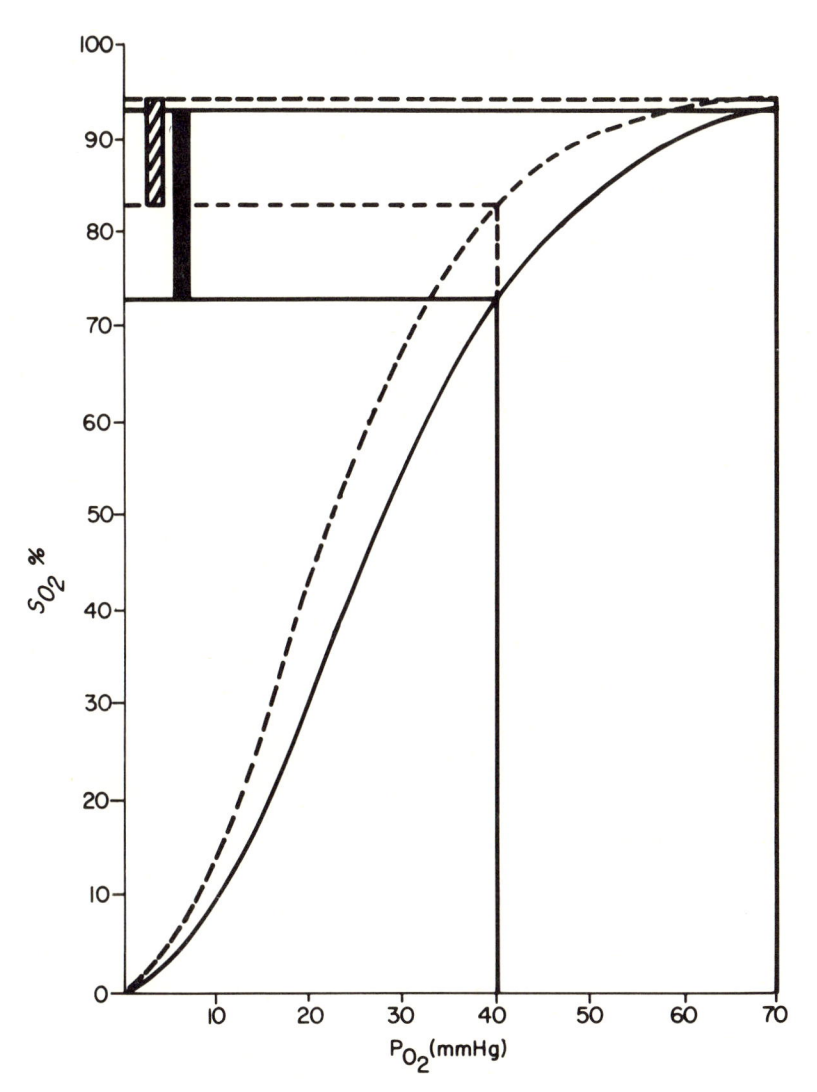

Fig 2 Postulated effect of a "left-shifted" oxygen-hemoglobin equilibrium curve: with a constant mixed venous O_2 tension, a rise in mixed venous O_2 saturation and consequent fall in A-V O_2 content difference would occur.

or extracting as much O_2 as with a normal curve, but reducing venous and, presumably, tissue PO_2 until a "critical" O_2 tension is reached and organ dysfunction occurs (Fig 3). Therefore,

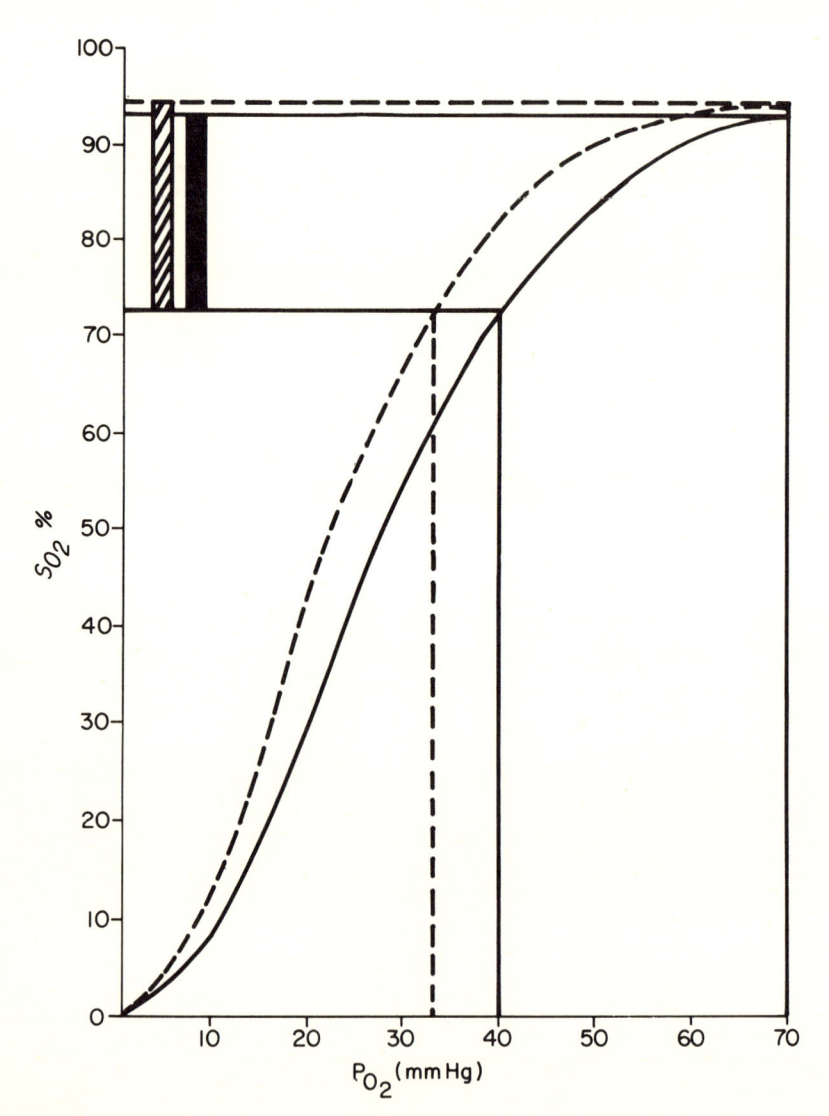

Fig 3 Postulated effect of a "left-shifted" oxygen-hemoglobin equilibrium curve: with a constant mixed venous O_2 saturation, there would have to be a decrease in mixed venous and, presumably, tissue O_2 tension.

the price of O_2 extraction with a "left-shifting" curve is probably a steadily falling tissue PO_2. The biologic significance of curve shifts was shown in 2 patients of approximately the same age and degree of anemia, one with a left-shifted curve (P_{50}=19) from a markedly reduced red cell DPG content due to a specific intraerythrocytic hexokinase deficiency, and one with a right-shifted curve (P_{50}=38) associated with an elevated red cell DPG due to a pyruvate kinase deficiency (15). Both patients were stressed by increasing the work load on a bicycle ergometer (16). The patient with the left-shifted curve, on minimal exercise, showed a prompt fall in central venous O_2 tension to 22 mm Hg (Fig 4). Oxygen delivery requirements, with an increasing work load, were met primarily by a steady increase in cardiac output since O_2 extraction became limiting at a low exercise load. In contrast, the O_2 tension of the right-shifted subject became progressively lower, never reaching 22 mm Hg, and he continued to extract increasing amounts of O_2 with a rising work load, needing to call forth only 50% of the increased cardiac output required by his "left-shifted" counterpart.

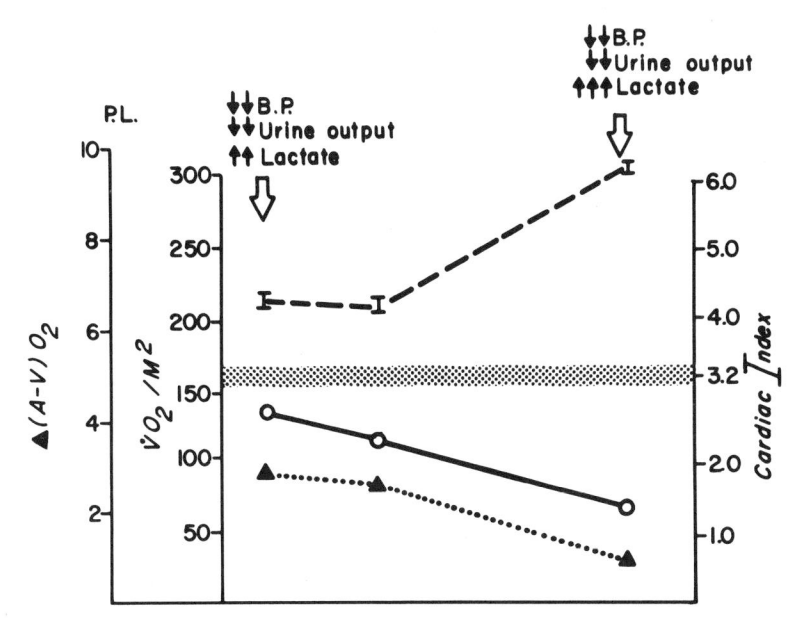

Fig 4 Changes in (A) O_2 saturation and (B) central venous O_2 tension (PvO_2) with exercise in Patient S.N. (red cell kexakinase deficiency), with a left-shifted oxygen-hemoglobin equilibrium curve, contrasted with changes with exercise in Patient D.S. (red cell pyruvate kinase deficiency), with a right shifted curve.
(Reproduced with permission from Oski et al. 12)

Another opportunity for studying the physiologic efforts of a left-shifted oxygen-hemoglobin equilibrium curve was afforded by the finding that patients on intravenous hyperalimentation with solution lacking inorganic phosphate rapidly developed hypophosphatemia and, due to a block in the dehydrogenation of glyceraldehyde-3-phosphate, developed marked reductions in red cell DPG and ATP and blood P_{50} (17,18). These patients manifest circumoral and peripheral extremity paresthesias, muscle weakness, mental obtundity, hyperventilation and diffuse slowing of the electroencephalogram. Simultaneous measurements of mixed venous O_2 saturation, mixed venous O_2 tension and cardiac output were made in one of these patients. The P_{50} was 16.5 mm Hg, in spite of a hemoglobin of 8.7 gms%. The mixed venous O_2 tension was 26.9 mm Hg, where normal is approximately 40; and she had a normal mixed venous O_2 saturation of 68%, producing an arterio-venous O_2 content difference of 3.6. The cardiac index was low normal: $2.8 1/min/m^2$. The O_2 consumption was low: $103/ml/min/m^2$. Her lactate was mildly elevated at 27 mg% and her lactate/pyruvate ratio was 20 to 1, where normal is approximately 10 to 1.

The patient failed to compensate for the shift to the left in the position of the curve, or the anemia, by increasing her cardiac output and, therefore, a low mixed venous and presumably, a low tissue O_2 tension resulted. For her to maintain a normal mixed venous O_2 tension of 40 mm Hg and an equivalent O_2 consumption, the arterio-venous O_2 content difference would, theoretically, have to fall from 3.6 to 1.2 volumes % and an increase in cardiac index from 2.8 to 8.6 $1/min/m^2$ would be required.

The pathophysiology of septic shock was found to be frequently dissimilar to hemorrhagic shock. The majority of our patients exhibited the so-called "hyperdynamic state" at one time or another in the course of their disease. The key findings are an inadequate O_2 consumption, in relation to the stressed patient's needs associated with an elevated cardiac index and a low peripheral resistance. Poor peripheral extraction of O_2, as shown by a small arterio-venous O_2 content difference, limits the total O_2 consumption. A steadily diminishing O_2 consumption often provided a clue to unresolved, or undrained, sepsis (Fig 5). The seemingly paradoxical observation of poor peripheral perfusion with an extremely high cardiac index has given rise to speculation on the cause of lowered peripheral O_2 extraction, loosely termed "shunting". Relatively little is known about possible intracellular defects produced by bacterial toxins that might contribute to this situation. The existence of peripheral anatomic "shunts" in this condition has never been shown and may be more apparent than real. It is here that the left-shifting equilibrium curve may play a significant role in the septic shock process and mimic the physiologic appearance of the elusive "shunts".

In all 15 patients, a direct correlation was noted between the red cell DPG concentration and the central venous O_2 tension (Fig 6). The correlation coefficient was 0.81. The lower the DPG, the lower was the central venous O_2 tension. The etiology of the fall in red cell DPG in septicemic shock is unknown.

Valtis and Kennedy (19) first demonstrated the progressive increase in O_2 affinity with blood storage. This has been subsequently shown to be secondary to a fall in red cell organic phosphate levels (20). After 3 days storage in ACD, DPG values fall to 50% of control; after 6 days, to 25% of control; and after 10 days, to 5% of the value in fresh blood (Fig 7) (21). Massive transfusion of stored blood, low in DPG, will lower a patient's P_{50} and decrease the delivery of O_2.

Five patients who received massive transfusions were studied. Again, the central venous O_2 tension was found to correlate with the red cell DPG concentration; the lower the DPG, the lower was the central venous O_2 tension (21). This increase in O_2 affinity in the severely ill, massively transfused patient may critically impair tissue oxygenation in an elderly patient with atherosclerosis. These patients should be given a generous amount of blood stored for less than 5 days.

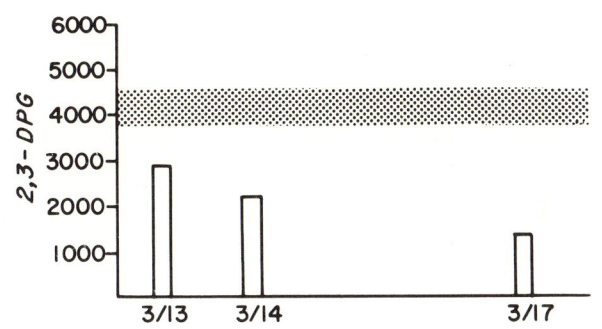

Fig 5 The "hyperdynamic state". In uncontrolled sepsis, the O_2 consumption continues to fall and clinical signs of peripheral perfusion deteriorate, despite a high, and rising, cardiac index. The 2, 3-DPG concentration falls as the clinical state worsens. The efforts to compensate, in terms of flow, in this patient were insufficient to salvage his life.
(Reproduced with permission from Miller et al. 14).

Attempts to correct this DPG deficit in stored blood have produced gratifying results. Blood stored for 28 days, with a DPG level of 285 nmoles/ml RBC's, as compared to a normal level of 4200 in fresh blood, was incubated with 0.1M inosine, pyruvate and phosphate (22). The DPG level reached 3916 in 1 hr and

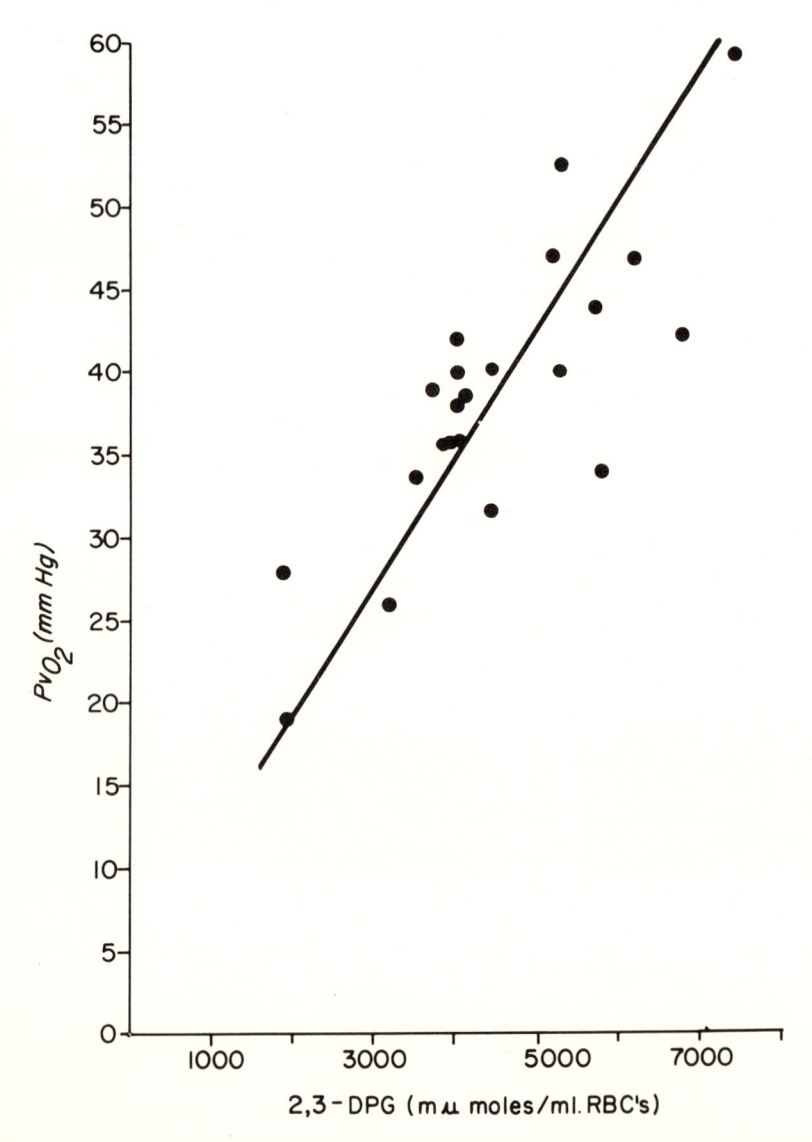

Fig 6 Correlation of the central venous O_2 tension (PvO_2) with the red-cell 2, 3-DPG concentration in patients with septic shock (r = 0.81, p<0.001).
(Reproduced with permission from Oski and Delivorio-Papadopoulos, 24).

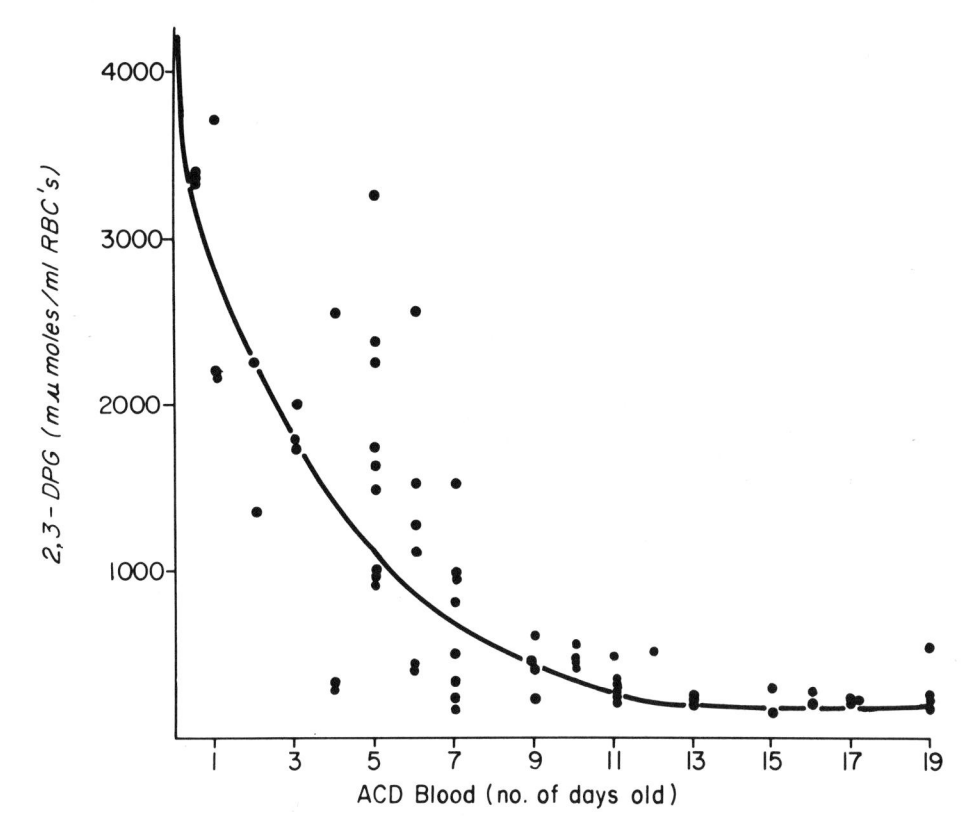

Fig 7 Rapid fall of 2, 3-DPG is depicted for blood stored in ACD. Semilogarithmic regression line for first 10 days of storage (r=0. 80, P ($\underline{t}$) <0.001).
(Reproduced with permission from Sugerman et al. 21)

6140 in 4 hr. If careful toxicity studies prove the relative safety of this mixture, it may prove to be extremely useful in improving the oxygen transport capability of stored blood, and in vivo, may be of value in clinical states, associated with poor O_2 delivery to tissues (23).

REFERENCES

1. MacLean, L.D., Mulligan, W.G., McLean, A.P.H. & Duff, J.H. (1967) Ann Surg. 166:543.

2. Siegel, J.H., Greenspan, M. & Del Guerico, L.R.M. (1967) Ann. Surg. 165:504.

3. Mela, L., Bacalzo, L.V. & Miller, L.D. (1971) Am J Physiol 220:571.

4. Sugerman, H.J., Miller, L.D., Oski, F.A., Diaco, J., Delivoria-Papadopoulos, M. & Davidson, D. (1970) Clin Res. 18:418.

5. Benesch, R. & Benesch, R.E. (1967) Biochem Biophys Res 26:162.

6. Chanutin, A. & Curnish, R.R. (1967) Arch Biochem. 96:121.

7. Oski, F.A., Miller, W., Delivoria-Papadopoulos, M. & Gottlieb, A.J. (1970) J Clin Invest 49:400.

8. Oski, F.A., Gottlieb, A.J., Delivoria-Papadopoulos, M. & Miller, W.W. (1969) N Eng J Med 200:1165

9. Edwards, M.J., Novy, M.J., Walters, C.L. & Metcalfe, J. (1968) J. Clin Invest 47:1851.

10. Torrance, J., Jacobs, P., Restrepo, A., Eschback, J., Lenfant, C. & Finch, C.A. (1970) N Eng J Med 283:165.

11. Miller, L.D., Sugerman, H.J., Miller, W.W., Delivoria-Papadopoulos, M., Diaco, J.F., Gottlieb, A.J. & Oski, F.A. (1970) Ann Surg 172:1051.

12. Mulhausen, R., Astrup, P. & Kjeldsen, K. (1967) Scand J Clin Lab Invest 19:291.

13. Lenfant, C., Torrance, J., English, E., Finch, C.A., Reynafarje, C., Ramos, J & Faura, J. (1968) J Clin Invest 47:2652.

14. Miller, L.D., Oski, F.A., Diaco, J., Sugerman, H.J., Davidson, D. & Delivoria-Papadopoulos, M. (1970) Surgery 68:187.

15. Delivoria-Papadopoulos, M., Oski, F.A. & Gottlieb, A.J. (1969) Science 165:601.

16. Oski, F.A., Marshall, B.E., Cohen, P.J., Sugerman, H.J. & Miller, L.D. (1971) Ann Int Med 74:44.

17. Travis, S.F., Sugerman, H.J., Ruberg, R.J., Delivoria-Papadopoulos, M., Miller, L.D. & Oski, F.A. N Eng J Med (in press)

18. Lichtman, M.A., Miller, D.R., Cohen, J. & Waterhouse, C. (1971) Ann Int Med 74:562.

19. Valtis, D.J. & Kennedy, A.C. (1954) Lancet 1:119.

20. Bunn, H.F., May, M.H., Kocholaty, W.F., & Shields, C.E. (1969) J Clin Invest 48:311.

21. Sugerman, H.J., Davidson, D.T., Vibul, S., Delivoria-Papadopoulos, M., Miller, L.D. & Oski, F.A. (1970) Surg Gynec Obstet 131:733.

22. Oski, F.A., Travis, S.F., Miller, L.D., Delivoria-Papadopoulos, M. & Cannon, E. (1971) Blood 37:52.

23. Pollock, T.W., Sugerman, H.J., Oski, F.A., Rosato, E.F. & Miller, L.D. (1971) Fed Proc 30:1965.

24. Oski, F.A. & Deliveria-Papadopoulos, M. (1970) J. Pediatr 77:941.

Supported in part by US Public Health Service Grants HD01919, GM15001 & GM1540 and a grant from the John A. Hartford Foundation, Inc.

COMPLETE OXYHEMOGLOBIN DISSOCIATION CURVES IN THERMAL TRAUMA

G. Arturson

Burn Center, University Hospital, Uppsala, Sweden

Practically all information available on the range of normal variation in the position and shape of the oxygen dissociation curve is based on multiple point determinations of O_2 content or saturation at a specific PO_2, constant pH and temperature. Less is known concerning the affinity of hemoglobin for oxygen following different kinds of trauma e.g. skin burns.

The aim of the present investigation was to establish the complete oxyhemoglobin dissociation curve in patients with severe skin burns. The position of the oxyhemoglobin dissociation curve at constant temperature (37°) and constant PCO_2 (40 mm Hg) was established taking into account the influence of blood pH, the concentration of hemoglobin (Hb) and carbon monoxide hemoglobin (HbCO), the intraerythrocytic concentration of 2,3-diphosphoglycerate (2,3-DPG), mean corpuscular hemoglobin concentration (MCHC), age, and sex of the patient. Comparisons were made with data from a control series.

Material and Methods

Controls

The position and shape of the complete oxyhemoglobin dissociation curve using blood samples from 91 healthy subjects both smokers and non-smokers, 32 men and 59 women, was investigated. The blood concentrations of Hb, HbO_2 and HbCO, the content of 2,3-DPG in the erythrocytes, MCHC and acid-base balance were also measured.

Patients

14 patients with skin burns were investigated in the same way as the control subjects. Four patients had second and third degree burns covering 20-35% of the body surface without severe clinical complications and 5 patients had 40-75% full-thickness burns with repeated, severe wound infection, transient ventilatory and cardiac insufficiency. Five fatal burns had 60-95% full thickness burns. All patients were treated similarly with dextran 70 (Pharmacia, Sweden) and electrolyte solutions during transportation to the Burn Center, and thereafter with plasma, albumin, Ringer-bicarbonate solutions, and erythrocyte concentrate. A combined enteral-intravenous dietary program of about 5,000 to 6,000 kcal/day was administered. Gamma-globulin solution (Kabi, Sweden) was given during the first 10 days after the accident. Antibiotics were administered as indicated by culture and sensitivity tests on the micro-organisms in the burn wounds. The local therapy was open, with early excision and skin-grafting with autogenous tissue except in 3 cases in which allografts were also used. For the treatment of respiratory insufficiency, an Engström respirator was used. All patients were treated in an isolated ward room, in which the humidity and temperature of the air surrounding the patient was kept constant at 30% and 34°C respectively. Intermittent drying of the wet burn wounds was performed.

Blood-sample procedure and methods

Arterial and central venous blood samples were taken with heparinized syringes in normal subjects and in patients at intervals after thermal trauma. The complete oxyhemoglobin dissociation curve was obtained in an oxyhemoglobin dissociation analyzer (13), in which blood PO_2 and O_2 content are registered continuously on an X-Y recorder as a known volume of O_2 gas diffuses into a known volume of deoxygenated blood. The pH, PCO_2 and PO_2 were determined by appropriate electrodes maintained in a separate system at 37°C before and after the completion of the run (20). The 2,3-DPG was determined by de Verdier and Killander's modification (9) of Barlett's chromatographic method (5). The values were expressed as mmoles per litre of packed erythrocytes (15). The concentrations of Hb, HbO_2 and HbCO were determined in an IL Model 182 CO-oximeter.

Results

Controls

As a result of multiple regression analyses using all available data from the normal material, regression coefficients were obtained and the values of log PO_2 at different values of O_2 saturation could be calculated. Thus the complete oxyhemoglobin dissociation curve was determined for men and women respectively at certain normal values of pH, 2,3-DPG etc. as shown in Fig. 1. The position of the oxyhemoglobin

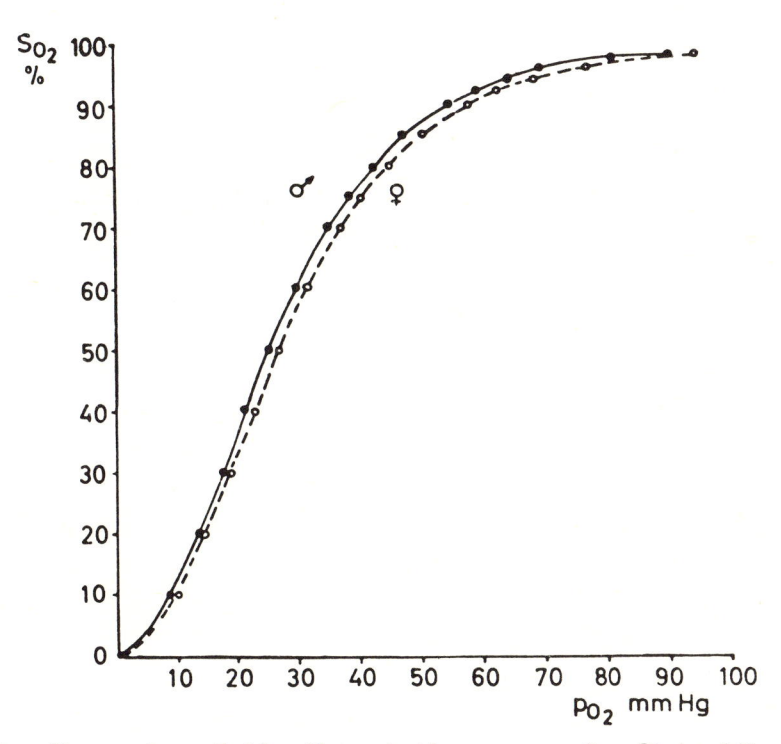

Fig. 1. The oxyhemoglobin dissociation curves for 91 healthy subjects (smokers and non-smokers), 32 men and 59 women at pH 7.40, PCO_2 40 mm Hg and temperature 37°C and the following normal values:

	men	women
2.3-DPG, m moles/l red cells	4,52 ± 1,15	5,28 ± 1,24
Hb, g/100 ml	15,1 ± 1,09	13,4 ± 0,86
MCHC, g/100 ml	34,1 ± 1,8	33,3 ± 1,0
Age, years	40,7 ± 8,8	27,1 ± 6,0
HbCO, %	4,3 ± 2,1	3,6 ± 1,7

dissociation curve is characterized in men and women by the P50 value (i.e. the PO_2 value giving 50% oxygen saturation of hemoglobin) defined according to the following equations:

Men:

$$\log PO_2 (50) = -0,4222 \cdot (pH) + 0,0463 \cdot (\log DPG) - 0,0028 \cdot (Hb) - 0,0001 \cdot (MCHC) + 0,0011 \cdot (Age) - 0,0082 \cdot (HbCO) + 4,53004. \quad \ldots (1)$$

Women:

$$\log PO_2 (50) = -0,3432 \cdot (pH) + 0,0279 \cdot (\log DPG) - 0,0082 \cdot (Hb) + 0,0046 \cdot (MCHC) - 0,0001 \cdot (HbCO) + 3,90156 \ldots (2)$$

The mean values and standard deviations of P50 are for 32 men = 25,1 ± 1,1 mm Hg and for 59 women = 26,6 ± 1,1 mm Hg at normal values of pH, 2,3-DPG, etc. according to Fig. 1. The affinity of hemoglobin for oxygen is lower in women compared to men and higher in smokers compared to non-smokers.

It is evident from the high coefficient of determination (70%) that practically all the variations in log PO_2 at the different values of saturation observed can be accounted for by the 6 factors represented in the regression equation for men. The corresponding figure for women was lower, 50%, indicating that factors other than those investigated were operating.

Patients

The affinity of hemoglobin for O_2 varied in different phases of the burn syndrome and with the condition of the patient. For these reasons, the data from the patients are mainly treated individually. The most important findings are illustrated in Figs 2-5.

A moderate increase in hematocrit and hemoglobin concentration was found directly following the trauma concomitant with a slight metabolic acidosis (Fig. 2).

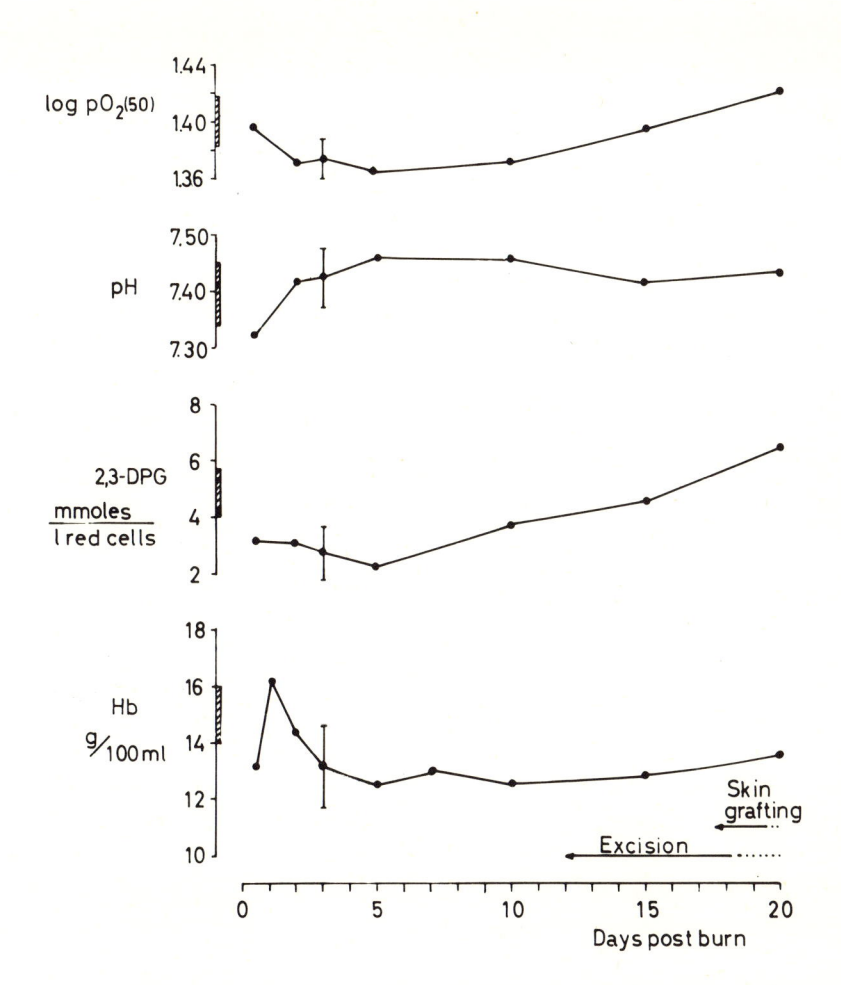

Fig. 2. Mean values and range of log PO_2 at 50 per cent saturation calculated from the actual values of pH, 2,3-DPG, Hb, MCHC, age and HbCO according to equation (1) in arterial blood from 9 male patients with second and third degree burns covering 20-75% of the body surface at different days after the thermal trauma.

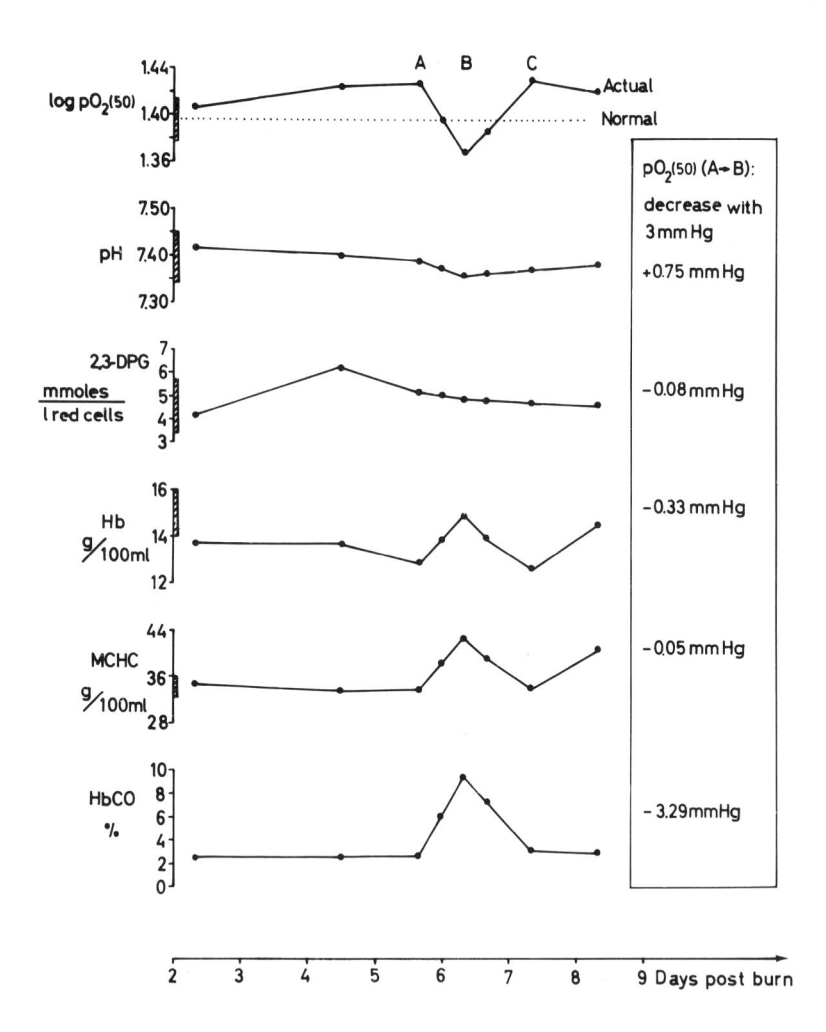

Fig. 3. The upper part of the figure shows log PO$_2$ values
calculated according to equation (1) using actual values
of pH, 2,3-DPG etc for the patient (unbroken line) as well
as using normal values (dotted line). Actual values of
pH, 2,3-DPG, Hb, MCHC and HbCO in arterial blood on different
days post burn are also given. A decrease in log PO$_2$ (50)
between A and B corresponding to a leftward shift of the
oxyhemoglobin dissociation curve at 50% saturation of 3 mm Hg
was calculated from the actual data of pH, 2,3-DPG, etc.
On the right hand side the contribution by the different
parameters to the displacement of the dissociation curve is
shown. The data are from a 36 year old man with a 75% full
thickness burn.

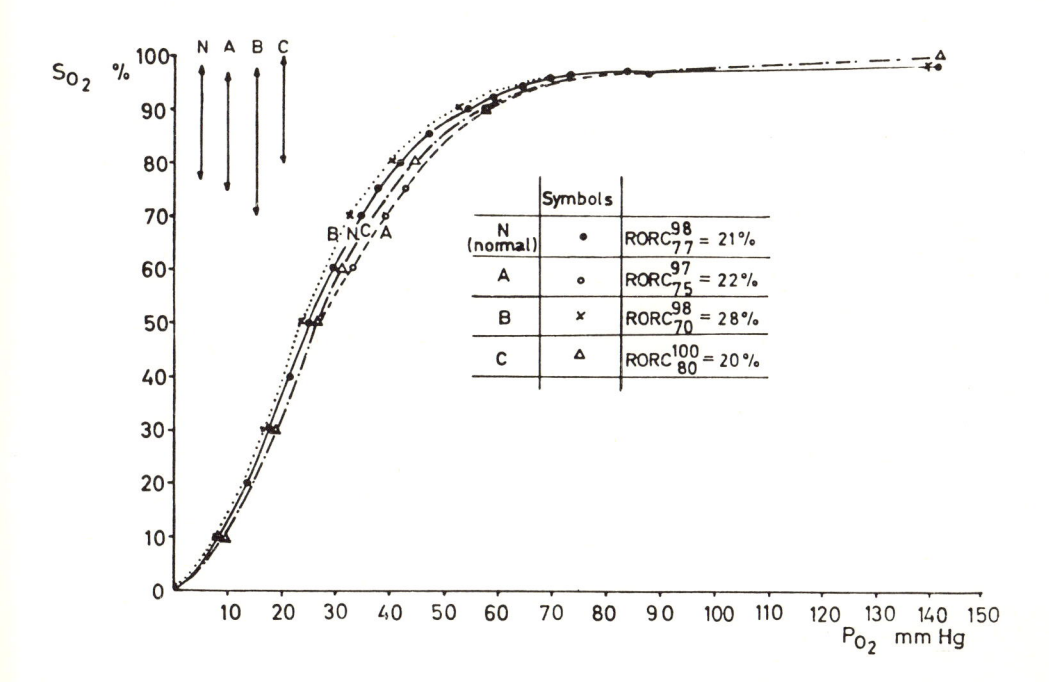

Fig. 4. Complete oxyhemoglobin dissociation curves from the same patient as in Fig. 3. Three different curves A B and C (as in Fig. 3) on days 5, 6 and 7 after the trauma have been calculated and compared with a calculated normal curve for a 36 year old man N. The relative O_2 releasing capacity for the different curves shows that the leftward shift of curve B concomitant with the maximum value of HbCO is associated with a high value of RORC due to a decrease in mixed venous blood PO_2.

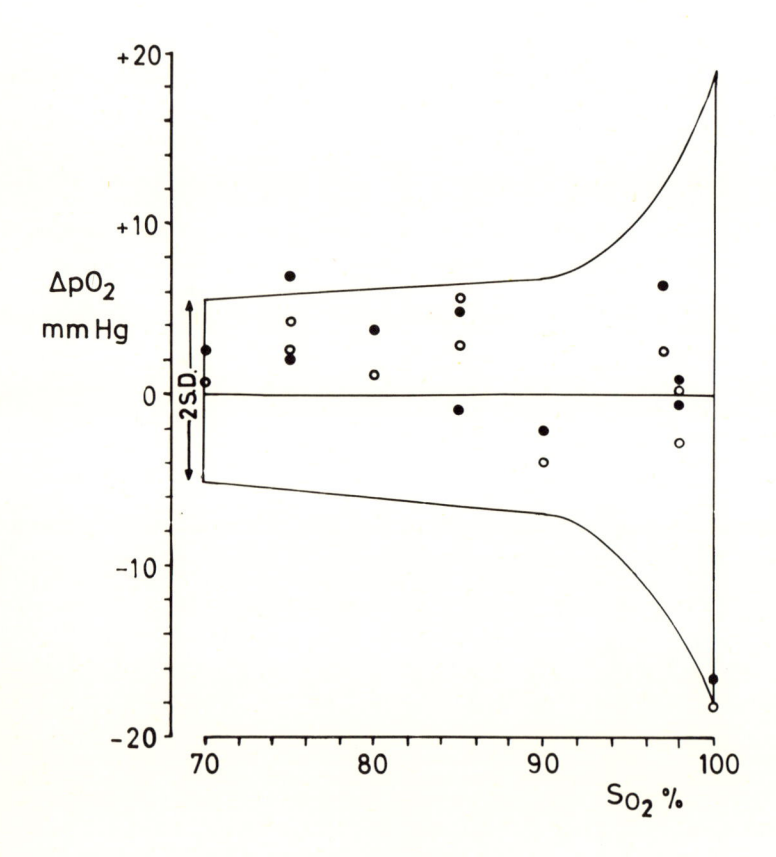

Fig. 5. Plot of differences between measured PO_2 and (1) calculated PO_2 using normal values (•) (2) calculated PO_2 using actual values for patient (O). The values are taken from the same patient as in Figs 3 & 4.

In all surviving patients, the affinity of hemoglobin for O_2 increased (a leftward shift of the oxyhemoglobin dissociation curve) with a simultaneous decrease in the 2,3-DPG content of the erythrocytes and a shift towards slight respiratory alkalosis during the first phase of the burn syndrome. This was followed by a decreased affinity of hemoglobin for O_2 and an increase in the 2,3-DPG concentration during the phase of excision and grafting of the wounds (Fig. 2). During this later phase, a slight chronic anemia as well as transient ventilatory insufficiency was often noticed.

The arterial and central venous blood PO_2 and O_2 saturations in these patients were in good agreement with those calculated from the normal as well as the actual values of pH, 2,3-DPG, Hb, MCHC, and HbCO according to equation (1). This indicates that the displacements of the oxyhemoglobin dissociation curves in different phases of the burn syndrome can be adequately accounted for by changes in the aforementioned factors of which the most important are pH, 2,3-DPG and similar results were obtained in patients with severe burns with periods of increased HbCO (See Figs 3-5). Thus in Fig. 3 a decrease in log PO_2 (50) between blood samples A and B, corresponding to a leftward shift of the oxyhemoglobin dissociation curve at 50% saturation of 3 mm Hg, could be calculated from the actual data of pH, 2,3-DPG, etc. according to equation (1). The displacements of the complete oxyhemoglobin dissociation curves in the same patient are shown in Fig. 4. The leftward displacement of the dissociation curve B, corresponding to the maximum value of HbCO, should imply a decrease in the relative O_2 releasing capacity (RORC) at the same PO_2. However, the decrease is compensated by a simultaneous decrease in the mixed venous blood PO_2. Fig. 5 shows that the difference between the experimental O_2 affinity (measured PO_2 and O_2 saturation) and the calculated normal and actual values of PO_2 at different saturations respectively are all within the normal range ($\pm$ 2 S.D.).

In patients with periods of pulmonary insufficiency as well as in the later stages of the burn syndrome in patients with a fatal outcome and persistent spontaneous hyperventilation, hypocapnia and respiratory distress with borderline oxygenation, the content of 2,3-DPG in the red blood cells was increased (8 to 15 mmoles per liter of erythrocytes) and the affinity of hemoglobin for O_2 decreased.

Discussion

Comparisons by different authors of the in vitro affinity of hemoglobin for O_2 (i.e. P50 values at PCO_2 = 40 mm Hg and pH = 7.40) from normal adults show fairly large variations (3,7,10,11,16). These variations are attributed to methodological errors and the influence of 2,3-DPG, carbon monoxide in smokers, age, sex, etc. The lower affinity of hemoglobin for O_2 in women compared to men in the present material might be due to the higher intra-erythrocytic 2,3-DPG concentration corresponding to the lower hemoglobin value. It thus seemed important to define the position and shape of the O_2 dissociation curve with respect to variations in 2,3-DPG, hemoglobin concentration, MCHC, HbCO as well as the age and sex of the subject. This has been done in the present material (cf Fig. 1, and the multiple regression analyses).

It is well known that the supply of O_2 to the different tissues in the organism is regulated primarily by the concentration of hemoglobin in the blood and the cardiac output. The uptake and release of O_2 by the hemoglobin is dependent on the concentration of oxygen in the surroundings of the erythrocytes and on certain other factors, especially the hydrogen - ion concentration and the intra-erythrocyte concentration of 2,3-DPG. When the 2,3-DPG concentration increases, the affinity of hemoglobin for oxygen decreases i.e. it is more difficult for the hemoglobin to take up and easier to release oxygen than previously. However, the hemoglobin is saturated to almost 100% at the partial oxygen pressures normally prevailing in the lungs. Thus, a reduced affinity of the hemoglobin for O_2 will, in general, favour the release of O_2 to the tissues.

In patients with severe thermal trauma, however, the hemoglobin is not always fully saturated. Periods of refractory respiratory insufficiency often occur, even in patients without any initial signs of pulmonary or facial burn trauma (1). The underlying pathological processes in the lungs are numerous and not fully understood. They will, however, cause a decrease in the available O_2. Moreover, the metabolic demands and the need for O_2 are known to increase considerably for several weeks after severe thermal injuries (1,6,8). In these circumstances the only mechanism, by which the impaired O_2 supply can be effectively corrected is by an improved delivery of O_2 from the lung to the tissue capillaries. This improvement can occur in 2 ways: (a) improved O_2 transport and (b) a decreased affinity of hemoglobin for O_2 in the tissue capillaries. It is well known that O_2 transport improves rapidly after an increase in the cardiac output but less rapidly after an increase in the hemoglobin level. In the present investigation a second mechanism for promoting the supply of O_2 to the tissues has been shown to operate, whereby the affinity of hemoglobin for O_2 is decreased. Furthermore, the displacement of the O_2 dissociation curve to the right is mediated by 2,3-DPG in the erythrocytes. By this mechanism the relative O_2 releasing capacity increases without any change in PO_2. It would appear of the utmost importance to keep the PO_2 high enough, especially in patients with severe thermal trauma in which generalized edema is often observed (1). The unexpectedly high PO_2 values found in mixed venous blood from the most severely burned patients is in accordance with this idea.

Two factors are of particular importance with respect to the increase in 2,3-DPG during O_2 deficiency following thermal trauma:
(a) changes in the blood pH and (b) changes in the oxygenation state of hemoglobin.
(a) It seems plausible that the acute hypoxic hypoxia in the early phase of the burn trauma induces a respiratory alkalosis which increases the blood pH and the red cell pH. The elevation of the intracellular pH promotes 2,3-DPG synthesis and decreases the rate of decomposition of 2,3-DPG which leads to an increase in 2,3-DPG concentration. This is in accordance with results from studies on the pH dependency of 2,3-DPG metabolism (2,4,14,17,18,19).
(b) According to Duhm and Gerlach (12) deoxygenation increases the rate of 2,3-DPG synthesis through elevation of the intracellular pH. In patients with periods of pulmonary insufficiency and very high intra-erythrocytic levels of 2,3-DPG these factors might be of importance. The increase in deoxyhemoglobin in chronic hypoxic hypoxia as well as in chronic anemia in the later phase of thermal trauma may also induce an increase in the 2,3-DPG concentration in the red blood cells. It is noteworthy that the directly measured arterial and central venous blood PO_2 and O_2 saturation in patients with thermal burns are in good agreement with those calculated using either appropriate normal values for the parameters in equation (1) or the actual values for the patient in question. This seems to indicate that the O_2 affinity in different phases of the burn syndrome can be predicted mainly from the values of pH, 2,3-DPG, Hb, MCHC, HbCO, age and sex of the patients.

Summary

Complete oxyhemoglobin dissociation curves have been recorded for 91 healthy subjects and 14 subjects with thermal burns. The position of the curves at constant temperature and PCO_2 was established taking into account the influence of blood pH, the concentrations of hemoglobin and carbon monoxide hemoglobin, the intra-erythrocytic concentration of 2,3-diphosphoglycerate, MCHC, age and sex of the subjects. The values of log PO_2 at different values of O_2 saturation were predicted by multiple regression equations. The affinity of hemoglobin for O_2 increased during the early phase and decreased during the late phase of the burn syndrome as well as in periods of refractory respiratory insufficiency. The O_2 affinity in normal subjects as well as in patients with thermal burn may be predicted from the recorded values of pH, 2,3-DPG, Hb and HbCO and to a lesser extent from MCHC, age and sex of the subject.

References

1. Arturson, G. (1970) Injury 1:226.

2. Asakuta, T., Sato, Y., Minakami, S. & Yoshikawa, H. (1966) Clin. chim. Acta 14:840.

3. Astrup, P., Engel, K., Severinghaus, H.W. & Munson, E. (1965) Scand. J. Clin. Lab. Invest. 17:515.

4. Astrup, P. & Thorshauge, C. (1970) Scand. J. clin. Lab. Invest. 26:46.

5. Barlett, G.R. (1959) J. Biol. Chem. 234:449.

6. Barr, P.O., Birke, G., Liljedahl, S.O. & Plantin, L.O. (1968) Lancet 1: 264.

7. Bartels, H., Bettec, K., Hilpert, P. & Riegel, K. (1960) Pflügers. Arch. 727:372.

8. Cope, O., Nardi, G.L. Ouijano, M., Rovit, H.L., Stanbury, J.B. & Wight, A. (1953) Ann, Surg. 137:163.

9. de Verdier, C.H. & Killander, J. (1962) Acta Physiol. Scand. 54:346.

10. Dill, D.B. & Forbes, W.H. (1941) Ann. J. Physiol.132:685

11. Duc, G. & Engel, K. (1970) Resp. Physiol. 8:118.

12. Duhm, J. & Gerlach E. (1971) Plügers. Arch. 326:254.

13. Duvelleroy, M.A., Buckles, R.G., Rosenkaimer, S., Tung, C. & Laver, M.B. (1970) J. Appl. Physiol. 28:227.

14. Gerlach, E., Lübben, K. & Deuticke, B. (1962) Nuc.-Med. (Stuttg.) 2:151.

15. Hjelm, M. (1969) Acta Univ. Upsal. Abstract of Upsala Dissertations in Medicine.

16. Naeraa, N. (1964) Scand. J. Clin. Lab. Invest. 16:630.

17. Rapoport, S. (1968) Biochemistry 4:69.

18. Rörth, M. (1970) In: Red cell metabolism and function, p. 57. Ed: G.J. Brewer. New York-London: Plenum Press,

19. Rörth, M. (1970) Scand. J. clin. Lab. Invest. 26:42.

20. Siggaard-Andersen, O. (1967) 3rd ed. Copenhagen, Munksgaard.

Acknowledgement

This work was supported by the Swedish Medical Research Council (Project No. B71-40X-676-06C).

BARO- AND CHEMORECEPTOR MECHANISMS IN HAEMORRHAGE

Joan C. Mott

Nuffield Institute for Medical Research

University of Oxford, United Kingdom

Introduction
Changes with age in the cardiovascular background
Haemorrhage
        Systemic arterial pressure
        Haematocrit
Response to haemorrhage after interference with some mechanisms of circulatory control
        Systemic arterial baroreceptors
        Systemic arterial chemoreceptors
        Efferent sympathetic activity
        Acute nephrectomy
Conclusion
        Importance of the renin-angiotensin system in the resistance of immature mammals to haemorrhage

INTRODUCTION

The observations summarized here arose out of a systematic examination, made for other purposes, of the effects of haemorrhage in animals of different ages, lightly anaesthetized with sodium barbitone. It became evident that in terms of the criteria detailed below, kittens and young rabbits withstood haemorrhage better than the adults of these species (10, 14). Further analysis showed that the contribution of arterial baroreceptor mechanisms in the maintenance of arterial pressure is relatively greater in adult rabbits and that mechanisms of renal origin have a predominant influence in immature rabbits (15). Both these species, especially the rabbit, are born in a somewhat immature condition. Developmental changes in cardiovascular function must be taken into account when designing appropriate tests of, and in assessing cardiovascular responses to, stimuli.

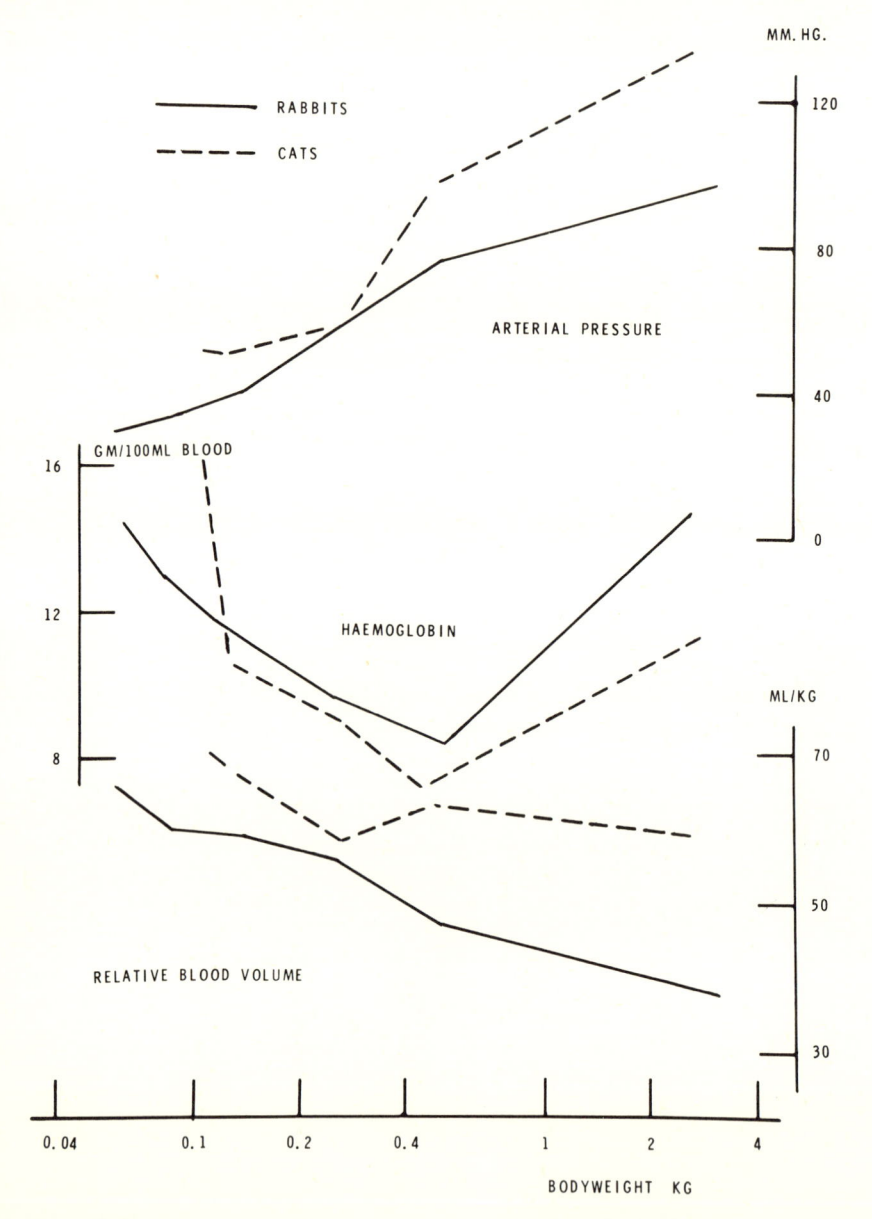

Fig. 1. Ordinates, arterial pressure (mm Hg), haemoglobin concentration (g/100 ml blood) and relative blood volume (ml/kg) in rabbits (————) and cats (- - - -) from birth to maturity. Abscissa, body weight in kg. (10, 13).

Changes with Age in the Cardiovascular Background

Resting arterial pressure rises during postnatal development and, in many species, relative blood volume (ml/kg) falls. Circulating haemoglobin concentration (g/100 ml blood) falls after birth to give a body concentration (g/kg) slightly below the adult level. Fig. 1.

Comparison of cardiovascular responses to various stimuli in animals of different ages and therefore of different resting arterial pressures and resting blood volumes is best made on a proportionate basis (5). In the present context, the question posed is what proportion of the blood volume must be removed to obtain a given proportionate fall of arterial pressure.

Haemorrhage. When 15-20% of the blood volume is removed from the circulation, arterial pressure falls but in the course of a few minutes, starts to rise again and reaches a plateau in about ten minutes. Repeated bleeding of rabbits and cats was carried out at intervals of not less than ten minutes until arterial pressure had fallen to one half or two-thirds of its initial value. Serial measurements of haematocrit or haemoglobin levels and recovery of

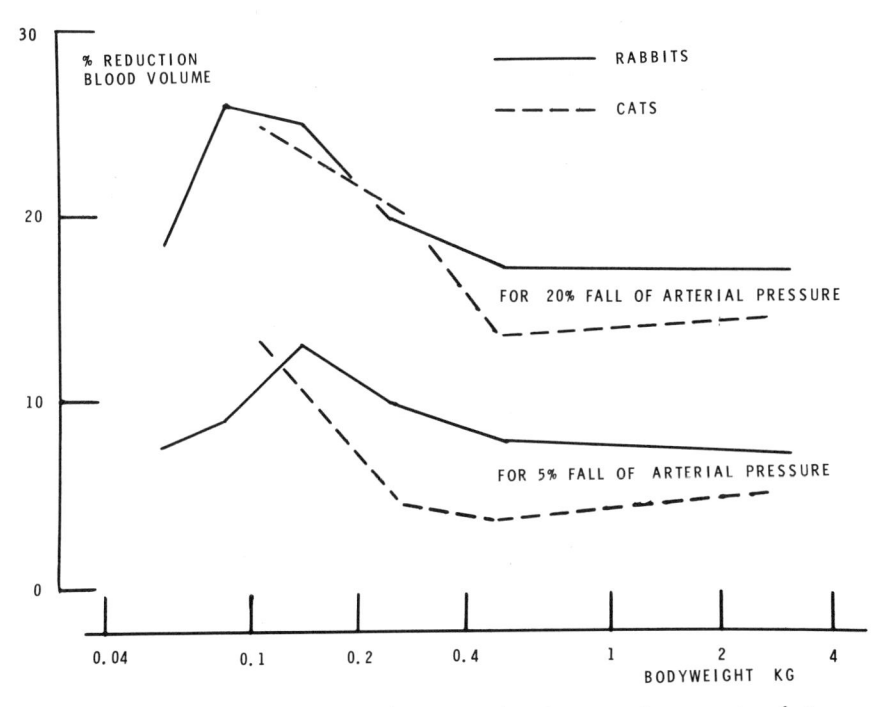

Fig. 2. Blood volume reduction required to reduce arterial pressure by 20% and by 5% in rabbits (———) and cats (- - - -) from birth to maturity. Each point is the mean value of measurements in 5-21 rabbits or 6-7 cats (10, 13).

the residual haemoglobin (10, 14) permitted calculation of the blood volume at each step in the bleeding procedure.

Arterial Pressure. The fall of arterial pressure during haemorrhage was broadly proportional to the fraction by which blood volume was reduced but varied significantly during development. Both kittens and infant rabbits maintained their arterial pressure better during stepwise haemorrhage than did cats and adult rabbits respectively. Fig. 2 summarizes the mean estimates (obtained by interpolation) of the blood volume reduction required to reduce arterial pressure by 5% and by 20%. Thus under the conditions of these experiments, superior resistance to haemorrhage was shown by kittens at birth and by rabbits aged about 4-14 days, when compared with the adults of these species.

Haematocrit. When kittens and rabbits of all ages were bled, the haematocrit fell but in some adult cats a rise in haematocrit followed stepwise haemorrhage. Analysis of the concurrent plasma protein changes showed that albumin-rich tissue fluid must have entered the vascular compartment during haemorrhage in kittens and rabbits but not in cats over 3 weeks old (13). Calculation showed that in rabbits haemodilution compensated for about one third of the blood removed by bleeding (10).

The Response to Haemorrhage after Interference with Some Mechanisms of Circulatory Control

Systemic Arterial Baroreceptors. It is well established that arterial baroreceptor reflexes are functional in immature rabbits and respond similarly to pressure changes of similar proportions at all ages (1, 7). That is to say that in the intact animal there is a degree of vascular tone such that a sustained rise of arterial pressure occurs when the carotid sinus and aortic depressor nerves are cut (10).

Resting blood pressure was inversely related to initial blood volume in rabbits in which the carotid sinus and aortic nerves had been cut but not in intact rabbits (15). A given fall of arterial pressure was produced by significantly less blood loss per kg. body wt. in adult rabbits so denervated than in intact animals. Rabbits 9-15 days old similarly treated however, behaved in the same way as intact control animals (Fig. 3).

In rabbits with denervated baroreceptors, the fraction of the blood volume which has to be removed to reduce the initial arterial pressure by 20% was significantly related ($P < 0.05$) to the initial blood volume. The practical consequence of this relationship was that a clear distinction in the responses to haemorrhage of denervated and intact rabbits is only to be expected in animals with blood volume of less than 42 ml/kg - that is on average in rabbits over 2.35 kg body weight (11, 12).

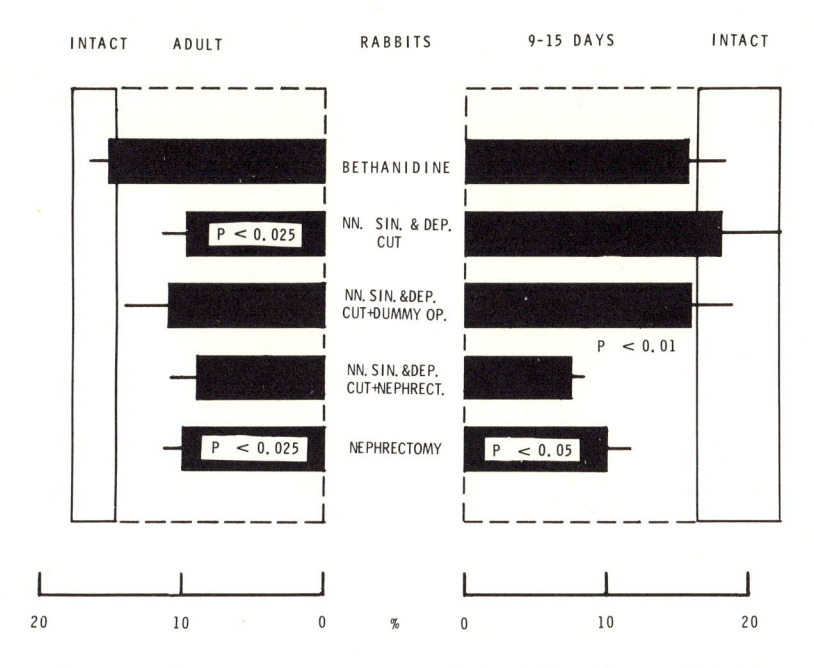

Fig. 3. % reduction of blood volume (black bars) required to cause a 20% fall of arterial pressure in rabbits treated as shown. The tail on each bar represents 1 S.E. (n = 5-9). The bars are superimposed on areas representing the corresponding values in 20 adult or 14 immature intact control rabbits. The solid rectangle covers the mean value ± 1 S.E. for the control animals. P values when superimposed on black bars are for differences from the appropriate control, otherwise for the difference between adjacent bars (13).

Arterial chemoreceptors. Section of the carotid nerves in the rabbit must interrupt chemoreceptor fibres but there seem to be few or no such fibres in the aortic nerves of the adult rabbit (for references see 8). Sympathetic tone of chemoreceptor origin is unlikely to be greater in adult than in immature rabbits. It follows that the excellent maintenance of arterial pressure during haemorrhage in denervated immature rabbits is achieved in spite of the absence of chemoreceptor activity.

Efferent sympathetic activity. Administration of bethanidine (1-3 mg/kg) which prevents the action of pressor amines released from sympathetic nerve terminals did not influence the response of adult or immature rabbits when bled, though the levels of arterial pressure throughout were somewhat lower than in untreated animals.

These observations made it seem likely that the mechanisms enabling immature rabbits to sustain haemorrhage so well were not to be sought in the autonomic nervous system.

More recently bioassay of the increments of arterial pressor amine activity, induced by removal of 25% of the blood volume, have shown that these are less in rabbits under 250g body weight than in older rabbits (4). However, since such immature rabbits are 3-6 fold more sensitive to the pressor actions of acute injections of adrenaline (2), they would not appear to be at any disadvantage in this respect.

Acute nephrectomy. Arterial pressure was reduced on average by 16 mm Hg by acute nephrectomy in rabbits of all ages. This reduction is proportionately very substantial in immature rabbits - 54% of resting arterial pressure at 0-1 day old and 27% at 9-15 days of age (2, 15). The falls seen after sham operation in immature rabbits were least in the youngest. It therefore seemed likely that the maintenance of resting arterial pressure in immature rabbits is partly determined by activity of renal origin.

Acute nephrectomy impaired the response of both adult and immature rabbits to haemorrhage (fig. 3). Since, however, any operative interference can lead to renin liberation (9) it seemed desirable to carry out further experiments in which changes of cardiovascular autonomic tone were minimized by previous section of the sinus and aortic nerves and in which sham operations were included as control. In both immature and adult rabbits, the addition of sham nephrectomy to the preparation of animals in which the carotid and aortic depressor nerves had been cut did not alter their response to haemorrhage from that seen in animals subjected to nerve section alone. (Fig. 3). However, immature rabbits subjected to nephrectomy in addition to baroreceptor nerve section sustained haemorrhage significantly less well than controls subjected to nerve section and sham operation only (Fig. 3). The corresponding experiments in adult rabbits showed no such difference (15).

Some recovery of arterial pressure after each withdrawal of blood in a series of bleeds occurred regularly in rabbits and was often particularly conspicuous in immature rabbits. Nephrectomy greatly impaired this recovery in immature but not in mature animals (15).

CONCLUSION

Importance of the Renin-Angiotensin System in the Resistance of Immature Mammals to Haemorrhage

The inference that the renin-angiotensin system is of relatively greater importance in cardiovascular control in immature than in adult mammals is supported by the recent finding of higher resting levels and of larger increments of angiotensin II-like activity during haemorrhage in immature than in adult rabbits (4).

Thus in the immature rabbit we have a situation in which demonstrably active arterial baroreceptor mechanisms (1, 7) seem unimportant in the maintenance of arterial pressure during haemorrhage. The most obvious cardiovascular difference between immature and adult rabbits is that the latter have relatively lower blood volumes and higher arterial pressures. In immature rabbits resting arterial pressure is partly dependent on the presence of the kidneys (2,15). It is possible that endogenous angiotensin-II is acting centrally to inhibit efferent parasympathetic activity (16). In these circumstances it appears that although baroreceptor (6, 7, 15) and chemoreceptor reflexes are readily demonstrable, nevertheless they appear to play little part in the steady state response of the immature whole animal to haemorrhage. Since recent experiments (3) show that lambs liberate more angiotensin II-like activity during haemorrhage than sheep, it may be generally true that an active renin-angiotensin system is a physiological concomitant of early postnatal life. Whether pathological activity of the system has comparable interaction with the nervous mechanisms of circulatory control remains to be directly tested.

REFERENCES

1. BLOOR, C. (1964). J.Physiol. (London) 174: 163.
2. BROUGHTON PIPKIN, F. (1971). Quart.J.exp.Physiol.56: 210.
3. BROUGHTON PIPKIN, F., S.M.L.KIRKPATRICK and J.C.MOTT. (1971) J.Physiol (London) 218: 61.
4. BROUGHTON PIPKIN, F., J.C.MOTT and N.R.C.ROBERTON. (1971) J.Physiol. (London) 218: 385.
5. DAWES, G.S., J.J.HANDLER, and J.C.MOTT. (1957). J.Physiol. (London). 139: 123.
6. DAWES, G.S., and J.C.MOTT. (1959) J.Physiol. (London).145:85
7. DOWNING, S.E. (1960) J.Physiol. (London). 150: 201.
8. KORNER, P.I., (1971) Physiol.Rev. 51: 312.
9. McKENZIE, J.K., M.R.LEE and W.F.COOK. (1967) Nature(London). 215: 542.
10. MOTT, J.C. (1965) J.Physiol.(London). 181, 728.
11. MOTT, J.C. (1966) J.Physiol. (London) 187: 28.
12. MOTT, J.C. (1967) J.Physiol. (London) 191: 131.
13. MOTT, J.C. (1968) J.Physiol. (London) 194: 35.
14. MOTT, J.C. (1968) J.Physiol.(London) 194: 659.
15. MOTT, J.C. (1969) J.Physiol. (London) 202: 25.
16. SCROOP, G.S., and R.D.LOWE. (1968) Nature,(London) 220:1331.

THE INFLUENCE OF BLOOD PRESSURE AND BLOOD FLOW ON THE LOCAL TISSUE Po$_2$ OF THE CAROTID BODY

D.W. Lübbers, H. Acker, H.P. Keller and E. Seidl

Max-Planck-Institut für Arbeitsphysiologie, Dortmund, W. Germany

In the warm-blooded animal systemic functions such as O$_2$ supply, respiration and circulation are monitored and regulated by means of specific sensors which co-ordinate different systemic parts by nervous or hormonal reflexes. In shock the oxygen supply is impaired, so the problem arises whether the sensor of blood oxygen plays an important role in the development or defence of shock.

There are different approaches to this question. Our approach was to find out about the operating principle of the sensor and from this the conditions which may produce respiratory and circulatory reflexes.

The best known O$_2$ sensor is the carotid body (3,7,8,16). There is experimental evidence that its nervous output depends on the Po$_2$ of the arterial blood. Since its blood flow is extremely high (about 2000 to 3000 ml/100g/min and the AVD-O$_2$ small (about 0.2 ml/100 ml) (4,15), it seemed reasonable to assume that the tissue O$_2$ tension of the carotid body should be high - but then it is difficult to predict what kind of stimulus could be responsible for the nervous output. I apologize that in the following I am presenting three hypotheses about the mechanism of the O$_2$ sensor with the data, but I hope this will stimulate the discussion.

By puncturing the carotid body with a Po$_2$ needle electrode we were able to measure the local Po$_2$ inside the carotid body (1). First the electrode reaches the periphery of the carotid body. We were very much surprised that the Po$_2$ in the very periphery was

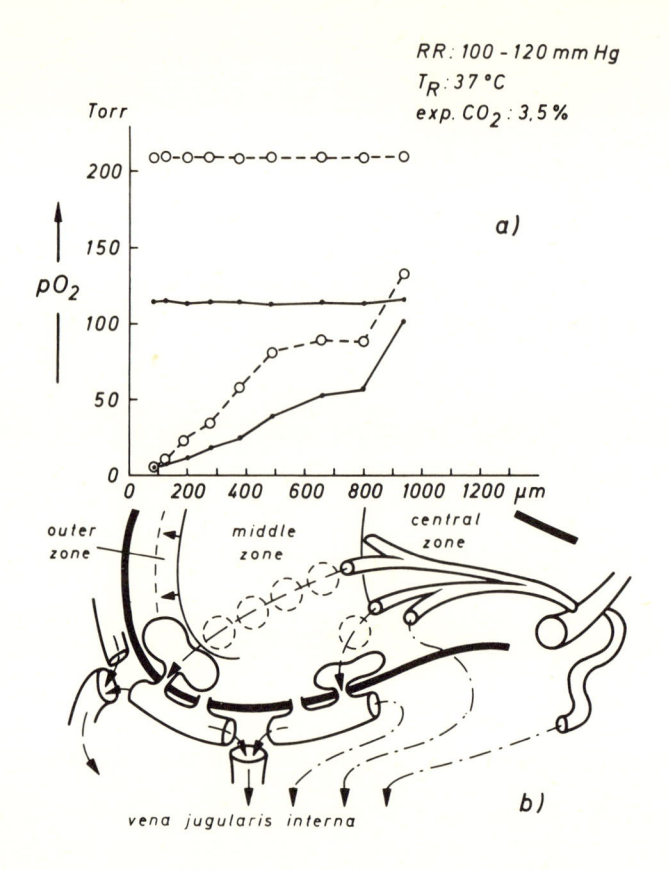

Fig 1 Oxygen Tension Field in the Carotid Body of the Cat (a) and a Schematic Drawing of the Corresponding Tissue (b).

a. (Abscissa: depth of the puncture measured from the surface; ordinate: O_2 tension in Torr; RR: mean arterial blood pressure; T_R: rectal temperature; exp. CO_2: expiratory CO_2; solid line: air respiration; broken line: respiration of 40% O_2 in N_2).

After the respiration of 40% O_2 in N_2 the tissue O_2 tension rises dependent on the depth of the puncture.

b. The drawing shows from the right to the left the inflowing artery with its ramifying parts (= central zone) and an anastomosis which bypasses the tissue. The capillaries of the middle and outer zone have different lengths and contact different numbers of glomoids. The blood is collected in sinusoids and leaves the carotid body entering the outside veins. These form two or more larger veins which flow into the v. jugularis. The dash and dot lines show in which way blood of arterial or almost arterial Po_2 could reach the veins.

close to zero. It could not be influenced by changing the P_{O_2} at the outer surface of the carotid body of the cat. Therefore, in the periphery the carotid body of the cat must have an outer layer of diameter 10 to 50 μm which O_2 cannot permeate. Pushing the electrode into the carotid body the electrode penetrates first the outer zone, then the middle zone and finally the central zone. Fig 1 shows that the tissue P_{O_2} increases continuously towards the central zone. Only in a few cases out of more than 500 punctures did we measure the arterial P_{O_2} which was monitored simultaneously by a P_{O_2} catheter electrode. This is understandable since the organ has rarely more than one or two inflowing arteries. Thus, the direction of the needle is important. The region in which the inflowing artery is found, we call central zone. The arteries ramify forming a capillary network in contact with the glomus cells (middle & outer zone). The capillary blood is collected into sinusoids. One of us (E. Seidl) has shown that the sinusoids form a broad-meshed net which mostly lies peripherally but is also found in the middle zone. It delivers the blood to more than one vein outside the carotid body.

If we change the arterial P_{O_2}, the tissue P_{O_2} changes correspondingly (Fig 2). Towards the centre the absolute changes become considerably greater than in the middle and outer zone (Fig 1). Decreasing arterial P_{O_2} therefore increases the breadth of the low P_{O_2} zone and vice versa (Fig 1, outer zone). Recent experiments of Purves in Bristol and of our group in Dortmund and Münster have shown that under normal conditions the nervous output follows the change of tissue P_{O_2} closely: a decrease of tissue P_{O_2} increases the nervous activity. These data led us to put forward the hypothesis - the first one - about the stimulation mechanism of the carotid body: The glomus cells in the low oxygen zone produce the nervous signals. The nervous output depends on the number of cells which are situated in the low P_{O_2} zone.

To answer the question of how the carotid body produces a P_{O_2} distribution showing a systematic P_{O_2} decrease towards the periphery and a high mixed venous P_{O_2} in the veins outside the carotid body, we will discuss four basically different models:

1) External shunting of blood

In this case the carotid body itself receives only a part of the total flow, the other part is shunted by an anastomosis.

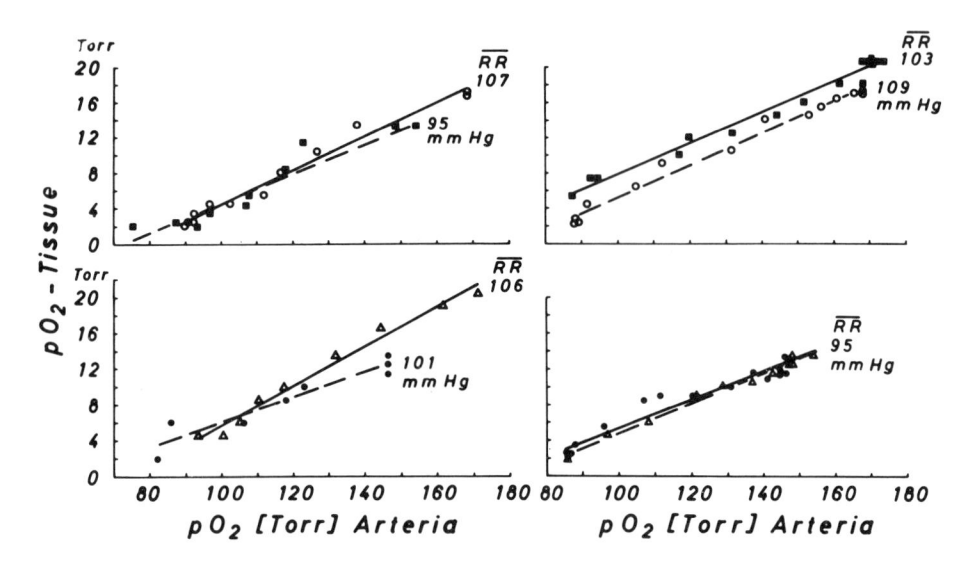

Fig 2 Tissue O$_2$ Response Curves to Changes in Arterial Po$_2$

(Abscissa: arterial Po$_2$ in Torr; ordinate: tissue Po$_2$ in Torr; RR: mean arterial blood pressure). The tissue Po$_2$ measurements were carried out in the same depth of the puncture (450 μm). After the height of the tissue Po$_2$ the measuring point is in the middle zone of the carotid body. A 50 Torr Po$_2$ change of the arterial blood changes the tissue Po$_2$ by about 12 Torr. The curves are obtained in a blood pressure range between 95 and 110 mmHg.

2) Internal shunting of blood

The capillaries inside the carotid body have different lengths which contact correspondingly different numbers of glomus cells. The high flow conductivity of the short capillaries would produce a high flow and a small Po$_2$ decrease. The small flow conductivity of the longer capillaries would produce a small flow and a great Po$_2$ decrease. Systematically increasing capillary lengths from the center to the periphery would result in the measured Po$_2$ decrease in the periphery.

The same situation would be brought about by other reasons for different flow conductivity (asymmetric flow system).

3) O_2 diffusion shunt

The capillaries form a countercurrent system as was found in the kidney (10), so that O_2 diffuses directly from the arterial to the venous part of the capillary.

4) O_2 diffusion barrier

The P_{O_2} distribution is brought about by a rather great diffusion barrier of a slowly respiring tissue.

The last possibility could be experimentally excluded. If the animal respires pure O_2, then the O_2 in the arterial blood and in the tissue of the carotid body changes with the same time constant. A diffusion barrier would cause a considerable delay in the change in the tissue P_{O_2} which was not observed. Therefore, only the three first possibilities remain.

Histologically a true countercurrent, as in the kidney, is not seen. There are regions in which vein and artery are close together, but only over relatively short distances. It is difficult to imagine that this contact could be sufficient to cause the P_{O_2} distribution found. In addition the countercurrent system would imply that almost the total flow perfuses the central, middle and peripheral zones. Therefore, we tried to measure the local blood flow in the different zones of the carotid body. This could be done by measuring the H_2 clearance with palladinized platinum microelectrodes (11,12,13).

After several experiments the best results were given by electrodes which were only 1 to 2 cm long and with the tip bent to an angle of 90°. The platinum wire of the electrode was soldered to a spiral of thin copper-wire close to the upper end of the electrode shaft. This part was covered and insulated by xantopren blue (self-curing catalyzed silicone liquid, Bayer, Leverkusen, Germany), so that the xantopren forms a little plate. After the preparation of the carotid body the field of operation was filled with thermostated saline. In the saline the electrode could float. An electrode suspended in this way follows all the movements of the carotid body without disturbing its circulation.

The small size of the electrode allowed the measurement of the flow in the 3 different zones of the carotid body (9). Fig 3 shows that the flow in the central zone is highest.

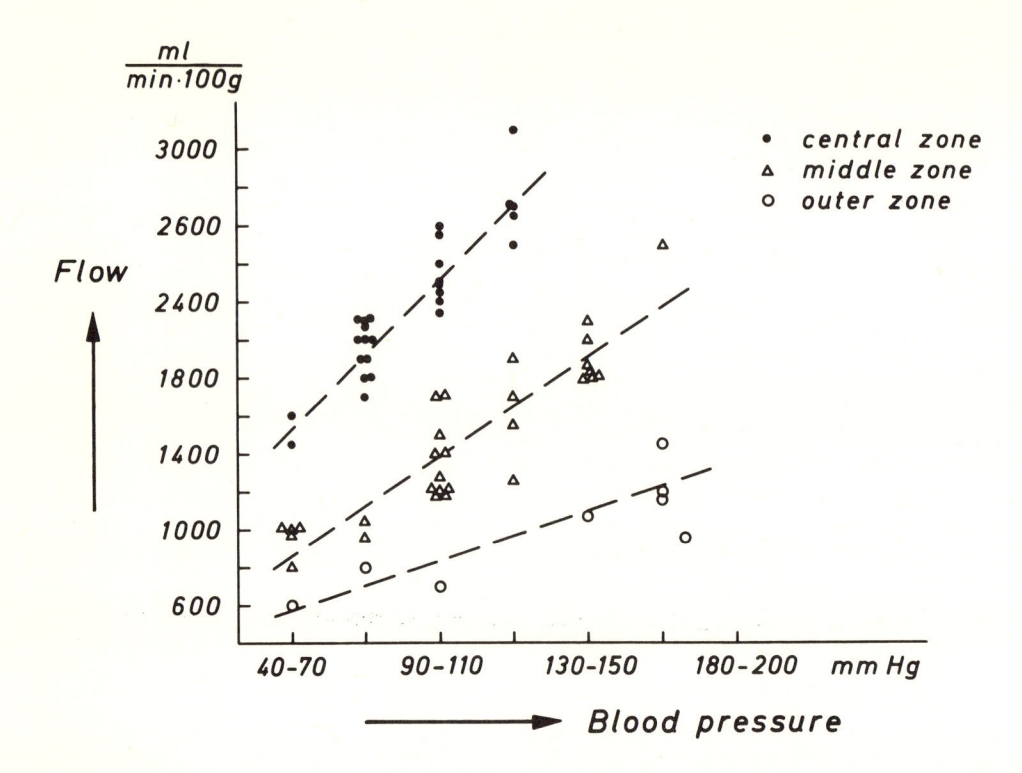

Fig 3 Flow Dependence of the Carotid Body on the Mean Arterial Blood Flow Measured by Microelectrode H_2 Clearance (Cat)

(Ordinate: blood flow in mm/100 g/min; abscissa: mean arterial blood pressure in mm Hg).
The three curves correspond to the flow of the central zone (highest flow), the middle zone (middle flow) and the outer zone (smallest flow).

In the middle zone the flow is about two thirds and in the outer zone one third that of the central flow. The values are taken from different experiments at normal blood Po_2 and Pco_2. Since the positioning of the electrode is very difficult, the scattering of the values was expected.

The relatively small flow in the periphery would contradict the countercurrent model, but since the perfused area is not very well known — the size of the zone increases from the centre to the periphery — the data of the flow/100g/min do not exclude the possibility of an equal <u>total</u> flow. One of us (H.P. Keller) succeeded in puncturing the carotid body with 2 electrodes so that the flow in different parts of the carotid body could be measured at the same time. There are only a few experiments of this type, but they show that it is possible to change outer zone blood flow without changing the central one and vice versa. This speaks strongly against the countercurrent system.

Only the first two possibilities remain. The existence of an anastomosis outside the carotid body described by De Castro (5) could be confirmed in the cat (Fig 1), but our regional flow measurements show that the flow in the central zone corresponds very well to the total flow values directly measured (4,15). We therefore assume as an heuristic hypothesis a combination of the second model with the first one: in this model the short capillaries should be situated in the centre, the longer ones towards the outer zone. The longer capillaries are in contact with more glomus cells. The glomus cells are sometimes gathered in small clusters which we call glomoids (Fig 1). The blood is collected from the capillaries in the sinusoids and leaves the carotid body into veins outside the carotid body. These veins receive blood from short and long capillaries, so it would be a mixed venous blood with a relatively high Po_2. The anastomosis can add also blood of high Po_2.

Such a flow system can be tested by changing the blood pressure (Fig 3). Under our experimental conditions (anaesthetized with Na pentobarbitone (30 mg/kg), paralyzed with gallamine triethiodide (15mg/kg) and surgical preparation of the carotid body with intact nerves) the flow increases almost linearly with increasing blood pressure. This holds for all 3 zones. A range in which blood flow was independent of the blood pressure could not be observed, in agreement with the literature (but 14).

The influence of blood pressure on the tissue Po_2 is different, depending on the way the blood pressure changes are brought about. In order to test the sensitivity of the sensor mechanism, we measured the way in which the tissue Po_2 response depended on the arterial Po_2 (2). In a range of a mean blood pressure of 95 to 110 mmHg the spontaneous changes of blood pressure gave reproducible results (Fig 2): The Po_2 response curves are almost straight lines, the slope of which depends on the distance of the measuring point from the centre and is almost independent of the blood pressure.

Blood pressure changes brought about experimentally by clamping the aorta showed different reactions. Sometimes blood presssure changes alter the tissue Po_2 as expected. Increasing blood pressure is accompanied by increasing tissue Po_2 and vice versa. But also there are often blood pressure changes without any change in tissue Po_2. Such behaviour is even more pronounced, when the blood pressure is changed to different levels with a blood reservoir as a pressure buffer. Here the slope of the tissue Po_2 response curve and its absolute value depends completely on the experimental situation. With stepwise increases in blood pressure above 110 mmHg the slope of the Po_2 response curve usually decreases. But on the return to the same pressure level other Po_2 response curves are always obtained. At the moment we are not able to predict the effect of blood pressure changes on the tissue Po_2 response curve. Knowing that the vessels in the carotid body are amply supplied by nerves, we think that these experiments show that the blood pressure has a strong influence on the local distribution of blood flow in the carotid body by means of nervous reflexes. The outside anastomosis could regulate the blood flow, so that the carotid body receives always the same amount of blood independent of blood pressure, but our flow measurements have not shown such a mechanism.

There remains another fact which is puzzling: If our Po_2 measurements are correct, then low Po_2 zones should have the highest tissue activity, since these cells produce the chemoreceptor signals, but there is no sign of anoxic tissue damage. May I risk another very early hypothesis – as the third one – in order to try to explain this phenomenon and the mechanism of O_2 stimulus: the high blood flow allows an appropriate anaerobic metabolism in the low Po_2 zone. Assuming that a transmitter substance released from cell type I stimulates nerve endings in the cell type II (which could be the so-called chemoreceptor cells of Purves) this reaction would proceed without energy deficiency in the low Po_2 zone. To stop this reaction the transmitter formation or release could be influenced and then the released transmitter destroyed. If we assume that the release can be modulated by reflexes and that destruction is brought about by a Po_2 dependent oxidase which itself can be influenced by intracellular pH or Pco_2, then our model would describe all the known facts about the chemoreceptor activity of the carotid body. The kind of transmitter substance involved (see 6) is an open question, for example monoamineoxidase has been demonstrated histochemically only in cell type II (17). To test the different hypotheses we digested the carotid body of the cat with a solution of collagenase and trypsin (pH 7,2) and succeeded in obtaining clusters of glomus cells type I and cells type II with visible nerve fibres. The growing of tissue cultures from these glomus cells has not so far been possible.

I must apologize for putting forward three hypotheses instead of just presenting our data, but I hope this can stimulate discussion of these phenomena.

REFERENCES

1. Acker, H., Lübbers, D.W. & Purves, M.J. (1971) Pflügers Arch. 329:136.

2. Acker, H. & Lübbers, D.W. In: Oxygen Supply. Theoretical and Practical Aspects of Oxygen Supply and Microcirculation of Tissue ed. by M. Kessler, D.F. Bruley, L.C. Clark Jr., D.W. Lübbers, I.A. Silver, J. Strauss. Urban & Schwarzenberg, München and Berlin University Park Press, Baltimore 1972 in print.

3. Biscoe, T.J. (1971) Physiol. Rev. 51:437.

4. Daly, M. deB., Lambertsen, C.J., Schweitzer, A. (1954) J. Physiol. (London) 125:67.

5. De Castro, F., Rubio, M. In: Arterial Chemoreceptors, ed. by R.W. Torrance Blackwell, Oxford 1968, p. 267.

6. Eyzaguirre, C. & Zapata, P. In: Arterial Chemoreceptors, ed. by R.W. Torrance Blackwell, Oxford 1969, p. 213.

7. Heymans, C. & Neil, E. (1958) Reflexogenic Areas of the Cardiovascular System Churchill, London 1958.

8. Joels, N. & Neil, E. (1963) Brit. Med. Bull. 19:21.

9. Keller, H.P. & Lübbers, D.W. In: Oxygen Supply. Theoretical and Practical Aspects of Oxygen Supply and Microcirculation of Tissue ed. by M. Kessler, D.F. Bruley, L.C. Clark Jr., D.W. Lübbers, I.A. Silver, J. Strauss. Urban & Schwarzenberg, München and Berlin. University Park Press, Baltimore 1972 in print.

10. Leichtweiss, H.-P., Lübbers, D.W., Weiss, Ch., Baumgärtl, H. & Reschke, W. (1969) Pflügers Arch. 309:328.

11. Lübbers, D.W. & Baumgärtl, H. (1967) Pflügers Arch. 294, R 39.

12. Lübbers, D.W. In: Blood Flow through Organs and Tissues ed. by W.H. Bain and A.M. Harper Livingstone, Edinburgh and London 1968, p. 162.

13. Lübbers, D.W., Baumgärtl, H., Fabel, H., Huch, A., Kessler, M., Kunze, K., Riemann, H., Seiler, D. & Schuchhardt, S. In: Oxygen Pressure Recording in Gases, Fluids and Tissues ed. by F. Kreuzer and H. Herzog. Karger, Basel and New York 1969, p. 136.

14. McCloskey, D.I., & Torrance, R.W. (1965) J. Physiol. (London) 179:37.

15. Purves, M.J. (1970) J. Physiol. (London) 209:395.

16. Arterial Chemoreceptors ed. by R.W. Torrance Blackwell, Oxford 1969.

17. Woods, R.I. (1967) Nature 213:1240.

SYMPATHETIC NERVOUS ACTIVITY AFTER HEMORRHAGE

L. Fedina, M. Kollai and A.G.B. Kovách

Experimental Research Department Semmelweis Medical

University, Budapest, Hungary

The role of the sympathetic nervous system in acute hemorrhage and hemorrhagic shock has been extensively investigated in recent years (Chien, 1967). Indirect evidence as well as direct recording of the action potentials in the different pre- and postganglionic sympathetic nerves showed, that there is a hyperactivity of the sympathetic system in hemorrhagic shock (Beck et.al. 1955; Chien, 1964; Corazza et.al.1963/a, 1963/b; Fedina et.al.1965; Floyd et.al.1952; Gernandt et.al.1947; Gootman et.al.1969; 1970; Thämer et.al.1969). Comparing the results obtained by direct recording of action potentials in sympathetic nerves, it can be seen, that there are no real discrepancies between the data of various authors in respect of the sympathetic hyperactivity and its time course in acute hemorrhage and hemorrhagic shock. Relatively little is known however, about the control of the sympathetic nervous activity in hemorrhagic shock.

The aim of this study to find out how the sympathetic nervous activity can be influenced by:
1.) stimulation of the afferent nerves,
2.) norepinephrine-induced blood pressure elevation, and
3.) asphyxia.

Methods. The experiments were performed on cats anesthetized with chloralose-urethan mixture (50 mg/kg and 200 mg/kg, respectively), immobilized with Flaxedil, and artificially ventillated. Rectal temperature was maintained between 37.0 and 39.0°C. For eliciting standard hemorrhagic shock we used the constant pressure technique of Wiggers as modified by Lamson et.al.(1945). Heparin was given to prevent blood coagulation. For recording of action potentials in sympathetic nerves the left renal sympathetic postganglionic nerve and the left cardiac sympathetic postganglionic nerve were dissected, cut peripherally and desheathed. Recordings from each nerve were taken either through bipolar hook electrodes (stainless steel or platinum), or by means of a small chamber electrode, applying conventional electrophysiological techniques. Arterial blood pressure was recorded simultaneously. The left sciatic nerve was stimulated with single square wave impulses, (see parameters in the Figures). Norepinephrine was injected intravenously in a dose of 3-30 µg/kg.

Results and discussion. In order to influence the sympathetic discharge we have used procedures which had well known effects in this respect. The sciatic nerve stimulation evokes through spinal and medullary structures a reflex response and a reflex inhibition of the sympathetic activity. Norepinephrine-induced blood pressure increase leads to a complete inhibition of the sympathetic activity through the baroreceptors and medullary structures. Asphyxia produces a marked increase in sympathetic activity by direct action on the sympathetic neurones in the medulla and spinal cord.

Fig. 1, as well as the others show the simultaneous recording of arterial blood pressure and the action potentials in a sympathetic postganglionic nerve; in Fig. 1 the activity of the cardiac sympathetic nerve is registered. The records show the events before and during hypovolemic hypotension and after retransfusion. The trace A is the control. Blood pressure was 120 mmHg. There was very little background activity in the cardiac nerve. Because of this very low activity only the reflex response evoked by sciatic nerve stimulation (marked by arrows), but not the reflex inhibition can be seen. The traces B and C show the events during the first and second bleeding period, when blood pressure was 60 and 35 mmHg, respectively. The activity of the cardiac sympathetic nerve increased markedly, and the stimulation of the sciatic nerve evoked a reflex response and a reflex inhibition in both the first and second bleeding period. The trace D shows the activity 20 minutes after retransfusion. Blood pressure was 110 mmHg, the sympathetic activity returned nearly to the control level. During posttransfusional deterioration of the circulation the fall in blood pressure led to an increased sympathetic activity in the cardiac nerve. We can see on the trace E that 105 minutes after retransfusion, when blood pressure had dropped to 60 mmHg, the sympathetic activity increased again, and the sciatic nerve stimulation evoked reflex response and inhibition. In the preterminal phase of hemorrhagic shock (trace F), when blood pressure was as low as 20 mmHg, the activity of the cardiac sympathetic nerve diminished but it was still somewhat higher, than the control. At the same time sciatic nerve stimulation could not evoke any reflex effects. It should be mentioned, that this type of reaction could be observed both in the cardiac and renal sympathetic nerves.

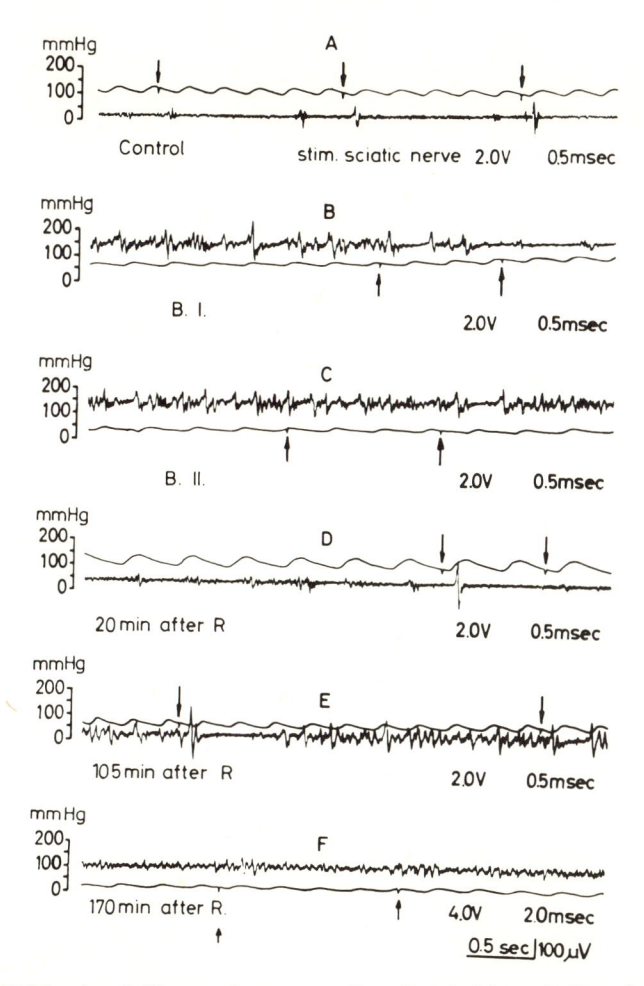

Fig. 1. The Effect of Hemorrhage on the Activity of the Cardiac Sympathetic Nerves.

The records in Fig. 2 demonstrate a somewhat different pattern of the sympathetic activity, that can be observed in some cases. In this case discharge in the renal sympathetic nerve was recorded before and during hypovolemic hypotension and after retransfusion. We can see the increased activity during the first and second bleeding period (traces B and C). However at the end of the second bleeding period the activity of the renal sympathetic nerve decreased despite the low arterial pressure (30 mmHg). At the same time there was an other conspicuous change, viz. the inhibition previously evoked by sciatic nerve stimulation disappeared (trace D). The absence of the reflex inhibition and the decline of the sympathetic activity may indicate the damage of the medullary sympathetic structures. However, this was reversible, because reflex

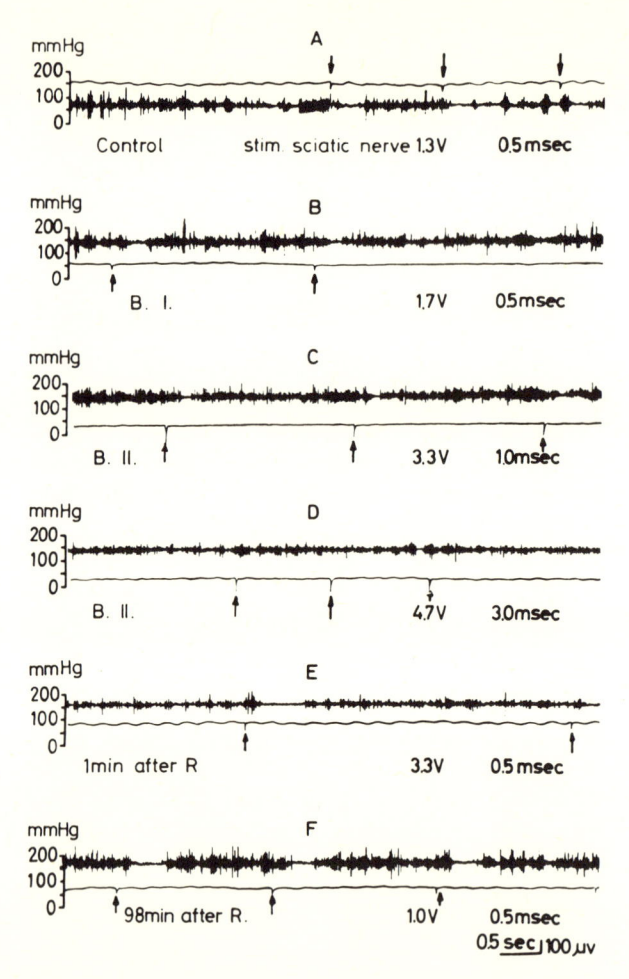

Fig. 2. The Effect of Hemorrhage on the Activity of the Renal Sympathetic Nerves.

inhibition reappeared after retransfusion, as it can be seen in the trace E. Further course of the activity resembled that seen in Figs 1 and 3.

In Fig. 3 we can follow the events during hypovolemic hypotension and after retransfusion. In this case the activity of the cardiac sympathetic nerve practically disappeared in the second bleeding period (trace B). After the retransfusion the activity recovered and there was a reflex inhibition evoked by the sciatic nerve stimulation too (trace C). In this animal the circulation deteriorated rather rapidly after the retransfusion. Together with the fall of the blood pressure there was a gradual increase

SYMPATHETIC NERVOUS ACTIVITY AFTER HEMMORRHAGE 477

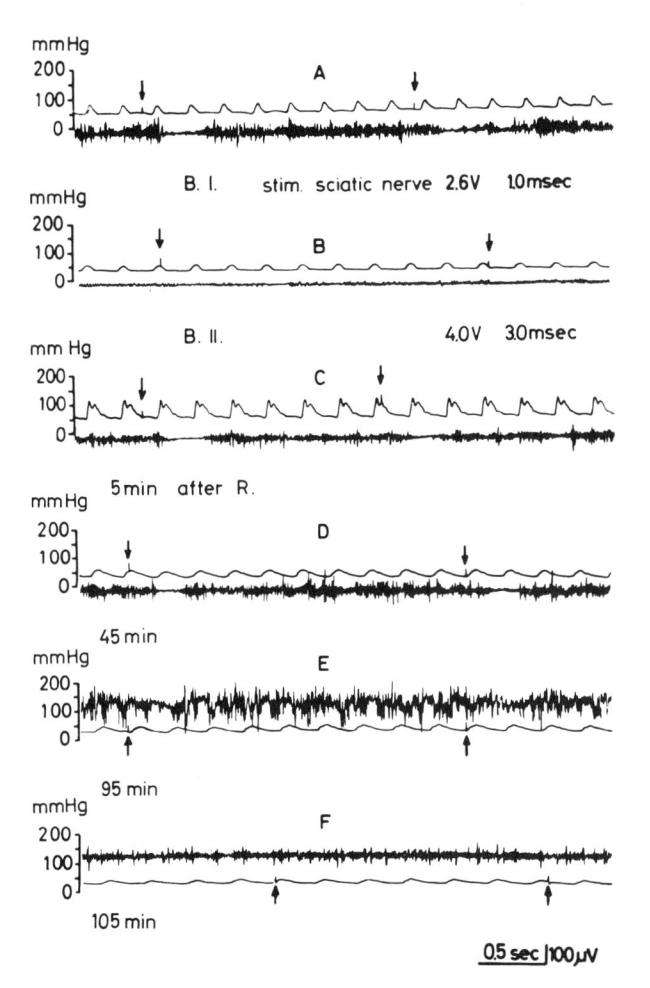

Fig. 3. Changes in the Activity of the Cardiac Sympathetic Nerves during Hypovolemic Hypotension and Posttransfusional Deterioration of the Circulation.

in the activity of the cardiac sympathetic nerve. The highest activity was recorded at a blood pressure level of 30-40 mmHg (trace E). At this time the reflex inhibition evoked by sciatic nerve stimulation disappeared. Ten minutes later, i.e. 105 minutes after retransfusion, the sympathetic activity decreased indicating the damage of the bulbar structures on account of ischemia (trace F). Now the question arises whether this damage is reversible or not.

The recordings in Fig. 4 illustrate the beneficial effect of saline infusion in the posttransfusional period. When comparing the first and second panel of recordings (A and B) it becomes evident that simultaneously with the drop in blood pressure the

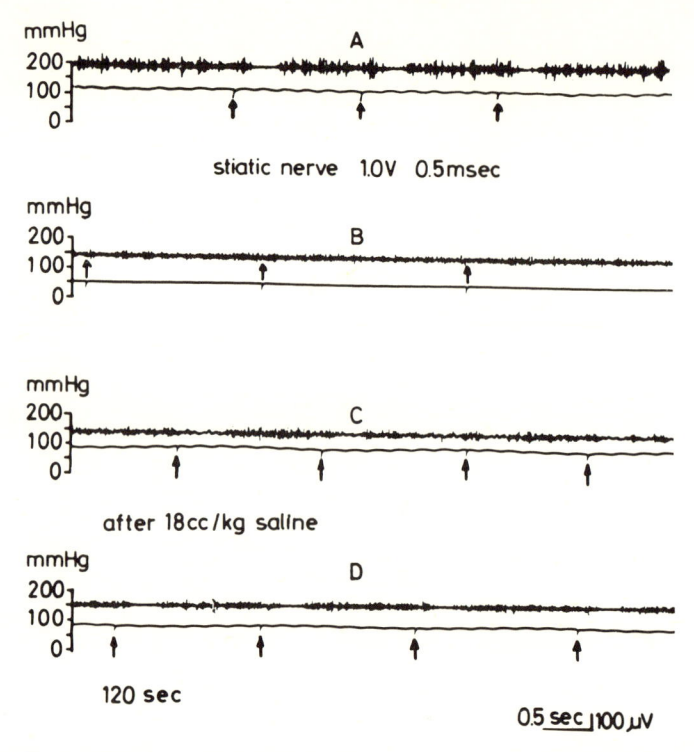

Fig. 4. The Effect of Saline Infusion in the Posttransfusional Period.

activity in the renal sympathetic nerve decreased, and reflex inhibition disappeared. In order to get a rise in blood pressure 18 ml/kg saline was infused (trace C). Blood pressure increased from 50 to 80 mmHg, and an increase in sympathetic activity and a reappearence of the reflex inhibition could be noted. This recovery was only temporary, and later, in the terminal phase of the hemorrhagic shock the saline infusion was not effective at all (trace D). We might ask whether is it possible to inhibit or increase the sympathetic activity by some other means? The recordings in Fig.5 illustrate the effect of the norepinephrine-induced blood pressure elevation and the asphyxia in the terminal phase of hemorrhagic shock. In the trace A blood pressure was 25 mmHg. Following the injection of 35 µg/kg norepinephrine blood pressure rose to 130 mmHg (trace B and C). We can see that simultaneously with the rise in blood pressure, there was a marked but not complete inhibition of the sympathetic activity. One minute later, when blood pressure dropped to 30 mmHg, the activity increased again, but it did not reach the level prior to norepinephrine injection (trace D). At the same time sciatic nerve stimulation was not effective at all. To answer the question, whether it is possible or not to increase the sympathetic activity in this late stage of hemorrhagic shock,

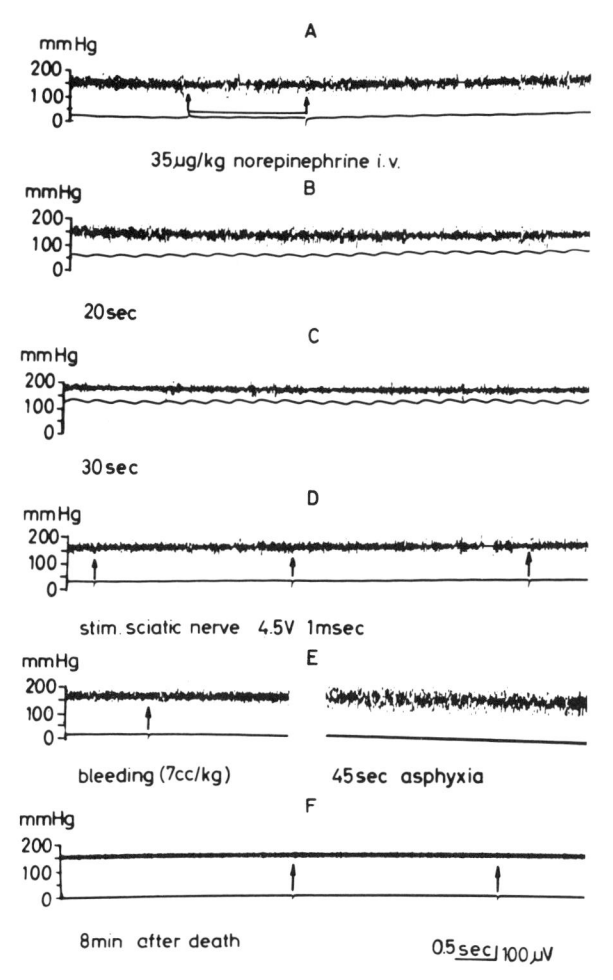

Fig. 5. The Effect of Norepinephrine and Asphyxia in the Terminal Phase of Hemorrhagic Shock.

first we removed 7 ml/kg blood from the animal in order to cause a more severe hypoxia. The record in the first half of the panel E shows, that the activity in the renal sympathetic nerve increased together with the fall of blood pressure. In the other part of this panel the effect of the asphyxia is shown. We can observe a further increase in the activity in the renal sympathetic nerve.

Summary. In our opinion there is a roughly inverse correlation between blood pressure and sympathetic nervous activity during hemorrhage and hemorrhagic shock except for the very late phase, when the activity usually declines, but it persists at a lower level until the death of the animal. In some cases there was a similar decline in the activity already at the end of the second bleeding period. Besides these facts a disappearence of the reflex inhibition evoked by sciatic stimulation could be observed indicating the damage of the medullary sympathetic structures. On the other hand, the activity could be partially inhibited by pressure elevation, or it could be increased by the hypoxia and asphyxia even in the terminal phase of the hemorrhagic shock.

Our results suggest that the control of the sympathetic outflow may be damaged after severe hemorrhage, and this damage may play a role in the development of the hemorrhagic shock.

REFERENCES

Beck, L. & Dontas, A.S. (1955) Fed. Proc. 14:318.

Chien, S. (1964) Amer. J. Physiol. 206:21.

Chien, S. (1967) Physiol. Rev. 47:214.

Corazza, R., Manfredi, L. & Raschi, F. (1963/a). Minerva med. 54:1691.

Corazza, R., Pinotti, O. & Raschi, F. (1963/b). Arch. Sci. Biol. (Bologna). 47:275.

Fedina, L., Kovách, A.G.B. & Vanik, M. (1965) Acta physiol. Acad. Sci. hung. Suppl. 26:35.

Floyd, N.F. & Neil, E. (1952) Arch. int. Pharmacodyn. 91:230.

Gernandt, B., Liljestrand, G. & Zotterman, Y. (1947) Acta physiol. scand. 11:230.

Gootman, P.M. & Cohen, M.I. (1969) Feder. Proc. 28:330.

Gootman, P.M. & Cohen, M.I. (1970) Amer. J. Physiol. 219:897.

Lamson, P.D. & De Turk, W.E. (1945) J. Pharmacol. exp. Ther. 83:250.

Thämer, V., Weidinger, H. & Kirchner, F. (1969) Z. Kreisl.-Forsch. 58:472.

HYPOTHALAMIC AND CORTICAL EVOKED POTENTIALS IN HEMORRHAGIC SHOCK

E. Dóra, A. G. B. Kovách, I. Nyáry

Experimental Research Department, Semmelweis Medical

University, Budapest, Hungary

In hemorrhagic hypotension the cortical electrical activity ceases almost entirely and after reinfusion the starting activity fails to return (Kovách and Fonyó, 1960). In the terminal phase of traumatic shock the energy-rich phosphate content of the brain decreases in rats (Kovách et.al. 1952, 1959). A significant hypercapnia was found in the hypothalamus, in the frontal cortex and in the sagittal sinus blood without marked changes in the thalamus, medulla oblongata and pons (Makláry and Kovách, 1968). Perfusion of the isolated head in ischemic and hemorrhagic shock or 50 ml/min intracarotid blood transfusion in hemorrhagic shock increased markedly the survival time of dogs (Kovách et.al., 1958).

In our present studies the neuronal function of the primary sensory cortex and hypothalamic ventromedial nucleus was investigated. Hypothalamic and cortical evoked potentials and spontaneous activity was studied during the development of hemorrhagic shock.

Materials and methods. Twenty-five adult cats of both sexes were used. The animals were anesthetized by intravenous chloralose (60-80 mg/kg immobilized by flaxedil (3-5 mg/kg) and artificially ventilated. Body temperature was kept between 36-38 C°. Heparin (1000 IU/kg) was administered as anticoagulant. Hemorrhagic shock was induced by a modified Lamson technique (Engelking and Willig, 1958). Arterial blood pressure was reduced to 50-60 mmHg and maintained at this level for 90 min, and subsequently at 30-40 mmHg for another 90 min. Following the hypovolemic period the shed blood was reinfused.

Hypothalamic and cortical evoked potentials were obtained by stimulating the sciatic and splanchnic nerves and the basal nucleus of amygdala. The responses were recorded unipolarly. The recording electrodes were made of glass insulated iridium-platinum wires (external diameter 0,25-0,35 mm). For recording cortical potentials silver ball was used. These electrodes served also to record spontaneous cortical activities. Deep electrodes were placed according to Szentágothai's stereotaxic atlas (1957). Arterial blood pressure was monitored continuously with a mercury manometer, the cortical and hypothalamic potentials with the usual electrophysiological methods (Dóra et.al., 1972). The bleeding volume was recorded every ten minutes during hemorrhagic hypotension.

The animals were divided into two groups depending on the severity of hemorrhagic shock; 15 cats suffered a severe, irreversible shock (Group A), in 8 animals the shock was of moderate severity (Group B).

Results. The majority of cats in Group A died within two hours following the reinfusion. The maximal bleeding volume was 17,8 ml/kg and before reinfusion 5,4 ml/kg blood remained in the reservoir. In Group B the maximal bleeding volume was 23 ml/kg and before reinfusion it was still 20 ml/kg. The cats in Group B were all alive two hours after reinfusion. The blood pressure declined slightly from the mean reinfusion value of 101 mmHg to 84 mmHg.

Hypothalamic and Cortical Spontaneous Activity and Evoked Potentials in Group A. A typical experiment is demonstrated in Figure 1. Thirty minutes after bleeding to 35 mmHg arterial blood pressure a weak desynchronization both in the hypothalamic and cortical spontaneous activities was observed. Rhythmic convulsive firing was in some cases later.

At the end of the hypovolemic period in the majority of the experiments the activity was reduced considerably or it disappeared totally and ceased completely after reinfusion.

In Figure 2, in the 30th minute of the second bleeding period hypothalamic and cortical responses could not be evoked by sciatic nerve stimulation. At the same time splanchnic and amygdala stimulation was effective both on the hypothalamus and on the cortex, but the hypothalamic responses differed considerably from the control curves concerning their shape. In the 90th minute only the amygdala stimulation was effective in evoking responses in the sensory cortex. After the reinfusion the evoked responses disappeared.

Hypothalamic and cortical spontaneous activity and evoked potentials in Group B. As it can be seen on Figure 3, 60 minutes after starting bleeding the frequency and amplitude of hypothala-

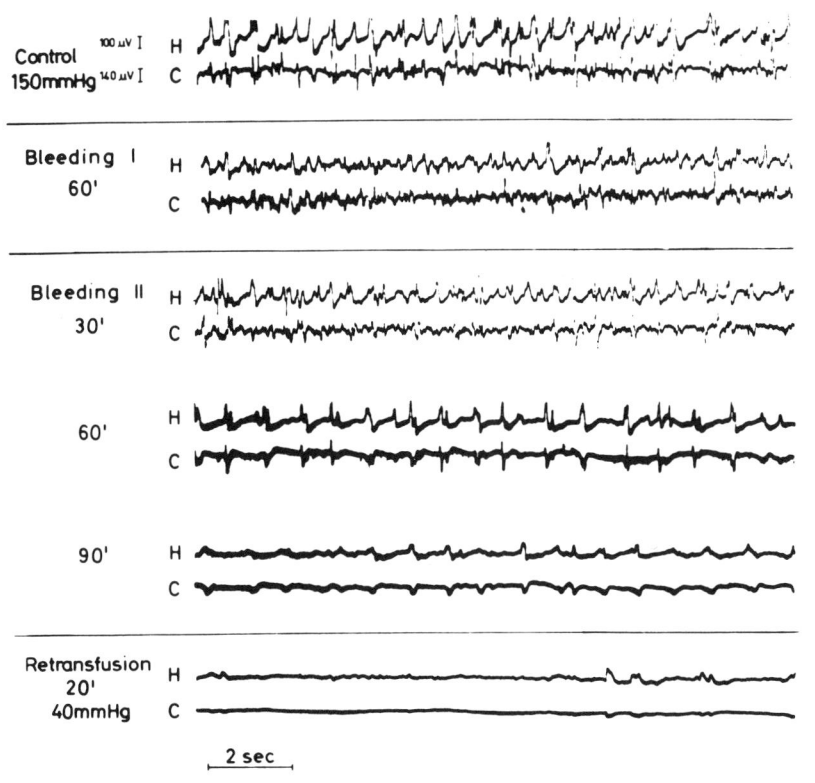

Fig.1. Hypothalamic (H) and cortical (C) spontaneous activity before, during hemorrhagic hypotension and after reinfusion in one single experiment of Group A.

mic and cortical spontaneous activity increased and the activity practically remained at this level until the end of the hemorrhage. Four hours following reinfusion definite synchronization could be observed.

Evoked potentials of a typical experiment of Group B is illustrated in Figure 4. In the 60th minute of the first bleeding period the amplitude of the potentials evoked by sciatic nerve stimulation decreased. In the 90th minute of the second bleeding period they almost disappeared, but after reinfusion the evoked potentials reappeared again. The hypothalamic and cortical responses evoked by splanchnic nerve and amygdala stimulation were only moderately influenced by the hemorrhage.

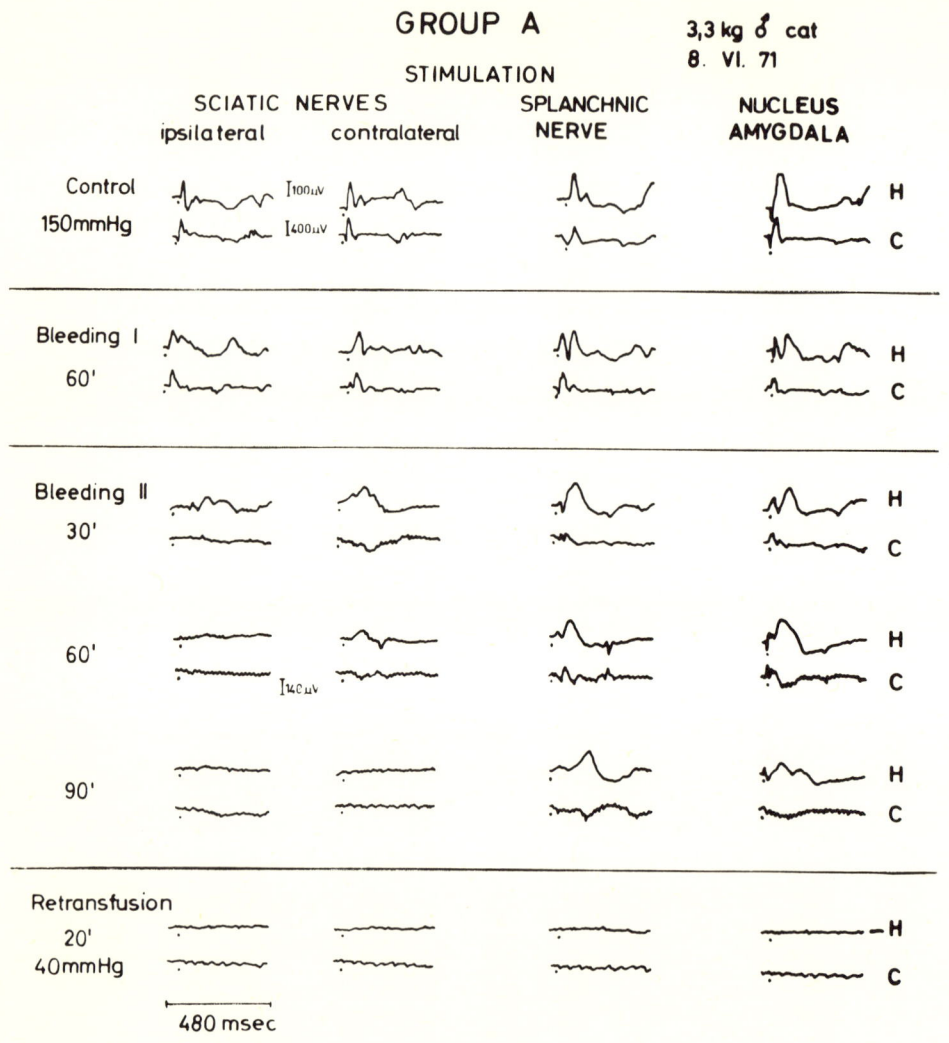

Fig. 2. Hypothalamic (H) and cortical (C) potentials evoked by sciatic and splanchnic nerve and amygdala stimulation before, during hemorrhagic hypotension and after reinfusion in one single experiment of Group A.

Discussion. In accordance with previous experiments (Kovách and Fonyó, 1960) the spontaneous electrical activity of hypothalamus and sensory cortex disappeared in the terminal phase of hypovolemic hypotension and failed to return after reinfusion in irreversible hemorrhagic shock. Before the disappearance of spontaneous electrical activity the amplitude of hypothalamic and cortical potentials evoked by sciatic stimuli decreased or in the majority of the experiments the evoked potentials vanished en-

GROUP B   2 kg ♂ cat
EEG   9. VI. 71

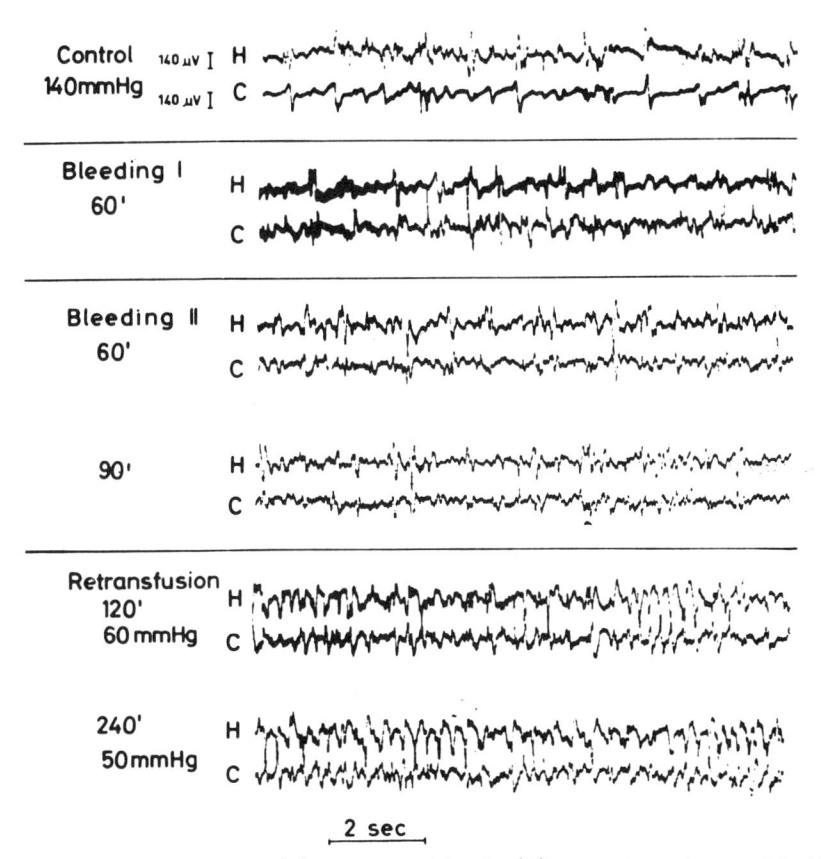

Fig. 3. Hypothalamic (H) and cortical (C) spontaneous activity before, during hemorrhagic hypotension and after reinfusion in one single experiment of Group B.

tirely. These changes ensued mostly between the end of first and beginning of second bleeding period. Brierley et.al. (1969) found that holding the arterial blood pressure at 25 mmHg the disappearance of somatosensory cortical evoked potentials was preceded by the escape of cortical spontaneous activity. To explain the difference between their and our results in this respect it is suggested that in our experiments the somatosensory stimuli could not reach the hypothalamus and sensory cortex because the neurones in the relay stations (nucleus gracilis, VPL thalami) were affected by the hemorrhage. This idea is supported by our

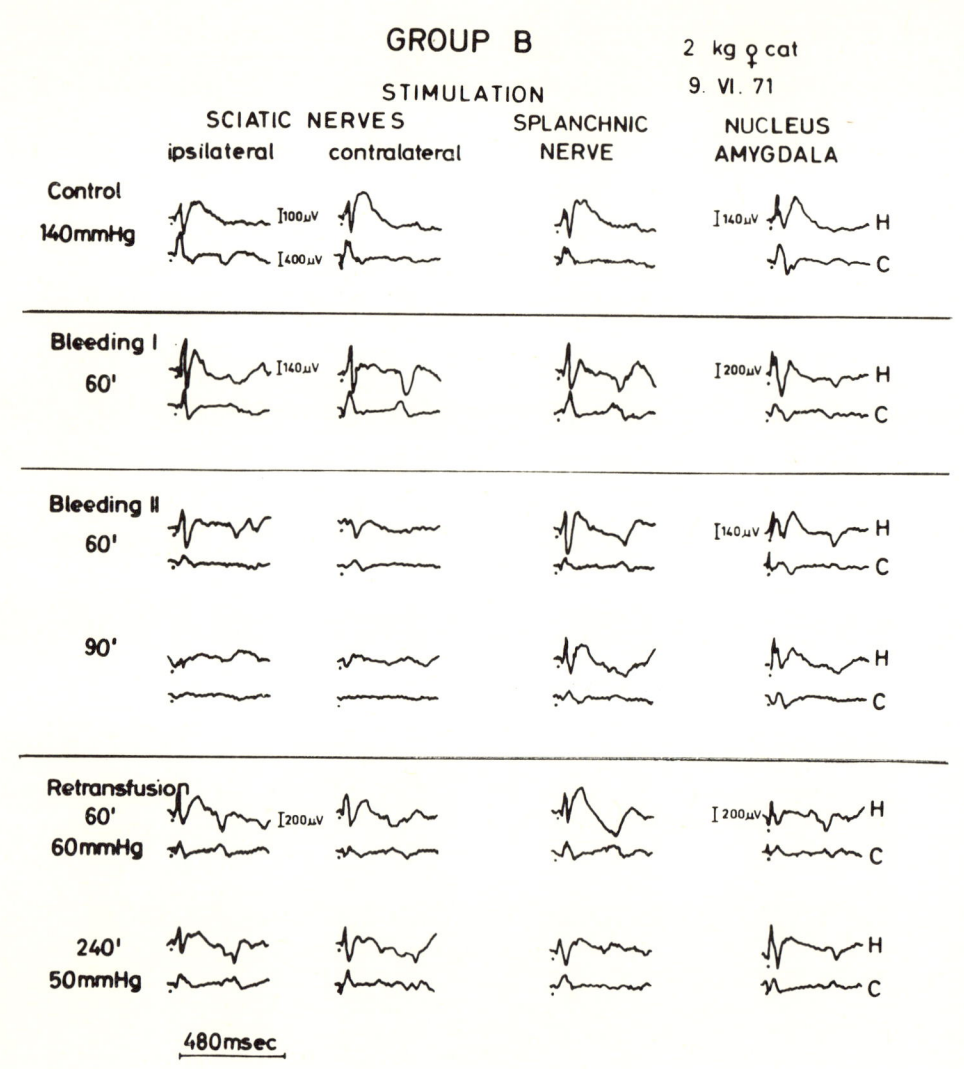

Fig 4. Hypothalamic (H) and cortical (C) evoked potentials produced by sciatic nerve, splanchnic nerve and amygdala stimulation before, during hemorrhagic hypotension and after reinfusion in one single experiment of Group B.

findings that at the time when somatosensory cortical and hypothalamic evoked potentials disappeared it was yet possible to get electrical responses by amygdala and splanchnic nerve stimulation.

Comparing the Group A and B considerable differences could be demonstrated in somatosensory hypothalamic and cortical evoked potentials between the irreversible and reversible group. On the basis of our experiments we suppose that the damage of the function is uneven in different brain regions during the development of hemorrhagic shock. It seems probable that the somatosensory system deteriorated earlier than the hypothalamus in hypovolemic shock.

Summary. In severe irreversible shock (Group A) hypothalamic and cortical potentials evoked by sciatic nerve stimulation decreased or disappeared, but considerable alterations were also obtained as regards the shape and amplitude of potentials evoked by splanchnic nerve and amygdala stimulation. All these changes ensued in the early phase of hemorrhage and proved to be irreversible.

In less severe shock (Group B) the changes were reversible which developed in the potentials evoked by sciatic nerve, splanchnic nerve and amygdala stimulation.

It is finally concluded that the electrical responsiveness of the ventromedial hypothalamic nucleus and the primary sensory cortex is significantly altered and depressed in irreversible hemorrhagic shock.

REFERENCES

Brierley, J.B., Brown, A.W., Excell, B.J. & Meldrum, B.S. (1969) Brain Res. 13:68.

Dóra, E., Nyáry, I., Kovách, A.G.B., Strausz, J. & Tulassay, T. (1972) Acta physiol. Acad. Sci. hung (submitted).

Engelking, R. & Willig, F. (1958) Pflugers Arch. ges. Physiol. 267:306.

Kovách, A.G.B., Bagdy, D., Balázs, R., Antoni, F., Gergely, J., Menyhárt, J., Irányi, M. & Kovách, E. (1952) Acta physiol. Acad. Sci. hung. 3:331.

Kovách, A.G.B., Fonyó, A. & Kovách, E. (1959) Acta physiol. Acad. Sci. hung. 16:157.

Kovách, A.G.B. & Fonyó, A. (1960) In: The Biochemical Response to Injury. Ed. H.B. Stoner and C.J. Threlfall. p.129. Oxford, Blackwell.

Kovách, A.G.B., Rőheim, P.S., Irányi, M., Kiss, S. & Antal, J. (1958) Acta physiol. Acad. Sci. hung. 14:231.

Maklári, E. & Kovách, A.G.B. (1968) Acta med. Acad. Sci. hung 25:13.

Szentágothai, J. (1957) In: A kisérleti orvostudomány vizsgáló módszerei, 3. Ed. A.G.B. Kovách, p. 65. Akadémiai Könyvkiadó, Budapest.

CENTRAL NERVOUS MECHANISMS IN RESISTANCE AND ADRENOCORTICAL REACTIONS TO TRAUMA

K. Murgaš, S. Németh and M. Vigaš

Institute of Experimental Endocrinology, Slovak Academy of Sciences, Bratislava, Czechoslovakia

In our laboratory, we are engaged in work on adaptation to trauma. For this purpose, Noble-Collip drum trauma (11) is one of the most convenient models.

We are using round, hollow drums 38 cm in diameter and 22 cm in depth, rotating at 60 revolutions per minute. Two rats with their two forelimbs and two hindlimbs tied together, are placed into these drums which are fitted with two opposing, fixed shelves. During rotation, the nearest of them raises the animals to a certain height before letting them tumble down when they are seized and raised by the second shelf and then tumbled down again, the whole process being repeated over and over again. A certain number of revolutions causes shock and eventually death to the animals. The critical number varies between 600 and 800, and without special protective treatment none of the animals survive more than 800 revolutions, shock producing circulatory failure being the explanation for this phenomenon(3). In our metabolic experiments, only 400 revolutions were applied which the animals survived without exception.

Regarding the nature of Noble-Collip drum trauma, the mechanical energy applied to the body of the animals during tumbling was considered to be the main factor causing injury (2). Some authors have regarded the procedure as a rather gruesome one which should not be used without anaesthesia. This may be true for the so-called "survival" experiments in which resistance against trauma is measured by counting the number of animals succumbing to a given dose after a given time. However, doses not exceeding 400-500 revolutions, even without anaesthesia, cause no substantial harm to the animals and are considered by us as a very suitable model for studying responses to stress. Regarding the "survival" experiments, some of them are simply impossible to perform under anaesthesia without loss of important information. However, their purpose should be carefully reconsidered again and again and the number of experiments restricted to the minimum. Moreover, we would like to show that physical injury may not be the main cause of death in traumatized rats and to point towards a rather active role played by the animals in creating their "destiny".

Our experiments were performed on adult male Wistar rats, which were kept at 24°C and fed a standard diet. Some groups called "fasted" were fasted 18 hr before the experiments.

I. Some metabolic responses of injured rats and the effect of medullectomy and adrenergic blocking agents

In fasted animals, immediately after injury (400 revolutions) glycogenolysis in liver and hypoglycaemia was observed, as well as a net increase in the blood pyruvic and lactic acid concentrations (1,12). This response of both acids remains unchanged in spite of a previous treatment with Dihydroergotamine (SPOFA, Prague), an α-blocking agent (100 µg/rat i.p. 1 hr before the experiments), Propanolol (INDERAL), a β-blocker (0,4 mg/0,2 ml/rat s.c. 1 hr before the experiments) and medullectomy performed more than 2 months before our experiments, or medullectomy combined with the depletion of catecholamine stores by treatment with reserpine (RESERPIN, GIULINI; 0,5 mg/1 ml/kg i.p. daily, 2 days before experimentation, always at 3 p.m.). We concluded from these results that catecholamine action could hardly be the cause for the observed increases in the blood pyruvate and lactic acid concentration (13). Instead, muscle work performed during an attempt to escape from the drum was proposed by us as their main cause. Moreover, the increased resistance of rats after adrenergic blockade (7,10) seems to be independent of changes in the blood pyruvic and lactic acids responses. Muscle work as the main cause of the post-traumatic increase in both acids also seems very probable in our experiments on rats subjected to trauma under anaesthesia.

II. Some metabolic responses of animals injured under Pentobarbital anaesthesia

In these animals, pentobarbital (SPOFA, Prague; 0,1 ml/100 g b.w. of a 3,3% solution in saline i.p.) was given about 5 min before injury, and the same amount of saline was given to the control group. In the anaesthetized rats, not only was the hypoglycaemia response to injury completely absent, but the post-traumatic hyperpyruvatemia and hyperlactacidemia were also substantially diminished. Moreover, 17 out of 18 such animals survived a dose of 900 revolutions whereas nearly 90% controls succumbed to it (14). From these experiments, our supposition regarding muscle work as the main factor evoking the increase in both acids, became even more probable and the opinion regarding the mechanical aetiology of death after Noble-Collip drum trauma rather uncertain. Unpublished experiments on rats treated with succinylcholiniodide (SPOFA, Prague; 0,1 mg/100 g b.w. i.p. before injury) proved unsuccessful in this respect, because doses sufficient for full muscle relaxation depressed the respiration of the animals (unpublished results). Interestingly enough, the post-traumatic increases in plasma corticosterone concentration were the same in both non-anaesthetized and anaesthetized animals.

III. Central nervous influences upon survival and plasma corticosterone levels

It seems that during injury extra muscle work is done because the animals are trying to escape from a new, unexpected, violent and evidently dangerous situation. Muscle work may be triggered by signals coming from the central nervous system. More than 10 years ago a paper was published in which the author (6) succeeded in dividing rats into animals with a low and those with a high central nervous system excitability. Animals of the first group showed increased resistance to injury compared with those in the second.

In our experiments, a somewhat different approach was chosen. Their aim was to get the animals well acquainted with the environment in which they would later be exposed to stress stimuli. Rats were put daily, for 30 min, into Noble-Collip drums, without turning them. After 20 days, these animals were subjected true Noble-Collip drum trauma which killed 60% of control animals. From the 15 tested animals previously "oriented" in the drum only one succumbed after 24 hr (5) and, in parallel experiments, such animals showed substantially smaller increases in their plasma corticosterone concentrations, than untrained controls (9).

Work performed on animals with electrocoagulative lesions or after electric or chemical stimulation of various brain structures showed that such procedures applied to some parts of the limbic system led to specific changes in aggressivity or irritability (4). After lesions of the septal area decreased resistance to trauma was observed, whereas lesions of the nucleus amygdalae markedly increased resistance, as compared with sham-operated controls. In these experiments as in those performed on anaesthetized rats, we failed to find any substantial difference in the response of the corticosterone concentration in plasma or adrenal homogenates.

In conclusion, it would seem that by accustoming the rat to the apparatus or by damaging central nervous structures alterations in the general level of non-specific excitability of the central nervous system were achieved. The consequence may be a more economical behaviour of the animals in the rotating drum.

Animals subjected to lethal doses of Noble-Collip drum trauma are dying from shock. Portal hypertension under the influence of increased catecholamine levels with consequent mesenteric pooling may be the underlying event (3). Central nervous influences may be exerting their action through changes in catecholamine secretion (8) or action, or by affecting the behaviour of the animals. From our experiments on anaesthetized rats at least, less muscle work seems to be beneficial by inhibiting the hypoglycaemic response, and the increase in the blood lactic acid level. Both hypoglycaemia and hyperlactacidemia, are adverse events and decrease resistance to injury.

SUMMARY

Resistance against Noble-Collip drum trauma depends not only on the mechanical energy applied to the animals, but is influenced by "psychic" factors and the active muscle work done in an attempt to escape from the unexpected and violent situation as well.

Rats subjected to Noble-Collip drum trauma under Pentobarbital anaesthesia showed substantially higher survival rates and lower hyperlactacidemia than unanaesthetized controls. Moreover, rats which were put daily into the drums without turning them, also showed a higher resistance and the post-traumatic increase in the plasma corticosterone concentration in these animals was less pronounced than that in previously not exposed animals.

Finally, changed resistance was also found in rats with lesions placed in their central nervous system; those destroying their nucleus amygdalae increased whereas those placed in the septal region decreased it.

The consequence of these interventions to the experimental situation or of certain nervous system lesions may be a more eceonomical behaviour of the animals in the rotating drum.

REFERENCES

Háva, O., Mráz, M. & Triner, L. (1959) Naunyn-Schmiedebergs Arch. Exp. Pathol. Pharmacol., 236, 81.
Hrůza, Z. & Poupa, O. (1964) In: Adaptation to Environment. Washington, D.C.: Amer. Physiol. Soc., p. 939.

Hrůza, Z. & Zweifach, B.W. (1969) J. Trauma, 9, 430.
Igič, R., Stern, P. & Basagič, E. (1970) Neuropharmacology, 9, 73.
Jonec, V. & Murgaš, K. (1969) Federation Proc., 28, 987.
Lát, J. (1958) Čs. fysiol., 7, 499.
Levy, E.Z., North, W.C. & Wells, J.A. (1954) J. Pharmacol. Exp. Ther., 112, 151.
Murgaš, K. & Kvetňanský, R. (1971) In: Proc. of the II. Internat. Meeting of Psychoneuroendocrinol. Soc., Budapest. (in press).
Murgaš, K., Dobrakovová, M. & Czako, M. Čs. fysiol., (in press).
Németh, S. & Vigaš, M. (1968) Endocrinologia Experimentalis, 2, 39.
Noble, R.L. & Collip, J.B. (1942) Quart. J. Exp. Physiol., 31, 187.
Vigaš, M. & Németh, S. (1968) Endocrinologia Experimentalis, 2, 91.
Vigaš, M., Németh, S. & Magdolenova, A. (1970) Physiol. Bohemoslov. 19, 354.
Vigaš, M. & Németh, S. Physiol. Bohemoslov. (in press).

THERMOREGULATION AFTER INJURY

H. B. Stoner

Experimental Pathology of Trauma Section, MRC Toxicology
Unit, Medical Research Council Laboratories
Woodmansterne Road, Carshalton, Surrey, England

It is now generally realized that the small mammal injured by limb ischaemia or scalding has difficulty in thermoregulation. The simplest evidence of this is the fact that while the core temperature does not fall when the injured animal is in a thermoneutral environment, at lower environmental temperatures it does and the lower the ambient temperature the faster it falls (Stoner, 1961; 1968). This is due to decreased heat production (Stoner & Pullar, 1963; Miksche & Caldwell, 1968) and is not due to any failure of O_2 transport or tissue hypoxia (Threlfall, 1970) at the particular time after injury under discussion, namely, the 'ebb' phase (Cuthbertson, 1942). More work is required on the effects of injury on thermoregulation particularly as surgeons now wish to manipulate the environment of their patients.

To investigate these changes O_2 consumption was determined at different environmental temperatures in rats after hind-limb ischaemia (Stoner, 1969). In the new steady-state shortly after the injury only the thermoregulatory component of O_2 consumption was affected. Without change in gradient the straight line relating O_2 consumption to ambient temperature was displaced to the left, the degree of displacement depending on the severity of the injury. In other words, the ambient temperature of the injured rat had to be reduced below the lower limit of the normal thermoneutral zone before O_2 consumption increased. At this stage O_2 consumption did not fall below the basal level. The injured rat behaved as if its environmental temperature was higher than it really was. A similar picture is seen shortly after scalding (Stoner, 1969).

In terms of the Hardy-Hammell movable set-point theory for

thermoregulation (Hammell, 1968) this could be described as a lowering of the set-point. On this theory the ambient and core temperature thresholds for the onset of shivering and thermoregulatory heat loss, which straddle the set-point and move with it, should also be displaced in the same direction, i.e. both should be lowered.

By recording the electromyogram of unanaesthetized rats in an environmental chamber (Stoner, 1971) it was found that the ambient temperature at which shivering began was lower than normal both during the second half of a 4 hr period of bilateral hind-limb ischaemia and 1 - 1½ hr later. Although the threshold was lower there was no significant alteration in the proportionality constant for the response. The response to cold stimulation of the deep thermoreceptors was also depressed as no shivering was seen when the core temperature fell after removal of the tourniquets to levels which provoked shivering in normal rats in a 20°C environment. Although piloerection occurs in injured rats from increased sympathetic activity it would be difficult to attribute these changes to increased insulation since they occurred in clipped rats and also their O_2 consumption-ambient temperature curves were parallel to those of the controls. Nor can they be attributed to decreased baroreceptor activity (Mott, 1963) since they preceded any fall in blood pressure.

Since the hypothalamus is cooled by the blood and since Kovách (1970) has shown, at least in the anaesthetized dog, that the hypothalamic blood flow is less well protected by autoregulation than that in other parts of the brain after haemorrhage, it was possible that the temperature in the hypothalamus differed from that in the colon after injury, that it might even rise and account for the changes (Stoner, 1970). The recent work of Cranston and Rosendorff (1971) makes this explanation unlikely and, in fact, it can be discounted since after 4 hr bilateral hind-limb ischaemia in a 20°C environment the temperature in the anterior hypothalamus fell more rapidly than in the colon (Stoner, 1972).

While the changes in the shivering thresholds seemed to go along with the idea of a lower set-point after injury, the fact that the main change occurred during the period of limb ischaemia was unexpected since the rat did not appear to have much difficulty with thermoregulation at that time. In a 20°C environment its core temperature does not fall. When exposed to successively lower ambient temperatures its core temperature did not rise as in controls but it did not fall as after limb ischaemia (Stoner, 1971).

The lower threshold for shivering during limb ischaemia was confirmed by stimulating the deep thermoreceptors by cooling the anterior hypothalamus directly with a thermode. In unanaesthetized uninjured control rats in a 20°C environment, shivering began at hypothalamic temperatures of 34.8 - 36.4°C. During limb ischaemia either no shivering was observed despite more severe cooling or slight shivering was seen between 31 - 32°C. Despite this, the core temperature rose in response to hypothalamic cooling much as in the controls (Stoner, 1972). This and other data suggested a possible dissociation of the effects of injury on the control of shivering and non-shivering thermogenesis. The O_2 consumption was therefore measured at different ambient temperatures during limb ischaemia (Stoner & Marshall, 1971).

O_2 consumption was increased when the tourniquets were in place at ambient temperatures above 5°C. The regression line relating O_2 consumption and ambient temperature for these rats intercepted the temperature axis at the same point as the corresponding line for the controls. This indicates that the increased O_2 consumption was due to a decrease in insulation. Loss of insulation when the tourniquets are in place can be attributed to the altered posture and diminished ability to huddle in response to cold. Nevertheless, in the ambient temperature range (5-30°C) the rat could meet a demand for extra heat by increasing non-shivering thermogenesis. Inhibition of non-shivering thermogenesis comes after removal of the tourniquets.

These results indicate that we are not dealing with a simple change in set-point. As already mentioned, if the set-point was lower the threshold for the onset of thermoregulatory heat loss should also be lower. Previously, Little and Stoner (1967) had found that the core temperature for the onset of increased heat loss from the tail of the normal rat in a 20°C environment was about 39.2°C and later (Stoner & Little, 1969) it was found that this threshold was unaltered during the period of limb ischaemia. The threshold after removal of the tourniquets was not determined as the core temperature could not be raised sufficiently without killing the rat. Heat has now been applied directly to the anterior hypothalamus with a thermode (Stoner, 1972). In controls in a 20°C environment the threshold temperature was between 39.0 and 40.3°C. During the second half of a 4 hour period of bilateral hind-limb ischaemia it rose slightly, the biggest change being from 39.0 to 41.5°C. In only one out of nine rats did the threshold fall; from 39.3 to 39.1°C. When the tourniquets were removed the response gradually disappeared so that $1\frac{1}{2}$ hr later it could not usually be elicited by hypothalamic temperatures as high as 44°C.

Changes also occur in the monoamine concentrations in the hypothalamus of the rat injured by limb ischaemia (Stoner & Elson, 1971). The 5-hydroxytryptamine concentration was unchanged but there was a decrease in the concentration of noradrenaline which was practically maximal at the end of 2 hr bilateral hind-limb ischaemia. Experiments with α-methyl-p-tyrosine showed that the fall was due to increased utilization of noradrenaline. Similar changes occurred after a full thickness scald of 20% of the skin area of the rat. The recovery of the concentration of noradrenaline after non-fatal injury took about 48 hr. Thermoregulation was impaired for a similar length of time (Stoner, 1968).

One can now begin to formulate an hypothesis to explain how non-thermal stimuli arising from the injury interfere with thermoregulation. A complete explanation is not yet possible but a neuronal model can be suggested which may help towards that goal.

Cremer and Bligh (1969) have suggested a neuronal model as an aide for correlating the phenomena of thermoregulation. In this model hot and cold stimuli excite primary hypothalamic neurones which in turn excite secondary neurones to give a cold to heat production pathway and a heat to heat loss pathway. In addition, there are crossed inhibitory neurones so that a cold stimulus will inhibit the heat loss pathway as well as stimulating heat production. In their original model there was little detail about the effector neurones. To accommodate the present results (Stoner, 1972) both effector pathways must be divided so that a cold stimulus will first cause an increase in non-shivering thermogenesis followed, with increasing stimulation, by recruitment of a pathway leading to shivering thermogenesis. Similarly, the heat loss pathway must be divided so as to give one path concerned with the arterio-venous anastomoses in the tail and another with the activation of the other heat loss mechanisms in the rat.

Given these modifications, the changes in thermoregulation produced by trauma can be visualized as due to a step-wise inhibition of the system, although other factors may be concerned in the changes in the heat-induced vasodilatation in the tail. Thus, the shivering pathway is inhibited while the tourniquets are in place. When they are removed, non-shivering thermogenesis is also inhibited by inhibition at a point central to the first.

The site of the neurones concerned in these responses to non-thermal stimuli and of their receptors is not known. Since one of the main events on removal of the tourniquets is loss of fluid from the circulation into the injured tissues the impulses inhibiting non-shivering thermogenesis may arise from volume receptors. The inhibitory terminals in the hypothalamus may be noradrenergic (Stoner & Elson, 1971) in keeping with the idea that noradrenaline is a general inhibitor in this system. Clearly much further work is necessary to arrive at the full explanation of these inhibitory mechanisms.

In conclusion, there seems little doubt that the difficulties in thermoregulation experienced by the injured animal during the ebb phase arise from central changes but their mechanism is still in doubt. A better appreciation of these changes is probably essential in understanding some of the metabolic changes after injury as they may be secondary to the hypothalamic disturbance.

REFERENCES

Cranston, W.I. & Rosendorff, C. (1971) J. Physiol. Lond. 215, 577.
Cremer, J.E. & Bligh, J. (1969) Brit. med. Bull. 25, 299.
Cuthbertson, D.P. (1942) Lancet i, 433.
Hammell, H.T. (1968) Ann. Rev. Physiol. 30, 641.
Kovách, A.G.B. (1970) J. clin. Path. 23, Suppl. (Royal Coll.Path) 4, 202.
Little, R.A. & Stoner, H.B. (1968) Quart.J. exp. Physiol. 53, 76.
Miksche, L.W. & Caldwell, F.T. (1968) Ann. N.Y. Acad.Sci. 150, 755.
Mott, J. (1963) J. Physiol. Lond. 166, 563.
Stoner, H.B. (1961) Scientific Basis of Med. Ann. Rev. p.172, London, Athlone Press.
Stoner, H.B. (1968) Ann. N.Y. Acad. Sci. 150, 722.
Stoner, H.B. (1969) Brit. J. exp. Path. 50, 125.
Stoner, H.B. (1970) In, 'Energy Metabolism in Trauma' Eds. R. Porter & J. Knight, p.1. Churchill, London.
Stoner, H.B. (1971) J. Physiol. Lond. 214, 599.
Stoner, H.B. (1972) J. appl. Physiol. At press.
Stoner, H.B. & Elson, P.M. (1971) J. Neurochem. 81, 1837.
Stoner, H.B. & Little, R.A. (1969) Brit. J. exp. Path. 50, 107.
Stoner, H.B. & Marshall, H.W. (1971) Brit. J. exp. Path. 52, 650.
Stoner, H.B. & Pullar, J.D. (1963) Brit. J. exp. Path. 44, 586.
Threlfall, C.J. (1970) In, 'Energy Metabolism in Trauma' Eds. R. Porter & J. Knight, p. 127, Churchill, London.

EFFECT OF STRESS ON VARIOUS CHARACTERISTICS OF NOREPINEPHRINE

METABOLISM IN CENTRAL NORADRENERGIC NEURONS

Anne-Marie Thierry

Groupe NB, Laboratoire de biologie moléculaire
Collège de France, 11, place Marcelin Berthelot, Paris 5e

Possible relationships between various physiological states, aggressive behaviour or emotions and biogenic amines have generated considerable scientific interest. Studies, in several species, reported during the past few years indicated that physical stress is associated with alterations in the metabolism of norepinephrine (NE) in the rat brain.

In earlier experiments, decrease in brain NE levels could be detected when animals were submitted to severe stress situations (Maynert & Levi, 1964). But formation and utilization of NE can be affected without altering the endogenous levels of the amine. New techniques and especially measurement of the turnover rather than levels better reflect possible changes in NE metabolism. Acceleration of NE turnover has been reported, without important changes in amine levels, in various stress situations induced by muscular exhaustion (Gordon et al., 1966), immobilization (Corrodi et al., 1971; Welch & Welch, 1968), changes in environmental temperature (Corrodi et al., 1968; Simmonds, 1969) and electric shocks applied to the feet (Bliss et al., 1968; Taylor & Laverty, 1969).

In all these experiments turnover rates of the amine could be obtained in normal or stressed animals by assuming a single compartmental model for the distribution of NE in noradrenergic neurons. However a few studies suggest that NE is stored heterogeneously in central noradrenergic terminals. For example ^{3}H-NE previously taken up (Glowinski et al., 1965) or synthesized from ^{3}H-tyrosine (Thierry et al., 1971a) in central noradrenergic neurons disappears in a multiphasic manner. Furthermore, the rate of NE synthesis estimated from the content of ^{14}C-NE accumulated at the end of a constant intravenous infusion of ^{14}C-

-tyrosine greatly exceeded the rate calculated from turnover studies (Sedvall et al., 1968). Moreover the existence of an heterogeneous disposition of catecholamines in central catecholaminergic neurons is also suggested from previous observations made in pharmacological (Weissman, 1966; Rech et al., 1968) endocrinological (Kordon and Glowinski, 1969) and behavioural studies (Stimus, unpublished observations). During the last few years, in order to understand better the regulatory processes of central NE metabolism, we have used a stress situation which was known to activate central noradrenergic neurons. The electric shock stress has the advantage of being controlled, in duration and intensity of the stimuli, within fairly narrow limits. It thus represents a good tool for investigating the properties of the regulatory processes of NE metabolism in central noradrenergic neurons. We would like to summarize and discuss two groups of results: the effects of various stress situations on the turnover of the transmitter, and the characteristics of the various forms of NE.

1. Effect of Electric Foot Shock Stress on the Activity of Central Noradrenergic Neurons

In our first experiments NE turnover has been used as an index of the biochemical activity of central noradrenergic neurons. It was estimated by following the decline in NE specific activity occurring 2 to 5 hr after the intracisternal administration of ^{3}H-NE. During this period ^{3}H-NE specifically taken up in central noradrenergic neurons disappeared in a single exponential phase. The stressful situation was induced by series of electrical shocks delivered to the paws of rats in individual boxes. The shocks were applied during 6 periods of 10 min each, alternating with rest periods of 20 min. The overall stress period lasted for 3 hr. Under these conditions, when the stress was given 2 hr after the intracisternal injection of ^{3}H-NE the NE specific activity was markedly reduced at the end of the stress session. This effect could be detected not only in the whole brain but also in various regions of the brain as well as in the spinal cord, indicating a general activation of the different systems of central noradrenergic neurons (Thierry et al., 1968). Higher current intensities of the shocks further enhanced the acceleration of NE turnover and increased the ^{3}H-methoxynoradrenaline accumulation in tissues (Fig. 1). This suggests that large amounts of ^{3}H-NE are released from nerve terminals under this experimental condition. Similarly an increased utilization of NE in central noradrenergic neurons under this stress could be shown by following the decline of endogenous NE blockade of its synthesis with the inhibitor of tyrosine hydroxylase: α-methyl-para-tyrosine. This general activation of central noradrenergic neurons was associated with a marked stimulation of the synthesis of the transmitter. In fact, increased synthesis of NE could even be detected in vitro by

measuring the quantities of ^{3}H-catecholamines formed from ^{3}H-
-tyrosine in brain stem slices of animals which had been just
previously exposed to a 3 hr stress situation (Thierry et al.,
1971a).
These data indicate that activation of noradrenergic neurons by
fluctuations of physiological states induce sustained modifications in the regulatory processes of NE metabolism which can be
visualized in vitro. This seems to be a general property of
aminergic neurons since circadian changes in the activity of
central serotoninergic neurons have also been demonstrated easily
in slices (Hery et al., 1972).

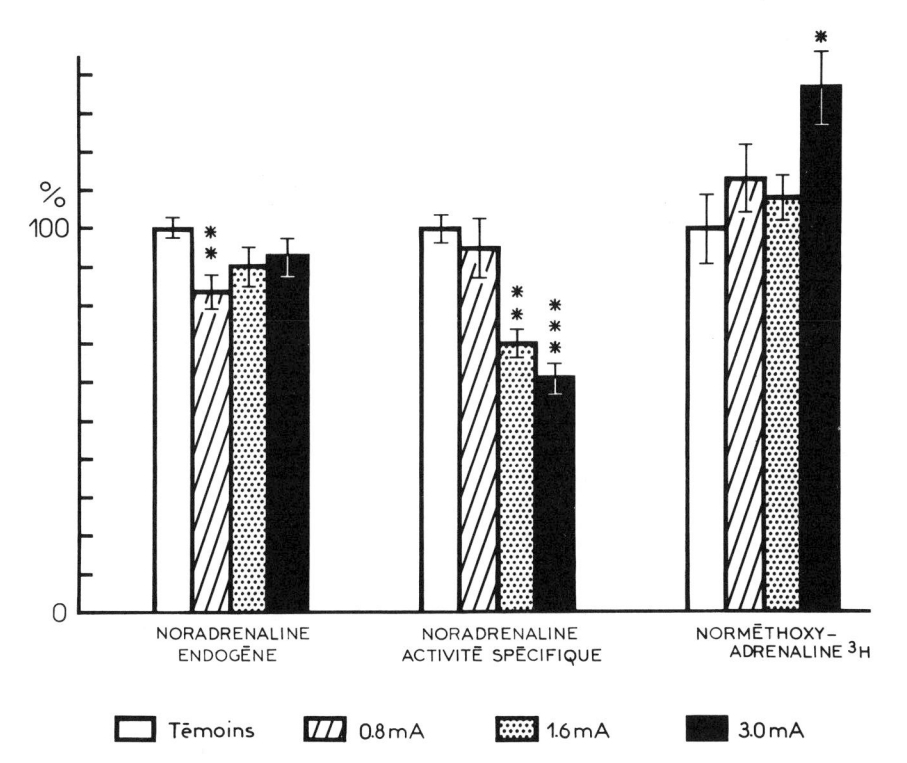

Fig. 1. Effect of variation in intensity of stimulation
on the turnover of NE and on the accumulation of ^{3}H-nor-
metanephrine in the brain stem.

Four groups of 8 rats received 4.7 µCi of ^{3}H-NE intracisternally. Two hr later, 3 groups were subjected to
electric shock stress for 3 hr, current intensity was
different in each group (0.8, 1.6 and 3.0 mA). All the
animals were killed 5 hr after ^{3}H-NE administration.
Results are expressed as percentage of control values $\pm$
S.E.M. *P < 0.02; **P < 0.01; ***P < 0.001 when compared
with control values.

Although the stress produced by a single session of mild electric shocks applied to the feet can induce an easily detectable increase in NE turnover, this change is not maximal. A greater acceleration of turnover was observed in animals submitted to the stress immediately after amphetamine administration (Javoy et al., 1968). Thus the individual effects of amphetamine and of the stress situation on the turnover of NE appear to be cumulative. A more potent activation of NE turnover was also observed after repeated exposure of the animals to the 3 hr stress once a day for 4 days (Thierry et al., 1968). This "chronic" stress particularly enhanced the turnover of NE in the telencephalon-diencephalon suggesting a sustained activation of the dorsal system of noradrenergic neurons. Similar observations were made after the application of electroshocks twice a day over a 10 day period. Twenty four hr after the last electroshock session, NE turnover was still markedly accelerated. The effect was again more pronounced in the telencephalon-diencephalon than in the brain stem mesencephalon (Kety et al., 1967). Furthermore, this sustained activation of NE turnover was associated with an increase in the activity of tyrosine hydroxylase in various parts of the brain (Musacchio et al., 1969) suggesting a stimulation of the enzyme synthesis.

2. Effect of Stress on the Disposition of NE localized in Various Intraneuronal Storage Forms

In subsequent studies the same stress but of shorter duration (15 min) was used to examine the changes in NE mobilization in its various storage forms during activation of central noradrenergic neurons. This mild stress markedly accelerated the utilization of ^{3}H-NE newly taken up or synthesized from ^{3}H-dopamine or ^{3}H-tyrosine (Fig. 2) in the brain stem of the rats; ^{3}H-NE levels were significantly decreased by 20 to 27% in 15 min. Endogenous NE levels were not affected. These results indicate, once more, that exogenous NE taken up in central noradrenergic neurons behaves as NE endogenously synthesized from its precursors. A very interesting observation was made when this stress of short duration was applied 3 hr after the initial accumulation of exogenous ^{3}H-NE or the initial formation of the labelled transmitter from ^{3}H-tyrosine. In contrast with that observed in the experiment described above, the 15 min stress had absolutely no detectable effect on the utilization of ^{3}H-NE stored for long time in tissues (Fig. 2). In conclusion, NE newly taken up or newly synthesized in central noradrenergic terminals is mainly localized in a "functional compartment" from which the amine is preferentially utilized on neuronal activation (Fig. 2). The amine stored for a longer time period is mainly localized in a "storage compartment", its utilization is only increased when animals are submitted to the stress of long duration (Fig. 1).

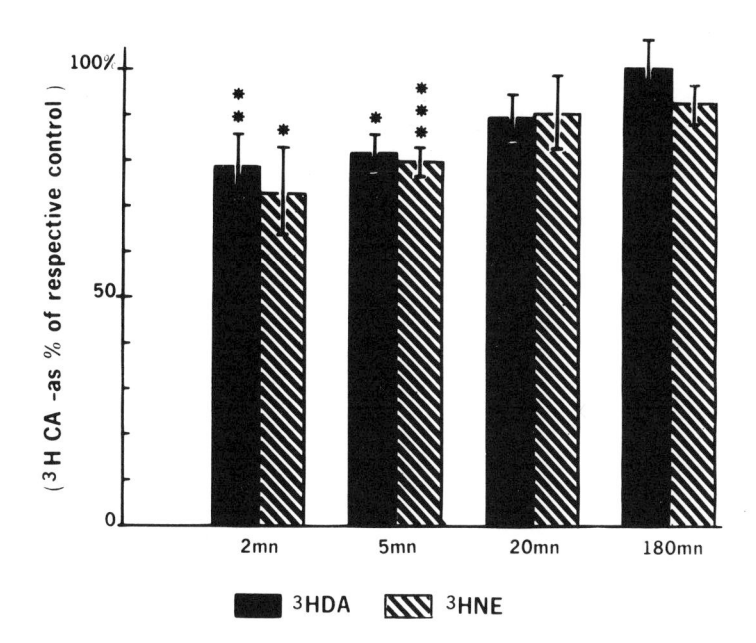

Fig. 2. Effect of 15 min stress on endogenously synthesized catecholamines stored for various time periods in the rat brain stem.

Eight groups of 8 rats received 23 µCi of ^{3}H-tyrosine intracisternally. Rats of 4 groups were submitted to electrical foot shocks for 15 min, 2, 5, 20 or 180 min after ^{3}H-tyrosine injection. Stressed and control rats were killed respectively 17, 20, 35 and 195 min after the labelled amino acid injection. ^{3}H-NE and ^{3}H-DA were estimated in brain stem. Results are expressed as percentage of respective control values ± S.E.M. *$P < 0.05$; **$P < 0.02$; ***$P < 0.01$ when compared with respective control values.

Surprisingly, we also observed in these experiments that dopamine (DA), the immediate precursor of NE, newly taken up or newly synthesized from ^{3}H-tyrosine was, as NE, utilized preferentially and at a faster rate than ^{3}H-DA stored for longer time in the brain stem of stressed animals (Fig. 2). This effect may be explained by a simultaneous enhanced release of newly synthesized DA and NE from noradrenergic terminals or by an acceleration of the conversion of DA into NE activation of noradrenergic neurons. Simultaneous release of DA with NE has already been shown in peripheral noradrenergic neurons (Thoenen et al., 1965). As most workers believe that NE is released by exocytosis, since vesicle-specific proteins together with NE have been detected extraneuronally during stimulation of peripheral sympathetic nerves (Geffen & Livett, 1971), the hypothesis of DA release from central noradrenergic neurons cannot be rejected. Recent experiments made with α-methyl para-tyrosine support the hypothesis of an acceleration of the DA-NE conversion during activation of noradrenergic neurons. Two hr after the administration of the inhibitor of catecholamines synthesis the 15 min stress stimulated both the untilization of DA and NE in the rat cortex, this was even associated with an increase on the NE/DA ratio. This observation suggests an acceleration of NE formation from the stored dopamine. Other results are also in favor of an important role for the last step of NE synthesis in the modulation of NE formation : an indirect interneuronal activation of central noradrenergic neurons by a neuroleptic has also been shown to increase the NE/DA ratio as a function of time after inhibition of tyrosine hydroxylase with α-methyl para-tyrosine (Thierry, et al., 1971).

CONCLUSION

Two types of regulatory process have been distinguished to explain the modifications of NE biosynthesis in noradrenergic neurons : 1) The rapid and mild increase in the transmitter untilization which reduces the intraneuronal level of the amine at the site of tyrosine hydroxylation, and thus limits the end product inhibitory mechanism of synthesis ; 2) The persistant and potent activation of NE untilization which induces an increased activity of tyrosine hydroxylase and dopamine hydroxylase : the two enzymes involved in the limiting steps of the amine formation. These two types of regulatory processes have been extensively studied in sympathetic nerves (Weiner, 1970) Some of the results presented in this report and other studies (Besson, et al., 1972) Reis & Molinoff, 1972) suggest that they also exist in central noradrenergic neurons. The stress experiments offer an interesting model for the further investigation these two main types of regulatory mechanisms whose characteristics may be slightly different from those observed at the periphery.

Another, and perhaps more interesting, aspect of the results described is represented by the new concept of the presence of a "functional pool" of NE in noradreneric terminals. It now seems clear that NE localized in the "Functional pool" is readily available for rapid extraneuronal untilization of the transmitter and is involved in any fluctuation of noradrenergic neuron activity. Preliminary studies indicate that the size of the functional pool in cortical noradrenergic terminals is a fifth of that of the "main storage compartment". On the other hand the turnover rate of NE in this compartment is five times higher than that observed in the storage compartment; finally the half life of NE in this "functional pool" is about 10 min.

These recent data as well as the selective effect of the stress of short duration on the amine metabolism in the functional compartment underline the particular attention which should be given to newly synthesized transmitters. Close correlations between subtle changes in behavioural states and biochemical changes in neurotransmitter metabolism can now be visualized by studying dynamic events occurring in this "functional pool".

REFERENCES

Besson, M.J., Cheramy, A., Gauchy, C. & Musacchio, J. 1972. To be published.
Bliss, E.L., Ailion, J. & Zwanziger, J. (1968). J. Pharmacol. 164. 122.
Corrodi, H., Fuxe, K. & Hokfelt, T. (1968). Life sci. 7, 107.
Corrodi, H., Fuxe, K., Lidbrink, P. & Olson, L. (1971). Brain Res. 29, 1
Geffen, L.B. & Livett, B.G. (1971) Physiol Rev. 51, 98.
Glowinski, J., Kopin, I.J. & Axelrod, J. (1965). J.Neurochem. 12, 25.
Gordon, R., Spector, S.,Sioerdsma, A. & Udenfriend, S.(1966). J. Pharmacol. 153, 440.
Hery, F.,Rouer, E. & Glowinski, J. Submitted for publication in Brain Res.
Javoy, F, Thierry, A.M., Kety, S.S. & Glowinski, J. (1968). Comm. Behavioural Biol. 1,43.
Kety, S.S., Javoy, F., Thierry, A.M., Julou, L. & Glowinski, J. (1967). Proc. Nat. Acad. Sci. 58, 1249.
Kordon, C. & Glowinski, J. (1969) Endocrinology, 85,924.
Maynert, E.W & Levi, R. (1964). J. Pharmacol. 143, 90.
Musacchio, J., Julou, L., Kety, S.S. & Glowinski, J. (1969). Proc. Nat. Acad. Sci.63,1117.
Rech, R.H., Carr, L.A. & Moore, K.E. (1968). J.Pharmacol. 160, 326.
Reis, D.J. & Molinoff, P.B. (1972). J.Neurochem. 19, 195.
Sedvall, G.C., Weise, V.K.Kopin, I.j. (1968). J.Pharmocol. 159, 274.
Simmonds, M.A.(1969). J.Physiol. 203, 199
Taylor, K.M.& Laverty, R. (1969). J. Neurochem. 16,1367.
Thierry, A.M., Blanc, G. & Glowinski, J. (1971a). J.Neurochem. 18, 449.
Thierry, A.M., Blanc, G. & Glowinski, J. (1971b). Eur.-J.Pharmacol. 14, 303.
Thierry, A.m., Javoy, F., Glowinski, J. & Kety, S.S.(1968). J.Pharmacol. 163, 163.
Theonen, H., Haefely, W., Gey, F.k. & Hurlimann, A.(1965). Life Sci. 4,2033.
Weiner, N. (1970). An. Rev. Pharmacol. 10. 273.

Weissman, A., Koe, K.B. & Tenen, S.S. (1966). J.Pharmacol.151,339.
Welch, B.L. & Welch, A.M. (1968) Nature, 218, 575.

HYPOTHALAMIC CATECHOLAMINES AND ADRENERGIC INNERVATION OF HYPO-

THALAMIC BLOOD VESSELS. POTENTIAL ROLE IN SHOCK.[1]

E. T. Angelakos, M. P. King and L. Carballo

Department of Physiology & Biophysics
Hahnemann Medical College & Hospital
Philadelphia, Pennsylvania

Previous studies by a number of investigators have established that the hypothalamus contains the highest concentration of norepinephrine (NE) among all regions of the brain (3,4,12,13,14, 24,26). This appears to be true of all species that have been studied. Results from our own studies in a number of laboratory species are shown in Table 1. We have also made extensive studies on the distribution of NE in various regions of the brain in the cat and squirrel monkey. These results are summarized in Table 2 and indicate that there are no striking differences in this respect between the feline and primate. The results obtained in the cat agree with previous findings (3, 18). Of all the tissues examined (Table 2) the hypothalamus had the highest concentration of norepinephrine (NE).

Table 1. Hypothalamic norepinephrine
μg/gm (mean ± S.D.)

CAT (9)	1.22 ± 0.29
DOG (4)	0.76 ± 0.11
Sq. MONKEY (19)	1.15 ± 0.30
Rh. MONKEY (3)	2.17 ± 0.32

---

[1] These studies were supported by a Themis Research Project Grant from the Department of the Navy and by USPHS NIH Grants HE 13008 and NB 08664.

Table 2. Norepinephrine concentrations in various regions of the cat and squirrel monkey brains.

μg/gm (mean ± S.D.)

Determ. (No. Animals)	CAT 3-6(9)	SQ. MONKEY 4-6(19)
Frontal Cortex	0.15 ± 0.05	0.08 ± 0.02
Parietal Cortex	0.13 ± 0.03	0.08 ± 0.02
Occipital Cortex	0.11 ± 0.03	0.08 ± 0.04
Cerebellum	0.10 ± 0.03	0.07 ± 0.03
Lenticular n.	0.40 ± 0.12	0.08 ± 0.02
Caudate n.	0.19 ± 0.03	0.11 ± 0.03
Hypothalamus	1.22 ± 0.29	1.15 ± 0.30
Pons	0.25 ± 0.09	0.36 ± 0.12
Medulla	0.20 ± 0.13	0.43 ± 0.16
Pia	0.75 ± 0.26	0.47 ± 0.09

The specific localization of NE within each region of the brain was studied using the fluorescence histochemical technique of Falck and Hillarp (9). Figure 1 shows a "stretch preparation" of pia membranes of the cat (1A and 1B) and squirrel monkey (1C and 1D). It became clear from these and subsequent studies (see below) that the NE found in the pia is predominantly localized in adrenergic nerve terminals innervating the pial vessels. These findings agree with previous observations and conclusions (10,11). It should be noted, however, that the adrenergic innervation of the pia vessels is quite extensive and as prominent as that found in regions which are known to be under extensive adrenergic vasomotor control (e.g. mesenteric, renal, splenic, etc.).

In systematic histochemical studies in cat and squirrel monkey brains it was found that in addition to the pia vessels, adrenergic innervation was present in many blood vessels of the brain especially in the gray matter of the brain stem. By contrast vessels within the cortex uniformly lacked any adrenergic innervation.

Among the regions examined, adrenergic innervation was found to be most prominent in the blood vessels found within the hypothalamus (Figure 2) (2). In the hypothalamus the adrenergic nerve terminals found in close association with the blood vessels appeared to be distinct from the adrenergic nerve terminals which are found distributed within the hypothalamic parenchyma and which have been described previously (1,5,10,12, 15,25). Adrenergic nerve terminals associated with blood vessels were found both in the parts of the hypothalamus where parenchymal adrenergic innervation was also present as well as in parts where such parenchymal innervation was lacking (Figure 2A and 2B).

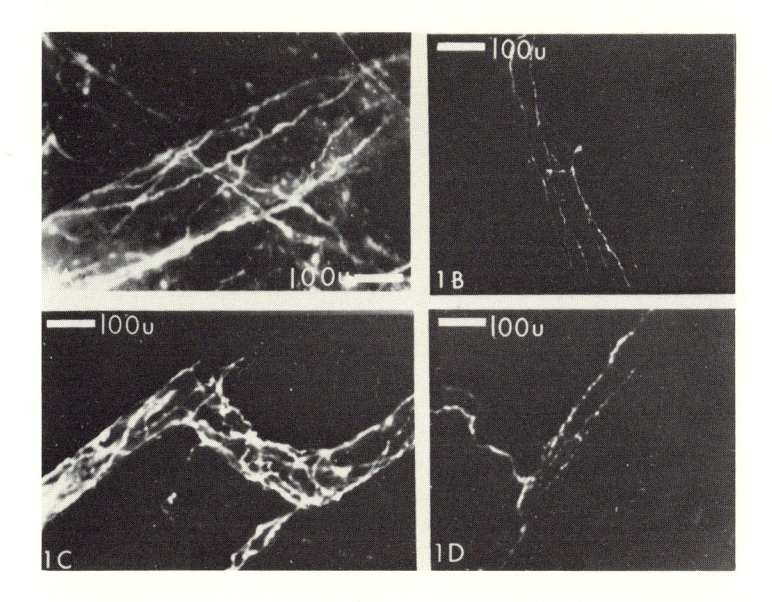

Figure 1: Fluorescence photomicrograph of the adrenergic plexus in a stretch preparation of the pial membranes from cats (A and B) and squirrel monkeys (C and D) demonstrating dense adrenergic innervation around the pial vessels of varying caliber.

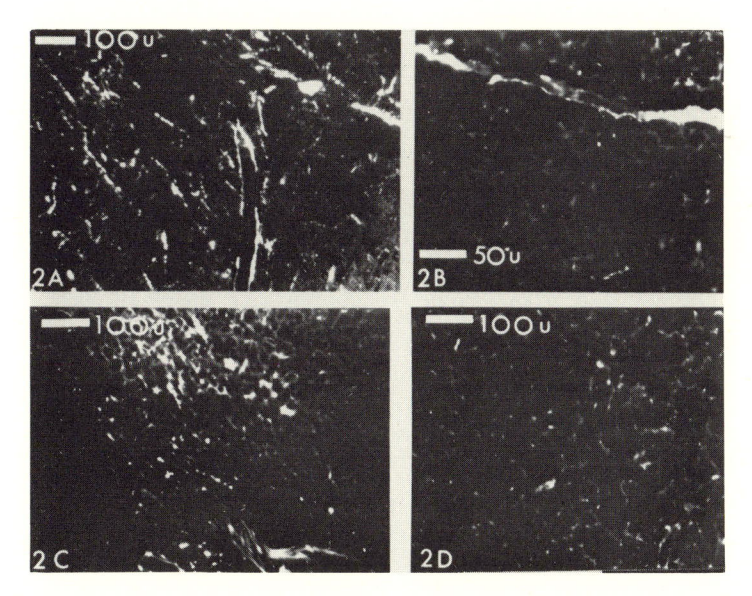

Figure 2: Fluorescence photomicrograph of a preparation from the hypothalamic region of the cat (A and B), baboon (C) and squirrel monkey (D) showing both vascular (A and B) and parenchymal innervation (A, C and D).

In a series of cats, bilateral superior cervical gnaglionectomy was performed and surviving animals were studied 7 to 14 days after the operation. Histochemical observation in these cats showed that the adrenergic innervation of the pia vessels was absent or greatly reduced. A similar decrease or complete disappearance was observed in the adrenergic nerve terminals normally present in association with hypothalamic blood vessels. By contrast, no discernible differences were found in the adrenergic nerve terminals present within the hypothalamic parenchyma. These observations confirm previous conclusions in which the sympathetic innervation of blood vessels of the brain (whenever this was found) was believed to originate from neurons located within the superior cervical ganglion (11, 16, 17, 20, 22).

Chemical analyses of the total NE present in various regions of the brain 7 to 14 days after superior ganglionectomy showed a decrease in hypothalamic NE by about 40% (Table 3). This suggests that a major fraction of NE found in the hypothalamus is localized in adrenergic nerve terminals innervating hypothalamic blood vessels.

Table 3. Superior cervical ganglionectomy in cats.

NE µg/gm (mean ± S.D.)

	Normal	Denervated*	Ratio
Cortex	0.13 ± 0.03	0.12 ± 0.04	0.92
Hypothalamus	1.22 ± 0.29	0.72 ± 0.26	0.59
Pia	0.75 ± 0.26	0.04 ± 0.03	0.05

(*) 7 - 14 days after operation. Confirmed histochemically.

Specific studies on the blood vessel and parenchymal innervation in various parts of the cat hypothalamus were made and the results are summarized in Table 4 which shows estimates of the presence and density of adrenergic nerve terminals. Similar results have been reported by Dahlstrom and Fuxe (6). In general the fluorescence _intensity_ of adrenergic nerve terminals found in association with blood vessels was greater than those found within the hypothalamic parenchyma.

Table 4. Adrenergic nerve terminations in various regions of the cat hypothalamus*.

	Parenchyma	Blood Vessels
Anterior		
Pre-optic n.	++	+
Supra-optic n.	++++	+++
Paraventricular n.	++	-
Suprachiasmal	++++	-
Middle		
Arcuate n.	+++++	-
Lateral n.	++	+
Ventro-medial n.	+++	+
Dorso-medial n.	+++	-
Ento-peduncularis	++++	++
Posterior		
Post. n.	+++	++
Mammillary n.	Variable	-

(*) Semi-quantitative subjective estimates from examination of a large number of sections using the fluorescence histochemical technique. (-) indicates little or no adrenergic innervation while (++++) indicates extensive innervation. The density of the fluorescent varicosities (nerve terminals) was taken as the major criterion although the intensity of fluorescence was also contributory in these visual estimates.

The results from these and previous studies show clearly that the NE found in the hypothalamus is associated with at least two distinct groups of nerve terminals. One group is associated with blood vessel innervation and apparently is derived from cells located in the superior cervical ganglion. A second group consists of the adrenergic nerve terminals found within the hypothalamic parenchyma and presumably synapsing with specific hypothalamic neurons. The latter group of terminals apparently originates from adrenergic neurons located in the medulla and pons. Clearly any evaluation of the changes in hypothalamic catecholamines, including estimates of turnover rates, must take into consideration possible differences between these two groups of nerve terminals. The two groups of adrenergic neurons (superior cervical and brain stem) are undoubtedly involved in different functions and are likely to respond differently. In addition histochemical observations suggest that the intraneuronal concentration of NE may be higher in the vascular than in the parenchymal adrenergic nerve endings.

The functional role of the adrenergic innervation of blood vessels of the brain is at present unclear. Early studies on total cerebral blood flow gave some conflicting results although the concensus indicated little or no effect of sympathetic nerve stimulation or injected adrenergic amines on total cerebral blood flow. These studies have been reviewed by Schmidt (22) and by Sokoloff (23). A lack of an effect on total cerebral blood flow would be consistent with the absence of blood vessel innervation of most of the vessels of the cortex. Nevertheless more recent studies suggest a substantial change of total cerebral blood flow following adrenergic stimulation (7,8), challenging this classical concept. On the other hand, decreases in hypothalamic blood flow following sympathetic stimulation have been reported by Schmidt as early as 1934 (21). More recently Rosendorff and Cranston (19) reported that local injection of NE into the rabbit hypothalamus produced vasodilation with small doses and vasoconstriction with large doses. The latter effect was blocked by dibenzyline, an alpha blocking agent.

The evidence that adrenergic innervation is present in the hypothalamus is of potential significance in conditions in which there is a generalized enhancement of adrenergic activity as it occurs in circulatory shock. The well known increase in circulating catecholamine levels as well as increased adrenergic nerve activity could produce direct effects on the blood supply of hypothalamic centers which in turn can affect markedly the various regulatory functions of this region. In this connection Dr. Kovach and associates have presented evidence in this symposium that there is marked decrease in blood flow through certain regions of the hypothalamus during hemorrhagic shock. Whether this decrease is related to the presence of adrenergic innervation in this region or perhaps to circulating catecholamines is now under study in this laboratory in collaboration with Dr. Kovach and associates.

Preliminary histochemical observations made in baboons in collaborative studies with Drs. Kovach, Reivich and associates indicated that the fluorescence of the adrenergic terminals of many hypothalamic blood vessels was markedly reduced as compared to controls. In addition the shocked animals had low hypothalamic NE levels. In general these preliminary observations favor the view that adrenergic factors may play a role in altering the hypothalamic circulation in shock and support the findings of Kovach and associates indicating a decrease of hypothalamic blood flow in shock. Clearly, further studies are needed to elucidate the potentially critical role of the adrenergic innervation of hypothalamic vessels in various forms of shock.

Summary

Available evidence indicates that the norepinephrine (NE) found in the hypothalamus is associated with two distinct groups of nerve terminals. One group originating from neurons in the medulla and pons with terminals in neurons and centers within the hypothalamic parenchyma and another group with neurons in the superior cervical ganglion. Terminals of the second group apparently innervate the hypothalamic blood vessels. Parenchymal and blood vessel innervation varies in different regions of the hypothalamus but has been found in all laboratory species examined including subhuman primates. Activation of the adrenergic neurons to the hypothalamic blood vessels under conditions of general adrenergic discharge, as occurs in circulatory shock, may alter the functional activity of the hypothalamic centers and thus play a significant role in the cardiovascular and hormonal alterations observed in shock.

References

1. Anden, N.E., Dahlstrom, A., Fuxe, K., Olson, L. & Ungerstedt, U. (1966) Experientia 22:44.

2. Angelakos, E.T., Irvin, J.D. & King, M.P. (1970) Fed. Proc. 29:416A.

3. Bertler, A. & Rosengren, E. (1959) Acta physiol. Scand. 47:350.

4. Carlsson, A. (1959) Pharm. Rev. 11:490.

5. Carlsson, A., Falck, B. & Hillarp, N-Å. (1962) Acta physiol. Scand. 56: suppl. 196.

6. Dahlstrom, A. & Fuxe, K. (1965) Acta physiol. Scand. 64: Suppl. 247.

7. D'Alecy, L.G. & Feigl. E.O. (1970) Fed. Proc. 29:520A.

8. Dollison, J.R., Bloor, B.M., Figueroa, A.F. & Loar, B.A. (1969) Surg. Forum 20:416.

9. Falck, B., Hillarp, N-Å., Thieme, G. & Torp, A. (1962) J. Histochem. Cytochem. 10:348.

10. Falck, B., Mchedlishvili, G.I. & Owman, Ch. (1965) Acta Pharmacol. Toxicol. 23:133.

11. Falck, B., Nielsen, K.C. & Owman, C. (1968) Scand. J. Clin. Lab. Invest. Suppl. 102: Sect VI, B.

12. Fuxe, K. & Hokfelt, T. (1970) Central monoaminergic systems and hypothalamic function. In: The Hypothalamus, ed.: L. Martini, M. Motta and F. Fraschini. New York, Academic, 123.

13. Fuxe, K., Hokfelt, T. & Ungerstedt, U. (1970) Morphological and functional aspects of central monoamine neurons. In: International Review of Neurobiology, ed.: C.C. Pfeiffer and J.R. Smythies. New York, Academic, 93.

14. Glowinski, J., Axelrod, J. & Iversen, L.L. (1966) J. Pharmacol. exptl. Therap. 153:30.

15. Hillarp, N-Å., Fuxe, K. & Dahlstrom, A. (1966). Pharmacol. Rev. 18:727.

16. Mitchell, G.A.G., (1956) Cardiovascular innervation. London: E. and S. Livingstone Ltd.

17. Nelson, E. & Rennels, M. (1970) Innervation of intracranial arteries. Brain 93:475.

18. Reis, D.J. & Wurtman, R.J. (1968) Diurnal changes in brain norepinephrine. Life Sciences 7:91.

19. Rosendorff, C. & Cranston, W.I. (1971) Circulation Res. 28:492.

20. Ross, G. The regional circulation (1971) Ann. Rev. Physiol. 33:445.

21. Schmidt, C.F. (1934) The intrinsic regulation of the circulation in the hypothalamus of the cat. Am. J. Physiol. 110:137.

22. Schmidt, C.F. (1950) The cerebral circulation in health and disease. Springfield, Illinois: Charles C. Thomas.

23. Sokoloff, L. (1959) Pharmacol. Rev. 11:1.

24. Vogt, M. (1954) J. Physiol. 123:451.

25. Vogt, M. (1957) Brit. Med. Bull. 13:166.

26. Vogt, M. (1959) Pharmacol. Rev. 11:483.

ACTIVITY OF ADRENAL CATECHOLAMINE-PRODUCING ENZYMES AND THEIR REGULATION AFTER STRESS

Richard Kvetňanský & Irwin J. Kopin

Institute of Experimental Endocrinology, Slovak Academy of Sciences, Bratislava, Czechoslovakia and Laboratory of Clinical Science, National Institute of Mental Health, Bethesda, Maryland 20014, U.S.A.

The pathway for biosynthesis of catecholamines from tyrosine was first proposed by Blaschko in 1939 (3), but only recently have separate enzymatic reactions been characterized. Conversion of tyrosine to dopa is catalyzed by the enzyme tyrosine hydroxylase (TH)(23), which is believed to be the rate-limiting step in the biosynthesis of the catecholamines (17). High levels of the relatively nonspecific aromatic L-amino acid decarboxylase (7,18) rapidly decarboxylate dopa to dopamine; decarboxylation therefore appears to play little or no role in regulation of catecholamine synthesis. Dopamine enters the chromaffin granules to be converted by the enzyme dopamine-beta-hydroxylase (DBH) to norepinephrine (16). Conversion of norepinephrine to epinephrine by phenyl-ethanolamine-N-methyltransferase (PNMT) is the final step in the biosynthesis of catecholamines in the adrenal medulla (1,8).

The last identified enzyme involved in the biosynthesis of catecholamines is tyrosine hydroxylase (23); its discovery was followed by a period of intensive investigation of the regulation of catecholamine biosynthesis. It was long believed that the adrenal medulla was controlled only by nerve impulses carried to the medulla by preganglionic sympathetic nerves. In 1953 Coupland (4) found that in species (such as man or rat) in which the adrenal cortex envelopes the medulla, epinephrine is the major medullary catecholamine; however little or no epinephrine is present in animals in which the adrenal chromaffin tissue lacks an envelope of cortex. Coupland therefore suggested that the adrenal cortex secretes a factor which influences the N-methylation of norepinephrine.

Kirshner and Goodall (8) and Axelrod (1) identified PNMT as

the enzyme responsible for the synthesis of epinephrine from norepinephrine in the adrenal medulla. Wurtman and Axelrod (35) showed that glucocorticoids, which are secreted from the adrenal cortex, increase the activity of PNMT in the adrenal medulla by inducing synthesis of the enzyme protein (36). Their observations suggest that the glucocorticoids are the factors alluded to by Coupland (4).

The normal rate of epinephrine biosynthesis requires that the adrenal medulla be surrounded by a normally functioning adrenal cortex (37) and that the pituitary gland be intact (12). Thus biosynthesis of adrenal catecholamines is regulated, at least in part, by humoral effects on an adrenal medullary enzyme, PNMT. A number of investigators have studied the humoral regulation of the rate of catecholamine biosynthesis and the activity of the synthetic enzymes other than PNMT. Mueller et at. (22) and Kvetnansky et al (10,12) found that after hypophysectomy rat adrenal TH activity diminishes with an apparent half-life of about 10 days. Administration of ACTH prevents the decrease in adrenal TH activity after hypophysectomy, but even a high dose of ACTH does not increase adrenal TH levels in normal rats (22,12). High doses of glucocorticoids or thyroxine fail to maintain levels of adrenal TH in hypophysectomized rats. Induction of adrenal TH in normal rats is seen, however, after drugs which cause reflex stimulation of the sympathetic nervous system [i.e. reserpine (21) or 6-hydroxydopamine (20)]. Drug-induced increases in adrenal enzyme activity are called "trans-synaptic" because they can be prevented by cutting the splanchnic nerve (27). An increase in TH activity has also been found after insulin-induced hypoglycemia (31) or repeated doses of 2-deoxy-D-glucose (26). The first physiologically induced elevation of adrenal TH was demonstrated by Kvetnansky et al. (10,11) in response to repeated immobilization. Exposure of rats to environmental cold (14,29) or psychosocial stimulation of mice also elevates the adrenal TH and PNMT activities (2).

DBH, the enzyme responsible for the formation of norepinephrine from dopamine, is decreased in the adrenal glands of hypophysectomized rats (5,10,32). Its level in the adrenals of hypophysectomized rats is restored by treatment with ACTH (5,10,32) or dexamethasone (5). Thus, adrenal DBH levels also appear to be dependent on an intact pituitary-adrenocortical system. The administration of reserpine also leads to a "trans-synaptic" increase in DBH activity in the rat adrenal (19). DBH levels are increased 4 days after a single dose of insulin (30) or after repeated administration of either insulin or 2-deoxy-D-glucose (26).

Repeated immobilization of rats (13) or chronic cold stress (14) elevates adrenal DBH activity by a physiological mechanism. Recently, Weinshilboum and Axelrod (33) have shown that DBH is present in the sera of both man and rat. Furthermore, Weinshilboum et al. (34) found that stimulation of the sympathetic nervous system (i.e. by repeated immobilization) leads to an elevation of serum DBH, presumably because of its release from the sympathetic nerves.

As indicated above, normal levels of adrenal medullary PNMT are maintained by glucocorticoids; elevation of PNMT activity above normal, however, is neuronally dependent and is prevented by adrenal denervation (12,28). The fact that TH, DBH and PNMT are similarly influenced raises the questions of whether the enzymes involved in the synthesis of adrenal catecholamines represent a functional unit and whether their synthesis is regulated by a common mechanism (28).

During stress there is an increase in the activity of both the sympathetic nervous and the hypothalamo-pituitary-adreno-cortical systems. Therefore, it was convenient to use the response to stress in the study of the regulation of adrenal catecholamine-synthesizing enzymes. Repeated daily immobilization (2.5 hr/day) has been used as a model for stress (11). After 7 or more repeated immobilizations, there is a threefold increase in activity of adrenal TH (11), a two- to three-fold increase in DBH (13) and a 50% increase in levels of PNMT (11). The increases in enzyme activity after repeated immobilization provided a means of examining factors regulating levels of catecholamine-synthesizing enzymes. Neuronal regulation of the adrenal medullary enzymes could be evaluated after denervation of the left adrenal gland by severing the left splanchnic nerve; the intact right gland served as a control. After repeated immobilization, there is a striking increase in TH and DBH activities in the intact right adrenal glands but no significant alteration of enzyme activities in the denervated adrenals (11,13) (Fig. 1). From these results it is clear that an intact nerve supply is necessary for an increase in the three enzymes in response to immobilization stress. Repeated immobilization results in an elevation of PNMT in both intact and denervated adrenals although the levels are higher in the intact than in the denervated gland (11).

Endocrine status also markedly influences stress-induced alterations in the levels of adrenal enzymes concerned with catecholamine biosynthesis. In normal animals, increases in adrenal TH and PNMT are evident after a single immobilization and become more marked after repeated immobilization (11). In hypophysectomized rats, adrenal levels of TH and PNMT are much lower than in sham-operated control animals. Relative increases of adrenal TH levels are the same (about threefold) in immobilized operated animals as in sham-operated animals, but the absolute levels remain much lower in adrenals of hypophysectomized rats than in adrenals of immobilized sham-operated controls (Fig. 2).

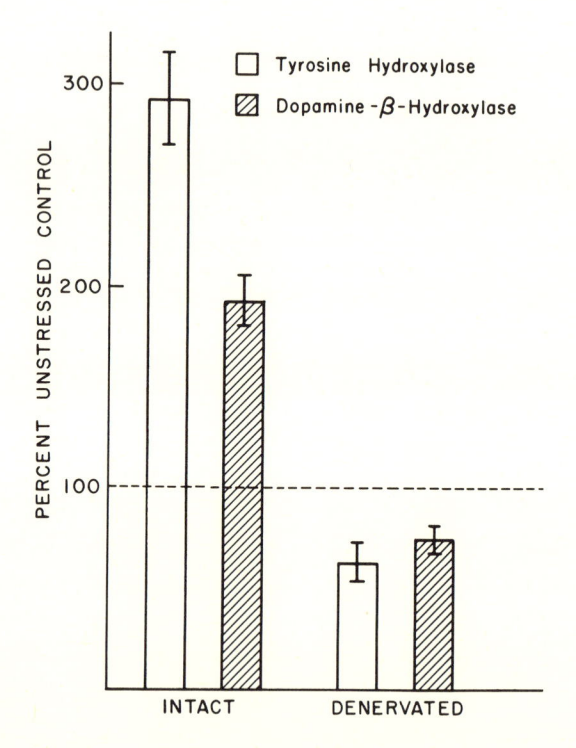

Figure 1. Effect of adrenal denervation on immobilization-
-induced elevation of tyrosine hydroxylase and dopamine-
-beta-hydroxylase activity. The left adrenal gland was
denervated by splanchnic nerve section 4 days prior to the onset
of the first of 7 daily 2·5 hr periods of immobilization.
Results are expressed as % of unstressed control and are mean
values ± S.E.M. for groups of 6 rats.

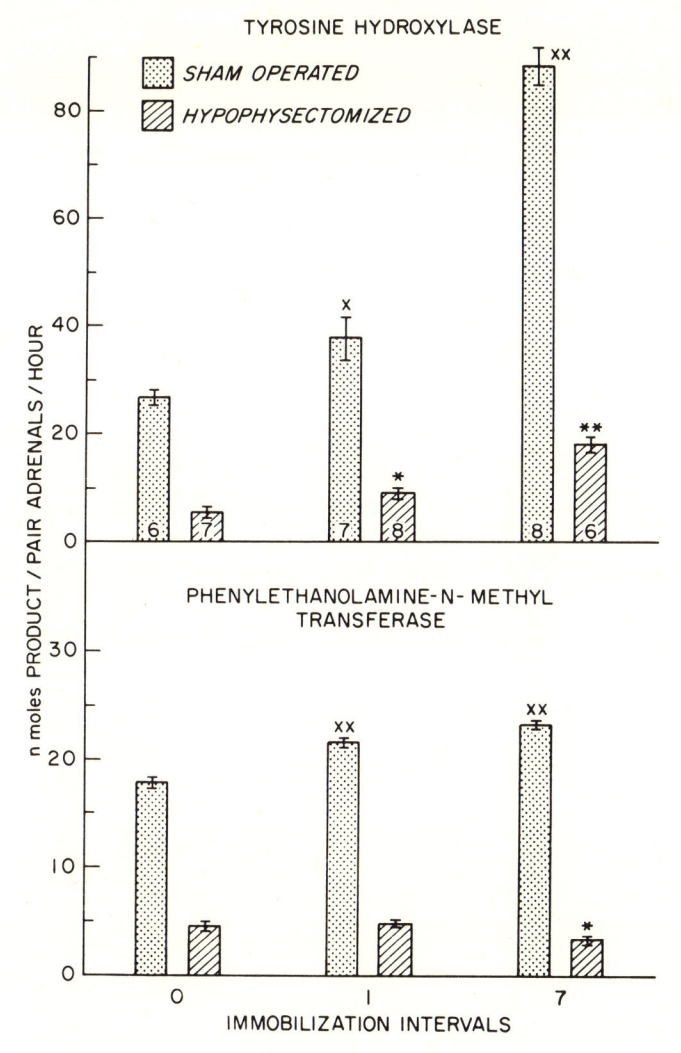

Figure 2. Effect of repeated intervals of immobilization on tyrosine hydroxylase and phenylethanolamine-N-methyltransferase levels in the adrenals of hypophysectomized and sham-operated rats. The number of daily 2·5 hr intervals of immobilization is indicated on the abscissa. Rats were killed 6 hr after the last immobilization when they were 2 weeks post-hypophysectomy. Results are the mean values ± S.E.M. for the number of rats indicated in the columns. Statistical significance compared to unstressed sham-operated group: x = $P < 0.02$; xx = $P < 0.001$. Statistical significance compared to unstressed hypophysectomized group: * = $P < 0.02$; ** = $P < 0.001$. (From Endocrinology 87: 1323, 1970.)

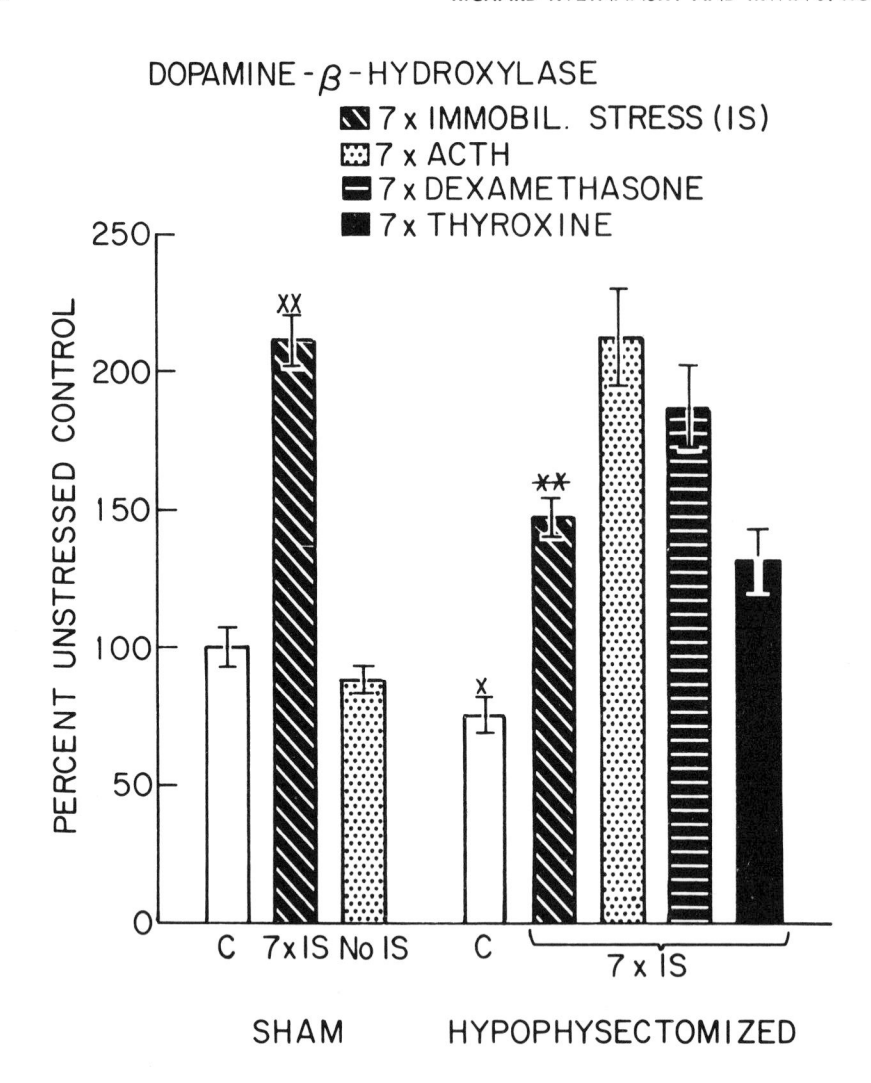

Figure 3. Effect of repeated intervals of immobilization and hormonal replacement on dopamine-beta-hydroxylase levels in the adrenals of hypophysectomized and sham-operated rats. Results are expressed as % of unstressed sham-operated animals and are mean values ± S.E.M. for groups of 6 to 7 rats killed 6 hr after the seventh daily immobilization for 2·5 hr and 37 days post-surgery. The doses administered daily (s.c. in 0·25 ml/animal) 60 min before each of the 7 immobilization periods were as follows: ACTH, 5 IU; dexamethasone, 1 mg; thyroxine, 10 µg. Control animals received saline injection. x = $P < 0.02$; xx = $P < 0.001$ compared to control sham-operated group. ** = $P < 0.001$ compared to control hypophysectomized group.

Repeated immobilization of hypophysectomized rats does not result in an increase in adrenal PNMT levels; instead, there is a small but significant decrease (Fig. 2).

The response to immobilization of adrenal DBH is similar to that of TH (13). Repeated immobilization results in a twofold increase in adrenal levels of DBH in both hypophysectomized and sham-operated rats, but the absolute level of the enzyme is much lower in the hypophysectomized animals (Fig. 3).

In unstressed hypophysectomized rats, treatment with ACTH partially restores the levels of adrenal TH (12). Repeated immobilization of ACTH-treated hypophysectomized rats results in elevation of adrenal TH levels to twice those found in untreated hypophysectomized animals (Table 1). Adrenal levels of DBH and PNMT in immobilized hypophysectomized animals are also partially restored by ACTH treatment. Dexamethasone treatment of hypophysectomized animals increases the levels of PNMT and DBH (5) but does not result in alteration of the TH levels (Table 1). Thyroxine does not appear to influence depressed levels of any of the 3 enzymes in the adrenals of hypophysectomized rats (Table 1).

After immobilization, there is a marked decrease in the epinephrine content of the rat adrenal because its synthesis fails to keep pace with its rapid release during stress (9). In hypophysectomized animals, the epinephrine levels in the adrenal gland are lower than normal and are further depressed by repeated immobilization (Fig. 4). The norepinephrine levels, however, are not decreased after repeated immobilization of hypophysectomized rats; only the rate of the last step, conversion of norepinephrine to epinephrine, seems to be diminished, which is similar to depressed levels of PNMT. Furthermore, treatment of hypophysectomized rats with either ACTH or dexamethasone prevents the marked immobilization-induced decrease in adrenal epinephrine (Fig. 4). **ACTH influences TH, DBH and PNMT, but treatment with dexamethasone restores only PNMT levels towards normal.** Thus, elevation of PNMT is associated with enhanced ability to synthesize epinephrine from norepinephrine and with maintenance of epinephrine levels in the adrenal glands of immobilized rats.

As indicated above, denervation of the adrenal gland prevents elevation of TH found after repeated immobilization. In hypophysectomized rats, denervation of the adrenal also prevents the immobilization-induced elevation of TH (Table 2). If ACTH is administered before each immobilization, the intact adrenal gland attains levels of TH comparable to those seen in normal rats which had been immobilized; but the content of TH in the denervated adrenal is only partially, but significantly ($P < 0.01$), elevated. Thus, both humoral and neuronal factors appear to influence the levels of TH in the adrenal medulla.

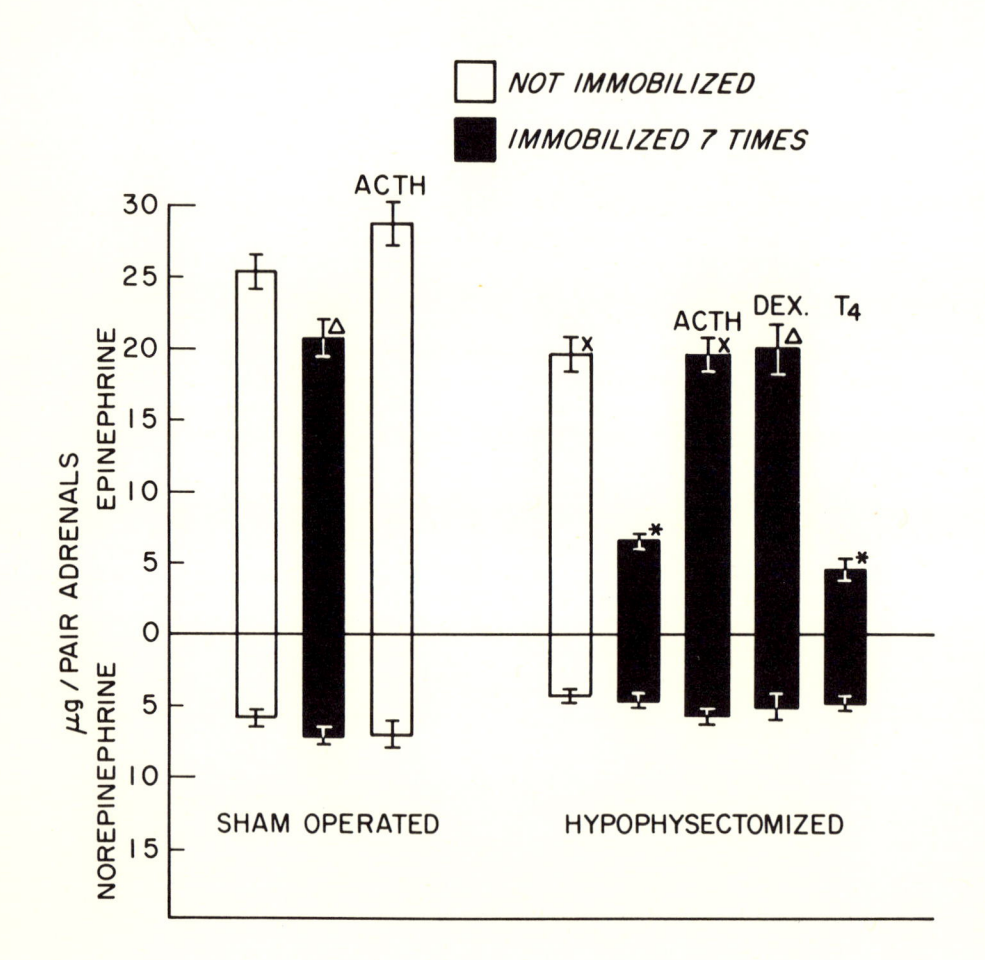

Figure 4. Effect of repeated intervals of immobilization on the catecholamine levels in the adrenal glands of sham-operated and hypophysectomized untreated or treated animals. Rats underwent surgery 1 week before the beginning of immobilization intervals (2·5 hr daily) and were killed 14 days after hypophysectomy. Hypophysectomized animals received either 5-IU ACTH, 1-mg dexamethasone or 10-μg L-thyroxine (s.c. in 0·25 ml/animal) 60 min before each of the 7 immobilization periods. Results are the mean values ± S.E.M. for groups of 6 to 7 animals. Statistical significance compared to sham-operated non-immobilized group: Δ = $P < 0.05$; x = $P < 0.01$; * = $P < 0.001$. (From Endocrinology 87: 1323, 1970.)

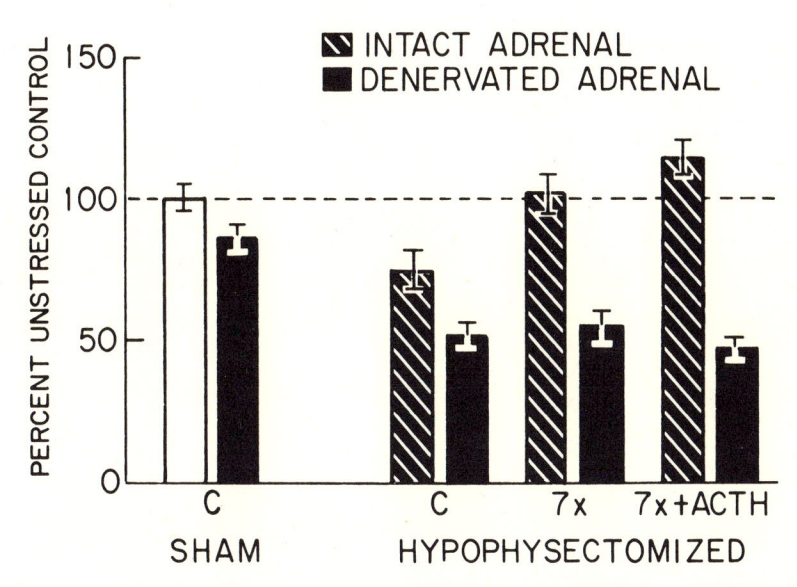

Figure 5. Effect of adrenal denervation on immobilization-induced elevation of dopamine-beta-hydroxylase in the adrenals of hypophysectomized rats. Results are expressed as percent of unstressed sham-operated group and are mean values ± S.E.M. for 6 to 7 rats. Animals were immobilized for 2·5 hr daily and killed 6 hr after the seventh immobilization (15 days after hypophysectomy and 12 days after denervation). ACTH (5 IU,s.c.) was administered to each rat 60 min before each immobilization period.

Table I. Effect of Hormonal Replacement on Tyrosine Hydroxylase (TH) and Phenylethanolamine-N-methyl Transferase (PNMT) levels in Adrenals of Hypophysectomized Immobilized Rats[a]

Treatment	Number of Immobilizations[b]	Hormone[c]	Adrenal Weight (mg/Pair)	TH (Units/Pair)	PNMT (Units/Pair)
Sham-operated	0	--	37.8 ± 1.4	31.9 ± 2.4	19.2 ± 0.6
	7	--	47.2 ± 1.7[d]	104.1 ± 7.5[d]	24.4 ± 1.4[d]
	0	ACTH	60.1 ± 2.0[d]	34.2 ± 3.0	24.4 ± 0.6[d]
Hypophysectomized	0	--	12.5 ± 0.7[d]	4.9 ± 0.9[d]	5.3 ± 0.3[d]
	7	--	14.5 ± 0.6[d]	29.6 ± 2.9[e]	4.5 ± 0.2[d]
	7	ACTH	29.4 ± 1.5[d,e,f]	66.7 ± 4.5[d,e,f]	17.9 ± 0.7[e,f]
	7	Dexamethasone	12.7 ± 0.8[d]	22.6 ± 2.5[e]	15.0 ± 0.7[d,e,f]
	7	Thyroxine	13.6 ± 1.0[d]	24.1 ± 4.0[e]	3.3 ± 0.3[d,e,f]

[a] Results are expressed as units (nmole product/hr) per pair of adrenal glands and are the mean values ± S.E.M. for groups of 6 or 7 animals. Rats were killed 14 days after surgery.

[b] Rats were immobilized for 2.5 hr daily.

[c] The doses administered daily 60 min before each of the 7 immobilizations (s.c. in 0.25 ml to each animal) were as follows: ACTH, 5 IU; dexamethasone, 1 mg; thyroxine, 10μg. Control animals received saline injection.

[d] Compared to sham-operated animals which were not immobilized, $P < 0.01$.

[e] Compared to untreated unstressed hypophysectomized animals, $P < 0.01$

[f] Compared to untreated stressed hypophysectomized animals, $P < 0.01$.

Table II. Effect of Adrenal Denervation on Immobilization-induced elevation of Tyrosine Hydroxylase in the Adrenals of hypophysectomized rats[a]

Treatment	Number of Immobilizations	Intact Side	Denervated Side
Sham-operated	0	13.8 ± 1.2	12.1 ± 1.1
Hypophysectomized	0	5.8 ± 0.8[b]	5.0 ± 0.5[b]
Hypophysectomized	7	14.4 ± 2.7	4.2 ± 1.9[b,c]
Hypophysectomized and ACTH	7	45.8 ± 1.9[b,d]	8.6 ± 0.9[c,d]

[a] Results are expressed as nmole product/one adrenal gland and are the mean values (± S.E.M.) for groups of 6 or 7 rats. Animals were immobilized for 2.5 hr daily and killed 6 hr after the seventh immobilization (15 days after hypophysectomy and 12 days after denervation). ACTH (5 IU/rat s.c.) was administered 60 min before each immobilization.

[b] $P < 0.01$ compared to sham-operated group

[c] $P < 0.01$ compared to intact side

[d] $P < 0.01$ compared to hypophysectomized, 7 times immobilized animals.

The decreased DBH activity in the adrenals of hypophysectomized rats is further diminished by denervation, and denervation prevents its elevation by immobilization (5). Denervation also appears to block the ACTH-induced elevation of DBH in the adrenals of hypophysectomized rats (Fig. 5). Thus, both DBH and TH are under dual hormonal-neuronal regulation. In hypophysectomized rats, severing the splanchnic nerve prevents the diminution of PNMT (12) produced by repeated immobilization (Table 3). Treatment with ACTH restores the elevation of PNMT produced by repeated immobilization in normal rats, but denervation partially interferes with restoration of the elevated enzyme level (Table 3). Adrenal corticoids or ACTH (presumably mediated by adrenal corticoids) are more important than neuronal factors in determining the level of PNMT.

ACTH appears to influence levels of PNMT in the adrenal medulla through its effects on steroid production in the adrenal cortex (36), but its action on adrenal TH appears to be mediated through a different mechanism. High doses of steroids, which reverse the effects of hypophysectomy on PNMT, do not restore levels of TH.

Table III. Effect of Adrenal Denervation on Immobilization-induced Changes of Phenylethanolamine-N-methyl Transferase[a] in the Adrenals of Hypophysectomized Rats*

Treatment	Number of Immobilizations	Intact Side	Denervated Side
Sham-operated	0	24.1 ± 0.9	21.6 ± 1.1
Hypophysectomized	0	8.3 ± 0.5[b]	6.9 ± 0.8[b]
Hypophysectomized	7	5.6 ± 0.4[b]	6.9 ± 1.1[b]
Hypophysectomized and ACTH	7	26.5 ± 2.1[d]	16.9 ± 1.1[b,c,d]

*Animals were immobilized for 2.5 hr daily and killed 6 hr after the seventh immobilization. ACTH (5 IU/rat s.c.) was administered 60 min before each immobilization.

[a] Results are expressed as nmole product/one adrenal gland and are the mean values (± S.E.M.) for groups of 6 or 7 rats.

[b] $p < 0.01$ compared to sham-operated group

[c] $p < 0.01$ compared to intact side

[d] $p < 0.01$ compared to hypophysectomized-immobilized animals

In the adrenal cortex, cyclic adenosine-monophosphate (cyclic-AMP) is thought to mediate the effect of ACTH on steroidogenesis. It is widely believed that cyclic-AMP acts as a secondary messenger which mediates the effect of ACTH on steroidogenesis in adrenocortical cells (6), and the possibility that it mediates the effects of ACTH on adrenal medullary TH activity during immobilization stress was considered (25). Immobilization for 30 min markedly increased the adrenal cyclic-AMP levels; with repeated immobilization, an additional rise in cyclic-AMP levels was noted. Hypophysectomy blocked the immobilization-induced rise in adrenal cyclic-AMP. Approximately 50% of the total adrenal cyclic-AMP level is in the medulla, and stress does not cause an increase in cyclic-AMP in the adrenal medulla. Denervation of the adrenal gland, however, reduced to about one-half the elevation of cyclic AMP found in the intact adrenal of immobilized rats (25). After denervation, cyclic-AMP levels were decreased only in the cortical part of the adrenal; there were no changes in the adrenal medulla although the medulla (not the cortex) is the target of most of the neurons in the splanchnic nerve. It is possible that either the splanchnic nerve or the adrenal medulla releases a factor which takes part in the regulation of adrenocortical cyclic-AMP. Because of the direction of blood flow from the cortex to the medulla, it seems unlikely that the medullary secretions are delivered to the cortex.

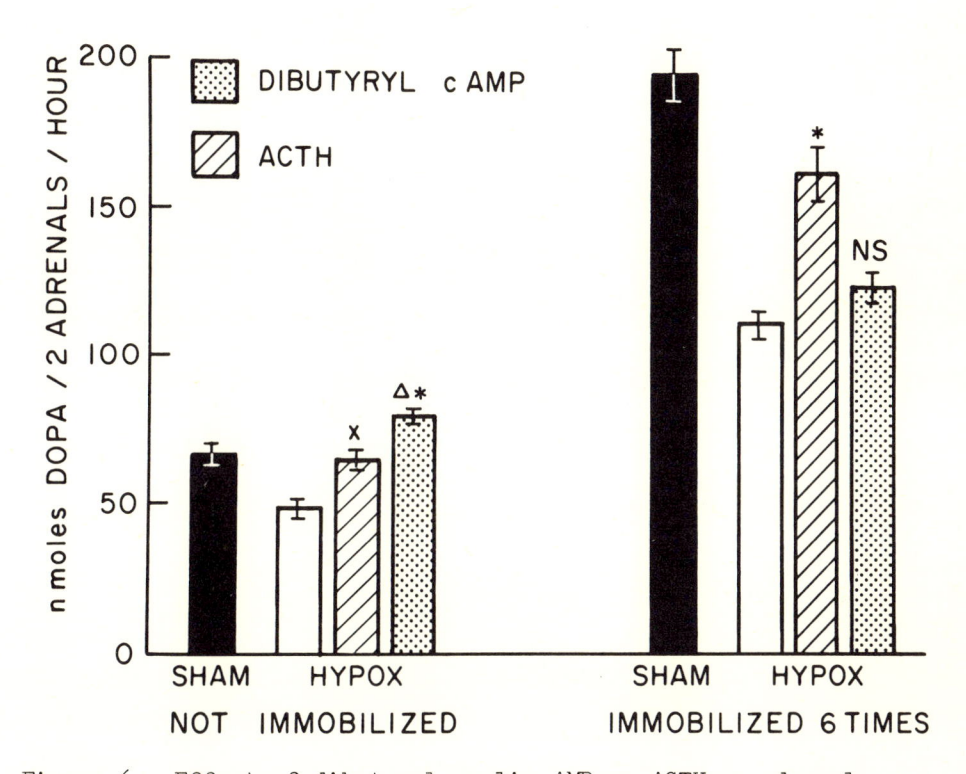

Figure 6. Effect of dibutyryl cyclic-AMP or ACTH on adrenal tyrosine hydroxylase levels in hypophysectomized rats. Hypophysectomized rats were treated for 6 days with either dibutyryl cyclic-AMP (25 mg s.c./rat twice daily) or ACTH (5 IU s.c./rat/day) Some rats were immobilized 6 times for 2·5 hr daily. Both drug treatment and immobilization were begun 2 days after hypophysectomy, and the animals were killed 8 days after hypophysectomy and 6 hr after the last immobilization interval. Results are expressed as mean ± S.E.M. for groups of 6 or 7 rats.
x = P < 0·01 compared to untreated hypophysectomized group;
✦ = P < 0·001 compared to untreated hypophysectomized group;
Δ = P < 0·02 compared to sham-operated control group; NS = not significantly different from untreated immobilized hypophysectomized group. (From Endocrinology 89:50, 1971)

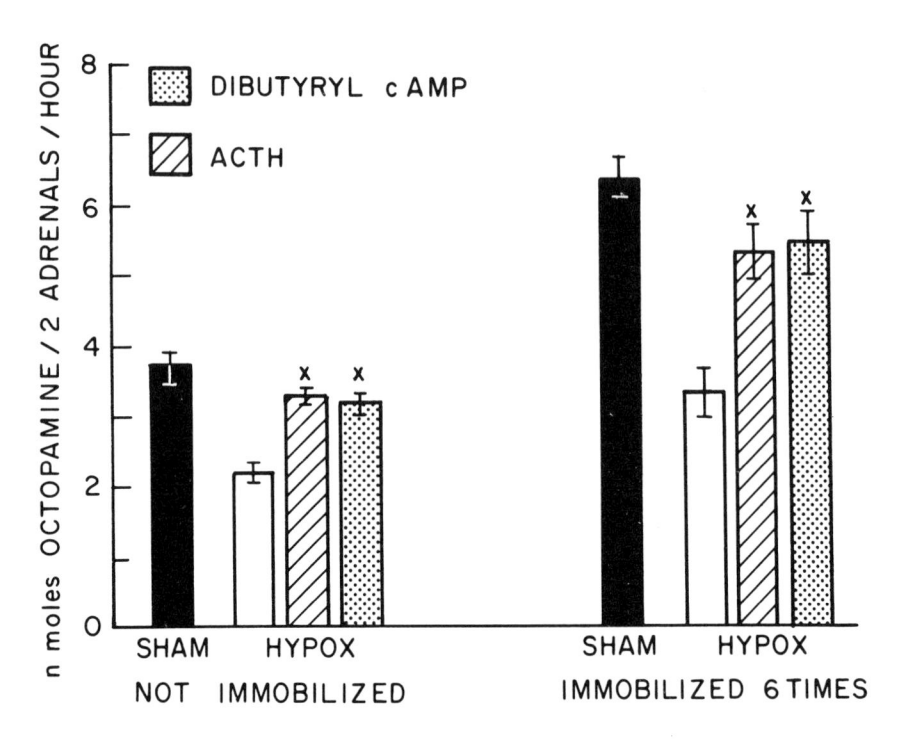

Figure 7. Effect of dibutyryl cyclic-AMP or ACTH on adrenal dopamine-beta-hydroxylase levels in hypophysectomized rats. Hypophysectomized rats were treated for 6 days with dibutyryl cyclic-AMP (25 mg, s.c./rat twice daily) or ACTH (5 IU,s.c./rat/day). Some rats were immobilized 6 times for 2·5 hr daily. Both drug treatment and immobilization were begun 2 days after hypophysectomy, and the animals were killed 8 days after hypophysectomy and 6 hr after the last immobilization period. Results are expressed as mean ± S.E.M. for groups of 6 to 7 rats. x = $P < 0.01$ compared to untreated hypophysectomized group. (From Endocrinology 89: 50, 1971)

Thus, it appears probable that some fibers from the splanchnic nerve may innervate the cortex and exert a synergistic influence on changes in cyclic-AMP induced by pituitary factors (25). Adrenals of both sham-operated and hypophysectomized animals repeatedly treated with ACTH contain elevated levels of cyclic-AMP (6,25).

The administration of dibutyryl cyclic-AMP to hypophysectomized rats results in both an increase in adrenal weights and adrenal content of DNA, RNA and protein and maintenance of adrenal cortical responsiveness to administered ACTH (24). To study whether ACTH allows immobilization-induced elevation of adrenal TH activity by direct or indirect action on cyclic-AMP production, dibutyryl cyclic-AMP was administered to hypophysectomized rats before each of 6 repeated immobilizations. Repeatedly immobilized hypophysectomized rats treated with ACTH show marked increases in adrenal TH levels (15), but treatment of the stressed animals with dibutyryl cyclic-AMP is not effective (Fig. 6). A slight elevation of adrenal TH activity induced by administration of dibutyryl cyclic-AMP in hypophysectomized unstressed animals may be mediated by an effect of the drug which increases splanchnic nerve activity.

Our results with dibutyryl cyclic-AMP administration (15) support the view that TH activity does not appear to be influenced by glucocorticoid secretion (12,22). Adrenal TH activity is therefore most probably regulated by a direct effect of ACTH on the adrenal medulla or by an unknown factor released from the adrenal cortex.

ACTH or dibutyryl cyclic-AMP treatment increase DBH activity in both hypophysectomized control and hypophysectomized immobilized rats (Fig. 7). Because dibutyryl cyclic-AMP elevates adrenal DBH activity without a marked increase in adrenal weights or elevation of PNMT levels, it is unlikely that the elevation of DBH activity is mediated by adrenal corticosteroids. ACTH may therefore have a direct effect on the adrenal DBH activity which is mediated by cyclic-AMP.

CONCLUSIONS

1. The first physiologically induced elevation of adrenal TH activity and DBH activity has been shown.
2. Regulation of adrenal TH activity requires both intact neuronal and pituitary-adrenal systems. The humoral regulation of adrenal TH is most probably mediated by a direct effect of ACTH on the adrenal medulla or by an unknown factor (not cyclic-AMP) released from the adrenal cortex following ACTH administration.
3. Adrenal DBH activity is also under both neuronal and humoral control, but the neuronal regulation appears to be the more important. Humoral regulation by ACTH is probably directly mediated to adrenal medullary DBH by cyclic-AMP.

4. Adrenal PNMT activity is also controlled by both humoral and neuronal factors, but the humoral/pituitary-adrenocortical system appears to be the more important.

Supported in part by Foundations' Fund for Research in Psychiatry, New Haven, Connecticut, Research Grant Number G 71-498

REFERENCES

1. Axelrod, J. (1962) J. Biol. Chem. 237, 1657.
2. Axelrod, J., Mueller, R.A., Henry, J.P. & Stephens, P.M. (1970) Nature, 225, 1059.
3. Blaschko, H. (1939) J. Physiol. 96, 50P
4. Coupland, R.E. (1953) J. Endocrinol. 9, 194.
5. Gewirtz, G.P., Kvetňanský, R.K., Weise, V.K. & Kopin, I.J. (1971) Molec. Pharmacol. 7, 163.
6. Grahame-Smith, D.G., Butcher, R.W., Ney, R.J. & Sutherland, E.W. (1967) J. Biol. Chem. 242, 5535.
7. Holtz, P., Heise, R & Ludtke, K. (1938) Arch. Exp. Path. Pharmak, 191: 87.
8. Kirshner, N. & Goodall, M. (1959) Biochim. Biophys. Acta 24, 658.
9. Kvetňanský, R.K. & Mikulaj, L. (1970) Endocrinology 87, 738
10. Kvetňanský, R.K., Weise, V.K. & Kopin, I.J. (1969) Pharmacologist 11, 274.
11. Kvetňanský, R.K., Weise, V.K. & Kopin, I.J. (1970) Endocrinology 87, 744.
12. Kvetňanský, R.K., Gewirtz, G.P., Weise, V.K. & Kopin, I.J. (1970) Endocrinology 87, 1323.
13. Kvetňanský, R.K., Gewirtz, G.P., Weise, V.K. & Kopin, I.J.(1971) Molec. Pharmacol. 7, 81.
14. Kvetňanský, R.K., Gewirtz, G.P., Weise, V.K. & Kopin, I.J. (1971) Amer. J. Physiol. 220, 928.
15. Kvetňanský, R.K., Gewirtz, G.P., Weise, V.K. & Kopin, I.J. (1971) 89, 50.
16. Levin, E.Y., Levenberg, B. & Kaufman, S. (1960) J. Biol. Chem. 235, 2080.
17. Levitt, M., Spector, S., Sjoerdsma, A. & Udenfriend, S. (1965) J. Pharmacol. exp. Ther. 148, 1.
18. Lovenberg, W., Weissbach, H. & Udenfriend, S. (1962) J. Biol. Chem. 237, 89.
19. Molinoff, P.B., Brimijoin, S., Weinshilboum, R. & Axelrod, J. (1970) 66, 453.
20. Mueller, R.A., Thoenen, H. & Axelrod, J. (1969) Science 163, 468.
21. Mueller, R.A., Thoenen, H. & Axelrod, J. (1969) J. Pharmacol. 169, 74.
22. Mueller, R.A., Thoenen, H. & Axelrod, J. (1970) Endocrinology 86, 751.

23. Nagatsu, T., Levitt, H. & Udenfriend, S. (1964) J.Biol. Chem. 239, 2910.
24. Ney, R.L. (1969) Endocrinology 84, 168.
25. Paul, M.I., Kvetnansky, R. Cramer, H., Silbergeld, S. & Kopin, I.J. (1971) Endocrinology 88, 338.
26. Silbergeld, S., Kvetnansky, R., Weise, V.K. & Kopin, I.J. (1971) Biochem. Pharmacol. 20, 1763.
27. Thoenen, H., Mueller, R.A. & Axelrod, J. (1969) J. Pharmacol. 169, 249.
28. Thoenen, H., Mueller, R.A. & Axelrod, J. (1970) Biochem. Pharmacol. 19, 669.
29. Thoenen, H. (1970) Nature 228, 861.
30. Viveros, O.H., Arqueros, L., Connett, R.J. & Kirshner, N. (1969) Molec. Pharmacol. 5, 60.
31. Viveros, O.H., Arqueros, L., Connett, R.J. & Kirshner, N. (1969) Molec. Pharmacol. 5, 69.
32. Weinshilboum, R. & Axelrod, J. (1970) Endocrinology, 87, 894.
33. Weinshilboum, R. & Axelrod, J. (1971) Circ.Res. 28, 307.
34. Weinshilboum, R., Kvetnansky, R., Axelrod, J. & Kopin, I.J. (1971) Nature (New Biology) 230, 287.
35. Wurtman, R.J. & Axelrod, J. (1965) Science, 150, 1464.
36. Wurtman, R.J. & Axelrod, J. (1966) J. Biol. Chem. 241, 2301.
37. Wurtman, R.J. (1967) In Recent Advances in Biological Psychiatry, 9, 359.

THE ROLE OF LYSOSOMAL HYDROLASES IN THE MECHANISM OF SHOCK

James J. Smith, Daniel J. Loegering, Daniel J. McDermott and Michael L. Bonin

Department of Physiology, Medical College of Wisconsin
Milwaukee, Wisconsin 53233

The early work of Zweifach, et al. (1) and Fine, et al. (2) suggested that the reticuloendothelial system might be importantly related to the widespread systemic failure in clinical and experimental shock.

In subsequent reports and reviews (3-5), increasing evidence accumulated that the reticuloendothelial system (RES) was functionally disturbed and hypoactive in shock, that artificially induced RES blockage or hypofunction increased the sensitivity of the animal to subsequent shock and also that an induced RES hyperfunction increased the tolerance to experimental shock. If it is true that the reticuloendothelial system is related and importantly concerned in shock survival, the question is what mechanism is involved in this RES effect.

Two main possibilities have been advanced in this regard; one is that the pathological effects in shock are induced by a toxin, such as endotoxin, histamine or other agent, whose progressive effect is inadequately counteracted by a now saturated and ineffective RES. A second possibility is that the RES, damaged in shock, releases toxic substances from the RES cells themselves. Particularly prominent in this latter concept has been the notion that lysosomal hydrolases are the responsible agents. If these intracellular enzymes are released through a cell membrane effect with perhaps an associated release of other toxic agents trapped in the cell, then the two theories are closely related and perhaps the two effects co-exist.

Janoff, Weissmann and their colleagues were among the earliest workers to report the association of lysosomal damage and mortality in experimental shock (6,7). Their reports and others indicated lysosomal involvement not only in trauma (6) and endotoxin (7) stress, but also in hemorrhagic shock (8,9). Horpacsy et al. (10) found that plasma levels of lactate, pyruvate and excess lactate were elevated early in shock and lysosomal enzymes only later and in severe shock and that the concentration of the latter paralleled the degree of shock. The RES has been implicated since it is known that RES tissues are rich in these hydrolases. Schumer, et al. (11) did not find a relationship between lysosomal breakdown and progression of symptoms in experimental oligemia in rats but there is a question in these experiments as to whether the stress imposed was adequate, that is, whether severe shock was actually produced.

In a series of interesting experiments, Lefer and his colleagues (12) found in the plasma of shock animals, a myocardial depressant factor (MDF) which appears to be a heat stable, dialyzable peptide of low molecular weight (88:1000). Lefer and Verrier (13) have recently reviewed the probable mechanisms of the corticosteroid effect in shock and have stated the more likely possibilities to be: (a) a direct effect on the heart or other part of the vascular system; (b) a possible neutralization of endotoxin or other vasoactive substance; (c) a metabolic effect, such as a decrease in lactate production or (d) a stabilizing action of corticosteroids on lysosomal membranes. Current evidence indicates that the last of these possibilities is the most plausible, thus adding additional credibility to the idea of a lysosomal hydrolase effect in shock pathology.

Site of Origin of Lysosomal Hydrolases

While it is known that lysosomes are heavily concentrated in RES tissues, it is also evident that they are widely scattered throughout the cells of other organs and tissues. In an effort to obtain more precise information on the role of lysosomal hydrolases in shock, increasing attention is being given to the site of origin of these enzymes. Bitensky et al. (8) emphasized the importance of the spleen as a source. Sutherland and his co-workers (14) gathered evidence from cytological studies of intestinal mucosa and lysosomal enzyme concentration during hemorrhagic shock in the dog that their primary source was small intestine and that the liver probably was not an important site of origin of these enzymes. A report that exclusion of the distal half of the small intestine substantially increased the survival rate of dogs in hemorrhagic shock also suggests the possible importance of the intestinal tract in the pathogenesis of shock (15).

Wiedmeier et al. (16) reported that the injection of blocking doses of certain inorganic colloids, such as silicon dioxide and colloidal carbon, resulted in marked hypotension and shock. These "toxic" colloids produced necrobiotic effects in RES macrophages of liver, spleen and lung with elevated plasma levels of acid phosphatase and β-glucuronidase; however, other colloids such as thorotrast and polystyrene latex caused no hypotension or toxicity and no elevation of lysosomal enzyme levels (16, 17). The prominence of the RES pathology and relative lack of changes in other tissues suggested that the cytoxic changes in RES tissues were closely related to the lysosomal enzyme release and to the hypotensive effects which resembled the circulatory collapse in shock (15).

Recently, Lefer and Martin (18) reaffirmed their belief that splanchnic ischemia and the resultant release of lysosomal hydrolases were important in the mechanism of shock and suggested further that the pancreas was the key source of the lysosomal enzymes; a later report from that laboratory indicated that glucocorticoids significantly stabilized the pancreatic lysosomes both in vivo and in vitro (19).

A current study in our own laboratory has the objective of analyzing further the relationship between lysosomal enzymes and hemorrhagic shock in the dog and the possible sources of these enzymes. Male dogs weighing 10-16 kg were dewormed, given distemper shots and observed for 10 days prior to use. The animals were anesthetized with sodium pentobarbital (30 mg/kg, iv) and the right femoral vein and artery and right jugular vein exposed and cannulated. The abdomen was opened, the portal vein cannulated with a polyethylene catheter and the abdomen closed. Catheters were then inserted into the aorta, inferior vena cava, and hepatic vein and hemorrhagic shock was induced by bleeding into a pressurized reservoir. A mean arterial pressure of 50 mm of mercury was maintained for 3 hr, which under our conditions results in a severe and usually irreversible shock.

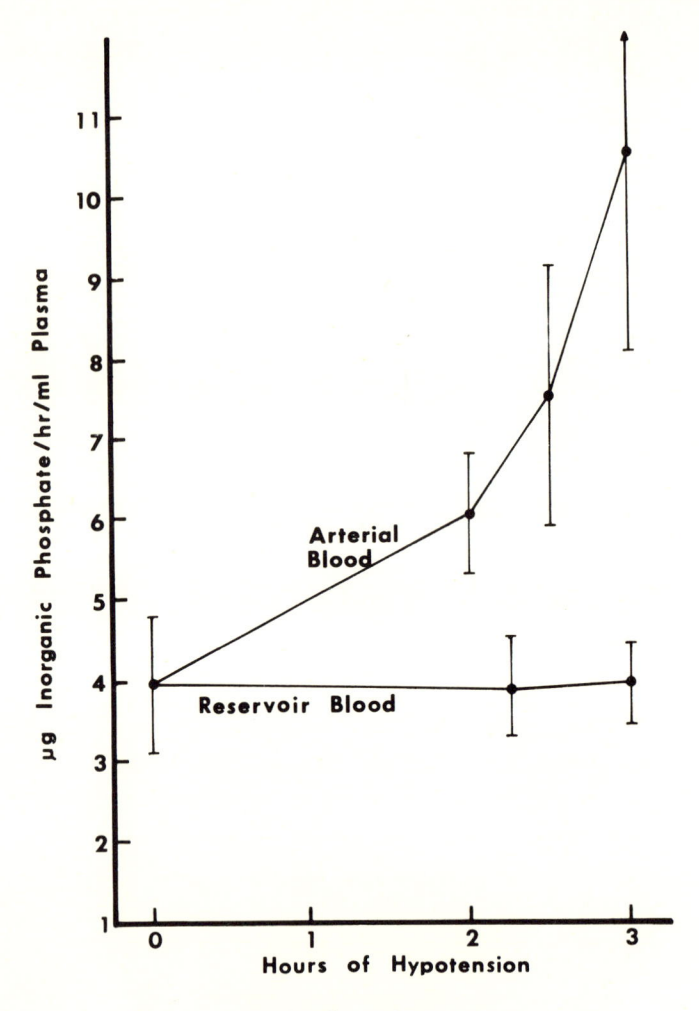

Figure 1 - Plasma Acid Phosphatase Activity
in Hemorrhagic Shock

In Fig 1 are shown the mean plasma acid phosphatase levsls and
in Fig 2 mean cathepsin levels at different stages in shock.
From a control value of approximately 4 µg phosphate/hr/ml plasma
of acid phosphatase per 100 ml of plasma, the concentration
increased after the 3 hr hypotension to an average value of
approximately 10.5; in the case of cathepsin there was likewise

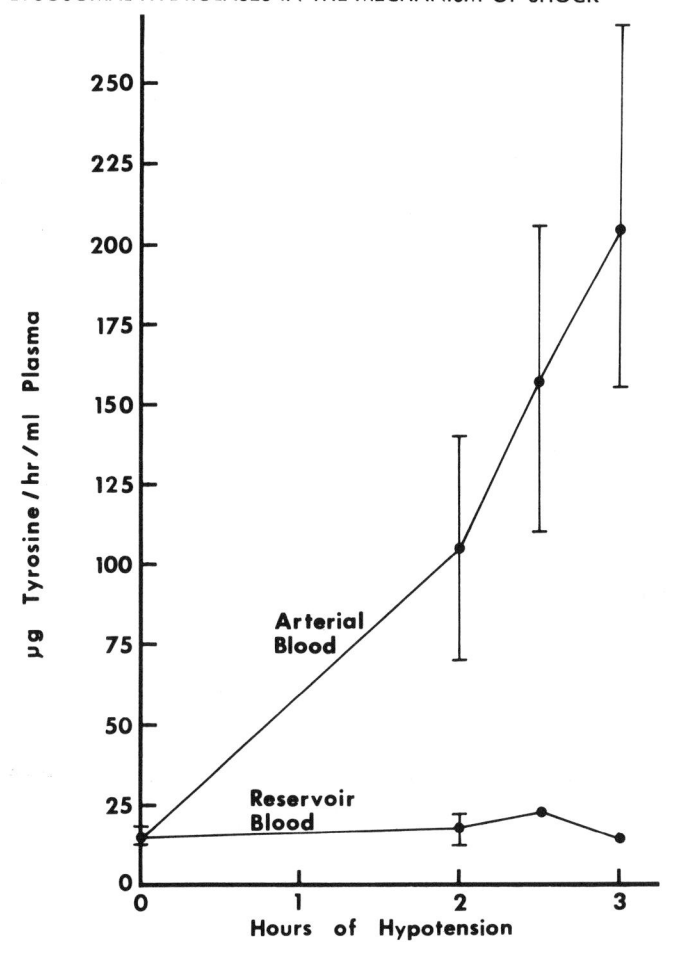

Figure 2 - Plasma Cathepsin Activity
in Hemorragic Shock

a sharp increase from a mean level of approximately 13 to about 205 at the end of the hypotension. There is considerable parallelism in the response of these two enzymes, both directionally and quantitatively and also a parallelism between enzyme concentration in the plasma and the degree of shock as characterized by the percent of blood uptake, the circulatory status of the animal and the level of arterial PO_2. These studies tend to confirm previous reports of the association of shock with lysosomal pathology.

A further objective is to obtain more precise information on the origin of these lysosomal enzymes through sampling the effluent blood from various regions of the body.

As is well known, the liver contains about 80 to 85% of the RES clearance capacity of the body. While hepatic RES tissue is rich in lysosomes, it is likely that the liver is also a key organ in clearing such enzymes from the blood. Thus, in severe stress the liver might simultaneously serve as a clearing agent for lysosomal enzymes as well as a source of these hydrolases. In view of this, it would seem that a determination of the relative enzyme concentrations in portal and hepatic venous blood might be of interest.

The plasma acid phosphatase values in portal and hepatic venous blood during shock in these dogs are shown in Fig 3.

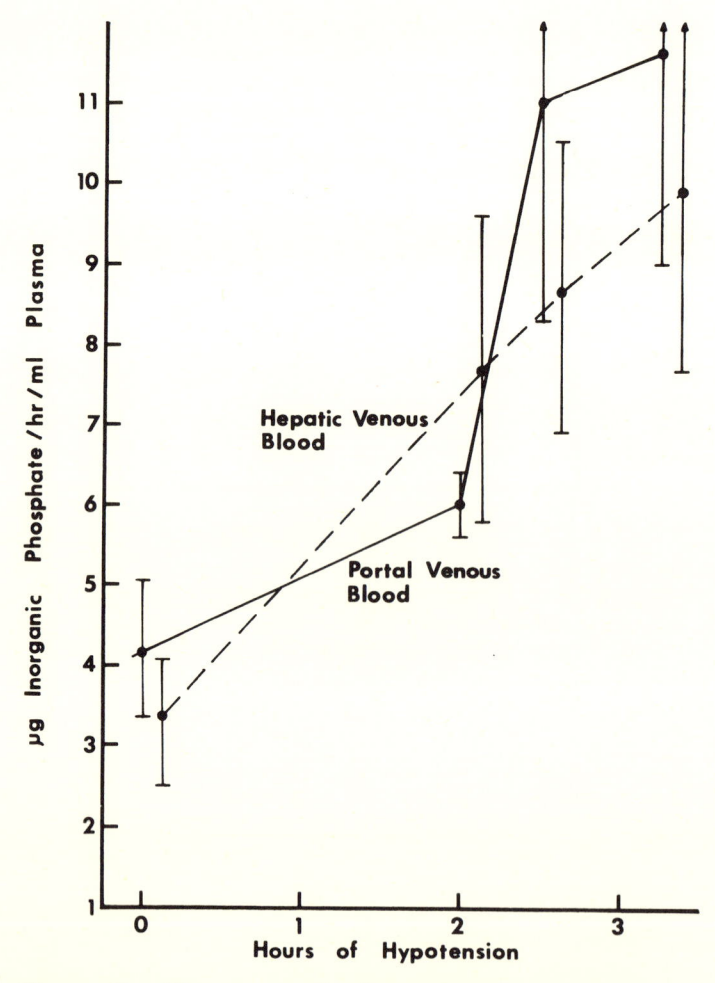

Figure 3 - Plasma Acid Phosphatase Activity in Hemorrhagic Shock

The mean values as well as results in individual animals studied thus far, show variability in portal-hepatic venous enzyme differences. The trend, however, seems to be toward an increase in portal concentration and an increase also in portal-hepatic venous gradient as shock progresses. This tendency was evident in the cathepsin values as well as in the acid phosphatase values. These results therefore do not indicate declining hepatic ability to clear the enzymes nor an increasing tendency for hepatic release. Further studies will be required in an effort to trace more specifically the sites of origin and sites of clearance of these enzymes.

A recent report on the induction of lysosomal enzymes in peritoneal macrophages indicates that the phagocytosis of red blood cells produced markedly increased lysosomal hydrolase levels, while in contrast, the phagocytosis of inert material such as polyvinyl or insoluble starch produced no such increase (20). Thus, the genesis of lysosomal enzymes was functionally related to the phagocytic process.

At the present time, specific deductions regarding the role of lysosomal hydrolases in irreversible shock are not possible. Van Lancker (21) has reiterated the fact that acid hydrolases are released in almost all forms of cellular necrosis but that this event is only questionably related to the mechanism which triggers cell death; he suggests that these hydrolases may not be involved in the sentence or execution of the cell but rather act as scavengers of the remains. In the case of hypofunction of an organ or death of an organism, the problem becomes even more complex. The lysosomal enzyme concept in shock seems therefore to be a promising one but as yet the evidence is largely circumstantial and a tangible cause-effect relationship is still unproven.

Summary

1) Research of the last 10 yr points increasingly to the role of the RES in the mechanism of death from shock; one of the more likely means through which the RES brings about its effects is via the release of lysosomal hydrolases into the plasma. These enzymes apparently bring about their effect either directly on systems of the body or through the formation of vasoactive agents from plasma precursors.

2) The evidence for the lysosomal concept rests primarily on 3 types of data:
   a) A close association found by a number of investigators between the increased concentration of lysosomal enzymes

in the plasma of shocked animals and the progression of shock. Cytological and cytochemical evidence regarding the site of origin of these enzymes in shock is fragmentary and not definitive.

b) Indication that the administration of certain colloids and toxins causes specific lysosomal changes in RES cellular components particularly liver and lung which are associated with an increase in plasma lysosomal enzyme concentration and with circulatory failure resembling the picture of experimental and clinical shock.

c) Increasing evidence that corticosteroids, which are perhaps the most effective specific agents in shock therapy, seem to produce their shock-protective action through stabilization of lysosomes.

3) While the lysosomal theory is one of the more promising explanations for the mechanism of irreversible shock, the evidence thus far is only indirect.

References

1. Zweifach, B.S., Benacerraf, B. & Thomas, L. (1957) J. Exp. Med. 106:403.

2. Fine, J., Rutenberg, S. & Schweinberg, F.B. (1954) J. Exp. Med. 110:547.

3. Zweifach, B.W. (1960) Ann. N.Y. Acad. Sci. 88:203.

4. Hershey, S.G. (1964) "Shock" Little, Brown, Boston.

5. Smith, J.J., McDermott, D.J. & Wiedmeier, V.T. In: "Shock-Metabolic Changes and Therapy." F.K. Schattauer, Stuttgard and New York, 1970, p. 403.

6. Janoff, A., Weissman, G., Zweifach, B.W. & Thomas, L. (1962) J. Exp. Med. 116:451.

7. Weissmann, G. & Thomas, L. (1962) J. Exp. Med. 116:433.

8. Bitensky, L., Chayen, J & Cunningham, G.J. (1963) Nature 199:493.

9. Lentz, P.E. & Smith, J.J. (1967) Proc. Soc. Exp. Biol. Med. 124:1243.

10. Horpacsy, G., Barankay, T., Tarnoky, K., Nagy, S. & Petri, G. (1971) Z. ges. exp. Med. 154:274.

11. Schumer, W., Kapica, S.H. & Teng, T. (1969) Surgery 99:325.

12. Lefer, A.M. & Martin, J. (1970) Circ. Res. 26:59.

13. Lefer, A.M. & Verrier, R.L. (1970) Clin. Pharm. Therap. 11:630.

14. Sutherland, N.G., Bounos, G. & Gurd, F.N. (1968) J. Trauma 8:350.

15. Gergely, M., Horpacsy, G. Barankay, T. & Petri, G. (1970) Acta Physiol. Hung. 37:301.

16. Wiedmeier, V.T., Johnson, S.A., Siegesmund, K.A. & Smith J.J. (1969) J. Reticuloendothelial Soc. 6:202.

17. Wiedmeier, V.T., Stekiel, W.J. & Smith, J.J. (1970) J. Reticuloendothelial Soc. 7:716.

18. Lefer, A.M. & Martin, J. (1970) Am. J. Physiol. 218:1423.

19. Jefferson, T.A., Glenn, T.M., Martin, J.B. & Lefer, A.M. (1971) Proc. Soc. Exp. Biol. 136:276.

20. Axline, S.G. & Cohn, Z.A. (1970) J. Exp. Med. 131:1239.

21. Van Lancker, J.L. In: Metabolic Conjugation and Metabolic Hydrolysis. Academic Press, N.Y. 1970.

These investigations were carried out with the assistance of research grants from the National Heart and Lung Institute (HE-06588) and from the Wisconsin Heart Association.

RETICULOENDOTHELIAL FUNCTION IN EXPERIMENTAL INJURY AND TOLERANCE

TO SHOCK

Burton M. Altura and S.G. Hershey

Departments of Anesthesiology and Physiology, Albert
Einstein College of Medicine of Yeshiva University,
New York City 10461

Although the morphology, ultrastructure and functions of the reticuloendothelial system (RES) are becoming well documented (64, 72), most of the information is immunologically oriented. The role of the RES in host defense mechanisms in injuries and systemic forms of stress is less understood.

Considerable interest in the relationship between the functional state of the RES and tolerance to experimental shock and trauma has developed (39, 42, 72, 83-85). Evidence suggests that tolerance to various types of experimental shock and trauma may be associated with the functional capacity of the phagocytic elements of the RES. Pretreatment of animals with various substances which stimulate RES phagocytic activity is, in many instances, associated with increased tolerance to shock (6,11,31 34,36,49,59,72,83-85,87), whereas materials which block or depress this system increase mortality (7,9,11,17,49,72,83-85,87).
It is interesting also that animals which are made tolerant to one form of shock are often cross-tolerant to other forms. Such evidence has been used (32,57,83-85,87) to suggest that the RES may represent the homeostatic system serving as a common pathway in the pathogenesis of and host adaptation to certain shock syndromes. Many of these chemicals and colloids are not without toxic effects on other organs (6,33,76,77,84) which makes RES studies in shock difficult to interpret. Furthermore, until the initiation of our work several years ago (2,4,5,9,37,38), the quantitative RES responses to experimental shock in normal animals were incompletely described (19,22,32,72,75,86).

It seemed logical to us to think that if there is a close relationship between reticuloendothelial function and tolerance to shock one should see the following: (a) animals subjected to various forms and degrees of experimental shock and injury should exhibit quantitative, sequential changes in RES activity; (b) animals which spontaneously recover from experimental shock should exhibit increased RES activity; (c) such animals might be expected to exhibit a cross-tolerance to other forms of experimental shock; (d) such cross-tolerant animals might not exhibit any signs of RES depression; and (e) it should be possible to manipulate RES activity pharmacologically during shock, i.e. increased RES activity should be seen in animals surviving shock after effective drug therapy. Furthermore, vasoactive drugs would be particularly appropriate test materials because their predominant actions are on the low-flow perfusion state (10,12,35,39,41,83). There is considerable evidence that the peripheral vascular insufficiency of shock can be attenuated or corrected by therapy which modifies the vasomotor behaviour of the microcirculation. Drugs which exert favourable actions on the different types of small vessels (1,3,10,13,35), should not only promote survival after shock but should also increase RES function.

The object of the present report is (1) to present some of our newer experimental studies, as well as pertinent portions of our older work on RES function in low-flow states and tolerance to experimental trauma (and shock) directly related to the above thesis, and (2) to discuss the future outlook and clinical implications of RES studies in stress and host defense.

Experimental

METHODS

    <u>Animals and Experimental Shock</u>. Female Wistar rats (150 ± 20 g) anesthetized with i.m. pentobarbital sodium were used. Different degrees of hemorrhagic, intestinal ischemic as well as Noble-Collip drum (NCD) shock were induced in groups of animals (6,7,11,37). Control animals through to LD_{95} shock procedures were utilized. Other animals were made tolerant to LD_{80} and LD_{100} NCD trauma. Combined injury shock was produced by superimposing hemorrhage 24 hr after intestinal ischemic shock (37). Animals were observed for survival for 48 hr or up to 5 days. All animals were grossly autopsied at death for signs of shock. Animals in the NCD groups with fractured skulls, subdural hematomata, lacerated viscera, or torn vessels in the thoracic or abdominal cavities were not included.

    <u>Hemorrhage</u> was induced by graded bleeding of anesthetized rats via a cannulated femoral artery over a 15-30 min period to

3% body weight and withholding the blood for different intervals e.g. an LD_{20} hemorrhage resulted when the blood was withheld for 1 hr. If the blood was withheld for 2 hr, 60% of the animals died. Intestinal ischemic (SMA) shock was produced by temporary ligation of the superior mesenteric artery; 20 min ligation gave a 15% mortality. Sham procedures were identical except that the arterial ligature was not tied in the SMA group and no blood was removed from the cannulated femoral artery in the hemorrhage group.

In other experiments, norepinephrine (0.13μg/min), angiotensin (0.10μg/min) and PLV-2 (0.03μg/min), a synthetic analogue of vasopressin, 2-phenylalanine-8-lysine vasopressin, made up to 2 ml in saline, or 2 ml saline alone, was infused over a 50 min period into rats subjected to mild form (LD_{15}) of SMA shock. The doses of these drugs were those previously determined to maintain blood pressure around normotensive levels (10,12,40,41).

Phagocytic indices were determined from the clearance of colloidal carbon, 4mg/100 g body weight, suspended in calf skin gelatin, by a slight modification (11) of the technique of Biozzi et al. (21). Blood samples were obtained 2,4,8,12 and 15 min after intravenous carbon injection, hemolyzed in 0.1% sodium carbonate and the carbon concentrations were measured photometrically at 675 mμ. Phagocytic indices were calculated:

$$K = (\log_{10} C_1 - \log_{10} C_2)/(t_2 - t_1)$$

where C_1 and C_2 are the colloidal carbon concentrations at times t_1 and t_2. The means and their standard errors were statistically analyzed by Student's *t* test.

RESULTS

RES Phagocytic Activity in Hemorrhage and Intestinal Ischemia

Influence of varying degrees of hemorrhage on survival and colloidal carbon clearance. Table 1 shows the survival rates and K values after varying degrees of hemorrhage together with those of control and sham-operated animals. (All these

TABLE 1

EFFECTS OF GRADED HEMORRHAGE ON SURVIVAL
AND CARBON CLEARANCE IN RATS*

Duration of Time Blood Withheld (Hr.)	Survivors/Total	% Survival	Phagocytic Index (K) (mean ± SE)
Controls	35/35	100	0.046 ± 0.002 (35)**
Sham***	12/12	100	0.043 ± 0.005 (12)
1.0	17/22	77	0.020 ± 0.002**** (16)
2.0	10/25	40	0.010 ± 0.002**** (8)
2.5	5/25	20	0.004 ± 0.002**** (8)
3.0	1/20	5	0.002 ± 0.003**** (6)

All of the hemorrhaged animals were bled 3.0% body weight. Test dose of colloidal carbon in calf skin gelatin = 4mg/100g body wt.
K values were obtained 1 hr after transfusion of shed blood.

   * Survival determined at 48 hr.
  ** Number of animals.
 *** Femoral arterial cannulation for 3 hr.
**** Significantly different from control and sham-operated animal (P<0.01).

phagocytic indices were obtained 1 hr after transfusion of the shed blood). It is apparent from the K values that the greater the degree of hemorrhage, the greater the early RES phagocytic depression.

TABLE 2

EFFECTS OF GRADED HEMORRHAGE ON CARBON CLEARANCE IN SURVIVING ANIMALS*

Duration of Time Blood Withheld (Hr.)	Phagocytic Index (mean K ± SE)		
	6 Hr.**	12 Hr.	24 Hr.
Controls	0.044 ± 0.005 (16)***	0.049 ± 0.006 (15)	0.041 ± 0.004 (16)
1.0	0.035 ± 0.008 (8)	0.060 ± 0.005 (10)	0.090 ± 0.003**** (9)
2.0	0.020 ± 0.006**** (9)	0.045 ± 0.005 (10)	0.080 ± 0.004**** (10)
2.5	0.014 ± 0.004**** (12)	0.035 ± 0.008 (8)	0.085 ± 0.005**** (10)

* Carbon clearances determined only on animals which survived hemorrhage.

** Time elapsed after transfusion of shed blood.

*** Number of animals.

**** Significantly different from controls (P<0.01).

Animals which survive these lethal hemorrhage procedures show progressively improved phagocytic indices (Table 2), and, within 24 hr, will exhibit K values more than 100% greater than in normal animals (Table 2). Usually these hyperfunctional RE systems return to normal by 96 hr.

TABLE 3

EFFECTS OF GRADED INTESTINAL ISCHEMIC (SMA) SHOCK
ON SURVIVAL AND CARBON CLEARANCE IN RATS[a]

Duration of SMA occlusion	Survivors/Total	% Survival	Phagocytic Index (K) (mean ± SE)
Controls	25/25	100	0.043 ± 0.002 (25)[d]
Sham SMA[b]	20/20	100	0.046 ± 0.003 (20)
20 min.	17/20	85	0.032 ± 0.002[c] (16)
45 min.	12/20	60	0.023 ± 0.003[c] (16)
75 min.	4/20	20	0.010 ± 0.004[c] (18)
90 min.	1/20	5	0.002 ± 0.003[c] (12)

Test dose of colloidal carbon in calf skin gelatin = 4mg/100g body wt. K values were obtained 3 hr after release of tie or sham procedure.

[a] Temporary ligation of superior mesenteric artery. Survival determined at 120 hr.

[b] Laparotomy followed by a loose, unoccluded tie around SMA for 75 min.

[c] Significantly different from control and sham-operated animal (P<0.001).

[d] Number of animals.

Influence of varying degrees of intestinal ischemia on survival and colloidal carbon clearance. Table 3 shows the survival rates and K values for animals subjected to varying degrees of SMA shock. It is apparent from these K values that the greater the degree of intestinal ischemia, the greater the early RES phagocytic depression.

Most of the animals surviving SMA shock exhibit, within 24-48 hr, K values which are more than 100% greater than in the controls (Table 4). By 96 hr these return to normal.

TABLE 4

INFLUENCE OF GRADED SMA SHOCK ON CARBON
CLEARANCE IN SURVIVING ANIMALS[a]

Duration of SMA occlusion	Phagocytic index (mean K ± S.E.)		
	24 hr.[b]	48 hr.	72 hr.
Controls	0.045 ± 0.004 (15)[c]	0.051 ± 0.006 (15)	0.042 ± 0.003 (18)
20 min.	0.080 ± 0.011[d] (20)	0.110 ± 0.018[d] (18)	0.105 ± 0.017[d] (16)
75 min.	0.073 ± 0.010[e] (15)	0.125 ± 0.020[d] (16)	0.095 ± 0.020[f] (16)
90 min.	0.090 ± 0.012[d] (14)	0.120 ± 0.015[d] (15)	0.105 ± 0.016[d] (14)

Test dose of colloidal carbon in calf skin gelatin = 4 mg/100g body wt.

[a] Carbon clearances determined only on animals which survived SMA shock procedures.
[b] Time elapsed after release of SMA occlusion.
[c] Number of animals.
[d] Significantly different from controls ($P<0.01$).
[e] Significantly different from controls ($P<0.02$).
[f] Significantly different from controls ($P<0.05$).

Influence of RES hyperphagocytosis in surviving animals on cross-tolerance to shock. SMA hyperphagocytic survivors were found to be cross-tolerant to subsequent LD_{60} hemorrhage procedures when challenged 24-36 hr after the original SMA procedure (Table 5).

TABLE 5

SURVIVAL OF RATS AFTER SHOCK
AND COMBINED INJURY

Group	Survivors/ Total Rats	% Survival
I. SMA shock*	45/53	85
II. Hemorrhage —* (Blood Withheld 2 hr.)	20/50	40
III. Combined Injury** (SMA + Hemorrhage)	24/25	96

* Animals in Group I (SMA shock) were subjected to a temporary ligation of superior mesenteric artery for 20 min. Animals in Group II (hemorrhage) were subjected to a 3.0% hemorrhage of body weight; shed blood transfused 2 hr. later.

** Animals in Group III (combined injury) were subjected first to the same SMA procedure as those in Group I, followed 24 hr. later by a 3.0% hemorrhage as in Group II. Animals were observed for survival for 5 days.

TABLE 6

PHAGOCYTIC INDICES (K VALUES)
AFTER COMBINED INJURY IN RATS*

Group	No. Rats	K value (mean ± SE)**
I. 1 hr. after combined injury		
Normal, unshocked controls	16	0.043 ± 0.005
Combined injury	12	0.045 ± 0.008
II. 3 hr. after combined injury		
Normal, unshocked controls	14	0.041 ± 0.006
Combined injury	15	0.046 ± 0.009
III. 24 hr. after combined injury		
Normal, unshocked controls	18	0.045 - 0.004
Combined injury	10	0.087 ± 0.011***

* Animals were first subjected to a temporary 20 min. ligation of the SMA, followed 24 hrs. later by a 3.0% hemorrhage by body weight. Two hr. later shed blood was transfused. Carbon clearances done 1, 3 and 24 hrs. after transfusion.

** Test dose colloidal carbon in calf-skin gelatin = 4mg/100g body weight.

*** $P<0.01$ (compared with paired controls).

RES phagocytic activity in combined injury shock.
Table 6 shows that, unlike normal animals subjected to acute hemorrhage, the combined-injury rats fail to show early RES phagocytic depression either 1 or 3 hr after reinfusion of shed blood. At 24 hr the combined-injury group showed significant RES hyperphagocytosis (Table 6) comparable to the groups subjected to hemorrhage or intestinal ischemia alone (Tables 1 & 3).

[Figure: Graph showing mean K-values vs time after shock in hours. Test dose colloidal carbon = 4mg/100gm i.v. Broken line indicates the normal range of activity.]

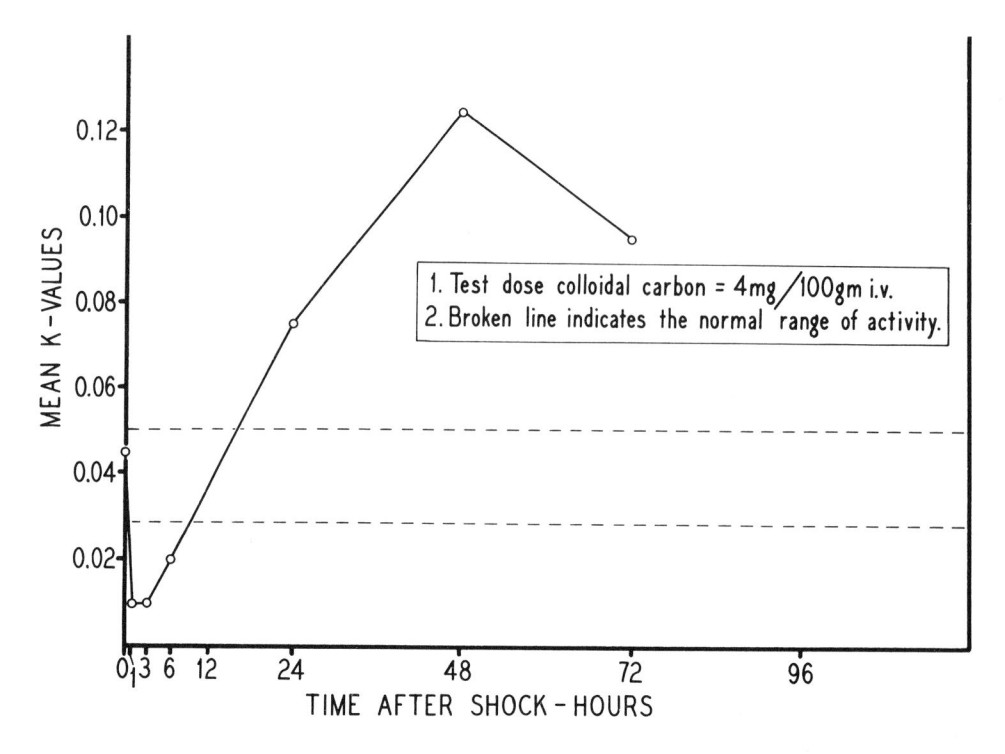

Fig. 1. Influence of NCD-trauma shock at 600 revolutions (LD_{25}) on colloidal carbon clearance in rats. Test dose of colloidal carbon 4 mg/100 g i.v. Broken line indicates the normal range of activity. Each mean K value represents 8-16 rats (Figure taken from ref. 5).

Influence of Varying Degrees of NCD Trauma on Colloidal Carbon Clearance

NCD Trauma: 600 revolutions. Fig. 1 compares the RE phagocytic function of rats after 600 revolutions drum trauma which gave a 25% mortality. The K values in these animals showed an 80% depression in carbon clearance for up to 3 hr after injury but reverted within 24 hr to values 65% above controls. By 48 hr the surviving rats exhibited carbon clearances which were almost 200% above the controls.

NCD Trauma: 850 revolutions. Fig. 2 shows the carbon clearance values of rats after 850 revolutions (80% mortality). These animals developed approximately 90% depression in RE

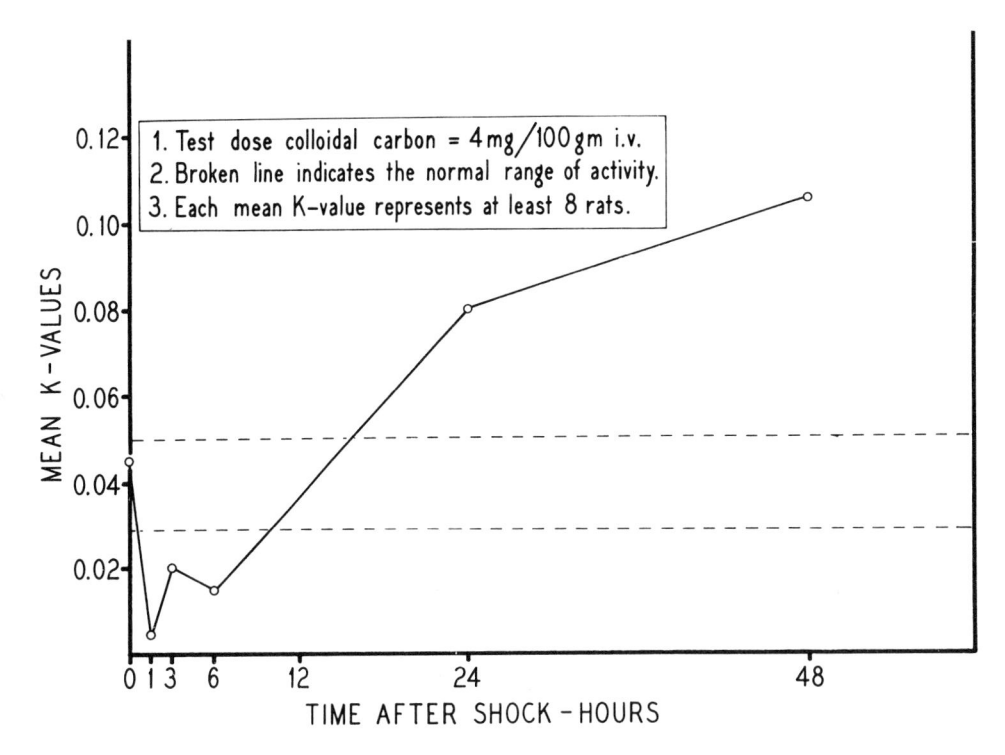

Fig. 2. Influence of NCD-trauma shock at 850 revolutions (LD$_{80}$) on colloidal carbon clearance in rats. Test dose colloidal carbon = 4 mg/100 g i.v. Broken line indicates the normal range of activity. Each mean K value represents 8-16 rats (Figure taken from ref. 5).

function after the first hr but after 24 hr showed carbon clearance rates which were almost 65% increased compared to untraumatized controls. Within 48 hr, the surviving animals had K values which were 133% stimulated.

Influence of Reticuloendothelial Hyperphagocytic Function on Tolerance to Trauma and Shock

Carbon clearances were determined in normal control rats. These rats were then subdivided into 2 groups depending on their carbon clearance. Group A contained animals with K values within one standard deviation of the mean, 0.045 ± 0.017. Group B comprised those animals with K values greater than 0.065. Both groups were then subjected, 18-24 hr later, to the LD$_{80}$ NCD trauma procedure; mortality was determined at

48 hr. The 15 rats with hyperphagocytic RE systems (Group B) were 5 times more resistant than Group A (Table 7). Similar results have been obtained with SMA shock (Table 8).

TABLE 7

INFLUENCE OF RES PHAGOCYTIC ACTIVITY IN NORMAL ANIMALS
IN PROTECTION AGAINST NCD-TRAUMA SHOCK*

Group	Phagocytic Index (Mean K Value ± SD)	Range of K Values	% Survival (No. Rats/Total)
Normal Population	0.045 ± 0.017 (94)**	0.010 – 0.180	21 (20/94)
A. Mean ± SD	0.033 ± 0.010 (75)	0.028 – 0.062	13 (10/75)
B. Hyperphagocytic	0.090 ± 0.025 (15)	0.065 – 0.180	65 (10/15)

* Carbon clearances determined 18-24 hr before animals were subjected to trauma (850 rev at 40 rpm). Dose colloidal carbon = 4mg/100g.

** Number of rats in group.

TABLE 8

INFLUENCE OF RES PHAGOCYTIC ACTIVITY IN NORMAL
ANIMALS IN PROTECTION AGAINST SMA SHOCK[a]

Group	Phagocytic Index (Mean K Value ± SD)	Range of K Values	% Survival (No. Rats/Total)
Normal population	0.043 ± 0.015 (98)[b]	0.010 − 0.175	20 (20/98)
A. Mean ± SD	0.032 ± 0.009 (76)	0.028 − 0.058	13 (10/76)
B. Hyperphagocytic	0.092 ± 0.027 (14)	0.062 − 0.175	71 (10/14)

[a] Carbon clearances determined 18-24 hr before animals were subjected to SMA procedure (75 min tie). Dose colloidal carbon = 4mg/100g.

[b] Number of rats in group.

Reticuloendothelial Function in Adaptation (Tolerance) to Trauma

Paired groups of rats were exposed to repeated drum-trauma to see if the phagocytic indices of adapted rats were correlated with susceptibility to subsequent lethal episodes of traumatic shock. All rats were first exposed to 400 revolutions followed 48 hr later by 600 revolutions and then challenged by an LD_{80}, 850 revolution procedure after various intervals. At 24 hr the rats not only had phagocytic indices which were 100% greater than control values, but they were completely resistant to lethal traumatic shock (Fig 3). At 48 hr the K values increased further and the rats remained completely tolerant to the

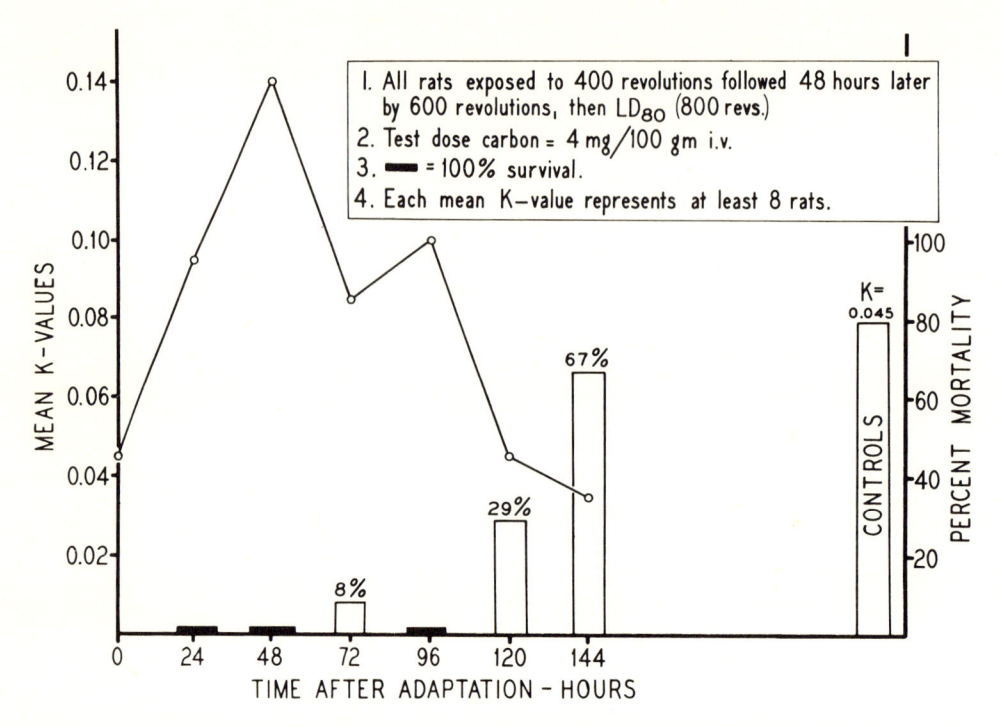

Fig. 3. Changes in RES phagocytic capacity and mortality in animals adapted to an LD_{80} trauma-shock procedure. All rats exposed to 400 revolutions followed 48 hr. later by 600 revolutions, the LD_{80} (850 rev). Test dose colloidal carbon = 4 mg/100 g i.v. Black solid horizontal bar (▬) = 100% survival. Each mean K value represents 8-16 rats. Each group of rats for mortality studies consists of 24-36 rats; control group consists of 85 animals (Figure taken from ref. 5).

LD_{80} Noble-Collip procedure. Although the phagocytic index was still above the control level at 72 hr, the K values were below those seen at 24 and 48 hr. This appeared to be correlated with a slight susceptibility of the rats to traumatic shock. Six days after adaptation the carbon clearance values showed slight RE phagocytic depression. This group of animals, at 144 hr, were now as susceptible as untreated controls to the LD_{80} shock procedure. Thus there is a very close correlation between phagocytic capacity of the RES and tolerance to experimental shock, and the degree of protection may be related to the degree of

RE hyperfunction. Other experiments show that rats made tolerant to completely lethal LD_{100} NCD-trauma shock maintain elevated RES phagocytic function during the period of adaptation (5).

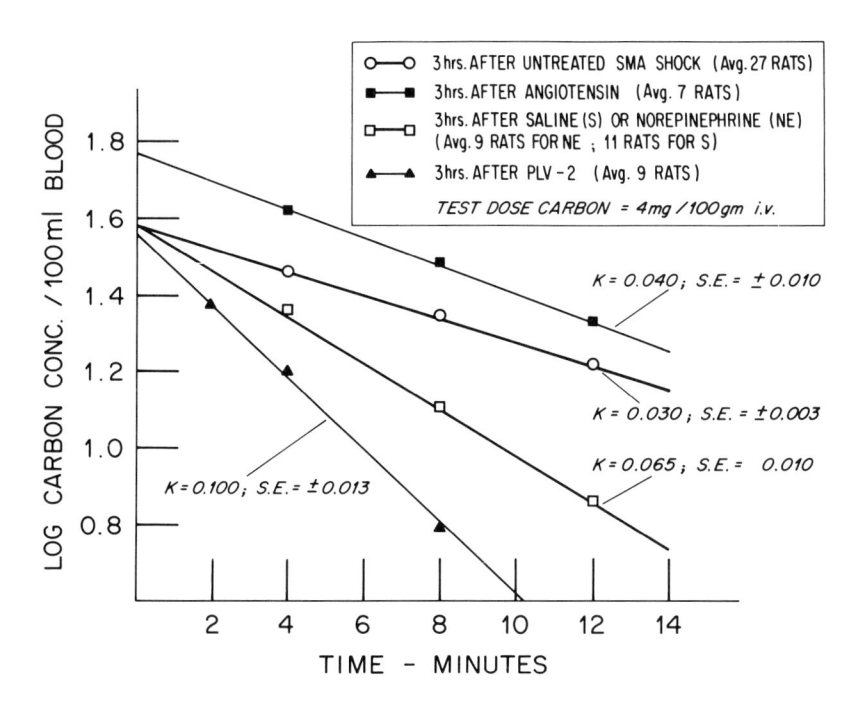

Fig. 4. Influence of vasopressors on RES phagocytic activity after intestinal ischemic (SMA) shock (Fig. taken from ref 2.). (Reproduced by permission from: Altura, B.M. & Hershey, S.G.: Use of reticuloendothelial phagocytic function as an index in shock therapy. Bull N.Y. Acad.Med.(1967) 43, 259.)

Pharmacologic Manipulation of RES Phagocytic activity in Acute Shock

Fig. 4 compares the RES phagocytic activity at 3 hr in vasopressor drug-treated and untreated SMA shock rats. Therapy with saline or norepinephrine resulted in > 100% increase in phagocytic activity when compared with the untreated shocked animals. These data also show that therapy with PLV-2 increased the carbon clearance rate more than 230% when compared with the untreated shocked rats. Angiotensin, which is not beneficial in shock (10,12,14,41), did not alter the clearance rate. We have obtained similar results with these drugs in NCD trauma shock (unpublished results). These data correlate well with earlier observations (8,10,12,14,35,40,41).

Discussion

Our findings indicate that phagocytic indices correlate well with the intensity of the trauma and the animal's subsequent response. The greater the degree of injury or shock, the greater the size and duration of early phagocytic depression. These experiments confirm and extend our previous studies (2,4,37,38). Similar findings have been reported by other laboratories (23,25,26,28,46,47,61,63,67,68).

The adaptive RES changes in the survivors and the NCD trauma tolerant rats probably result in more efficient handling of blood-borne tissue mediators, metabolites, and/or other noxious tissue products (5,68,85). This may account for the greater resistance of drum-tolerant animals and animals subjected to a second lethal injury. Support for this comes from histological studies on rats adapted to Noble-Collip drum trauma shock, showing enlarged Kupffer cells (85). Although reduced blood flow may partially account for the early depression in RE phagocytic activity (15,18,27), it is probably not the complete explanation (5,63,67-69). Recent work (63,67-69) strongly suggests that depletion of plasma factors (opsonins), essential for RES phagocytosis (18,62,64,77), is probably the main cause.

It has often been demonstrated that adaptation to specific types of experimental shock can result in cross resistance to other types of stress (5,7,42,83,87,88). Although such tolerance has been partly attributed to age of the animals (43), it can be induced in adrenalectomized, chemically sympathectomized as well as anesthetized animals (43). The cause of this tolerance to trauma is not known. It has, however, been shown (9,49,84,85) that induced adaptation to trauma can be attenuated and even abolished by blockade of the RES.

Of further interest are our findings which show that some of a normal rat population exhibit hyperfunctional RE systems which are correlated with protection against lethal NCD trauma as well as against lethal SMA shock (7). Although this correlation may be coincidental, this is unlikely (2,4-7,9,11,17, 23,26,28,32,34,36-38,49,57,59,63,67-69,83-85).

Although dissociation of hyperphagocytosis and tolerance has been reported in certain types of experimental shock (6,11,20, 31,59,60,72,74,78,79), these differences may be related to the method used to induce RES stimulation. When dissociation was found, the RES was stimulated by toxic materials, such as thorotrast, saccharated iron oxide, zymosan, certain triglycerides, etc., which may have induced pathological changes.

Although the earlier studies of Zweifach (84,86) and Blattberg and Levy (22) did not indicate a close correlation between survival and RES phagocytic function, the difference from our results may be related to experimental circumstances and methodology (37). The chief difference between their studies and ours is the test dose of colloidal carbon. We used 4.0 mg/100 g of colloidal carbon (over 90% cleared in liver), whereas the previous investigators used higher doses. The small test dose was chosen to avoid depletion of opsonins or other plasma factors necessary for complete phagocytosis of carbon particles. The recent observations of Saba (63) and Schildt (68,69), indicating depletion of opsonins during surgery and trauma, support this contention. Furthermore, these latter investigators used RES test substances other than colloidal carbon. Our data, together with that reported previously, seem not only to support, but also strongly reemphasize, the existence of a close, functional relationship between tolerance to experimental shock (and trauma) and the phagocytic elements of the RES. Our studies also provide quantitative data which could possibly be used as a basis for diagnostic indices of the course of shock and its response to therapy. Our data also strongly support the concept of Fine (32) and Zweifach (83-85) that the RES may represent the homeostatic system serving as a critical common pathway in the pathogenesis of and in host adaptation to certain shock syndromes. Although the mechanisms responsible for the RE changes in experimental trauma and shock remain to be elucidated, the available data indicate that failure of the RES can account for the etiology and progression of the syndrome in at least one form of shock, namely bowel ischemia (7).

Future Outlook and Clinical Implcations of RES Studies Related to Stress and Host Defense

Recent studies involving a variety of stressful situations 16,29,45,46,47,56,61,65,66,70,71,77,80,81) suggest that RES changes are involved in these stresses.

RES phagocytosis in man. The information developed in animals now warrants studies of the RES-shock relationship in man. Several investigators (16,28,44,48,61,65,66,70,80) have measured RES function in man. The recent appearance of several safe RES test substances makes these determinations practicable. These materials could be used to develop reference values in normal man and to determine whether RES patterns similar to those in stressed animals exist clinically. Where feasible, RES measurements should be made in conjunction with other cardiovascular measurements.

Chemically-induced stimulation of the RES with materials suitable for man. Most of the materials used to stimulate the RES in animals are not suitable for human use. However, we have been exploring materials which are relatively safe for human use. Preliminary studies indicate that three substances, choline chloride, microaggregated human serum albumin and glyceryl trioleate stimulate RES phagocytic capacity and enhance the survival of rats exposed to several types of shock (6,11,34,36). However, each of these substances, given individually, does not protect against all the four types of shock used (hemorrhage, SMA, ischemia, NCD-trauma and endotoxin). Some recent experiments show that, in rats, extremely low doses of the pure synthetic estrogen, 17-beta estradiol, are not only a potent RES stimulant (52,53), but also increase survival after endotoxin shock (authors' unpublished data), the most refractory of the shock models we have been using. We consider it worthwhile to extend these types of experiments to use combinations of the above materials, to explore other synthetic estrogens (52,53) and other RES stimulating lipids (24) for their prophylactic effects. The most promising of these newer materials have to be subjected to standard batteries of acute and chronic toxicity studies. Such materials might be used to pre-treat patients scheduled for massive elective surgical procecures.

Relationship of RES function to microcirculatory behaviour in shock. There is some evidence that the RES may participate in regulation of blood flow in other tissues besides the liver and spleen. Zweifach (85) found that blockade or stimulation of the RES altered vascular reactivity in the mesenteric and skin vessels. Certain chronic infections and endotoxins which affect RES phagocytic capacity also alter microcirculatory vasomotor activity (9,57). Vasoactive drugs which improve survival after several types of experimental shock not only alter microcirculatory dynamics (10,12,41) but also stimulate RES phagocytic capacity (2,4,38). No microcirculatory studies, however, are yet available with RES stimulating materials suitable for man.

Nature of plasma factors promoting RES phagocytosis of test particles. Although the mechanism(s) whereby RES phagocytic activity is acutely and adaptively modified in shock and trauma is not clear (5,26,64,67) one must consider the

possibility that a depletion and subsequent increase in
phagocytosis-promoting plasma proteins (P-PPP) are primarily
responsible for the early RES depression and late RES
hyperphagocytosis. If so, several questions must be answered:
(1) Are there similar biphasic changes in P-PPP in other forms
of trauma and shock as in hemorrhage and bowel ischemia?
(2) And if so which of the plasma proteins are responsible?
(3) Do such P-PPP play a role in mediating choline chloride-,
glyceryl trioleate- and microaggregate human serum albumin
induced- RES stimulation? The answers to these questions could
have clinical importance.

SUMMARY

Experiments were designed to explore the possibility of a
close relationship between the functional capacity of RES and
tolerance to several types of injury. Sequential changes in
RES activity developed after hemorrhage, bowel ischemia and
trauma. Recovery and cross-tolerance were accompanied by marked
stimulation of the RES. Phagocytic indices were depressed
early after mild and lethal forms of hemorrhage, bowel ischemia
and trauma, with development of hyperphagocytosis (>100%) in
survivors at 24 hr. Further, data indicate that RES phagocytic
indices correlate well with both the intensity of the trauma or
shock episode and the subsequent progression towards recovery.
The results strongly suggest that the greater the degree of
injury or shock, the greater are both the magnitude and duration
of early phagocytic depression. Animals made tolerant to lethal
trauma (LD_{80} - LD_{100}) show marked stimulation of the RE phagocytic
elements. In addition, animals which survived mild forms of
intestinal ischemia became cross-tolerant to a second stress.
Unlike animals exposed to bowel ischemia alone, animals subjected
to the combined injury failed to show early depression in RES
phagocytic function, and on recovery had increased K values
similar to those after the single stresses. The marked RES
phagocytic depression seen early in shock could be pharmacologically manipulated and reversed by vasoactive drugs. The
experimental evidence firmly supports the existence of a close
interrelationship between tolerance to experimental injury and
the status of the RES. The data support our previous suggestion
that RES clearance measurements may be useful indices of the
course of shock and its therapy. The outlook and clinical
implications of future RES studies related to stress and hose
defense is discussed. A review of the literature raises the
question as to whether the RES is a mediating system not only in
trauma and shock but also in systemic stress and host defense in
general.

REFERENCES

1. Altura, B.M. In: Physiology and Pharmacology of Vascular Neuroeffector Systems, edited by J.A. Bevan, R.F. Furchgott, R.A. Maxwell & A.P. Somlyo. (1971) Basel: Karger, p.274.

2. Altura, B.M. & Hershey, S.G. Bull. N.Y. Acad. Med. (1967) 43:259.

3. Altura, B.M. & Hershey, S.G. (1967) Angiology 18:428.

4. Altura, B.M. & Hershey, S.G. In: Intermedes Proceedings 1968: Combined Injuries and Shock, edited by B. Schildt and L. Thorén. (1968) Stockholm: Almqvist and Wiksell, p.185.

5. Altura, B.M. & Hershey, S.G. (1968) Am. J. Physiol. 215:1414.

6. Altura, B.M. & Hershey, S.G. (1970) J. Pharmacol. Exptl. Ther. 175:555.

7. Altura, B.M. & Hershey, S.G.(1971)J. Reticuloendothelial Soc., 10, 361.

8. Altura, B.M. & Hershey, S.G. In: Proc. Symp. on Vasoactive Polypeptides (Florence, July 1971), edited by N.Back and F. Sicuteri, New York, Plenum Press, in press.

9. Altura, B.M., Hershey, S.G., Ali, M. & Thaw, C. (1966) Res: J. Reticuloendothelial Soc. 3:447.

10. Altura, B.M., Hershey, S.G., & Altura, B.T. In: Advances in Experimental Medicine and Biology, vol. 8, Bradykinin and Related Kinins: Cardiovascular, Biochemical and Neural Actions, edited by F. Sicuteri, M. Rocha e Silva and N. Back (1970) New York: Plenum Press, p. 239.

11. Altura, B.M., Hershey, S.G. & Hyman, C. (1966) J. Reticuloendothelial Soc. 3:57.

12. Altura, B.M., Hershey, S.G. & Mazzia, V.D.B. (1966) Am. J. Surgery 111:186.

13. Altura, B.M., Hershey, S.G. & Zweifach, B.W. (1965) Proc. Soc. Exptl. Biol. Med. 119:258.

14. Altura, B.M., Hsu, R., Mazzia, V.D.B. & Hershey, S.G. (1965) Proc. Soc. Exptl. Biol. Med. 119:389.

15. Asiddao, C.B., Filkins, J.P. & Smith, J.J. (1964)   J. Reticuloendothelial Soc. 1:393.

16. Attié, E. & Drouet, M. In: Proc. 6th Intl. Meeting of RE Soc. Freiburg, 1970, in press.

17. Beeson, P.B. (1947) Proc. Soc. Exptl. Biol. Med. 64:146.

18. Benacerraf, B., Biozzi, G., Halpern, B.N. & Stiffel, C. In: Physiopathology of the Reticuloendothelial System, edited by B.N. Halpern (1957) Oxford: Blackwell, p.52.

19. Benacerraf, B. & Sebestyen, M.M. (1957) Federation Proc. 16:860.

20. Benacerraf, B., Thorbecke, G.J. & Jacoby, D.J. (1959) Proc. Soc. Exptl. Biol. Med. 100:796.

21. Biozzi, G., Benacerraf, B. & Halpern, B.N. (1953) Brit. J. Exptl. Pathol. 34:441.

22. Blattberg, B. & Levy, M.N. (1962) Am J. Physiol. 203:409.

23. Carteny, G. (1965) 11 Policlinico 72:228.

24. Cooper, G.N. (1964) Res: J. Reticuloendothelial Soc. 1:50.

25. D'Amico, G.D. & Cambria, S. (1963) Giorn. It. Chirurq. 19:501.

26. Dhawan, I.K., Balkrishna, B.N., Mohapatra, L.N. & Venkataswamy, B. (1970) J. Surgical Res. 10:67.

27. Dobson, E.L. In: Physiopathology of the Reticuloendothelial System, edited by B.N. Halpern. (1957) Oxford: Blackwell, p. 80.

28. Donovan, A.J. (1967) Arch. Surg. 94:247.

29. Douglas, B.H. & Grogan, J.B. (1970) Am. J. Obstet. Gynecol. 107:44.

30. Filkins, J.P. & Diluzio, N.R. (1966) Proc. Soc. Exptl. Biol. Med. 122:548.

31. Filkins, J.P., Lubitz, J.M. & Smith, J.J. (1964) Angiology 15:465.

32. Fine, J., Rutenberg, S. & Schweinburg, F.B. (1959) J. Exptl. Med. 110:547.

33. Fitzpatrick, F.W. & Di Carlo, F.J. (1964) Ann. N.Y. Acad. Sci. 118:233.

34. Hershey, S.G. & Altura, B.M. (1966) Proc. Soc. Exptl. Biol. Med. 122:1195.

35. Hershey, S.G. & Altura, B.M. (1966) Schweiz, Med. Wschr. 96:1467, 1516 (2 parts).

36. Hershey, S.G. & Altura, B.M. In: Intermedes Proceedings 1968: Combined Injuries and Shock, edited by B. Schildt and L. Thorén. (1968) Stockholm: Almqvist and Wiksell. p. 195.

37. Hershey, S.G. & Altura, B.M. (1969) Anesthesiology 30:138.

38. Hershey, S.G. & Altura, B.M. (1969) Anesthesiology 30:144.

39. Hershey, S.G. & Altura, B.M. In:Textbook of Veterinary Anesthesia, edited by L.R. Soma. Baltimore: Williams and Wilkins Company, 1971, p.529.

40. Hershey, S.G., Altura, B.M. & Orkin, L.R. (1968) Anesthesiology 29:466.

41. Hershey, S.G., Mazzia, V.D.B., Altura, B.M., & Gyure, L. (1965) Anesthesiology 26:179.

42. Hershey, S.G. & Zweifach, B.W. (1961) Canad. Anaesth. Soc. J. 8:529.

43. Hruza, A., & Poupa, O. In: Handbook of Physiology. Adaptation to Environment. (1965) Washington, D.C. (Am. Physiol. Soc.) P.939.

44. IIO, M., & Wagner, H.N. (1963) J. Clin. Invest. 42:417.

45. Juhlin, L. (1958) Acta Physiol. Scand. 43:262.

46. Lemperle, G. (1967) J. Infect. Diseases 117:7.

47. Lemperle, G. (1970) Plastic and Reconstr. Surg. 45:435.

48. Lemperle, G., Reichelt, M. & Denk, S. In: Proc. 6th Intl. Meeting of RE Soc., Freiburg., 1970, in press.

49. McKenna, J.M. & Zweifach, B.W. (1965) Am. J. Physiol. 187:263.

50. Megirian, R., Laffin, R.J. & Warrington, D. (1970) J. Reticuloendothelial Soc. 7:529.

51. Najjar, V.A., & Nishioka, K. (1970) Nature 228:672.

52. Nicol, T. & Bilbey, D.L.J. In: Reticuloendothelial Structure and Function, edited by J.H. Heller. (1960) New York: Ronald Press, p.301.

53. Nicol, T., Quantock, D.C. & Vernon-Roberts, B. In: Advances in Experimental Medicine and Biology, vol. 1. The Reticuloendothelial System and Atherosclerosis, edited by N.R. DiLuzio and R. Paoletti. (1967) New York: Plenum Press, p. 221.

54. Noble, R.L. (1943) Am. J. Physiol. 138:346.

55. Noble, R.L. & Collip, J.B. (1942) Quart. J. Exptl. Physiol. 31:201.

56. Old, L.J. Clarke, D.A., Benacerraf, B. & Goldsmith, M. (1960) Ann. N.Y. Acad. Sci. 88:264.

57. Palmerio, C. & Fine J. (1970) Arch. Surgery 98:679.

58. Pisano, J.C. & DiLuzio, N.R. (1970) J. Reticuloendothelial Soc. 7:386.

59. Reichard, S. (1968) Radiology 91:132.

60. Ringle, D.A., Herndon, B.L. & Bullis, H.R., Jr. (1966) Am. J. Physiol. 210:1041.

61. Rittenbury, M.S. & Hanback, L.D. (1967) J. Trauma 7:523.

62. Rowley, D. Phagocytosis. In: Advances in Immunology, vol. 2, edited by W.H. Taliaferro and J.H. Humphrey. (1962) New York: Academic Press, p. 241.

63. Saba, T.M. (1970) Nature 228:781.

64. Saba, T.M. (1970) Arch. Intern. Med. 126:1031.

65. Salky, N.K., DiLuzio, N.R., Levin, A.G. & Goldsmith, H.S. (1967) J. Lab. Clin. Med. 70:393.

66. Salky, N.K., Mills, D. & DiLuzio, N.R. (1965) J. Lab. Clin. Med. 66:952.

67. Schildt, B.E. (1970) Acta Chir. Scand. 136:359.

68. Schildt, B.E. (1971) Dissertation: Uppsala University.

69. Schildt, B.E. & Bouveng, R. (1971) Life Sci. Part 1, 10:397.

70. Sheagren, J.N., Block, J.B. & Wolff, S.M. (1967) J. Clin. Invest. 46:855.

71. Šljivič, M.S. (1970) Brit. J. Exptl. Pathol. 51:130.

72. Smith, J.J. In: Shock and Hypotension, edited by L.C. Mills and J.H. Moyer. (1965) New York: Grune and Stratton, p.327.

73. Smith, J.J., Grace, R.A., Wiedemeier, V.T. & Chase, R.E. (1967) Res: J. Reticuloendothelial Soc. 4:433.

74. Stiehm, E.R. (1962) J. Appl. Physiol. 17:293.

75. Stoner, H.B. (1961) Brit. J. Exptl. Pathol. 42:523.

76. Stoner, H.B. In: Intermedes Proceedings 1968: Combined Injuries and Shock, edited by B. Schildt and L. Thorén (1968) Stockholm: Almqvist and Wiksell, p.181.

77. Stuart, A.E. (1970) The Reticuloendothelial System, Edinburgh Livingstone.

78. Stuart, A.E. & Cooper, G.N. (1962) J. Pathol. Bacteriol. 83:245.

79. Trejo, R.A., Crafton, C.G. & DiLuzio, N.R. (1971)   J. Reticuloendothelial Soc. 9:299.

80. Wagner, H.N., Iio, M. & Hornick, R.B. (1963) J. Clin. Invest. 42:427.

81. Wagner, G.F. & Dobson, E.L. (1954) Am. J. Physiol. 179:93.

82. Wooles, W.R. & DiLuzio, N.R. (1963) Science 142:1078.

83. Zweifach, B.W. (1958) Brit. J. Exptl. Anaesthesia 30:466.

84. Zweifach, B.W. (1960) Ann. N.Y. Acad. Sci. 88:203.

85. Zweifach, B.W. In: Shock, edited by S.G. Hershey. (1964) Boston: Little, Brown and Co., p.113.

86. Zweifach, B.W. & Benacerraf, B. (1958) Circulation Res. 6:83.

87. Zweifach, B.W., Benacerraf, B. & Thomas, L. (1957) J. Exptl. Med. 106:403.

88. Zweifach, B.W. & Thomas, L. (1957) J. Exptl. Med. 106:385.

This study was supported in part by Public Health Service Grants HE-12462 and HE-11391 from the National Heart and Lung Institute.

B.M. Altura is a recipient of Public Health Service Research Career Development Award 5-K3-GM-38, 603.

MECHANISMS OF HYPOTHERMIC PROTECTION AGAINST ANOXIA

James A. Miller, Jr. and Faith S. Miller*

Department of Anatomy, Tulane University

New Orleans, La.

About 10 p.m. one night some years ago, a junior in a southern medical school was directed to place a just-delivered stillborn infant in the refrigerated storage pending the arrival of the mortician on the following morning. Twelve hr later, when he pulled out the tray and unwrapped the child he was confronted by a cold but pink, very-much-alive infant. The infant was placed immediately in a warm incubator; almost as quickly developed respiratory problems, and within 2 hr was dead (Stallings, personal communication). Similar types of experiences can be recalled by almost any obstetrician who has had wide experience. Are these merely clinical observations for which no explanations can be offered or do they constitute evidence that there may be a neglected approach to the problem of resuscitation of the asphyxiated neonate?

Our interest in hypothermia for hypoxic or anoxic states dates back 25 yr when it occurred to us that, according to van't Hoff's rule, it should be possible to reduce metabolic requirements (and thereby prolong survival during hypoxia) 50% or more by a 20°C reduction in body temperature. Our early experiments confirmed this expectation for neonatal guinea pigs and demonstrated that this was not merely a postponement of eventual death since cooled animals recovered spontaneously (without artificial respiration) from exposures which were lethal for littermates maintained at 37°C. None of the animals which recovered showed evidence of brain damage (Miller, 1949; Miller & Miller, 1954). In this species the percentage depression in O_2 uptake was the reciprocal of the survival time (Miller & Miller,

* Deceased June 11, 1971

1962). Since guinea pigs are far more mature at birth than human infants, most of our recent experiments have been on neonates which are less mature. In 8 different species 37°C proved to be far from ideal during asphyxia. Indeed, for most species the longest survival was at 15°C body temperature. The response of puppies to hypothermia was outstanding. At 15°C body temperature they tolerated 7 times the exposure to asphyxia which was lethal for littermates at 37°C. This was true protection. Animals of all species at 15°C recovered spontaneously from exposures which were twice as long as those which were lethal for their warm littermates. The majority of puppies was able to recover from exposures 8 times longer than those which killed the warm littermates. None of the spontaneously recovered cold animals showed evidence of brain damage. In contrast none of the warm animals which had been asphyxiated for 2 times TLG (time of last gasp) recovered even though given artificial respiration with O_2.

Since clinically asphyxiation begins in utero, a series of experiments was performed in which the onset of cooling was delayed 5, 7, 10 and 15 min after the beginning of asphyxiation and the animals were removed from the asphyxiation chamber at 15 min (Table 1). Although all warm littermates succumbed,

TABLE 1

EFFECTS OF COOLING AFTER ONSET OF ASPHYXIA (PUPPIES)

Min of Asphyxia before cooling	No.	Deaths	Spontaneous Recoveries No.	%
Control (kept at 37°C)	15	15	0	0
5	10	0	10	100
7	5	0	5	100
10	10	2	8	80
15	10	0	10	100
Total	35	2	33	94

TABLE II*

RESUSCITATION OF CESAREAN-DELIVERED
PUPS ASPHYXIATED FOR 2 TIMES TLG [+]

Body Temp.	No.	Stages in Recovery			
		1st Gasp	1st Breath	Righting Reflex	Complete Recovery
37°C	16	9	4	0	0
15°C	16	16	15**	15**	15**

* Condensed from Table II of Aufderheid & Miller (1972)

[+] TLG = Time of Last Gasp. The total time of asphyxia for both groups was 37.7 min.

** The cold animal which failed to take a breath was found at autopsy to have been inadequately resuscitated. Its lungs were completely atelectatic.

94% of the cooled animals recovered without assistance. More recently pups have been delivered by cesarean section, asphyxiated for two times the TLG (time of last gasp) of the warm littermates and then resuscitated by clearing the airways and artificial respiration with O_2. As seen in Table II only 4 of the warm puppies could be induced to take a breath and none recovered sufficiently to exhibit the righting reflex. By contrast 15 out of 16 cooled puppies recovered completely, and were successfully suckling the following day with no behavioural evidence of sequelae. In the exceptional case it was found that the artificial respiration had been inadequately performed and the lungs were completely atelectatic (Aufderheid & Miller, 1972).

At this point it might be appropriate to discuss the apparent discrepancies between our results and those obtained when unanesthetized, air breathing non-hypoxic animals were cooled. Under these circumstances cold-exposure is highly stressful for adult mammals and for most neonates (except the newborn opossum, cf. Rink & Miller, 1967). They react by liberating catecholamines, increasing motor activity, shivering, burning brown fat, vasoconstricting their arteries, and generally increasing metabolism in an attempt to prevent the fall in body temperature. These activities deplete anaerobic energy stores, and as in the case of vasoconstriction, themselves induce hypoxia in vital organs (e.g. heart and brain). Asphyxia blocks these responses to cold exposure. Asphyxiated animals which are exposed to cold exhibit no motor activity except gasping, no shivering, no thermogenesis from brown fat (Dawkins & Hull, 1965) and no liberation of catecholamines (Carmichael, 1972). Instead they behave like poikilotherms, cool rapidly, and consequently, greatly reduce their metabolic requirements. Thus asphyxia blocks reflexes which are deleterious for non-hibernating mammals and permits the more basic effects of cold on chemical and enzymatic activity to express themselves. Although hypothermia alone may be hazardous, when combined with asphyxia it becomes a life-saving treatment (Miller, 1965).

In guinea pigs measurements of O_2 uptake showed a close parallel between depression in O_2 requirements and prolongation of survival during anoxia. However, in puppies the prolongation of gasping was twice that which was anticipated from O_2 uptake studies. Puppies used 3.7 times as much O_2 at 37°C as they did at 15°C, but cold animals lived 7 times, not 3.7 times as long as did their warm littermates (Table III).

A clue to the reason for this is seen in the blood chemistry studies in the lower part of Table III. Blood acidity changed half as fast and serum K^+, glucose and pyruvate 1/3 as fast in the cold as in the warm animals. Lactate which according to Villee is an excellent indicator of tissue anoxia, paralleled the TLG data exactly. The sparing action of hypothermia on cardiac glycogen during asphyxia is even more striking. This was confirmed by the

TABLE III*

BIOCHEMICAL EFFECTS OF HYPOTHERMIA
DURING ASPHYXIA (PUPPIES)

	Coenothermic (37°C)	Hypothermic (15°C)	Ratio Coeno./Hypo
O_2 Uptake (ml/100g/hr)	174	47	3.7:1
TLG when asphyxiated (min)	15	105	1:7

Changes during 15 min Asphyxia			
pH (units)	-0.55	-0.25	2:1
Serum Na^+ (mEq)	+0.5 (N.S.)	+2.4 (N.S.)	N.S.
K^+ (mEq)	+1.7	+0.6	3:1
Glucose (mg %)	+67.4	+23.5	3:1
Lactate (mg/100ml)	+11.5	+1.6	7:1
Pyruvate (mg/100ml)	-1.8	-0.6	3:1
Cardiac Carbohydrate (mg/g)	-11.7	-1.0	11:1

* Modified from Table 4 (Miller & Miller, 1969)

cardiovascular studies which showed a damping of the catecholamine phase and the maintenance of above normal pulse pressures in the cold anoxic heart for about 20 min beyond the time of last gasp (Fig. 1).

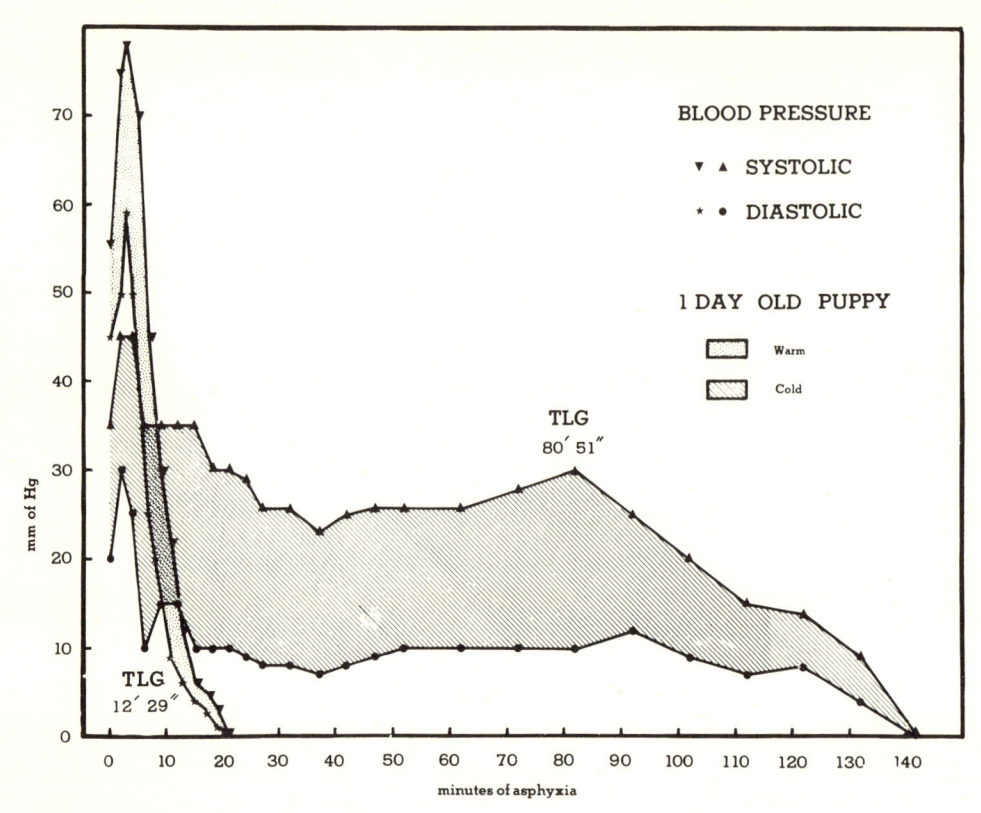

Fig 1 (from Miller, 1971)
Systolic and diastolic blood pressures in a warm (37°C) and a cold (15°C) puppy during asphyxiation in 95% N_2+5% CO_2. Note the higher initial and nearly double maximal systolic and diastolic pressures in the warm animal. Note especially the very large pulse pressures in the cold littermate and extended period (over 100 min) during which greater than normal pulse pressures were maintained in the total absence of O_2. The cold animal gasped 6 times as long as the warm.

      Other evidence for the protection provided by hypothermia is seen in an electron microscope study by Crow (1969). The unasphyxiated puppy myocardium shows a rich supply of glycogen, undamaged mitochondria and firmly attached intercallated discs. The warm asphyxiated heart shows severe depletion of glycogen, many swollen and some destroyed mitochondria and dehiscence of the discs. In the cold asphyxiated heart glycogen is still present. There are many normal and few abnormal mitochondria and intercallated discs are intact (Fig 2a, 2b, 2c).

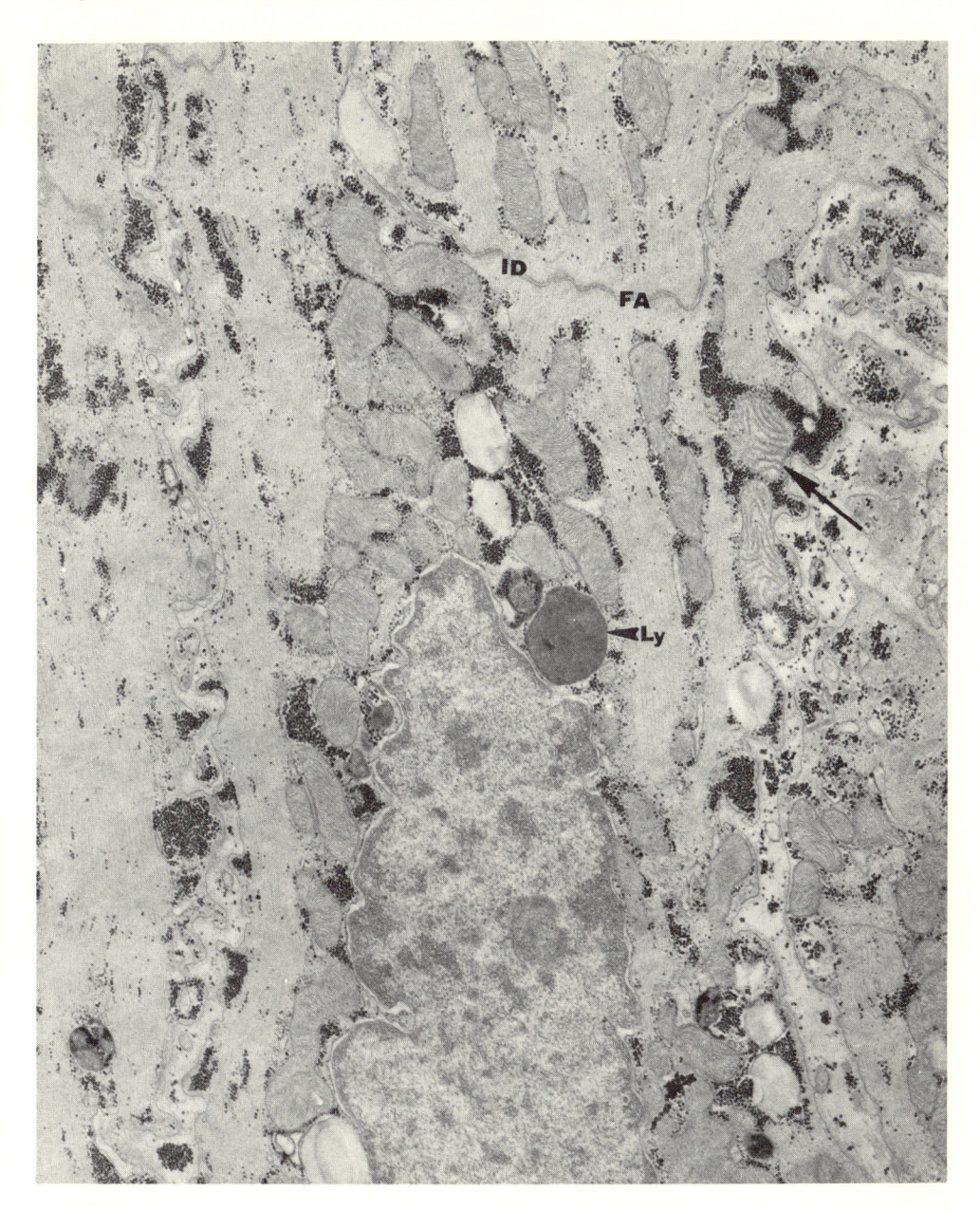

Fig 2a. Longitudinal section of a myocardial cell of an
unasphyxiated warm puppy showing the nucleus with normal
chromatin distribution, mitochondria with dense matrices,
an abundance of glycogen at one pole of the nucleus, and an
intercalated disc (ID) with fascia adherens (FA) tightly
applied. (From Crow, 1969).

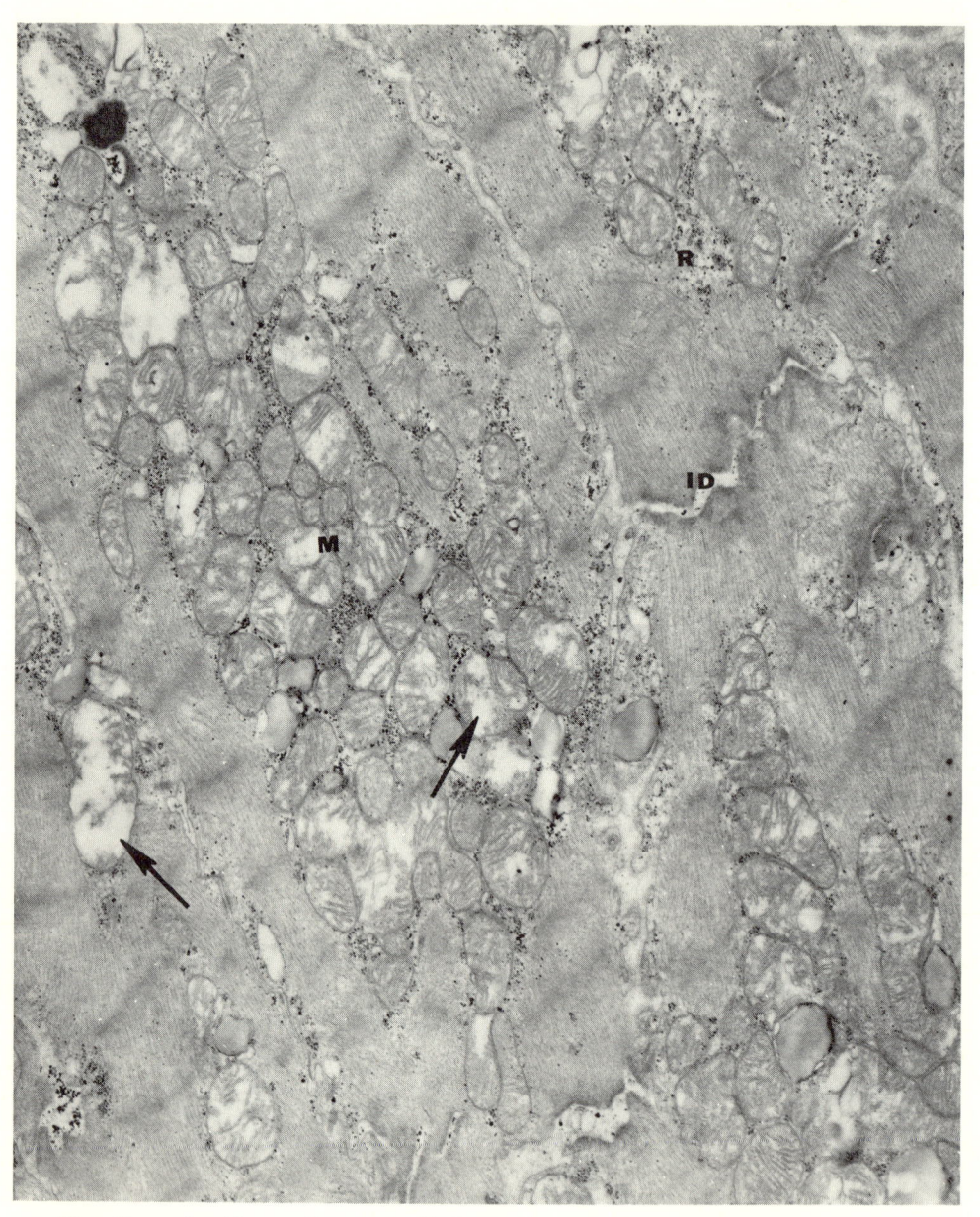

Fig 2b. Section of myocardial cells from coenothermic littermate of 2a asphyxiated at 37°C body temperature until TLG. Note the virtual loss of glycogen with only ribosomes remaining. The mitochondria (M) are swollen, and their matrices are rarified. Many show nearly total destruction of mitochondrial structure (arrows). The intercalated discs (ID) show dehiscences with even the fascia adherens pulled apart. (From Fig 15, Crow, 1969).

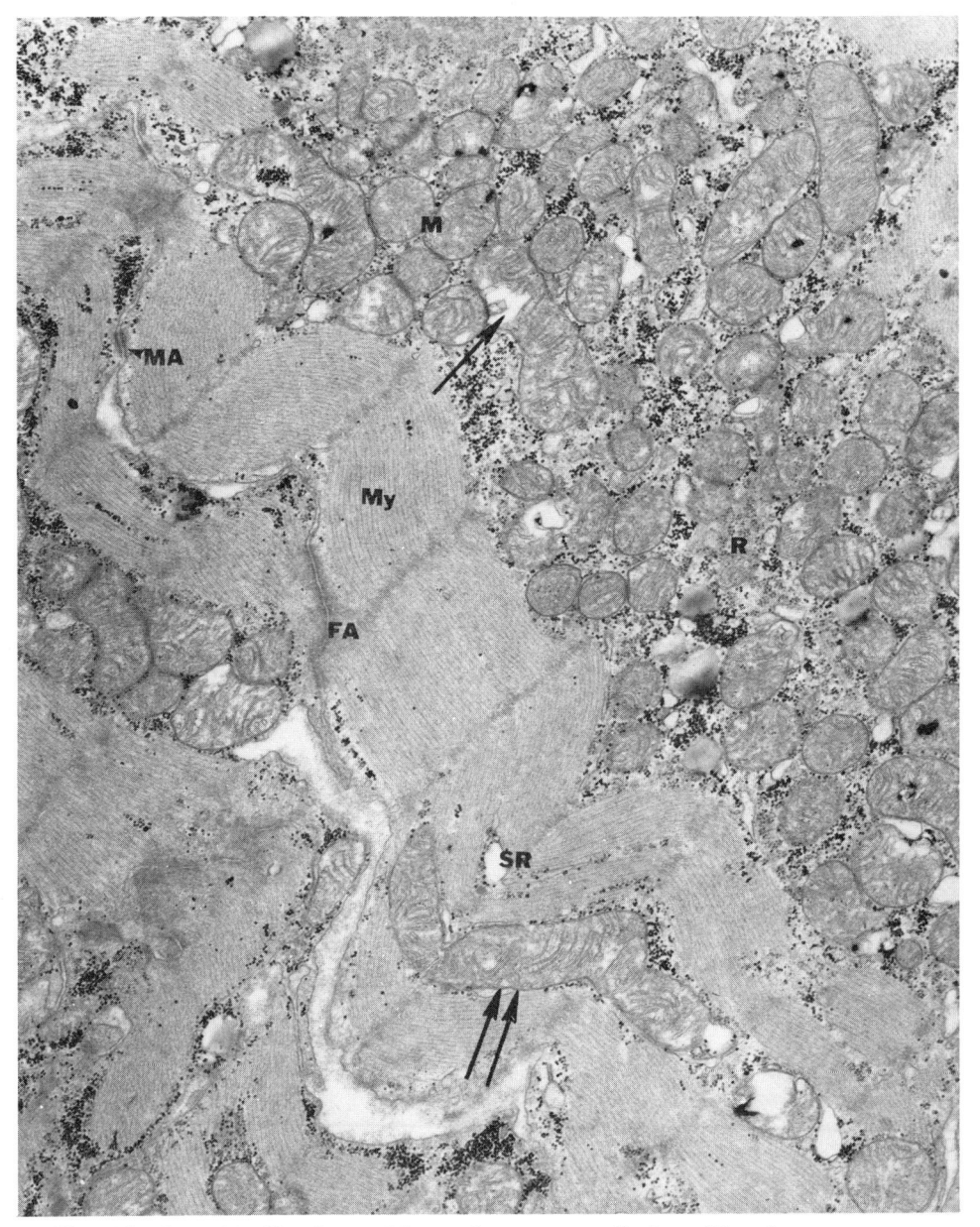

Fig 2C. Longitudinal section of a myocardial cell of an hypothermic littermate which had been asphyxiated for the same length of time as was 2b. Note the presence of glycogen, somewhat less than in 2a but much more than in 2b. There are some swollen mitochondria with areas of rarefaction (arrow) but many appear uninjured. The intercalated disc (ID) shows tightly adherent maculae adherentes (MA) and fasciae adherentes (FA). An occasional giant mitochondrion is seen (arrows) (From Fig. 21, Crow, 1969).

Since in adults hypoxia-hypercapnia is beneficial during cooling, we tested whether newborns that we cooled in a sealed vessel or breathing 10% O_2 + 5% CO_2, would be more or less tolerant of asphyxia than were littermates cooled to the same temperature while breathing air (Table IV). The table shows that although pretreatment with hypoxia-hypercapnia did not affect the survival of the warm controls, it did prolong that of the cooled animals. Hypoxia-hypercapnia (H-H) exerts its

TABLE IV

EFFECTS OF HYPOXIA-HYPERCAPNIA DURING COOLING

| Body Temp. | TLG in 95% N_2 + 5% CO_2 ||||
| | Control || Hypoxia-Hypercapnia ||
	TLG min	%	TLG min	% of Control
37°C	15	100	14.8	98
15°C	105	100	120	114
10°C	85	100	114	134
5°C	45	100	70	156

benefit during the cooling process, since it is effective in isolated heads which are thereby deprived of all circulation during asphyxia (c.f. Figure 3 in Miller, 1971). In Fig 3 it is seen that the H-H treatment doubles the blood volume in the heads of animals at 15°C and at 10°C, but not at 5°C. In other words - in the temperature range in which it is effective (at 15°C and 10°C) it counteracts the vasoconstriction which occurs at very low temperatures.

During asphyxia blood pH falls steadily and at last gasp is about 6.85 in puppies. Since this is associated with a number of undesirable sequelae, the control of acid-base changes

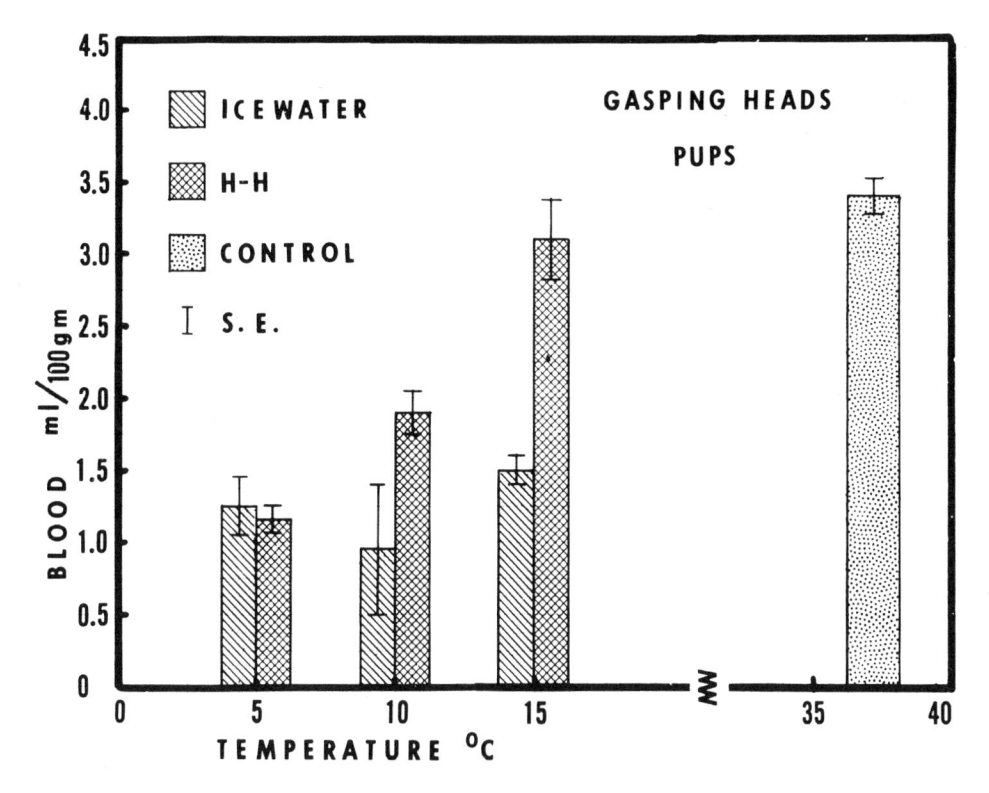

Fig 3. Influence of Hypoxia-Hypercapnia on Blood Volume of Heads of Pups at Different Temperatures.

is part of many resuscitation programs. We have analysed the effect of one of the more promising buffers (THAM= Trishydroxymethyl-aminomethane).

When tested on warm littermates, maintenance of pH within the normal range only increased survival 17%. Glucose, which lowered pH below control levels, increased it 19% and the combination of THAM + glucose gave a 69% increase. Cooling to 15°C gave a 6.5 times increase in TLG and cooling plus THAM plus glucose gave an 11 times increase. These are the longest survivals yet achieved. In addition they exemplify the fact that has been observed on numerous occasions that hypothermia, when added to an effective treatment for asphyxia, enhances its efficacy.

There is another aspect of blood pH which has prognostic significance. Because it is unaffected by fluctuations in blood O_2 and can be measured in less than a minute and because, as shown in Fig 4, it falls at a regular rate during asphyxia, blood pH offers a means of prognosis of the outcome of resuscitation even if the past asphyxial history is unknown. As shown on the right hand side of the graph, puppies at 6.85 can be completely resuscitated with O_2 in 100% of the cases, using artificial respiration. At 6.84 only 60% were completely resuscitated and at 6.75 only 40%. We were 100% unsuccessful in resuscitating those at 6.67. Therefore with no knowledge of

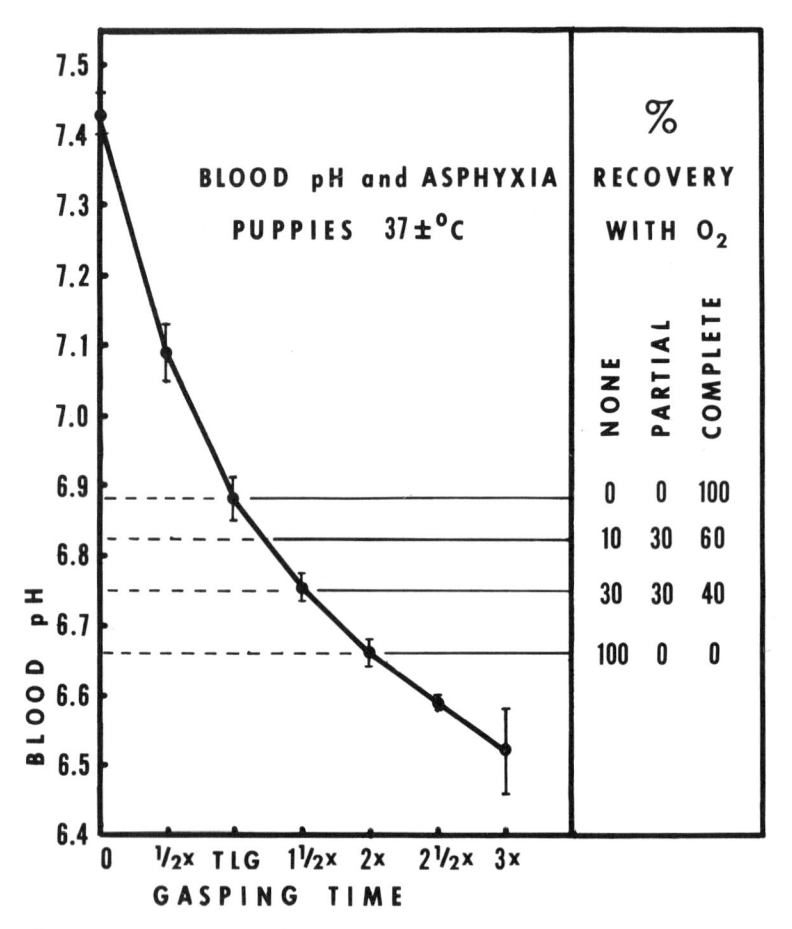

Fig 4 (from Miller, 1971). Blood PH during asphyxia in warm puppies and its relationship to resuscitatability. Note that at last gasp (TLG) it is 6.85 and all animals can be completely resuscitated when respired with O_2. At pH 6.75 only 40% can be completely resuscitated and at 6.67 none can be resuscitated

its history excepting that a pup has been acutely asphyxiated, we can predict the outcome of resuscitative efforts merely by checking the pH of its arterial blood.

Hypothermia has been used in the treatment of over 200 severely asphyxiated newborn infants. As seen in the table the survival rates are not inferior to those reported for 175 similarly depressed infants in the collaborative study (Drage et al., 1964). Sixty-nine of Westin's cooled infants have been followed up. One case with prolonged intrauterine asphyxia has been found to have cerebral palsy and one child which which was believed to have suffered from an intracranial haemorrhage just before birth shows lack of concentration and psychological retardation (Westin, 1971). Since 7.5% of infants with 5 min Apgar scores of 1 to 3 show patent neurological symptoms at one year of age and more develop them later (Drage & Berendes, 1969), this is a remarkably good record for the cooled infants. In addition Cordey has 10 cases in which he has combined hypothermia with bicarbonate treatment to correct the acid-base status of the infant. He reports excellent results from the combined therapy (Cordey, 1972 personal communication).

Thus, although it is virtually impossible to prove by clinical trials whether or not a treatment is effective, the clinical work suggests that the human infant does not differ materially from the infants of other mammals in which hypothermia has been shown to be effective in preventing death or brain damage from asphyxia.

Summary

We have demonstrated that hypothermia saves the lives of asphyxiated newborn animals and that it is effective if applied after 5 min., 10 min or even 15 min in pups which gasp for approximately 15 min when warm. It is effective in part because it reduces overall O_2 uptake. In part it is effective because it reduces strikingly the rates of biochemical degradation which occur during asphyxia. However, an important aspect is the cardiovascular effects which result in better heart action and increased circulating blood volume by combatting the generalized vasodilation which occurs in deep shock.

There is a fortunate synergy between asphyxia and hypothermia. Each counteracts deleterious effects of the other. Hypoxia-hypercapnia blocks the shivering and

TABLE V

SURVIVALS OF WARM AND COOLED INFANTS WITH 5-MINUTE APGAR SCORES OF 3 OR LESS

	Weights at Birth								
	2,501 gm & over			2,001–2,500 gm			2,000 gm & under		
	No.	Survivals	Deaths	No.	Survivals	Deaths	No.	Survivals	Deaths
	"COLD"–(ACTIVELY COOLED. BODY TEMPERATURES AS LOW AS 21°C)								
Westin et al. (1962, 1964)	54	50	4	5	4	1	10	2	8
Krokfors et al. (1961)	4	3	1	1	1	0	0	0	0
Cordey+ (1961,1964,1967,1972)	31	30	1	2	1	1	–	–	–
Auld et al. (1962)	–	–	–	1	1	0	–	–	–
Dunn & Miller+ (1969)	18	17	1	3	3	0	5	3	2
Ehrstrom et al.++ (1969)	21*	14	7*	3	2	1	1	0	1
Cedercreutz (1971)	22	21	1	–	–	–	–	–	–
Totals	150	135	15	15	12	3	16	5	11
% Deaths		10.0%			20.0%			68.7%	
	"WARM"–(NO ACTIVE COOLING. BODY TEMPERATURES NOT RECORDED)								
Drage et al. (1964)	175	148	27	27	19	8	86	19	67
% Deaths		15.4%			29.6%			77.9%	

* One case omitted because the infant was dead before cooling was initiated (5 min Apgar 0, 15 min Apgar 0)
+ Two cases omitted because initial Apgar scores were 4 or above
++ Four cases omitted because initial Apgar scores were 4 or above. In addition, these cases are not strictly comparable since as long as 1 hr intervened before the induction of hypothermia.

non-shivering thermogenesis and the extreme vasoconstriction which is caused by cold exposure whereas cold counteracts the vasodilating effects of anoxia as well as increasing tolerance of asphyxia by reducing metabolic needs.

Hypothermia can be combined with other effective treatments and in each case tested (THAM + Glucose or nembutal for example) it has been found to be additive.

The results with human trials suggest that the responses of deeply asphyxiated human infants do not differ substantially from those of experimental animals.

REFERENCES

Aufderheide, W. & Miller, J.A., Jr. (1972) Unpublished results.

Auld, P.A., Nelson, N.M., Nicopoulos, D.A., Helwig, F. & Smith, C.A. (1962) New Engl. J. Med. 267:1348.

Carmichael, S.W. (1971) Dissertation, Tulane Library.

Carmichael, S.W. (1971) Submitted to Anat. Rec.

Cedercreutz, C. (1971) Personal communication quoted by Westin, B. (1971) Amer. J. Obstet. and Gynecol. 110:1136.

Cordey, R. (1961) Bull. Fed. Soc. Gynecol. Obstet Langue Franc. 13:507.

Cordey, R. (1964) Obstet. & Gynecol. 24:760.

Cordey, R. & Chausse, J.M. (1967) Gynaecol. 163:124.

Cordey, R. & Chiolero Réanimation du nouveau-né par hypothermie: données acido-basiques et evolution de 33 enfants à long terme. Manuscript 1972

Crow, C.A. (1969) Dissertation, Tulane Library.

Dawkins, M.J.R. & Hull, D. (1965) Scientif. Amer. 213:62.

Drage, J.S., Kennedy, C. & Schwartz, B.K. (1964) Obstet. Gynecol. 24:222.

Drage, J.S. & Berendes, H. (1966) Pediat. Clin. N.Am. 13:635.

Dunn, J.M. & Miller, J.A., Jr. (1969) Amer. J. Obstet. Gynecol. 104:58.

Ehrstrom, J., Hirvansalo, M., Donner, M. & Hietalahti, J. (1969) Ann. Clin. Res. 1:40.

Krokfors, E., Pitkänen, H., Hirvensalo, M. & Rhen, K. (1961) Evipanos Suomen Lääkaril 17:987.

Miller, J.A., Jr. (1949) Science 110:113.

Miller, J.A., Jr. (1965) Bull. Tulane Univ. Med. Faculty 24:193.

Miller, J.A., Jr. (1971) Amer. J. obstet. Gynecol. 110:1123.

Miller, J.A. & Miller, F.S. (1954) Surgery 36:916.

Miller, J.A., Jr. & Miller F.S. (1962) Amer. J. Obstet. Gynecol. 48:44.

Rink, R. & Miller, J.A., Jr. (1967) Cryobiol. 4:24.

Villee, C.A. (1967) In "Brain Damage in the Fetus and Newborn from Hypoxia or Asphyxia". Eds. L.S. James & R.E. Myers. Ross Labo Pub. 1967 p. 47.

Westin, B. (1971) Amer. J. Obstet. Gynecol. 110:1134.

Westin, B., Nyberg, R, Miller, J.A., Jr. & Wedenberg, E. (1962) Acta. Paediat. Suppl. 139:1.

RESISTANCE OF THE NEW-BORN TO INJURY

R. A. Little

Experimental Pathology of Trauma Section, MRC Toxicology Unit, Medical Research Council Laboratories Woodmansterne Rd, Carshalton, Surrey, England

Many experimental models of injury have as their acute effect a loss of fluid from the circulation (Millican, 1960; Stoner, 1969). This fluid loss will be compensated, at least partially, by the movement of fluid into the circulation from the extravascular fluid compartments in the uninjured parts of the body (Tabor, Rosenthal & Millican, 1951).

As very little work has been done on the effect of age on post-traumatic fluid loss and its compensation, the ability to withstand post-traumatic fluid loss has been studied in the rabbit during the first weeks of life. During the experiments the animals were kept in a thermoneutral environment (Hull, 1965).

Three models of injury have been studied:- (a) a period of bilateral hind limb ischaemia (Rosenthal, 1943); (b) a full thickness scald of the dorsum (Bailey, Lewis & Blocker, 1962) and (c) the intraperitoneal implantation of lyphogel crystals. Lyphogel is a polyacrylamide hydrogel which expands in aqueous solutions to absorb exactly five times its own weight of water and low molecular weight solutes while excluding proteins and other substances with a molecular weight of 20,000 or more. The first two models (i.e. hind limb ischaemia and scalding) involve the loss of protein containing fluid from the circulation while the implantation of the lyphogel crystals results in the withdrawal of protein-free fluid from the circulation.

Consider first the simplest injury - the withdrawal of water and electrolytes from the circulation. Isosmotic protein-free fluid was withdrawn from the circulation of 1 day-old and 22 day-old rabbits by the intraperitoneal implantation of the lyphogel crystals. The 'dose' of lyphogel was calculated to give a 60% reduction in plasma volume in both groups of animals. Serial blood samples were taken from the retro-orbital plexus and the plasma volume was calculated at each time interval and expressed as a percentage of the initial volume. This loss of protein-free fluid was partially compensated in both groups of animals and the compensation was greater in the 1 day-old rabbits (Little, 1971a).

Thus the new-born rabbit is better able to compensate a loss of protein-free fluid from the circulation than the older animal. The loss of protein-free fluid from the circulation is a rather artificial situation and it is perhaps more relevant to consider a realistic situation where proteinaceous fluid is lost.

Post-traumatic fluid loss was most clearly localised after a period of limb-ischaemia and this injury has been studied in most detail, although the same pattern of response was seen after scalds. The fluid loss into the injured limbs after tourniquet removal was very rapid, the maximum increase in limb volume occurring within 1 hr of the end of 1, 2, 3 or 4 hr periods of ischaemia (Little, 1972). The fluid loss was calculated from measurements of the excess water content of the injured limbs 1 hr after tourniquet removal. The slope of the regression line (calculated by the method of least squares) relating the excess fluid in the injured hind-limbs to the period of ischaemia was significantly greater ($P < 0.05$) in the 1 - 2 day old than in the 22 day-old rabbits (Little, 1972).

From the regression lines the LD50 fluid loss, i.e. the fluid loss necessary to kill 50% of a group of rabbits, could be found for the LD50 periods of ischaemia calculated by probit analysis in earlier experiments (Little, 1971a). The LD50 fluid loss (expressed as ml/100 g body weight) was calculated for 1 - 2, 5, 10 and 22 day-old rabbits and found to be significantly higher in the 1 - 2 day-old than in the older animals (Little, 1971a).

At this point changes in body composition with age have to be taken into account. The total body water content and the plasma volume are both greater in the neonate. Thus although the LD50 fluid loss expressed as ml/100 g body weight was greater in the new-born rabbit it is possible that if this LD50 fluid loss was expressed as a percent plasma volume or per cent total body water the difference with age would no longer be apparent. When the LD50 fluid was expressed as a percentage of the total body water or of the 'true' intravascular plasma volume (Little, 1970) it was significantly higher during the first 48 hr of life than at any other age studied. The new-born rabbit is, therefore, able to tolerate a greater loss of proteinaceous fluid from the circulation than the older animal and the mechanism of this resistance must now be considered.

When calculating the 'LD50 fluid loss' the loss was estimated from the excess water content of the injured limbs. Another estimate of the fluid loss can be made from the changes in the great vessel haematocrit value. The calculation assumes that there has been no change in the red-cell volume and this has been confirmed (Little, 1971a). If there was no compensation of the circulatory fluid loss, the two measurements will be the same, but if compensation is occurring, the value calculated from the changes in haematocrit will be less than the value calculated from the wet and dry weights.

The 'apparent' fluid loss, calculated from the haematocrit values, was less than the 'real' fluid loss at all ages studied - in other words compensation was occurring at all ages. When the 'apparent' LD50 fluid loss was expressed as a percentage of the 'true' plasma volume there was no significant change with age. This means that in order to kill half of a group of rabbits at any age, the plasma volume must be reduced by the same amount (20-30%) but to achieve this 1.4 times as much fluid had to be removed from the circulation of the 1 - 2 day than of the 22 day-old rabbit.

The volume of fluid moved into the circulation (measured 1 hour after tourniquet removal) was significantly greater ($P < 0.001$) in the 1 - 2 day-old rabbit than in the older animal. The new-born rabbit being able to restore nearly five times as much fluid to the circulation as the 22 day old animal.

The compensatory movement of fluid across the vascular endothelium into the circulation is occurring in the uninjured tissues of the body. The factors influencing this movement are summarised in the Landis and Pappenheimer (1965) equation and it is relevant to see whether any changes with age in these factors can explain the greater fluid movement into the circulation found in the new-born rabbit.

The tissue component (tissue interstitial fluid pressure and the osmotic pressure of the proteins in the interstitial fluid immediately outside the vessel walls) is very difficult to measure and as its contribution to the equilibrium is probably rather small it has not been considered. Post-traumatic hypovolaemia and fall in arterial blood pressure will lead to vasoconstriction and a fall in capillary hydrostatic pressure in the uninjured tissues of the body (Oberg, 1964). It is difficult to predict how the age of the animal will affect these changes in capillary pressure but as the cardiovascular reflexes of the new-born rabbit appear to be well-developed (Mott, 1966) one can assume that the changes in capillary pressure will probably not differ markedly with age.

The other important factors are the plasma protein osmotic pressure and the capillary filtration coefficient. The loss of proteinaceous fluid into the injured tissues can change the plasma protein concentration and thereby the osmotic pressure exerted by the plasma proteins. After injury the plasma protein concentration rose significantly in the 1 day-old rabbit, fell significantly in the 40 day-old animals and showed no change at intermediate ages (Little, 1971a). The plasma-protein osmotic pressures were measured using a Hansen osmometer (Hansen, 1961) fitted with an Amicon UM10 semi-permeable membrane (synthetic hydrated polymer impermeable to solutes with a molecular weight over 10,000). Blood samples were taken 15 min before the end of an LD50 period of ischaemia and 1 hr after tourniquet removal. In the 1 day-old animals the samples were taken from separate animals whereas in the older animals both samples were taken from the same animal. At all ages studied the 'control' plasma protein osmotic pressure was higher than the values previously reported for the adult rabbit by Intaglietta & Zweifach (1971). The most striking result was the significant increase ($P < 0.02$) in the plasma protein osmotic pressure found in the 1 day-old rabbit.

From the Landis & Pappenheimer equations it can be deduced that this post-traumatic increase in plasma protein osmotic pressure will favour the movement of fluid into the circulation.

Whereas a change such as that found in plasma protein osmotic pressure will largely determine the direction of net fluid movement, the rate of movement will be determined by the capillary filtration coefficient. Thus for a unit change in pressure across the vascular endothelium, the movement of fluid will be greatest in the tissue with the highest capillary filtration coefficient.

The capillary filtration coefficient has been measured in the hind-limb of the 1 day-old and the 5 day-old rabbit using a water filled occlusion plethysmograph. The capillary filtration coefficient in the 1 day-old rabbit (0.033 ± 0.004 ml/min/mm Hg/100 g tissue) was significantly greater ($P < 0.01$) than in the 5 day-old rabbit (0.013 ± 0.003 ml/min/mm Hg/100 g tissue) (Little, 1971b). The fact that the capillary filtration coefficient in the neonate is 2.5 times larger than in the older animal means that a change in the transmural pressure equilibrium will cause a greater movement of fluid in the new-born.

When considering the compensatory movement of fluid into the circulation some account must be taken of where the fluid is coming from. A factor very relevant to this is the finding of a readily available extravascular albumin pool during the first 48 hr of life (Little, 1970).

Thus one can conclude that the greater ability of the new-born rabbit to withstand post-traumatic fluid loss is a result of its ability to compensate the loss. The more efficient compensation can be explained by the high capillary filtration coefficient, the post-traumatic increase in plasma protein osmotic pressure and the freely available extravascular albumin space found at this age. The correlation between differences in capillary filtration coefficient and resistance to circulatory hypovolaemia has been clearly demonstrated previously. The increased ability of flying birds, as compared to non-flying birds, to compensate post-haemorrhagic circulatory hypovolaemia has been shown to be a result of the high capillary filtration coefficient characteristic of the flying birds (Folkow, Fuxe & Sonnenschein, 1966; Djojosugito, Folkow & Kovách, 1968; Kovach, Szasz & Pilmayer, 1969; Kovách & Balint, 1969.

Although anoxia is not important in the ebb phase of the response to non-haemorrhagic injury it is important in the terminal necrobiotic phase and when considering the total resistance of the new-born to injury its increased resistance to anoxia (Mott, 1961) must also be taken into account.

REFERENCES

Bailey, B.N., Lewis, S.R. & Blocker, T.G. (1962) Tex. Rep. Biol. Med. 20, 20-29.
Djojosugito, A.M., Folkow, B. & Kovách, A.G.B. (1968) Acta physiol. scand. 74, 114-122.
Folkow, B., Fuxe, K. & Sonnenschein, R.R. (1966) Acta physiol. scand. 67, 327-342.
Hansen, A.T. (1961) Acta physiol. scand. 53, 197-213.
Hull, D. (1965) J. Physiol. 177, 192-202.
Intaglietta, M. & Zweifach, B.W. (1971) Microvasc. Res. 3, 72-82.
Kovách, A.G.B. & Balint, T. (1969) Acta physiol. Acad. Sci. hung. 35, 231-243.
Kovách, A.G.B., Szasz, E. & Pilmayer, N. (1969) Acta physiol. Acad. Sci. hung. 35, 109-116.
Landis, E.M. & Pappenheimer, J.R. (1963) Handbook of Physiology. Section 2: Circulation. Volume II. p.975. American Physiological Society, Washington.
Little, R.A. (1970) J. Physiol. 208, 485-497.
Little, R.A. (1971a) Ph.D. Thesis, University of London.
Little, R.A. (1971b) Proc. XXV Int. Congress of Physiol. Sci., 349.
Little, R.A. (1972) Brit. J. exp. Path. 53, 180-189.
Millican, R.C. (1960) The Biochemical Response to Injury. Ed. Stoner, H.B. & Threlfall, C.J. Oxford, Blackwell.
Mott, J.C. (1961) Br. Med. Bull. 17, 144-148.
Mott, J.C. (1966) Br. Med. Bull. 22, 66-69.
Oberg, B. (1964) Acta physiol. scand. 62, suppl. 229.

Rosenthal, S.M. (1943) Publ. Hlth Rep., Wash., $\underline{58}$, 1429-1436.
Stoner, H.B. (1969) Ann. R. Coll. Surg. Engl. $\underline{44}$, 308-323.
Tabor, H., Rosenthal, S.M. & Millican, R.C. (1951) Am.J.
    Physiol. $\underline{167}$, 517-522.

COMPARATIVE PHYSIOLOGY OF HEMORRHAGE. A STUDY OF ADRENERGIC

RECEPTOR STIMULATION AND BLOCKADE IN PIGEON AND MAMMAL

G. Rubányi, L. Huszár, and A.G.B. Kovách

Experimental Research Department, Semmelweis Medical

University, Budapest, Hungary

It has been shown (Kovách & Szász, 1968) that pigeons are more resistant to hemorrhage than mammals; some pigeons survived a slow blood loss of 8 ml per 100 g b.w., i.e. one equal to their total blood volume. When different avian and mammalian species bled at the rate of 1% body weight per hr, were compared it was found that they differed in tolerance to blood loss (Figure 1).

The flying and diving birds like the pigeon and duck are resistant to blood loss. The hen and the crow occupy an intermediate position and the domesticated pheasant together with the mammals belongs to the sensitive group (Kovách et al.1969).

Different explanations may account for the difference between birds and mammals: differences in blood volume, sensitivity to hypoxia, more efficient regulation of blood volume by different compensatory mechanisms. Folkow et al.(1966) have shown that the capillary filtration coefficient is higher in the duck than in mammals. Kovách and Bálint (1969) demonstrated that hemodilution after bleeding occurs more rapidly and to a greater extent in the pigeon than in the rat. Djojosugito et al.(1968) have found that after hemorrhage in the duck the capillary filtration coefficient rose threefold, close to the maximal level for duck muscle; the speed of fluid transfer is much higher from the extravascular space than in mammals.

There could be another explanation for the resistance of birds; difference in the activation of the sympathetic nervous system. It is known that in hemorrhagic shock the endogenous catecholamine concentration in blood increases (Bedford, 1917; Watts, 1956; Fowler, 1964). This elevation of blood catecholamine level and

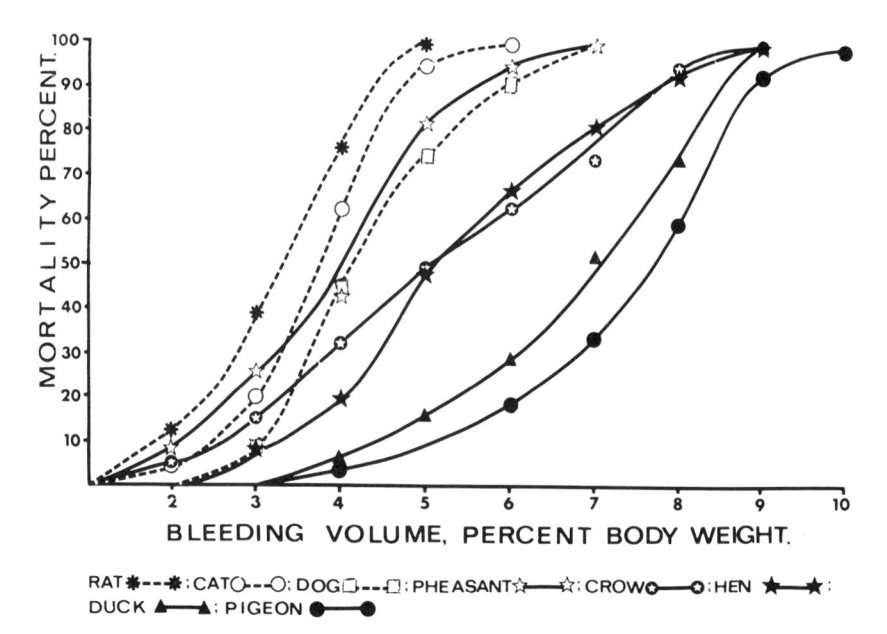

Fig.1.: Mortality percent of various mammals and birds under the effect of continuous bleeding. Every hour 1% body weight of blood was withdrawn. The percentage of animals lost during bleeding has been recorded and plotted (ordinate). (A.G.B. Kovách, E. Szász and N. Pilmayer: Mortality of Various Avian and Mammalian Species Following Blood Loss. Acta physiol.Acad.Sci.hung. 1969.35.109).

sympathetic activity (Fedina et al.1962) may be one important factor making the shock condition irreversible (Kovách et al.1957; Nickerson, 1962). There are several possible mechanisms: a. permanent vascoconstriction of certain blood vessel sections (Malcolm 1905; Cannon 1923; Jourdan 1949); b. plugging of the microcirculation (Baeckström et al.1971); c. breakdown of tissue metabolism.

Comparing the catecholamine content of different tissues, the vascular sensitivity to catecholamines and the effect of adrenergic blocking agents in different species the following data are available. On basis of the work of Östlund (1954), Prosser (1961) and Duke (1971) the sequence of norepinephrine content of adrenal medulla of different species is as follows: invertebrates (the presence of catecholamines is very uncertain); whale, pigeon, newborn rabbit (80-90% norepinephrine contents); mouse, man, guinea pig, adult rabbit (2-25% norepinephrine contents).

The transmitter released at the terminals of sympathetic postganglionic nerves contains 80-90% norepinephrine in mammals and 60-70% epinephrine in pigeons (Sturkie, 1970).

Epinephrine administered i.v. diminishes the blood pressure of earth-worm contrary to vertebrates (Prosser 1961.). Only large doses stimulate the heart rate in invertebrates. Phenoxybenzamine failed to block the vasoconstrictor effect of epinephrine or norepinephrine in pigeons (Kovách et al.1969). According to Thompson and Coon (1948) Dibenamine does not block the pressor effect of epinephrine in the chicken in contrast to other adrenolytics such as dihydroergotamin or Tolazoline which exert an inhibitory effect. In contrast to this, Barger and Dale in 1907 found a response to ergot alkaloids in birds similar to that to Dibenamine. It is not known whether this lack of a pressor blocking effect of Dibenamine is characteristic of other avian species.

In our present experiments we studied the catecholamine sensitivity, the properties of the alpha and beta adrenergic receptors in the pigeon and compared them with data on mammals.

METHODS

Forty-five pigeons of either sex, weighing 400 g on an average, were used for the experiments. The birds were anaesthetized by intramuscularly injected urethane (1-1,5 g/kg), immobilized by flaxedyl and subjected to artificial respiration. To monitor arterial blood pressure the left basilar artery was prepared, cannulated and the pressure recorded on a polygraph. The heart rate was measured from an ECG lead and indicated on an Analyser-Gamma ratemeter. To provoke the cerebral ischaemic reflex the carotid arteries were prepared. The substances were injected into the left basilic vein.

Norepinephrine (10 µg/kg) and the effect of carotid occlusion (15-20 sec) were used for alpha-adrenergic receptor stimulation. Phenoxibenzamine (10+10+20 mg/kg) and dimethoxy-benzyl-aminoethyl-pyridin (DBAP 5 mg/kg) were administered as blocking agents. Isoprenaline for beta stimulation (0,5, 1,0, 5,0 µg/kg) and propranolol (1 mg/kg) for beta receptor blockade was used.

Student's t-test was used for statistical evaluation.

RESULTS

<u>Sensitivity to exogenous catecholamines.</u> Pigeons are considerably less sensitive to exogenous catecholamines than dogs. This is proved by the fact, that for the same pressure change (60 mmHg arterial mean blood pressure elevation) pigeons need much larger doses: sixfold for norepinephrine (dog:2,5 µg/kg; pigeon:15 µg/kg) and threefold for epinephrine (dog:2,5 µg/kg; pigeon:7,5 µg/kg). To elicit a 30 mmHg drop in blood pressure a seven and a halffold dose of isoprenaline has to be used in pigeons (dog:1 µg/kg; pigeon:7,5 µg/kg).

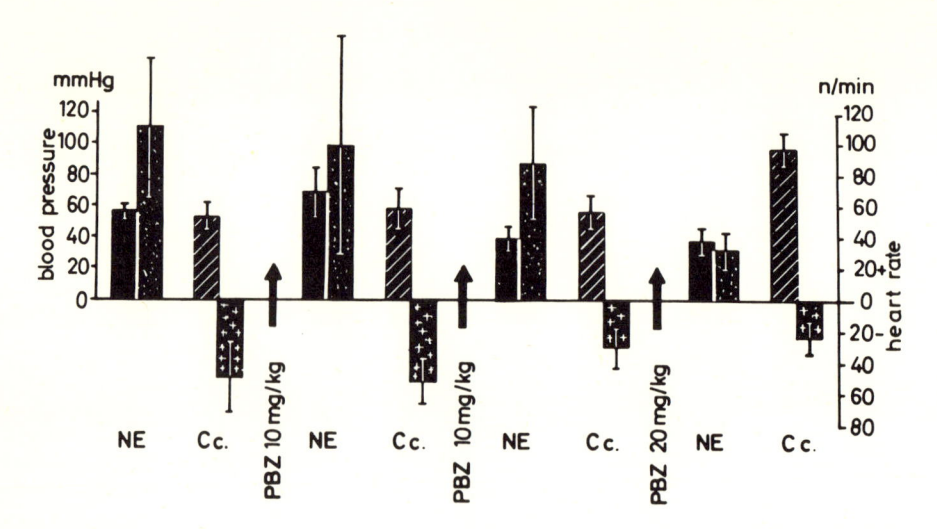

Fig.2.: The effect of norepinephrine (NE) and Carotid clamping (Cc) on mean arterial blood pressure and heart rate of the pigeon before and after administration of increasing doses of phenoxybenzamine (PBZ). (Please note that in this and in all further figures the columns indicate the variations relative to the initial values)

The effect of phenoxybenzamine on pigeons. Phenoxybenzamine in a 10 mg/kg dose evokes a biphasic effect on the systemic blood pressure: an initial short increase from 90.1±7.2 to 108.1±3.6 mmHg (P < 0.05) and following this a decrease for half an hour to 70.6±4.8 mmHg (P < 0.05). The heart rate rises from 347.5±31.2 to 408.6±29.5 beat per min (P < 0.05).

Figure 2 shows that the blood pressure increasing effect of both norepinephrine and carotid clamping rises after administration of 10 mg/kg phenoxybenzamine in pigeons. Norepinephrine: effect on arterial blood pressure in pigeons before phenoxybenzamine from 103.6±3.3 to 158.7±14 mmHg (P < 0.001) and after phenoxybenzamine from 78.0±27.8 to 147.3±19.5 mmHg (P < 0.01). Carotid occlusion: effect on arterial blood pressure in pigeons before phenoxybenzamine from 98.5±3.3 to 152.0±11.0 mmHg (P < 0.001) and after phenoxybenzamine from 85.0±7.0 to 143.0±18.5 (P < 0.01).

After the second administration of 10 mg/kg phenoxybenzamine the effect of norepinephrine is still significant, from 83.3±6.5 to 121.8±14.4 mmHg (P < 0.01). The pressor response following carotid occlusion remained unchanged from 81.0±11.4 to 137.0±17.0 mmHg (P < 0.01). The administration of a further 20 mg/kg phenoxybenzamine to the pigeon reduces the effect of norepinephrine. The blood pressure rises significantly from 68.0±9.0 to 106.0±18.0 mmHg (P < 0.05).

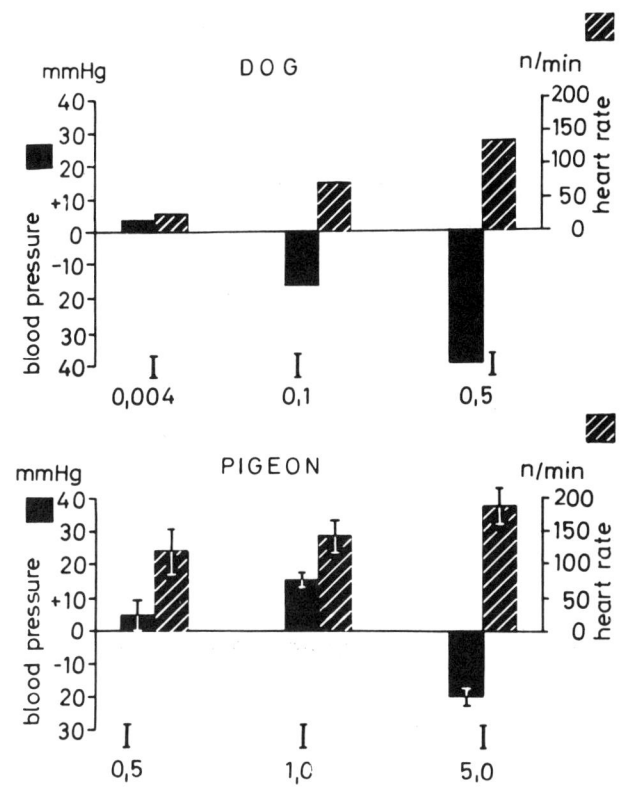

Fig.3.: The effect of different doses of isoprenaline on mean arterial blood pressure and heart rate of dogs and pigeons. (I= isoprenaline µg/kg).

The effect of dimethoxy-benzyl-aminoethyl-pyridin (DBAP) on the pigeon. DBAP in a 5 mg/kg dose decreased the blood pressure from 83.1±3.2 to 72.4±4.0 mmHg (P< 0.05), but it did not significantly influence the heart rate of pigeons. We have found that an adrenolytic dose of DBAP in mammals (5 mg/kg) inhibits the blood pressure increasing effect of norepinephrine. The blood pressure response to norepinephrine was from 86.4±5.0 to 115.0±6.5 mmHg (P< 0.02) before DBAP and from 77.7±9.0 to 87.2±11.6 mmHg (P< 0.2) after DBAP. DBAP does not block the pressor response following carotid clamping. The blood pressure rose from 88.0±8.0 to 130.0± 14.2 mmHg (P< 0.01) before and from 75.0±14.0 to 138.0±10.3 mmHg (P< 0.01) after DBAP.

The effect of isoprenaline in the pigeon. Figure 3 shows an unexpected result. Isoprenaline in a dose of 0.5 µg/kg which causes a considerable drop in blood pressure and increases heart rate in

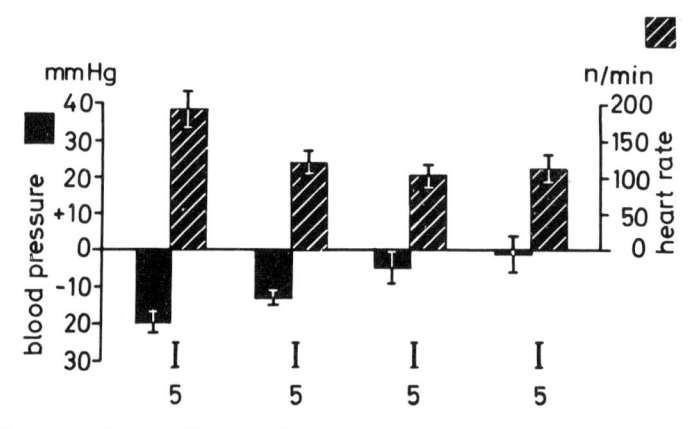

Fig.4.: Tachyphylaxis-like effect of successive administration of isoprenaline on mean arterial blood pressure of pigeons. (I= isoprenaline μg/kg).

dogs has no significant influence on the mean arterial pressure in pigeons (96.6±5.1 before, 100.8±4.9 mmHg after [P> 0.8]), but increases the heart rate to the same degree as in dogs, from 186.4±14.3 to 303.5±59.9 beat per minute (P< 0.02). On the other hand instead of decreasing, isoprenaline in a dose of 1 μg/kg increases blood pressure in pigeons from 92.1±6.5 to 102.2±4.8 mmHg (P< 0.01).

To achieve a blood pressure decrease by isoprenaline ten times the mammal dose (5.0 μg/kg) is necessary and in this case the blood pressure decreases from 88.2±8.4 to 68.5±6.4 mmHg (P< 0.05). Having administered isoprenaline to pigeons in such doses several times at regular 15 minute intervals, we observed a tachyphylaxis-like effect, illustrated in figure 4. In some experiments the administration of a third and fourth dose brought about a rise in blood pressure.

The effect of beta blocking propranolol on pigeons. Propranalol in 1 mg/kg dose decreases the mean arterial pressure from 96.2± 6.5 to 82.8±8.1 mmHg (P< 0.05) and the heart rate from 241.2±25.0 to 153.2±18.2 (P< 0.05) in the pigeon. Isoprenaline in 1 μg/kg dose, causing a rise in blood pressure, provokes a pressor effect of the same degree also after 1 mg/kg propranolol (Fig.5). Using isoprenaline in a pressure reducing dose (5 μg/kg) after propranolol a considerable blood pressure rise takes place from 78.7± 4.1 to 91.3±2.3 mmHg (P< 0.05). Similarly to mammals, the heart rate accelerating effect of isoprenaline decreases after administration of propranolol. The values before propranolol were 258.2± 20.6 and isoprenaline increased them to 442.7±12.6 beats per minute

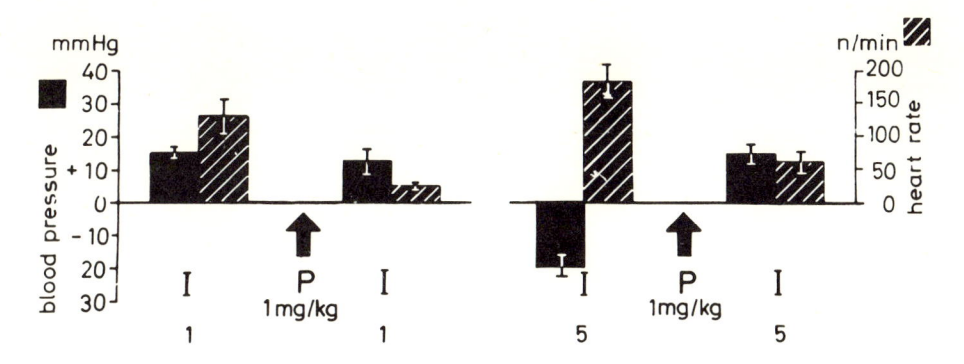

Fig.5.: The effect of isoprenaline on mean arterial blood pressure and heart rate of the pigeon before and after administration of propranolol. (I =isoprenaline μg/kg); (P= propranolol).

($P<0.05$). After propranolol pretreatment the heart rate was 168.2±10.5 and isoprenaline only elevated it to 191.0±18.7 beats per minute.

DISCUSSION

Discussing our results first we have found a difference in the sensitivity of exogenous catecholamines. The lower sensitivity of pigeons could be explained by numerous causes, on which we have no data. The adrenergic receptors are different in pigeons; the catecholamines link in other places; this link is different from that in mammals; the catecholamines undergo destruction in the circulation of pigeons; the catecholamine level in the blood of pigeons is originally so high that the exogenous catecholamines administered in mammal doses are not efficient. We may suppose however that this low sensitivity can be one further important factor in the mechanism of the high tolerance to hemorrhagic shock of that species. We wish to point out the difference in sensitivity between epinephrine and norepinephrine on pigeons. In this species the epinephrine increases the arterial blood pressure twice as efficiently as norepinephrine, just the opposite of the findings in mammals.

Broughton Pipkin (1971) and Mott (1971) found that the sensitivity of newborn rabbits to catecholamines is much higher (3-6 fold) than that of adult ones, though the newborn tolerate the different forms of shock better. (Miller (1971); Little, (1971)). On the other hand Broughton Pipkin (1971) and Mott (1971) found that the concentration of catecholamines in blood in the course of a hemorrhagic shock increases much less in newborn rats, than in adult ones. Further investigations are needed both in phylogenetic and in ontogenetic fields to understand these contradictions.

Phenoxybenzamine does not block the pressure effect of norepinephrine in pigeons except in subtoxic doses. This dose can cause such unphysiological changes that it is very hard to say whether the specific adrenolytic effect of phenoxybenzamine is alone responsible for diminishing the blood pressure increasing effect of norepinephrine. The arterial blood pressure raising effect of carotid clamping does not change even after administration of subtoxic doses. DBAP in pigeons has a significant blocking effect on norepinephrine, but it cannot block the vasoconstriction elicited by carotid clamping. From these findings it seems to be possible that pigeons have either two different types of alpha receptors on the smooth muscles of the peripheral blood vessels, or the transmitter substance released by the sympathetic nerve terminals following carotid clamping is different from norepinephrine.

During the studies of beta stimulant isoprenaline and beta blocking agent propranalol we came to the following conclusions. First it is most probable that isoprenaline stimulates both the alpha and the beta adrenergic receptors of pigeons. This is proved by the results that the administration of low (0.5-1 µg/kg) doses of isoprenaline to pigeons causes blood pressure increase, and the blood pressure decreasing effect following the injection of larger doses (5 µg/kg) turns into pressor effect after propranolol administration. The second conclusion is that the beta receptors in the heart and in peripheral blood vessels seem to be different from each other because small doses of isoprenaline stimulate only the beta-receptors of the heart, while having an alpha-receptor stimulating effect on the peripheral blood vessels.

We can conclude that there are several differences between pigeons and mammals concerning the alpha- and beta- adrenergic receptor properties. As to the high tolerance to hemorrhagic shock of this species the low sensitivity to exogenous catecholamines seems to be an important factor.

REFERENCES

Baeckström, P., Folkow, B., Kovách, A.G.B., Löfving, B. & Öberg,B. (1971) In: 6th European Conference on Microcirculation, Aalborg, 1970. Eds. J. Ditzel & D.H. Lewis, Basel, Karger p.16.
Barger, G. & Dale, H.H. (1907). Biochem.J. 2 240.
Bedford, E.A. & Jackson, G. (1916). Proc. Soc. exp. Biol. Med. 13, 85.
Braunwald, E. (1966) Symposium on β-Adrenergic Receptor Blockade. Editorial Introduction. Amer. J. Cardiol. 18, 303.
Broughton Pipkin, F. (1971). Quart.J.exp. Physiol. 56, 210.
Broughton Pipkin, F., Mott, J.C., & Roberton, N.R.C. (1971) J. Physiol.(London) 218, 385.
Cannon, W.B. (1923) Traumatic Shock. Appleton, New York.
Djojosugito, A.M., Folkow, B., & Kovách, A.G.B. (1968). Acta physiol. scand. 74, 114.
Dukes, H.H. (1970) Physiology of Domestic Animals. 8th Edition. Ed. Swenson, M.J., Cornell University Press, Ithaca.
Fedina, L., Kovach, A.G.B. & Vanik, M. (1965) Acta physiol. Acad. Sci. hung. Suppl. 26, 35.
Folkow, B., Fuxe, K. & Sonnenschein, R.R. (1966)Acta physiol. scand. 67, 327.
Jourdan, F., Chatonet, J., Collet, A., & Pazat, P. (1949) J. Physiol. (Paris) 41, 193A.
Kovách, A.G.B. (1966) Orvosképzés, 5, 321.
Kovách, A.G.B. & Bálint, T. (1969) Acta physiol.hung. 35, 231.
Kovách, A.G.B. & Szász, E. (1968) Acta physiol. hung. 34, 301.
Kovách, A.G.B., Szász, E. & Pilmayer, N. (1969) Acta physiol. hung. 35, 109.
Kovách, A.G.B., Takács, L., Mohácsi, A., Káldor, A. & Kalmár, Z. (1957) Acta physiol. hung. 11, 181.
Little, R.A. (1971). This Symposium, p.
Marshall, A.J. (1961) Ed. Biology and Comparative Physiology of Birds. Vol. 1-2, London-New York, Academic Press.
Miller, J.A. (1971) This Symposium, p.
Nickerson, M. (1962) In: Schock, Pathogenese und Therapie. Berlin-Göttingen-Heidelberg, Springer Verl. p. 434.
Östlund, E. (1954) Acta physiol. scand. 31, Suppl. 112, 1.
Prosser, C.L. & Brown, F.A. Jr., (1961) Comparative Animal Physiology. 2nd Edition W.B.Saunders Co. Philadelphia-London.
Reader, R. (1967) Foreword to the Symposium on the Catecholamines in Cardiovascular Physiology and Disease. Circ. Res, 21, Suppl.3 1.
Sturkie, P.D. (1970) Fed. Proc. 29, 1674.
Sturkie, P.D. (1965) Avian Physiology. Ithaca, Cornell Univ.Press, p.766.
Thompson, R.M. & Coon, J.M. (1948) Fed. Proc. 7, 259.
Watts, D.T. (1956) Amer. J. Physiol. 184, 271.

BIOSYNTHESIS OF ADRENAL CATECHOLAMINES DURING ADAPTATION OF RATS TO IMMOBILIZATION STRESS

Richard Kvetňanský

Institute of Experimental Endocrinology, Slovak Academy of Sciences, Bratislava, Czechoslovakia

The study of the reactions of the organism to chronic, long term, repeated stress, attains an extraordinary importance, mainly because nowadays civilisation is accompanied by new factors, to which the organism has to be adapted, i.e. crowds, traffic, noise, rush, human relations, etc. The sympatho-adrenal system is activated in the first place by stress and immediately secretes catecholamines - epinephrine and norepinephrine. Up to the present time we have little information about the function or reaction of these very active biological compounds during adaptation to daily repeated stress.

This is the reason why we have recently turned our attention in our laboratory to the problem of catecholamines in stress and during the adaptation of rats to daily repeated stress. For a stress model, on which our experiments were mainly performed, we chose immobilization, which produces both psychological stress (struggling) and physical stress (muscle work) (24,14). Daily repeated immobilization for 150 min causes a significant increase in the weight of the adrenal and a many-fold increase in plasma corticosterone, which are good indicators of the activation of hypothalamo-pituitary-adrenocortical system (19,11). Beside the adrenal cortex the adrenal medulla also participates in the increase in weight of adrenals after repeated immobilization (12). We measured the cell nuclei of the adrenal medullary cells by the karyometric method (21) in histological sections and they appear to be larger in the adrenals of repeatedly immobilized rats (12). These findings suggested an increase in the activity of the adrenal medulla after repeated immobilization.

The histologically demonstrated increase in the activity of adrenal medulla during repeated immobilization had to be shown in the biochemical indicators too. Therefore we measured the content of catecholamines in the adrenal medulla and their urinary excretion in rats which were immobilized only once or repeatedly. To estimate the catecholamines we used our modification of the method of von Euler and Lishajko (5).

First we studied the effect of varying intervals of the first immobilization on adrenal and urinary catecholamines (13). A single period of immobilization did not result in marked alterations in the adrenal catecholamine content although there were marked increases in catecholamine excretion during the period of immobilization. A significant decrease in adrenal epinephrine was seen after 90 min or longer periods of immobilization. This decrease persisted during the 24 hr following release of the rats from immobilization (13). Adrenal norepinephrine levels remained almost unchanged. Urinary epinephrine excretion (24 hr collection) was significantly increased for each interval of immobilization from 15 to 240 min. There was no increase in norepinephrine excretion following periods of immobilization up to 1 hr. After 150 or 240 min of immobilization, however, significant increases in norepinephrine excretion were apparent. Pearl et al. (22) also showed a small decrease in total adrenal catecholamines in restrained rats and Leduc (18) showed a decrease in adrenal epinephrine after 24 hr cold stress in rats without a marked change in adrenal norepinephrine. Bygdemann and Euler (3) have reported that after splanchnic nerve stimulation or chemically induced release of epinephrine from the cat adrenal medulla there is a rapid resynthesis of this catecholamine. Gordon et al. (7) have demonstrated a small decrease of rat adrenal epinephrine after exercise which was more apparent when α-methyl-p-tyrosine was used to block the biosynthesis of catecholamines. Exercise was shown to increase catecholamine formation from tyrosine $-^{14}C$ (7). This acceleration of catecholamine synthesis by stress can explain why immobilization did not cause a marked decrease in adrenal catecholamines. Marked increases in urinary catecholamine excretion after immobilization of rats in small cages for 16 hr were found (8). While in our experiments urinary epinephrine was significantly increased at each interval of immobilization studied, an increase in urinary norepinephrine was apparent only after 150 min of immobilization. Urinary norepinephrine largely reflects the activity of sympathetic nerve endings. In humans increased excretion of urinary norepinephrine occurs during and after physical stresses or exercise (4). In our experiments, after the first 2.5 hr of immobilization, rats were not able to walk and they appeared physically exhausted. Perhaps this is why a significant increase of urinary norepinephrine was seen only after longer intervals of immobilization. In other experiments we found that the greatest increase in excretion rates

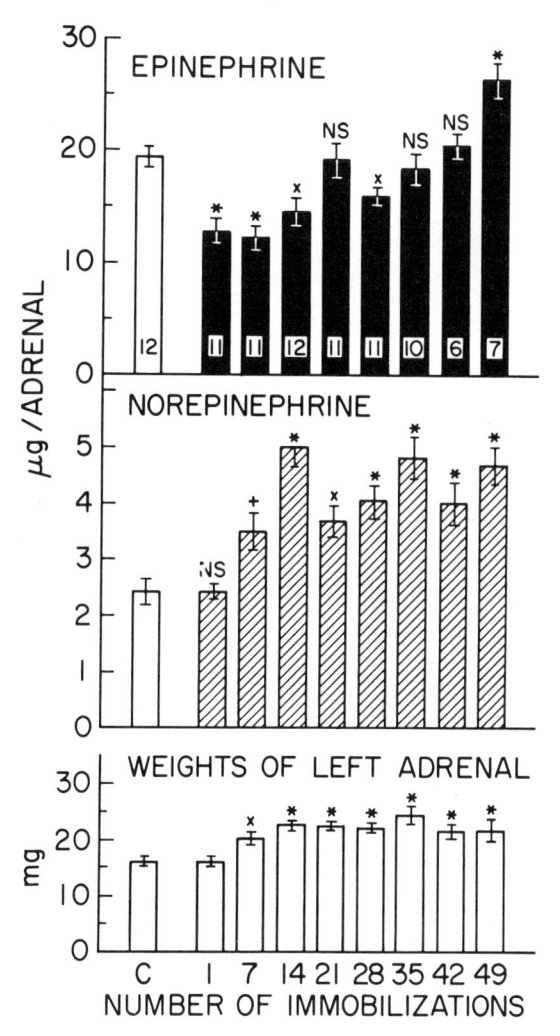

Fig. 1. Effect of daily repeated immobilization for 2.5 hr on adrenal catecholamines and adrenal weights. Values are expressed as the mean (±SEM). The number of animals is indicated in each column. Statistical significance compared with unstressed group (C): (NS=not significant, +=p<0.02, x=p<0.01, *=p<0.001).
From : Endocrinology, 87, 738, (1970).

of epinephrine and norepinephrine in 24 hr urine collection occurred during the interval of immobilization (more than half the amount of epinephrine and a third of the amount of norepinephrine excreted)(13).

The most convenient interval for study of repeated immobilization was 2·5 hr. After the first 2·5 hr immobilization, levels of adrenal epinephrine were decreased for at least 24 hr. Thus, animals immobilized again on the second day had a deficit in the adrenal epinephrine. Immediately after repeated daily immobilization (7-14 times), levels of adrenal epinephrine were still decreased (13) (Fig. 1). After about 40 days of repeated immobilization, however, levels of adrenal epinephrine were no longer depressed. Levels of adrenal norepinephrine appeared to be elevated after one week of repeated intervals of immobilization and remained elevated. As indicated earlier, 24 hr after the first interval of immobilization, levels of adrenal epinephrine were still diminished. Immediately after the ninth daily immobilization, epinephrine levels were diminished, but 24 hr later (at the time that the tenth stress would have been given), levels of adrenal epinephrine returned to normal (13) (Fig. 2). After the 70th or 350th immobilization no decrease in adrenal epinephrine was apparent either immediately or 24 hr after immobilization.

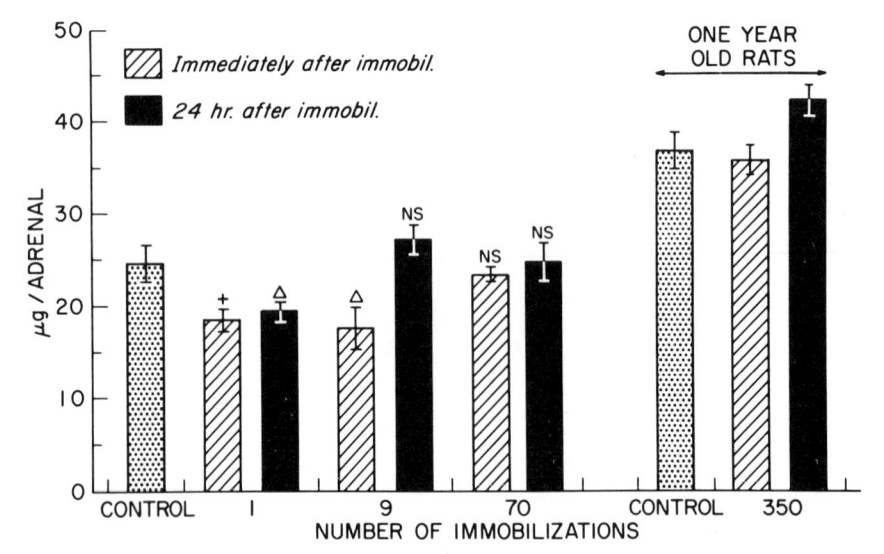

Fig. 2. Effect of repeated immobilization on adrenal epinephrine content. Rats were immobilized daily for 2·5hr and killed immediately after the last immobilization or 24 hr later. Results are mean values (± SEM) for groups of 7-10 rats. Rats immobilized 350 times (5-6 animals/group) were compared with control rats of the same age. Levels of adrenal epinephrine in 1 yr-old rats were not significantly changed. Statistical significance compared with unstressed group (control):

(Legend to Fig. 2 continued)
(△ P < 0.05, += P < 0.02). From: Endocrinology 87: 738 (1970)

Urinary excretion of epinephrine was increased on the first day of immobilization and became maximal after seven days of repeated immobilization (13) (Fig. 3). After this, there seemed to be a decrease from this peak rate but excretion of epinephrine continued to be markedly elevated above the levels obtained after the first immobilization. Twenty-four hr after the ninth daily immobilization the levels of adrenal epinephrine approximated those in the control animals (Fig. 2) but excretion of urinary epinephrine was maximal at this time of repeated immobilization (Fig. 3). This suggests that probably at this time there is increased biosynthesis of adrenal catecholamines. Repeated immobilization continued to cause a decrease in adrenal epinephrine (Fig. 1), however, presumably because the rate of release of epinephrine during the short interval of immobilization (2.5 hr) still exceeded synthesis. In rats subjected to 40-350 repeated immobilizations, however, there was no decrease of adrenal epinephrine immediately or 24 hr after immobilization (Fig. 2), although urinary excretion of epinephrine was still increased (Fig. 3). Leduc (18) showed that there was neither

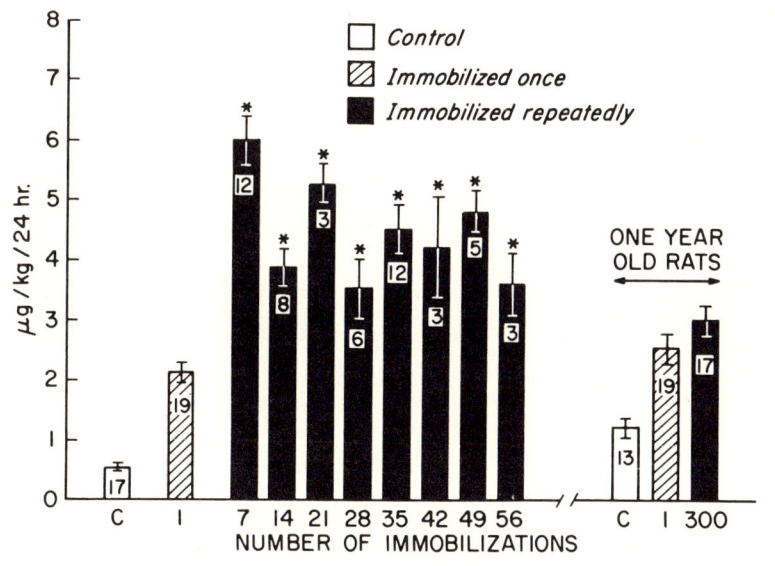

Fig. 3. Effect of repeated immobilization on urinary epinephrine excretion. Rats were immobilized daily for 2.5 hr. Values are the mean (± SEM) of 24 hr epinephrine excretion for the number of pairs of rats indicated in the columns. Statistical significance compared with the group immobilized only once: (* = P < 0.001). Rats immobilized 300 times were compared with control rats of the same age and had significantly higher levels of urinary epinephrine (P < 0.001) but did not differ significantly from rats of this age immobilized only once. From Endocrinology 87: 738, (1970).

inhibition nor activation of the destruction of administered catecholamines in rats during adaptation to cold. If metabolism of epinephrine in our studies on repeatedly immobilized rats is similarly unaffected, the increased excretion of free epinephrine is probably caused by its increased biosynthesis rather than by changes in degradation of this catecholamine.

We have already published a paper about increased levels of total adrenal catecholamines in 1966 (12). Similar increases in the levels of adrenal medullary catecholamines after repeated stress were found by Welch (31) in mice that had been made aggressive by several weeks of individual housing and then were allowed to fight daily for 5-10 min, and by Hruza (9) in rats adapted to trauma in the Noble-Collip drum. Welch concluded that the large increases in adrenal catecholamines found after repeated daily exposure to fighting probably indicate that the adrenal medulla had undergone adaptive changes in its capacity for synthesis and storage of catecholamines (31). During adaptation of rats to chronic cold stress increases in adrenal epinephrine and norepinephrine content and in urinary catecholamine excretion have also been found (18).

Our results suggest that the "adaptation" of the adrenal medulla to repeated immobilization stress in rats was the result of an enhanced ability to replace the catecholamines which were released, rather than to a diminished release of catecholamine in response to the repetition of immobilization.

To obtain confirmation and see if in "adapted" animals the biosynthesis of adrenal catecholamines had been really increased, we studied the levels of radioactive catecholamines after in vivo administration of labelled catecholamine precursors (17). Tyrosine-^{14}C conversion to catecholamines has been used as an index of the rate of synthesis of these compounds in vivo (27). Formation of labelled amines from dopa-^{3}H can be used to assess changes in the rate of catecholamine synthesis and to allow for change in rate of catecholamine release which do not involve the tyrosine hydroxylase step (10) Tyrosine-^{14}C (50 μci/rat) and dopa-^{3}H (37.5 μci/rat) were simultaneously administered intravenously either 1.5 hr after the beginning of the last immobilization or 24 hr after terminating the last immobilization. The animals were decapitated 1 hr after injection of the radioactive compounds and the adrenals after homogenization applied to Dowex 50 for separation of norepinephrine and epinephrine in the first fraction and dopamine in the second (23). Tritium and carbon-14 content was assayed in both fractions by liquid scintillation spectrometry.

TABLE 1. Effect of immobilization on dopamine synthesis from tyrosine-14C and dopa-3H in the rat adrenal

No. of immobilizations	Hr after last immobilizations	Dopamine cpm/pair adrenals Carbon-14	Tritium	Carbon-14/tritium
0	—	405 ± 48	247 ± 18	1.99 ± 0.30
1	0	634 ± 85a	223 ± 44	3.19 ± 0.25b
1	24	831 ± 86c	232 ± 44	3.47 ± 0.31c
7	0	1491 ± 64d	243 ± 13	6.32 ± 0.33d
7	24	880 ± 97c	205 ± 34	4.49 ± 0.50c
40	0	1710 ±177d	470 ± 93a	4.68 ± 0.43d
40	24	1471 ±354a	331 ± 68	4.36 ± 0.22

Rats were given tyrosine-14C (50 µCi) and dopa-3H (37.5 µCi) iv 1 hr before decapitation at the times indicated. Each group consisted of 5-7 rats except for the last group, which contained only 3 rats.
a$p<0.05$, b$p<0.02$, c$p<0.01$, d$p<0.001$.
From: Endocrinology 89 : 46, (1971)

TABLE II. Effect of immobilization on β-hydroxylated catecholamine synthesis from tyrosine-14C and dopa-3H in the rat adrenal

No. of immobilizations	Hr after last immobilizations	Norepinephrine and epinephrine cpm/pair adrenals Carbon-14	Tritium	Carbon-14/tritium
0	—	450 ± 67	265 ± 41	1.99 ± 0.26
1	0	621 ± 50	254 ± 11	2.47 ± 0.19
1	24	795 ± 67[a]	246 ± 32	3.36 ± 0.25[a]
7	0	3127 ± 194[b]	489 ± 55[a]	6.72 ± 0.53[b]
7	24	3192 ± 498[b]	715 ± 87[a]	4.42 ± 0.27[b]
40	0	2369 ± 269[b]	494 ± 36[a]	4.29 ± 0.15[b]
40	24	1994 ± 558[c]	420 ± 132	4.97 ± 0.35[b]

Rats were given tyrosine-14C (50 μCi) and dopa-3H (37.5 μCi) iv 1 hr before decapitation at the times indicated. Each group consisted of 5-7 rats except for the last group, which contained only 3 rats.

[a] $p<0.01$, [b] $p<0.001$, [c] $p<0.05$.
From: Endocrinology 89 : 46, (1971)

From the Table I (17) it is apparent that dopamine synthesized from tyrosine-^{14}C was significantly increased by immobilization in all groups. Dopamine synthesized from dopa-^{3}H was not changed after the first or seventh immobilization, only after the 40th immobilization was there a small increase. The ratio of carbon-14 to tritium in the catecholamine provides an index of the rate of tyrosine hydroxylation. These results therefore indicate that there is an increase in the rate of conversion of tyrosine to dopa catalysed by tyrosine hydroxylase and that the rate of dopamine formation from dopa, catalysed by l-amino acid decarboxylase, is unchanged, except in the group of animals immobilized 40 times. Synthesis of the β-hydroxylated catecholamines (norepinephrine and epinephrine) from tyrosine -^{14}C is also markedly increased particularly after repeated immobilization (17) (Table II). The ratio of carbon-14 to tritium in these compounds is increased to about the same extent as in dopamine. Synthesis of β-hydroxylated compounds from dopa-^{3}H is not changed after the first immobilization but is significantly increased after repeated immobilization. Since dopamine-^{3}H levels do not appear to be increased (Table 1), the increase in β-hydroxylated compounds formed from dopa-^{3}H may be explained as an increase in the rate of conversion of dopamine to norepinephrine, catalysed by dopamine-β-hydroxylase.

These results clearly show that there is an increase in the biosynthesis of adrenal catecholamines after repeated immobilization of rats. Catecholamine synthesis from tyrosine-^{14}C has also been shown to be increased in the adrenals of rats subjected to forced exercise (7). The rate of catecholamine biosynthesis after immobilization was mainly changed at steps tyrosine-dopa and dopamine-norepinephrine which are catalysed by the enzymes tyrosine hydroxylase and dopamine-β-hydroxylase. Therefore in our next work we measured in vitro the levels of these enzymes in the adrenals of immobilized rats. Tyrosine hydroxylase (TH) was assayed using the method of Nagatsu et al. (20). The levels of this enzyme in adrenals of immobilized rats are shown in Fig.4 (14).After a single immobilization interval of 2.5 hr, the adrenal glands had significantly increased tyrosine hydroxylase activity. After 7 such daily immobilization intervals the TH activity had risen to over 3 times the control levels. No further increase in TH activity was found in the adrenals of rats which had been immobilized daily for 6 weeks. The enhanced synthesis of the enzyme protein is most probably responsible for the elevation of adrenal TH after immobilization (14).

EFFECT OF REPEATED IMMOBILIZATION STRESS ON RAT ADRENAL TYROSINE HYDROXYLASE

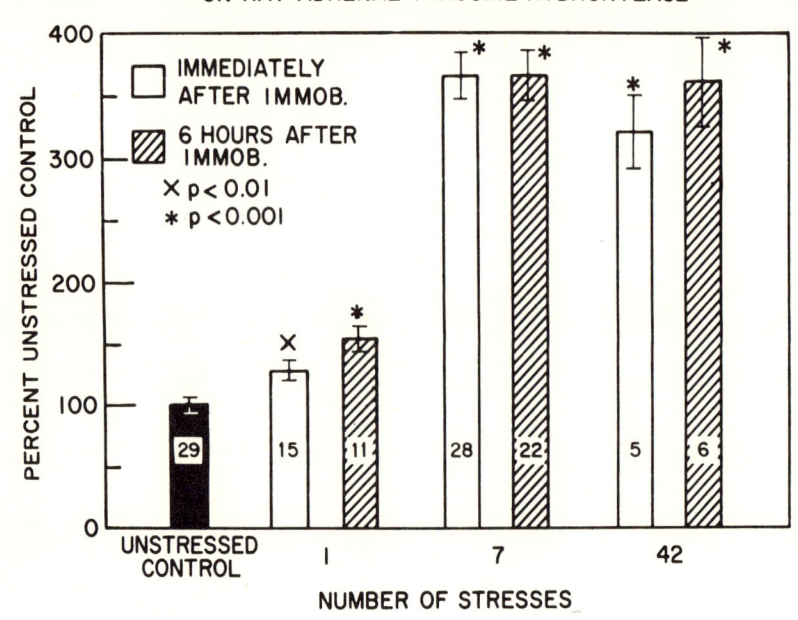

Fig. 4. Effect of repeated immobilization on tyrosine hydroxylase levels in the rat adrenal. Rats were immobilized for 2.5 hr daily for the indicated number of times. The levels of tyrosine hydroxylase are expressed as per cent of unstressed control (mean ± SEM) for the number of rats indicated in the columns. Control levels of tyrosine hydroxylase were 58.2 ± 3.1 nmoles tyrosine hydroxylated per pair of adrenals per hr.
From: Endocrinology 87 : 744, (1970).

Dopamine-β-hydroxylase was assayed by the method of Friedman and Kaufman (6) as modified by Viveros et al. (29). Adrenal dopamine-β-hydroxylase (DBH) levels in immobilized rats are shown in Fig. 5 (15). After repeated immobilization levels of DBH were significantly higher than those found in the adrenals of the control animals. Six hr after the 7th or 42nd immobilization the elevation of the enzyme was more marked. The increment in activity of this enzyme during the 6 hr after the 1st, 7th and 42nd immobilization progressively increased (Fig. 5).

Viveros et al. (28) showed that DBH is released from the perfused adrenal medulla along with epinephrine. Levels of DBH found 6 hr after the cessation of repeated immobilization were higher than immediately after immobilization suggesting that

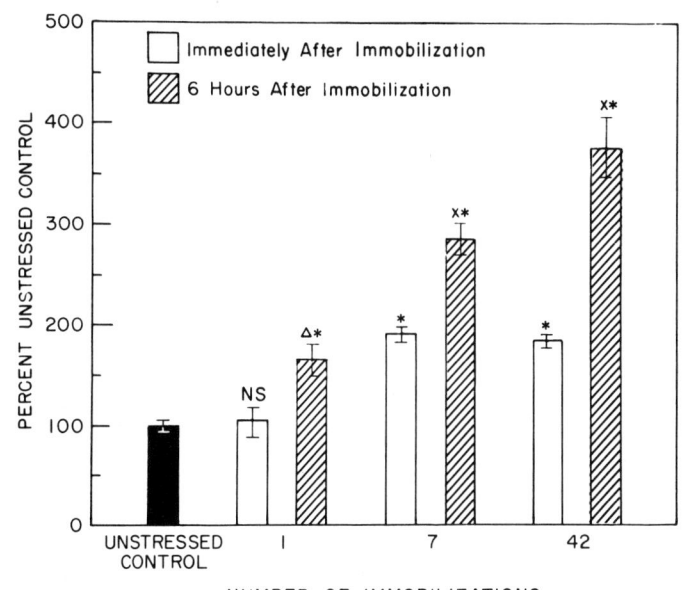

Fig. 5. Effect of repeated immobilization on rat adrenal
dopamine β-hydroxylase activity.
Rats were immobilized for 2.5 hr daily for 1, 7, or 42 days and
were killed either immediately after or 6 hr after the last
immobilization. Results are the mean levels ± standard errors
for groups of six rats and are expressed as percentages of the
dopamine β-hydroxylase activity found in the unstressed control
group. The level of dopamine β-hydroxylase in the unstressed
control group was 2.54 nmoles of octopamine-^{3}H formed per hour
per pair of adrenals.
*P<0.001 compared to unstressed controls.
△*P<0.01 compared to the group killed immediately after
immobilization.
✕*P<0.001 compared to the group killed immediately after
immobilization.
From: Mol. Pharmacol., 7 : 81, (1971).

the enzyme had probably been released in vivo during immobilization
and replenished during 6 hr interval after immobilization was
terminated (15). Weinshilboun et al. (30) found an increase in
plasma DBH activity after immobilization of the rats and assumed
that most of the DBH was released from sympathetic nerve endings.
The increase in adrenal DBH during 6 hr after immobilization was
prevented by inhibitors of protein or messenger RNA synthesis (15).
These results suggest that the turnover rate of dopamine-β-
hydroxylase had increased during repeated immobilization.

EFFECT OF REPEATED IMMOBILIZATION STRESS ON RAT ADRENAL PHENYLETHANOLAMINE-N-METHYLTRANSFERASE

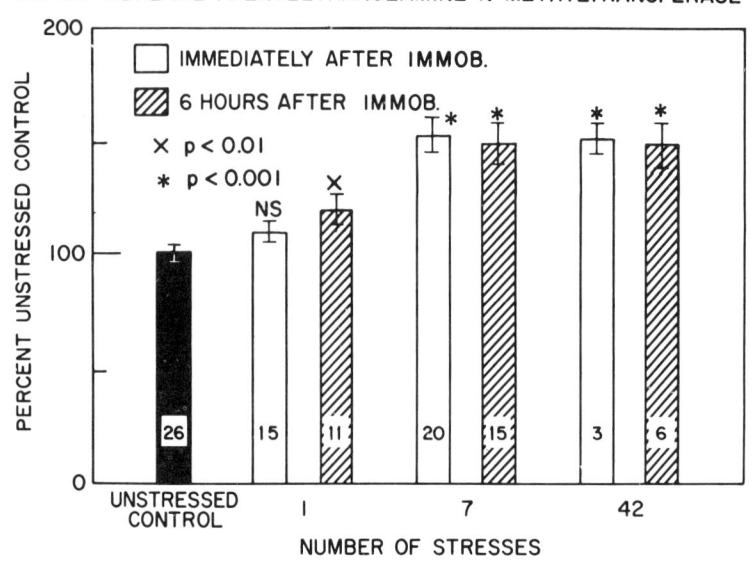

Fig. 6. Effect of repeated immobilization on phenylethanolamine-N-methyl transferase (PNMT) levels in the rat adrenal. Rats were immobilized for 2.5 hr daily for the indicated number of times. The levels of PNMT are expressed as per cent of unstressed control (mean ± SEM) for the number of rats indicated in the columns. Control levels of PNMT were 54.5 ± 1.3 nmoles metanephrine formed per pair of adrenals per hr.
From: Endocrinology 87 : 744, (1970).

In addition to these 2 enzymes we also measured the activity of phenylethanolamine-N-methyl transferase (PNMT), the enzyme catalysing conversion of norepinephrine to epinephrine. We used the technic of Axelrod (1). In Fig. 6 are shown the PNMT levels after immobilization (14). PNMT levels are also increased after repeated immobilization but not as strikingly as the TH or DBH levels. Recently increases in adrenal TH and PNMT activity after exposure to cold (26,16), psychosocial stimulation in mice (2), and TH and DBH activity after repeated 2-deoxy-D-glucose administration (25) were demonstrated.

Thus, all these 3 adrenal catecholamine-synthesizing enzymes - TH, DBH ,nd PNMT - are increased as a consequence of repeated immobilization stress. Levels of adrenal TH measured in vitro (14) and the rate of tyrosine hydroxylation estimated in vivo from the carbon-14/tritium catecholamine ratio (17) show a good general agreement (Table III). The rate of dopamine-β-hydroxylation also parallels the levels of DBH after repeated immobilization of rats (Table III).

TABLE III. Effect of immobilization on enzyme levels and catecholamine synthesis in rat adrenals.

Immobilization	Tyrosine hydroxylation Enzyme level[a]	Rate in vivo[b]	Dopamine-β-hydroxylation Enzyme level[a]	Rate in vivo[c]
Control None	100 ± 4	100 ± 12	100 ± 4	100 ± 16
During 1st	127 ± 8	139 ± 10	105 ± 14	96 ± 4
After 1st	152 ± 10	177 ± 7	168 ± 15	93 ± 12
During 7th	365 ± 19	317 ± 20	191 ± 7	185 ± 21
After 7th	364 ± 19	219 ± 15	288 ± 11	270 ± 33
During 40th	320 ± 30	218 ± 24	186 ± 5	186 ± 14
After 40th	360 ± 34	231 ± 2	377 ± 30	158 ± 50

Rats were treated as indicated in Tables 1 and 2, and all results are expressed as per cent control.
[a] Enzyme levels were obtained from experiments reported in Fig. 4 and Fig. 5.
[b] The relative rate of tyrosine hydroxylation was estimated from the relative magnitudes of the carbon-14/tritium ratio in the adrenal catecholamines (sum of dopamine, norepinephrine and epinephrine).
[c] The relative rates of dopamine-β-hydroxylation were calculated from the tritiated β-hydroxylated catecholamines found in the adrenals (see Table II).
From: Endocrinology 89 : 46, (1971).

In conclusion our results showed that there is an essential increase in biosynthesis of adrenal medullary catecholamines during adaptation of rats to repeated immobilization stress. This conclusion was supported by the increased levels of adrenal and urinary epinephrine and norepinephrine, by increased synthesis of adrenal catecholamines measured in vivo after administration of labelled precursors of catecholamines and by increases in the activity of catecholamine-synthesizing enzymes: tyrosine hydroxylase, dopamine-β-hydroxylase, and phenylethanolamine-N-methyl transferase measured in vitro.

REFERENCES

1. Axelrod, J. (1962) J. Biol. Chem. 237: 1657.
2. Axelrod, J., Mueller, R.A., Henry, J.P. & Stephens, P.M. (1970) Nature, 225: 1059.
3. Bygdemann, S. & Euler, U.S.V. (1958) Acta Physiol. Scand, 44: 375.
4. Euler, U.S.V. (1964) Clin. Pharmacol. Therap. 5: 398.
5. Euler, U.S.V. & Lishajko, F. (1961) Acta Physiol. Scand. 51: 348.
6. Friedman, S. & Kaufman, S. (1965) J. Biol. Chem. 240: 4763.
7. Gordon, R., Spector, S., Sjoerdsma, A. & Udenfriend, S. (1966) J. Pharmacol. Exp. Therap. 153: 440.
8. Graham, L.A. (1966) Acta Physiol. Scand. 68: 18.
9. Hrúza, Z., Albrecht, I., Jelínková, M. & Erdösová, R. (1966) Physiol. Bohemoslov. 15, 434.
10. Kopin, I.J., Weise, V.K. & Sedvall, G.C. (1969) J. Pharmacol. exp. Ther., 170, 246.
11. Kvetňanský, R., (1967) Dissertation Thesis, Bratislava.
12. Kvetňanský, R., Mitro, A., Mikulaj, L. & Hocman, G. (1966) Brat. Lek. Listy 46: 35.
13. Kvetňanský, R. & Mikulaj, L. (1970) Endocrinology 87: 738.
14. Kvetňanský, R., Weise, V.K. & Kopin, I.J. (1970) Endocrinology 87: 744.
15. Kvetňanský, R., Gewirtz, G.P., Weise, V.K. & Kopin, I.J. (1971) Mol. Pharmacol., 7, 81.
16. Kvetňanský, R., Gewirtz, G.P., Weise, V.K. & Kopin, I.J. (1971) Amer. J. Physiol. 220, 928.
17. Kvetňanský, R., Weise, V.K., Gewirtz, G.P. & Kopin, I.J. (1971) Endocrinology 89, 46.
18. Leduc, J. (1961) Acta Physiol. Scand. 53, Suppl. 183.
19. Mikulaj, L. & Csiba, J. (1963) Csl.Physiol. 12: 330. Abstract
20. Nagatsu, T., Levitt, M. & Udenfriend, S. (1964) Anal.Biochem. 9: 122.
21. Palkovits, M. (1961) Z.Mikr.Anat.Forsch. 67, 343.
22. Pearl, W., Balazs, T. & Buyske, D.A. (1966) Life Sci. 5: 67.
23. Sedvall, G.C., Weise, V.K. & Kopin, I.J. (1968) J. Pharmacol. Exp. Ther. 159, 274.

24. Selye, H. (1950) Acta, Inc., Medical Publishers, Montreal, Canada, 1950.
25. Silbergeld, S., Kvetňanský, R., Weise, V.K. & Kopin, I.J. (1971) Biochem.Pharmacol. 20: 1763.
26. Thoenen, H. (1970) Nature, 228: 861.
27. Udenfriend, S., Zaltzman-Nirenberg, P., Gordon, R. & Spector, S. (1966) Mol. Pharmacol. 2: 95.
28. Viveros, O.H., Arqueros, L. & Kirshner, N. (1968) Life Sci. 7: 609.
29. Viveros, O.H., Arqueros, L., Connett, R.J.& Kirshner, N. (1969) Mol. Pharmacol. 5: 60.
30. Weinshilboun, R., Kvetňanský, R., Axelrod, J. & Kopin, I.J. (1971) Nature - New Biol. 230, 287.
31. Welch, B.L. & Welch, A.S. (1969) Proc. Nat. Acad. Sci. 64: 100.

PHYLOGENETIC ASPECTS OF ADRENOCORTICAL ACTIVITY DURING THE PROCESS OF ADAPTATION

M. Juráni, L. Mikulaj* and K. Murgaš*

Department of General Zoology and Animal Physiology

Comenius University, Bratislava, and the

*Institute of Experimental Endocrinology

Slovak Academy of Sciences, Bratislava, Czechoslovakia

A multitude of experiments on mammals have shown that adrenal glands react to stress by secretion of corticosteroids from the adrenocortical tissue (7,11,19,25,27,46-48,54,58). This reaction of adrenal glands has been interpreted as one of the manifestations of the process of adaptation of the organism to stress (1).

Our task was to find out if there was a similar reaction to stress in other species of phylogenetic lower classes of the subphylum Vertebrata. For the sake of comparison of some of the classes we chose the following representatives: From the class Teleostomi, species Salmo gairdneri irideus, from Amphibia, species Rana esculenta, from Aves, species Anas platyrhynchos, Gallus domesticus and Columba livia domestica, from Mammalia, species Rattus norvegicus (Wistar).

It was known from literature that the interrenal tissue of fish increased its activity at the time of migration and spawning and persisted even after spawning (23,28,43-45), that activation was also provoked by different kinds of stress, e.g. holding in a confined space and repeated bleeding (24), anaesthesia (18) and forced exercise provoked by swimming in a tank (26), trapping in a handnet or agitating in shallow water (15,18,57). With frogs the activity of the interrenal tissue depended on the season and temperature of environment (42,50,55), and similarly, bleeding led to an increase in its activity (31). With birds an increased activity of interrenal tissue was observed after surgical operation (21,35,41), after exposure to cold (5,10) and

Fig.1. Basal plasma corticosteroids level in some species of vertebrates. B = corticosterone, F = cortisol.

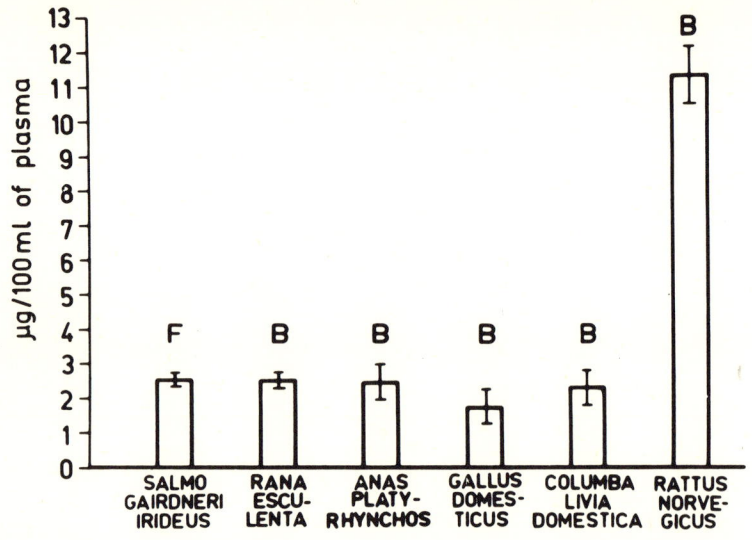

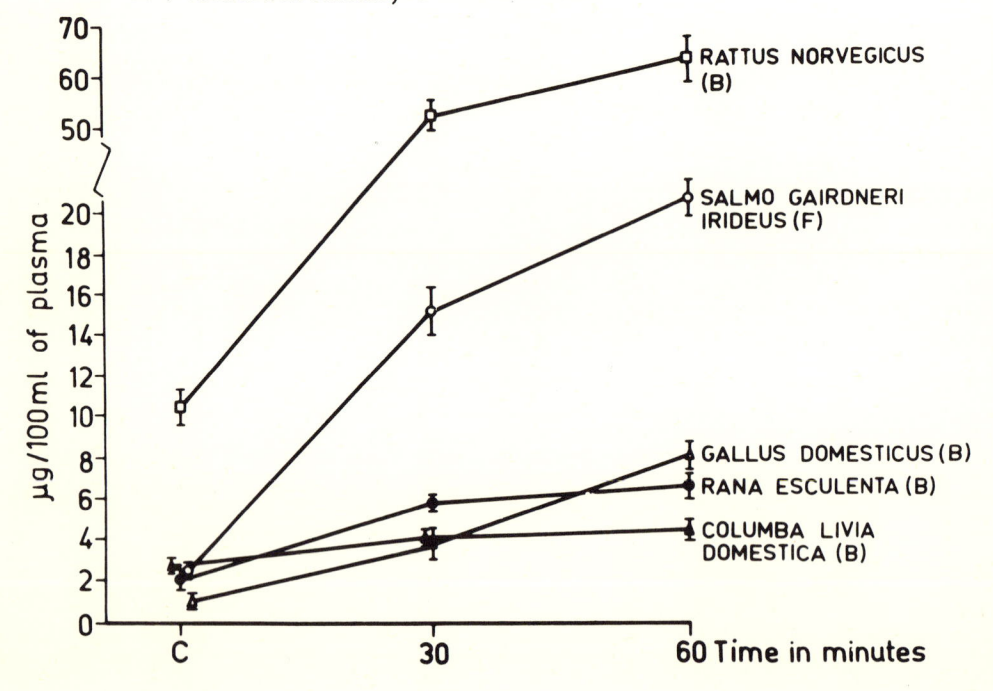

Fig. 2. Changes of plasma corticosteroids levels in some species of vertebrates after application of ACTH (5 I.U./kg i.p.) B = corticosterone, F = cortisol.

Fig.3. A comparison of plasma corticosteroid levels in some vertebrates following 1 min. electric stimulation (60 V.A.C.) B = corticosterone, F = cortisol.

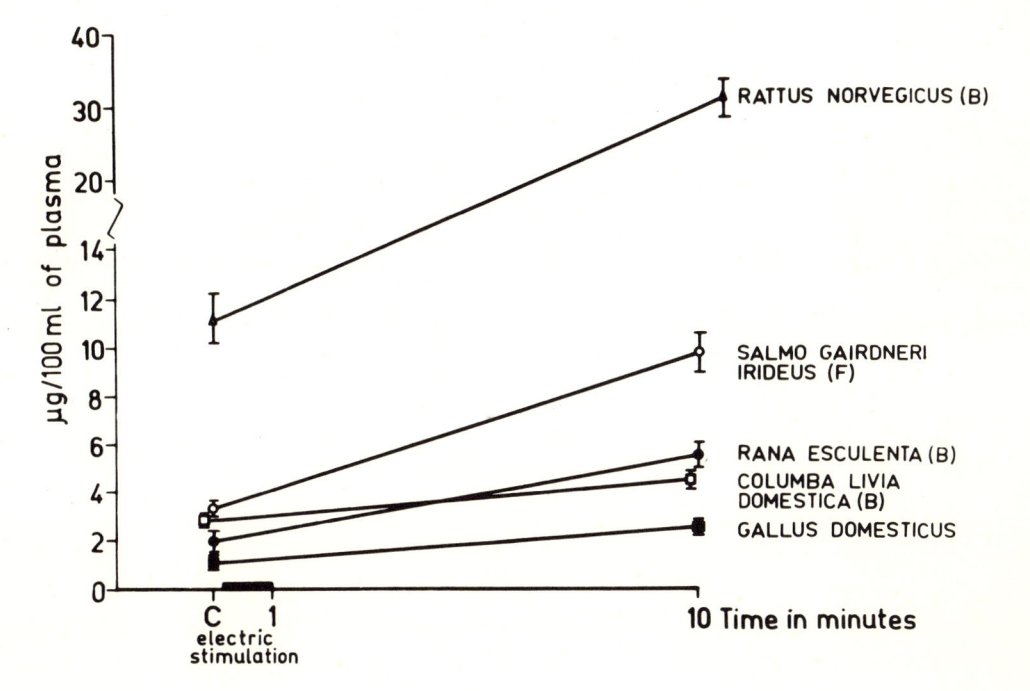

water deprivation (10). Activity was measured by histological or biochemical methods.

In our experiments we examined the reaction of interrenal tissue by measuring the levels of free plasma corticosterone and free plasma cortisol. Recently we have been measuring corticosterone or cortisol in adrenorenal or adrenal homogenate according to the dominant corticosteroid in the respective species. In trout it is cortisol (23,29), in other animals corticosterone (2,4,12,14, 22,31,52).

Our first task was to obtain intact animals in order to find out the basal level of corticosteroids, a problem that is essentially less complicated with common laboratory animals. The catching of fish, frogs and birds had to be performed in such a way that the animals would not be frightened and they have to be obtained from their natural environment, without transportation or keepin g in laboratory conditions. In the case of trout the fish were caught by a fishing rod directly from a fish breeding pond. We also used a fishing rod to catch frogs. Ducks and hens were captured from a big breeding flock by hand or by special hooks, doves were caught in nets. Rats were taken from breeding cages. The animals were decapitated immediately after catching, their blood was collected in tubes with heparin and centrifuged. Cortisol was measured by a fluorimetric method (Donaldson et al., 16), corticosterone was measured by a modification of the fluorimetric method of Van der Vies (53). The basal levels are shown in Fig. 1. These basal levels of plasma corticosteroids are lower than reported in literature in the same, or related species of these animals (2,3,8,15,31,58). The differences are evidently caused by differences in the biochemical methods (2,6,20,30) and by the different ways of capturing, keeping or killing the animals.

Our first experiments with short immobilization stress showed less activation of the interrenal tissue in animals phylogenetically lower than mammals. The problem looked as if it was really a question of a lower reactivity of adrenal glands. It is known that exogenous application of ACTH provokes the highest activation of adrenocortical tissue in mammals. A similar activation of interrenal tissue could be presumed, though a non-homologous type of ACTH could not be avoided. A positive adrenostimulatory effect of mammalian ACTH in lower groups of vertebrates has been observed by several authors (9,24,26,31,35,52). In our experiment (Fig.2.) we gave 5 I.U. ACTH/kg of body weight intraperitoneally and observed a difference in the reaction of interrenal tissue. The most marked stimulation occurred in the adrenal glands of rats and fish. The response in other species was much less. In some experiments we used more drastic forms of stress: The effect of electric stimulation by a 60 V.A.C. source in one second intervals for one min is shown in Fig.3. Looking for a more appropriate stimulus from the standpoint of ecology, in hens we chose as additional kinds of stress, tarsometatarsus fracture, dipping into cold water,

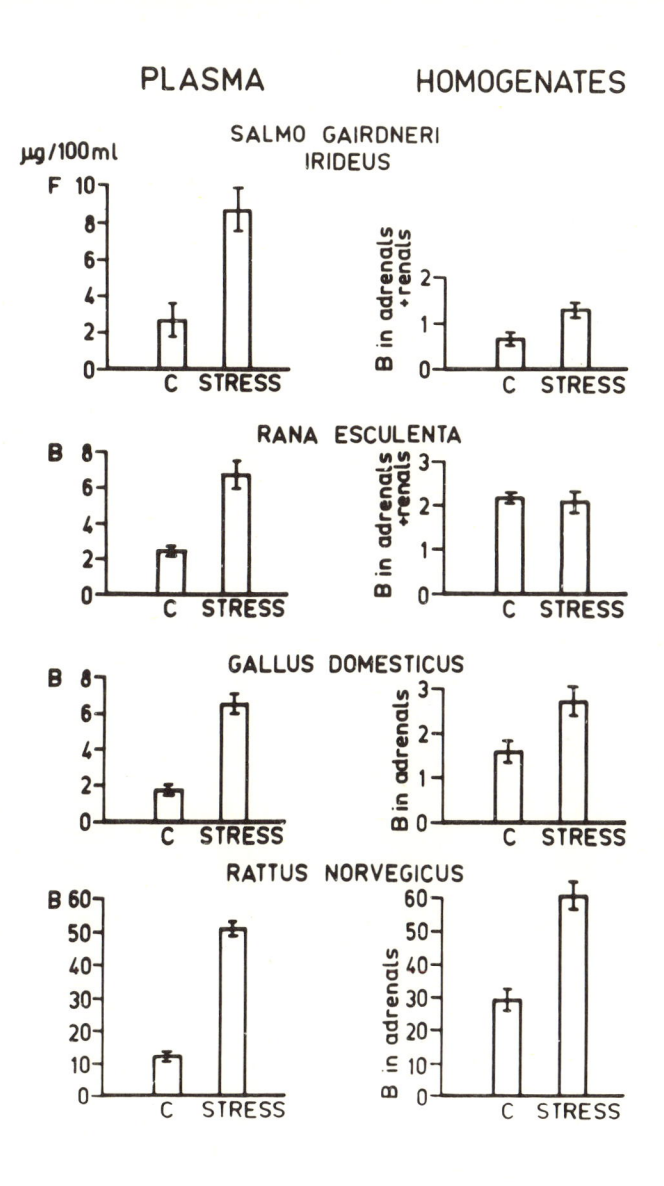

Fig.4. Effect of 30 min immobilization on corticosteroids in plasma and adrenorenal or adrenal tissue homogenates in some vertebrates.

Fig.5. Effect of 30 min immobilization on in vitro corticosterone production of adrenal glands in some vertebrates.

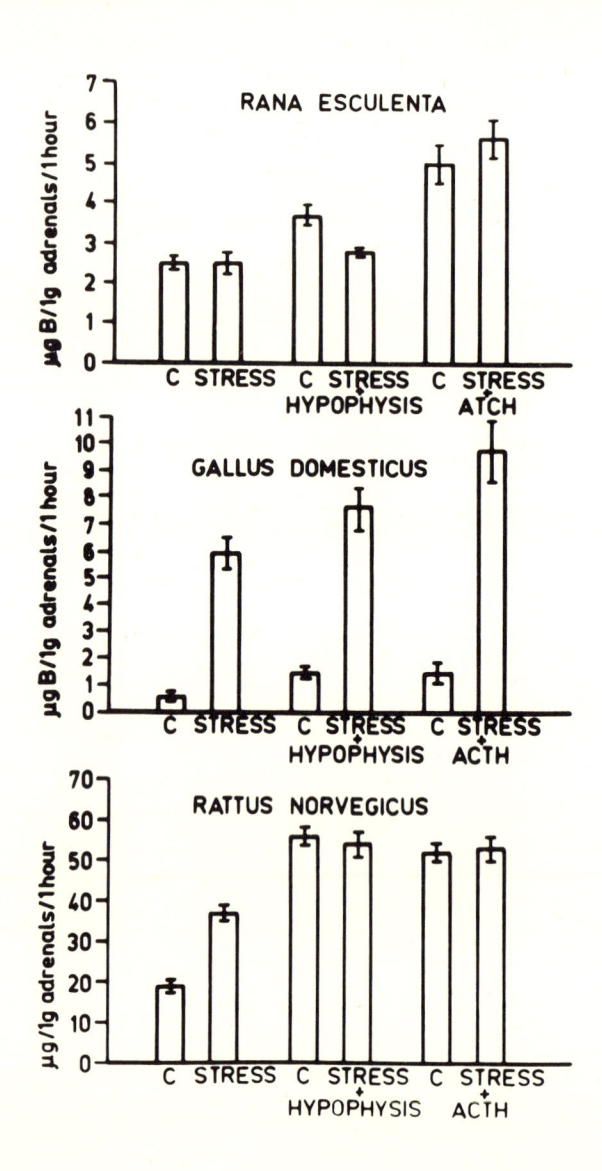

Fig.6. Adrenal stimulation by ACTH (1 I.U./animal) at various environmental temperatures in Rana esculenta.
B = corticosterone.

Fig.7. Adrenal stimulation by stress (0 - 120 min holding in a confined space) at various environmental temperatures in Salmo gairdneri irideus. F = cortisol.

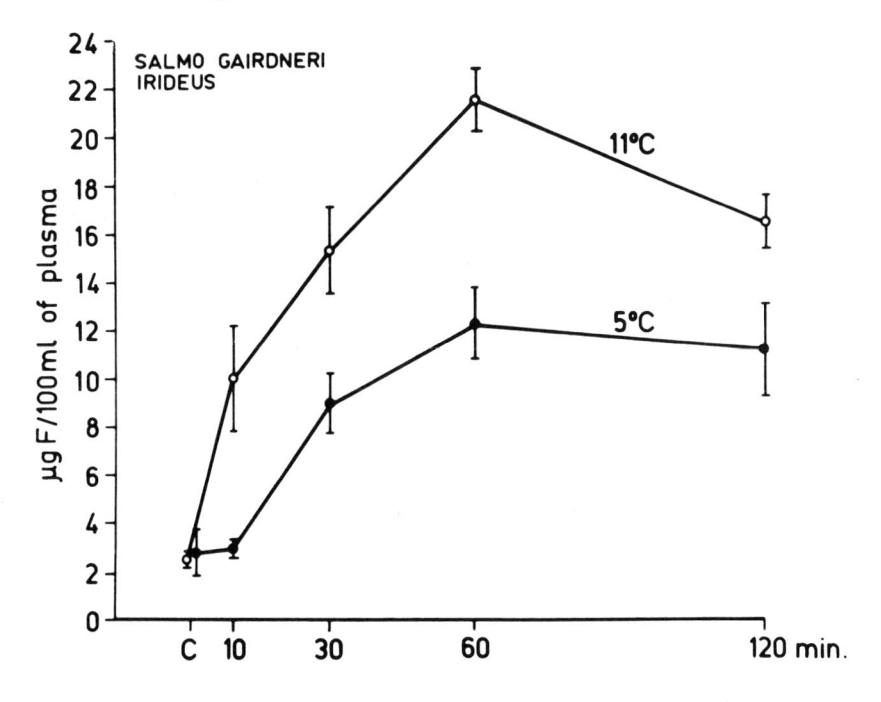

frightening by a dog and a bird of prey. None of these stimuli led to a greater increase in plasma corticosteroid concentrations than immobilization, so we could continue to use that original kind of stress as the one with maximal activation effect. The immobilization of trout was achieved by the limitation of their living space, keeping them in a small aquarium with stable water exchange, the other animals had their legs bound.

Before studying the interrenal reaction to repeated stress stimuli we examined the reaction of interrenal tissue of different species to an acute stress stimulus. To make the stimuli approximately identical in quantity and quality we chose a standard immobilization of 30 min. Besides the corticosteroid concentration in the plasma, the adrenorenal or adrenal homogenate content was also measured. Observing the two indexes used in mammals (11,58) it is possible to define more precisely the activity of interrenal tissue. Comparing the two indexes of interrenal activity in the separate species (Fig.4.) again the most intensive stimulation occurs in the adrenal glands of rats. As an index in vivo adrenocortical activity in mammals the in vitro production of corticosterone was often also measured (13,49,53,54,56). This is why post-stress extirpation of adrenal glands of the animal was performed for the sake of incubation. One group was incubated in pure medium, another together with the hypophysis of the stressed animals and a third one with 0,3 I.U. ACTH/1,5 ml incubation medium. The results of the in vitro measurements are summarised in Fig.5. The latter two figures show that if the stimulating effect of stress is evaluated solely on the basis of the plasma corticosteroids it is about the same for frogs and hens. However, the in vitro techniques show strong stimulation in hens which is absent in frogs. On a percentage basis it is even greater than that observed in rats.

From these results it is clear that it is necessary to investigate the mechanisms of the species differences in the responses of the hypophyseal-interrenal system to stress. Other authors have also reported considerable variability in the adrenocortical response to stress in mammals, even in species or strains related to each other (17,33,34,51).

With poikilothermic animals it is also necessary to take into account the temperature of the environment and the season (42,50,55). Fig.6. shows the dependence of the reaction of the interrenal tissue of the frog on the temperature of the environment after the administration of 1 I.U. ACTH/animal. Trout exposed to stress at different seasons showed a quicker increase in the circulating cortisol level in November with a water temperature of 11°C which reached a higher value than in February when the water temperature was 5°C (Fig.7.) The non-reactivity of the

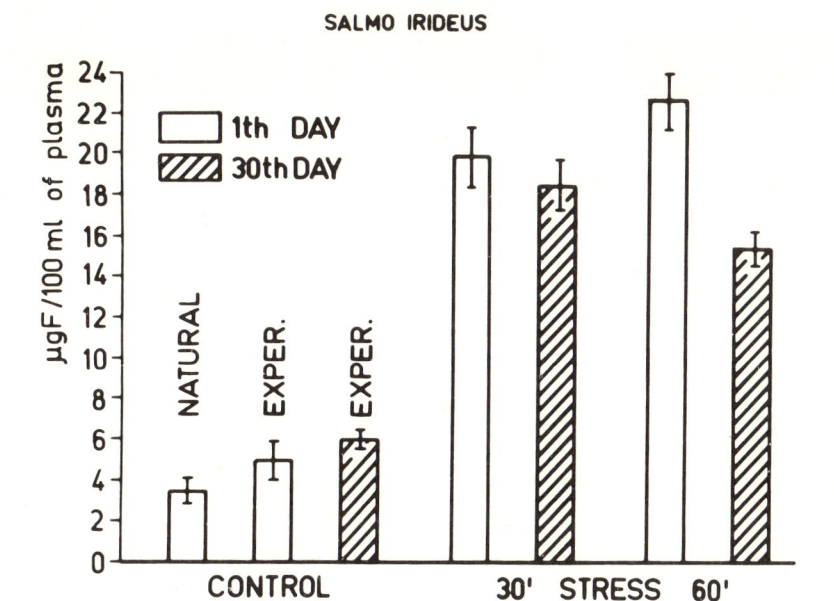

Fig. 8. Adrenal stimulation by repeated stress (holding daily for 4 hr in a confined space) in Salmo irideus. On the last day decapitated at 30th or 60th min. F = cortisol.

interrenal tissue to exogenous ACTH in trout during winter months was also observed by Hill and Fromm (16).

Since the study of changes in adrenal secretion of mammals under chronic intermittent stress is the chief theme of our work, we concentrated on the same effect in lower species of vertebrates. We tried to discover if the interrenal tissue was activated like the mammalian adrenal by repeated stress stimuli (32,36-40). The first experiment was performed on trout. The trout were placed into cages submerged in breeding ponds. For 30 days the cages were raied for 4 hr each day so that there was only a minimum water-level. On the last day after the decrease in the water level the trout were decapitated either at the 30th or 60th min (Fig.8.) We found out that the effect of chronic stress could be observed in fish as in mammals (37), expressing itself as a lower activation of interrenal tissue in the fish repeatedly exposed to stress situations. Further experiments showed a similar reaction in the other species observed. Thus, change in the response of the interrenal tissue to repeated stress is probably the same in all the species having a developed regulation system of hypothalamus-hypophysis-adrenal glands.

REFERENCES

1. Bajusz, E. (1969) In: Physiology and pathology of adaptation mechanisms, edited by E. Bajusz and collaborators. Pergamon Press, Oxford.
2. Baylé, J.D., Boissin, J. & Assenmacher, I. (1967) Compt. Rend., 265, 1524.
3. Boissin, J., Baylé, J.D. & Assenmacher, I. (1966) Compt. Rend., 263, 1127.
4. Boissin, J., Soulé, J. & Daniel, J.Y. (1966) Compt. Rend. Soc. Biol., 160, 1624.
5. Boissin, J. (1967) J. Physiol. Paris, 59, 423.
6. Bouillé, C., Daniel, J.Y., Boissin, J. & Assenmacher, I. (1969) Steroids, 14, 7.
7. Boulonard, R. (1966) Federation Proc., 25, 1195.
8. Bradshaw, S.E. & Fontaine-Bertrand, E. (1970) Comp. Biochem. Physiol., 36, 37.
9. Breitenbach, R.P. (1962) Poultry Sci., 41, 1318.
10. Brown, K.I. (1961) Proc. Soc. Exp. Biol. Med., 107, 538.
11. Brudieux, R. (1969) Compt. Rend. Soc. Biol., 163, 2300.
12. Crabbé, J. (1961) Endocrinology, 69, 673.
13. De Weid, D., Van der Wal, B. & Van Goch, J.J. (1964) In: Proc. II. Intern. Congr. Endocrinol., London, p.64; Intern. Congr. Ser. 83, Excerpta Medica Foundation.
14. Donaldson, E.M. & Holmes, W.N. (1965) J. Endocrinol., 32, 329.
15. Donaldson, E.M. & McBride, J.R. (1967) Gen. Comp. Endocrinol., 9, 93.
16. Donaldson, E.M., Fagerlund, U.H. & Schmidt, P.J. (1968) J. Fish Bd., Canada, 25, 71.
17. Eleftheriou, B.E. (1964) J. Endocrinol., 31, 75.
18. Fagerlund, U.H.M. (1967) Gen. Comp. Endocrinol., 8, 197.
19. Feldman, S., Conforti, N., Chowers, I. & Davidson, J.M. (1969) Neuroendocrinol., 5, 290.
20. Frankel, A.I., Cook, B., Graber, J.W. & Nalbandov, A.V. (1967) Endocrinology, 80, 181.
21. Frankel, A.I., Graber, J.W. & Nalbandov, A.V. (1967) Endocrinology, 80, 1013.
22. Guillemin, R., Clayton, G.W., Smith, J.D. & Lipscomb, H.S. (1958) Endocrinology, 63, 349.
23. Hane, S. & Robertson, O.H. (1959) Proc. Nat. Acad. Sci., U.S. 45, 886.
24. Hane, S., Robertson, O.H., Wexler, B.C. & Krupp, M.A. (1966) Endocrinology, 78, 791.
25. Hess, J.L., Denenberg, V.H., Zarrow, M.X., & Pfeifer, W.D. (1969) Physiol. Behav., 4, 109.
26. Hill, C.W. & Fromm, P.O. (1968) Gen. Comp. Endocrinol., 11, 69.
27. Hodges, J.R. & Mitchley, S. (1970) J. Endocrinol., 47, 253.
28. Idler, D.R., Ronald, A.P. & Schmidt, P.J. (1959) Can. J. Biochem. Physiol., 37, 1227.

29. Idler, D.R., Freeman, H.C. & Truscott, B. (1964) Can. J. Biochem., 42, 211.
30. James, V.H.T., Townsend, J. & Fraser, R. (1967) J. Endocrinol., 37, xxviii.
31. Johnston, C.I., Davis, J.O., Wright, F.S. & Howards, S.S. (1967) Am. J. Physiol., 213, 393.
32. Jonec, V., Murgaš, K, & Kvetňanský, R. (1966) Federation Proc., 25, 1200.
33. Levine, S. & Treiman, D.M. (1964) Endocrinology, 75, 142.
34. Levine, S. & Treiman, D.M. (1969) In: Physiology and pathology of adaptation mechanisms, edited by E. Majusz and collaborators. Pergamon Press, Oxford, p. 171.
35. Macchi, I.A., Phillips, J.G. & Brown, P. (1967) J. Endocrinol., 38, 319.
36. Mikulaj, L., & Csiba, J. (1963) Cš. fysiol., 12, 330.
37. Mikulaj, L., Lichardus, B., Mitro, A. & Csiba, J. (1964) J. Physiol., Paris, 56, 612.
38. Mikulaj, L. & Kvetňanský, R. (1965) Physiol. Bohemoslov., 15, 439.
39. Mikulaj, L., Javorka, K., Krahulec, P., Kvetňanský, R. & Spišák, S. Cš. fysiol., 15, 99.
40. Murgaš, K. (1968) Doctoral thesis, Inst. Experiment. Endocrinol., Slovak Acad. Sci., Bratislava.
41. Nagra, C.L., Birnie, J.G., Baum, G.J. & Meyer, R.K. (1963) Gen. Comp. Endocrinol., 3, 274.
42. Nakamura, M. (1967) Endocrinol., Japan, 14, 43.
43. Robertson, O.H. & Wexler, B.C. (1959) Endocrinology, 65, 225.
44. Robertson, O.H., Krupp, A.M., Favour, C.B., Hane, S., & Thomas, S.F. (1961) Endocrinology, 68, 733.
45. Schmidt, P.J. & Idler, D.R. (1962) Gen. Comp. Endocrinol., 2, 204.
46. Selye, H. (1950) Stress, Montreal, Acta. Inc. Med. Publish.
47. Smookler, H.H. & Buckley, J.P. (1969) Int. J. Neuropharmac., 8, 33.
48. Solomon, G.F., Merigan, C. & Levine, S. (1967) Proc. Soc. Exp. Biol. Med., 126, 74.
49. Sólyom, J., Sturcz, J., Spät, A., Mészáros, I. & Ludwig, E. (1969) Acta Physiol. Acad. Sci., Hung., 36, 371.
50. Suzuki, N. (1968) Endocrinol., Japan, 15, 82.
51. Treiman, D.M. & Levine, S. (1969) Endocrinology, 84, 676.
52. Urist, M.R. & Deutsch, N.M. (1960) Proc. Soc. Exp. Biol. Med., 104, 35.
53. Van der Vies, J., Bakker, R.F.M. & de Weid, D. (1960) Acta Endocrinol., 34, 513.
54. Van Goch, J.J., de Weid, D. and Schönbaum, E. (1963) Am. J. Physiol., 205, 1083.
55. Van Kemenade, J.A.M., Van Dongen, W.J. & Van Oordt, P.G.M.J. (1968) Z. Zellforsch., 91, 96.
56. Vecsei-Weisz, P. & Kemény, V. (1964) Acta Physiol. Acad. Sci. Hung., 24, 237.
57. Wedemeyer, G. (1969) Comp. Biochem. Physiol., 29, 1247.
58. Zimmermann, E. & Critchlow, V. (1967) Proc. Soc. Exp. Biol. Med., 125, 658.

ENDOCRINE FUNCTIONS DURING ADAPTATION TO STRESS

L. Mikulaj and A. Mitro

Institute of Experimental Endocrinology

Slovak Academy of Sciences, Bratislava, Czechoslovakia

Among the countless observations on acute stress, only few data can be found that deal with the behaviour of adrenocortical responses to long-term, chronic, intermittent or frequently repeated stress (11,14,15,28,29,46,52,58). This is somewhat surprising in view of the fact, that chronicity or intermittent repetition is one of the basic properties common to the majority of stress factors, be they climatic, physical, sociological, industrial or other, which man encounters most frequently and to which he becomes adapted whether in the ontogenetic or phylogenetic sense of the term. In this respect only the effect of hypoxia (3,10,13,16,24,50, 56,57,62,64), and the influence of cold (1,2,9,12,20,25,26,27,49,63) has been relatively well studied, though it should be borne in mind that cold represents a rather specific situation. Most reports dealing with these problems have appeared in recent years and this proves that many physiologists do not accept the application of results obtained in acute stress experiments to the whole process of adaptation (4,5,6,7,19,21,47,48,51,53,54,55,59,60,61). We were of the same opinion when starting, in 1964, to study the changes in the endocrine system during the course of long-term repeated exposure to stress. The main stimulus for our work was the conviction that a knowledge of this function of the organism in its dynamics, in its time course, and in its interglandular and neuroendocrine relationships, may be a condition for a better understanding of the role and the significance of adrenal hormones in metabolic changes and in the whole adaptation process.

Methods and Results

Our work so far has involved mainly rats, but some preliminary experiments were also performed on human beings. Using animals we selected such stress stimuli which are capable of high stimulation effect on adrenocortical activity and of simultaneously inducing marked metabolic changes. Most commonly we used immobilisation (120 - 180 min), Noble-Collip drum trauma (60 rotations/min), or the subcutaneous injection of formalin (0.5 ml 2% solution). Adrenal activity was measured by fluorometric determinations of corticosterone in plasma or in excised and homogenized adrenals (8). Blood was obtained by decapitation of the animals immediately after stress.

In rats exposed to a stress of 150 min immobilisation daily for 42 days, we observed a 5 - 7 fold increase in plasma corticosterone concentrations immediately following the stress during the first days of immobilisation. Groups of animals were decapitated at weekly intervals. The large initial stress response gradually decreased during the following weeks so that at the end of the experiment the response was only twice the resting value even though the daily stressing stimulus remained unchanged. A similar sequence of changes was found in the corticosterone values in adrenal homogenates (30,31). The adrenal weight increased about 30% in the first week and this increase persisted throughout the experiment. When plasma corticosterone concentration was expressed per unit adrenal weight, it fell markedly. When corticosterone production by adrenal slices was followed in vitro it was found to fall similarly (35). In another experiment we showed that with a daily subcutaneous injection of formalin the response of the adrenal cortex became inhibited after the 20th day (33). A change in the stress response also becomes apparent after such extreme stress as traumatisation in the Noble-Collip drum. Following 28 days of repeated drum stress the level rises in the same way as in animals stressed for the first time, but it returns faster to the resting values (34).

We have performed many similar tests over the past few years and in our view, the finding of gradual inhibition of adrenocortical response during the course of adaptation to a repeated intermittent stress, may be a general phenomenon. Nevertheless, there was one exception: after daily faradic stimulation for 30 days, the resulting corticosterone elevations did not show any declining trend (45).

The fundamental question is, of course, whether or not these changes result from some functional or morphological alteration in the adrenal cortex during the intensive repeated stimulation. However, testing the secretory capacity of the gland with an exogenous standard dose of 1 IU ACTH i.p. has shown that the adrenals retain their secretory activity and the latter is essentially directly proportional to the mass of the secretory tissue (30,32). The dynamic changes in adrenocortical activity were also demonstrated morphologically. Exposure of the animals to an immobilisation stress in the early stage of the experiment was followed by the reduction of the positive lipid reaction in the cells of the fascicular zone. This reduction failed to occur in the later stage of the experiment, despite the fact that the duration and intensity of the stress had not changed (42). Further examination of adrenocortical morphology by usual staining techniques confirmed the previous functional changes (41).

Summarizing this part of our study, it seems that the changes in adrenocortical activity are not primarily caused by the adrenal glands themselves.

At the present time we cannot give a definite answer to the question whether the above-mentioned changes in the plasma corticosterone values result solely from altered adrenocortical secretion and to what extent changes in metabolism in the liver and in other tissues, or the changes in excretion participate in them. The disappearance rate of tritium-labelled corticosterone remained unchanged when animals stressed for the first time were compared with those stressed 45 times (39). Similarly, no difference was found in the in vitro hepatic reduction of the Δ^4-3-keto group in the corticosterone molecule. On the other hand, a significant decrease in the reduction of the α-keto side chain by liver homogenates was found in vitro in male rats subjected to repeated intermittent stress (39). The decreased response to stress still persists in the adapted animal as long as a month after the last exposure to immobilisation stress. In the case of traumatic stress in the Noble-Collip drum, we found even 60 days following the last exposure a significantly lower decline of plasma corticosterone in animals that had previously been adapted to this type of stress (37,38).

In the further study of adaptation to repeated stress we turned our attention to the higher regulatory regions, in particular, to the adenohypophysis. It was possible in two ways in our laboratory; either the morphological and histochemical, or indirect, functional indicators could be used. Unfortunately, we have no immunobiological method for the direct measurement of ACTH available.

One of our experiments was based on the presumption that if in the course of a repeated intermittent stressing adrenocorticotrophic pituitary activity is weakened, the contralateral hypertrophy following unilateral adrenalectomy should be more pronounced at the beginning, in the first weeks of the experiment, and attenuated in its later stages. And, indeed, this was confirmed (36).

We shall give a concise outline of the histological picture of adenohypophysis. Acidophils were found to be differentiated into large and small; among the basophils, the so-called thyreotrophs, i.e. angular cells, lost their staining properties in the course of the experiment, then finally became normal again; gonadotrophs, i.e. round cells, gradually lost their staining ability and in the 5 - 6th week, numerous degenerating gonadotrophs could be observed. Chromophobes did not show any major changes (41). There are many objections against the concept of the existence of a specific relation between the classical division of adenopituitary cells and the production of particular trophic hormones (22). Be that as it may, the above finding concerning thyreoptrophs and gonadotrophs is in good agreement with the results of thyroid and testicular function-indicators which we followed in other tests (31,40,41) and which are in good agreement with results of other authors.

A histochemical investigation of the pituitary has shown that the activity of 5 dehydrogenases-, glucoso-6-phosphate-, lactate-, succinate-, isocitrate-, and malate-dehydrogenase, in the adenohypophysis was increased on the 7th day of the experiment, but by the end of the 5th week had returned to the control values, despite the unchanged repeated stimulation (44).

If we assume all these indirect experimental and morphological tests to reflect the activity of the pituitary, then we may presume that the pituitary adrenocorticotrophic function in adapted animals is also gradually inhibited like the adrenocortical secretion. This would imply that during the course of adaptation the pituitary, too, is but a mediator of the regulatory integrating changes that take place in the higher planes of neuroendocrine relationships.

At this stage, the study of structure and mechanism responsible for the changes in adrenocortical secretion during the process of adaptation to the repeated stress is getting to the level of central nervous system. This should be expected if we consider the well-known results of endocrinologists in the past 15 yr. In our laboratory we have observed the nuclei ventromedialis, dorsomedialis and arcuatus of the hypothalamus, which are believed to have close relationship to pituitary-adrenocortical function. In NDM and NAr, using a karyometric method of tracing the volume of the neural cell nuclei (18), we found a diminution of the nuclei in the first experimental phase, but in the second phase a tendency to return to the values of the control group was seen. The changes in the NVM did not show the same type of changes and in the course of time they only become more marked (43). The diminution of cell nuclei can signify either an increasing active inhibition or it might be the result of a passive inhibition.

Discussion

There are many definition of adaptation, but generally, it is defined as "transformation first of function and finally of structures effected by long-term changes in external environment of the organism" (46). Thus, "adaptation results from a prolonged or repeated exposure to a stress condition and so it represents the summation of modification in the original responses to stress, which results in more economical and broader homeostasis and reduces the strain on body and mind"(27). It follows that adaptation is conditioned by the function of an analysing and integrating mechanism in the course of time. Therefore, adaptation should be considered as a process and not as a reaction.

Since Selye's discovery of adrenocortical activity as a basic component of the body's reaction to a stressor, countless observations on adrenals and the stress have appeared in literature. Unfortunately, a very simplified deduction has often appeared: the more intensive the adrenocortical reaction, the better the adaptation of an organism to the stress. Now, if we take as our starting point the forementioned definition on one hand, and on the other hand the fact that adrenal cortical activity during the course of adaptation becomes inhibited, the basic question appears to be: Is the activity observed in the initial exposure to stress the most economical or optimal? It is well known that the exaggerated adrenal reaction with its metabolic consequences shows a negative, catabolic balance. The more excessive the adrenal reaction, the more negative the balance. And it always appears to be excessive except when it takes place under conditions of real danger to life, during fight, flight, hunger, injury, etc., i.e. during states for which it had formerly been phylogenetically encoded as a highly logical and adequate reaction. Today, on its first exposure to stress, the organism is not able to analyse and foresee the consequences of stress and hence it mobilises its capacities to the maximum, preparing itself for every possibility. Leter, when it has taken account of the metabolic consequences of such excessive mobilisation, it begins to adapt them, i.e. usually to inhibit them. This is the case in the majority of modern stresses, particularly in man, where usually there is only a symbol of the real original stress situation (17). Under such circumstances this primaeval phylogenetically fixed neurohumoral reaction appears to be rather a stereotype and must gradually be inhibited to prevent it from being detrimental to the organism.

Thus, the inhibition of adrenocortical secretory activity during the course of a long-term repeated exposure to stress appears highly logical. This conclusion is not surprising and Selye already pointed it out in 1936: "This defense reaction is most marked when a damaging influence acts on the organism for the first time. We saw repeatedly that, after an initial enlargement, the adrenal decreases in size and returns to normal even though the injections of the drug, such as morphine, atrophine or adrenaline, which produces the initial enlargement are continued" (52). In recent years this statement has been enlarged and enriched by many new aspects of functional and regulatory inter-relations, but many questions remain still unsolved. However, it should be underlined, that, particularly in interpreting the hormonal and metabolic relationships, we should always bear in mind that the findings obtained in acute stress cannot be simply and schematically applied to the whole adaptation process. The adaptive reaction becomes significantly altered during the intermittent or chronic action of the stress. Further study of neuroendocrine regulatory mechanisms is needed before this field of physiology can contribute to the solution of practical problems in the everyday life of modern civilisation.

REFERENCES

1. D'Angelo, S.A. (1960) Fed.Proc.Suppl.,II, 51.
2. Boulouard, R. (1966) Fed.Proc., 25, 1195.
3. Darrow, D.C. & Sarason, E.L. (1944) J.Clin.Invest.23, 11.
4. Fiorica, V. & Iampietro, P.F. (1966) Proc.Soc.Exp.Biol.Med., N.Y. 122, 647.
5. Fregly, M.J. (1969) Environ.Res., 2, 435.
6. Frenkl, R. & Csalay, L. (1970) Acta.Phys.Acad.Sci.Hung.,36, 365.
7. Furner, L.R. & Stitzel, R.E. (1968) Biochem.Pharmacol., 17,121.
8. Guillemin, R., Clayton, G., Smitty, J. & Lipscomb, H.(1959) J.Lab.Clin.Med., 53, 830.
9. Hart, J.S., (1958) Fed.Proc. 17, 1045.
10. Hartroft, P.M. & Bischoff, M.B. (1969) Fed.Proc.,28, 1233.
11. Henkin, R.I. & Knigge, K.M. (1963) Amer.J.Physiol., 204,710.
12. Heroux, O. (1960) Fed.Proc., 19, suppl., 5, 82.
13. Hornbein, Thomas, F. (1962) J.Appl.Physiol., 17,(2),246.
14. Hruza, Z., & Chytil, F. (1959) Can.J.Biochem.Physiol., 37, 61.
15. Hruza, Z. & Poupa, O. (1964) Am.Physiol Soc., Washington, D.C. p. 939.
16. Hurtado, A. (1966) Amer.Industr.Hyg.Assoc.J., 27, 313.
17. Charvat, J. (1964) Čas.lék.českých., 103, 761.
18. Jakoboj, W. (1935) Z.Mikr.Anat.Forsch., 38, 161.
19. Jakoubek, B. & Semigninovský, B. (1970) Life Sc.PT.1.,9,1169.
20. Jonec, V. & Murgaš, K. (1969) Fed.Proc., 28, 987.
21. Kotby, S. & Johnson, D.Harold (1966) Fed.Proc., 25, 552.
22. Baker., B.L. (1970) J.Histochem.Cytochem., 18, 1.
23. Kevetňansky, R. & Mikulaj, L. (1970) Endocrinology.,87,738.
24. Langley, L.L. & Clarce, R.W. (1942) Yale,J.Biol.Med.,14,529.
25. Le Blanc, J. (1967) Amer.J.Physiol., 212, 530.
26. Le Blanc, J., Robinson, D. & Sharman, D.F. (1967) Amer.J. Physiol., 213, 1419.
27. Le Blanc, J. (1966) Ann.N.Y.Acad.Sci., 134, 721.
28. Máslinski,Cz., & Nieldzielski,A.(1964) Experientia,20, 264.

29. Matthews, B.H.C. (1953) J.Physiol. (Paris), 122, 31.
30. Mikulaj, L. & Csiba, J. (1963) Čsl.fysiol., 12, 330.
31. Mikulaj, L., Kutka, M., Štolc, V. & Sámel, M. (1964) Čsl. fysiol., 13, 501.
32. Mikulaj, L., Lichardus, B., Mitro, A. & Csiba, J.(1964) J. Physiol. (Paris), 56, 612.
33. Mikulaj, L., Javorka, K., Krahulec, P., Kvetňansky, R. & Spišák, S. (1966) Čsl.fysiol., 15, 99.
34. Mikulaj, L. & Kvetňanský, R. (1966) Physiol.Bohemoslov.,15, 439.
35. Mikulaj, L., Bartová, A., Kolena, J. & Kvetňanský, R. (1966) Brat.lek.listy., 46, 29.
36. Mikulaj, L., et al. (1966) Čsl.fysiol., 15, 98.
37. Mikulaj, L. & Kvetnansky, R. (1967) Čsl.fysiol.,16, 246.
38. Mikulaj, L. et al. (1967) Proceedings of Hung.Endocrin.Soc., Pecs, p. 35.
39. Mikulaj, L., Kvetňanský, R. & Kolena, J. (1968) III Int. Congress of Endocrinology, Mexico.Abstracts Expt.Medica, I.C.S. 157, p. 142.
40. Mikulaj, L. Háčik, T. & Mitro, A. (1970) Physiol.Bohemoslov., 19, 334.
41. Mikulaj, L., Mitro, A., (1972) Brat. Lek. Listy, 58,282
42. Mitro, A. & Mikulaj, L. (1965) Endocrinologie, 49, 18.
43. Mitro, A. & Mikulaj, L. (1965) Biologia., 20, 856.
44. Mitro, A., Sokolová, V. & Mikulaj, L. (1966) Čsl.Soc.Hist. Cytol., 1, 83.
45. Murgaš, K., Mikulaj, L. & Juráni, M. Physiol.Bohemoslov., in press.
46. Poupa, O. (1961). The development of homeostasis. Publ.House Cs.Academy of Sciences, Prague, p. 23.
47. Rinne, U.K., Kivalo, E. & Kunnas, R. (1964) Ann.Med.Exptl. Biol.Fenniae, 42, 12.
48. Rosecrans, J.A. & Watzman, N. (1966) Biochem.Pharmacol., 16, 1707.
49. Scaria, K.S. & Premalatha, L.S. (1967) Ind.J.Exp.Biol., 5,256.
50. Slater, J.D.H., Tuffley, R.E., Williams, E.S. & Beresford, C.H. (1969) Clin.Sci ., 37, 327.
51. Smookler, H.H. & Buckley, J.P. (1969) Int.J.Neuropharmacol., 8, 33.
52. Selye, H. (1936) Brit.J.Exp.Path., 17, 234.
53. Smookler, H.H. & Buckley, J.P. (1970) Fed.Proc.,29,1980.
54. Stark, E., Fachet, J. & Mihaly, K. (1965) Endocrinologie,49,1 & 27.
55. Stark, E. & Fachet, J. (1968) Acta.Med.Acad.Sci.Hung., 25,251.
56. Stickney, J.C. & Van Liere, E.J. (1953) Physiol.Rev.,33,13.
57. Timiras, P.S., Pace, N. & Hwang, C.A. (1957) Fed.Proc.,16,340.
58. Vecsei, P. (1962) Endocrinologie, 42, 154.
59. Vecsei, P. & Kessler, H. (1970) Experientia., 26, 1015.
60. Welch, B.L. & Welch, A.S. (1969) Proc.Nat.Acad.Sci.,64,100.

61. Weltman, A.S., Sackler, A.M. & Owens, H. (1968) Physiol. Behav., 3, 281.
62. Williams, E.S. (1966) Proc.Roy.Soc., B. 165, 266.
63. Wilson, O. (1966) Acta.Univ.Lund.Sect. II, No. 21.
64. Zapata-Ortiz, V., Castro, R., Farnandez, E., Geu, A. & Batalla, L. (1966) Acta.Physiol.Lat.Amer.,16,66.

METABOLIC AND ENDOCRINE FUNCTIONS OF RATS CONDITIONED TO NOBLE-COLLIP DRUM TRAUMA

Š. Németh and M. Vigaš

Institute of Experimental Endocrinology, Slovak Academy of Sciences, Bratislava, Czechoslovakia

A great deal of research work has already been done on the changes in endocrine and metabolic function in rats adapted to Noble-Collip drum trauma (9,16,19,24,25). However, final conclusions about these problems cannot yet be made. This is understandable as there is lack of knowledge about the basic nature of this type of trauma. Mechanical energy applied to the body of the animals was considered to be the main factor causing injury (9). However, some of our recent work (30,31) made it clear that a more active part is taken by the injured animals themselves. They are trying to escape from the unexpected and apparently dangerous situation by doing active muscular work. Both central nervous influences and afferent signals from the periphery may lead to increased catecholamine (33) and corticosterone secretion (17). Vasoconstriction of the portal bed (13) with subsequent mesenteric pooling (27) loss of plasma and hemoconcentration (1) may be the consequences of the former. Muscular work is done under rather unfavourable conditions. There exists hypoxia which, though facilitating glucose entry and phosphorylation in muscle (18), leads to an excessive accumulation of lactate, considered by some authors to be an important factor contributing to the irreversibility of shock (28).

Our paper on changes in rats adapted to Noble-Collip drum trauma deals with various functions. We wonder if a common denominator can be found for them, even to-day. Adaptation consists in tumbling rats repeatedly starting with low doses and subsequently increasing them (26). Our own drums turn at 60 revolutions per min and we begin with 300 revolutions (5 min), repeating the tumbling each 3rd or 4th day and increasing the dose by 180 revolutions (3 min) until a dose of 1200 revolutions (20 min) is reached. Experiments are always performed after an interval of at least 9 days. Our results presented here concern adult male rats of the Wistar strain kept at 24°C and fed a standard laboratory diet. Some groups called "fasted" were fasted approximately 18 hr before performing the experiments. The trauma dose used in experiments on changes in metabolism never exceeded 400 revolutions (6 min 40 sec).

SOME METABOLIC AND HORMONAL RESPONSES OF NON-ADAPTED AND ADAPTED RATS OBSERVED IMMEDIATELY AFTER INJURY AND THEIR POSSIBLE MECHANISM

In fasted, non-adapted rats, immediately after injury, hypoglycemia and nearly total depletion of the liver glycogen stores were observed together with a net increase in blood pyruvate and lactate levels (4,29). In fed rats, during this period, the same changes take place with the exception that instead of hypoglycemia, a hyperglycemic response was found (2,4). In adapted rats, a less pronounced hypoglycemia in fasted (19) or hyperglycemia in fed animals (2) was reported together with a substantially reduced hyperlactacidemic response in both groups (9,19). Regarding post-traumatic hyperpyruvatemia, its diminution has been reported (2). However, in our own experiments an exaggeration of post-traumatic hyperpyruvatemia was found (21) and may be, together with the smaller increase in blood lactate, the consequence of an improved O_2 supply to the tissues in adapted animals. This is very probable from our experiments on rats under pentobarbital anesthesia in which partial O_2 tension, O_2 saturation, partial CO_2 tension, standard bicarbonate, base excess, pH and hematocrit were measured in arterial blood from the aorta and venous blood from the posterior vena cava. In non-adapted rats, immediately after injury, net hypoxic changes and lowered blood pH values were found not only in venous blood, but as a consequence of insufficient compensation by the lungs also in the arterial blood. In adapted animals subjected to injury, the O_2 saturation of the arterial blood remained undiminished, and also the acidotic changes, fairly obvious in the non-adapted group, were less marked in the adapted animals (20). These results fit well with those reported by other authors who found decreased O_2 consumption after injury in non-adapted and normal, or even increased, basal metabolic rates in adapted animals (16).

Regarding other metabolites in blood, increased post-traumatic levels of non-esterified fatty acids, amino nitrogen and urea, with decreased triglyceride levels were observed. In adapted rats a diminution or even absence of these responses was found, with the exception of the blood amino nitrogen changes (9).

It has already been pointed out that hypoxia in the traumatized animals arises as a direct consequence of changes in the circulation and distribution of blood. The vasoconstrictor action of catecholamines upon portal vessels with consequent mesenteric pooling seems to be the explanation for them. Concerning the levels of catecholamines in blood, their increase after injury has been observed (33). Moreover, the post-traumatic increase in the blood catecholamine levels seems to be reduced in adapted animals (3). However, these results should be reconfirmed as no completely reliable methods for blood catecholamine determination in rats seem to exist. Concerning the catecholamine content of homogenates from the adrenals, no differences were found between non-adapted and adapted rats without trauma (12) and these results were confirmed by us. Moreover, in our laboratory, the same post-traumatic increases in adrenaline were found in both groups (21). On the other hand, there is ample evidence in favour of a decreased reactivity toward catecholamines in adapted animals with regard to their vasoconstrictor (13), lipolytic (10) and perhaps glycogenolytic effects (12). In our own laboratory, decreased pyruvate and lactate acid hyperproduction was found in fasted, adrenalectomized animals if adapted, 1 hr after 20 µg adrenaline / 100 g s.c. (21).

However, there is not only less hypoxia in the adapted rats, but their sensitivity to hypoxia also seems to be altered. Enhanced lactate production was found in diaphragms from traumatized, adapted rats as compared with traumatized, non-adapted animals, in the Warburg apparatus, in Krebs-Ringer-bicarbonate medium with glucose as substrate and under conditions of absolute anoxia, the diaphragms having been excised immediately after injury. An increased resistance of glycolytic pathway enzymes against the effect of injury may be the underlying factor. Furthermore, in Krebs-Ringer-phosphate medium with 100% O_2, very low pyruvate levels were observed after incubation of diaphragms from traumatized, adapted rats. They should be ascribed to enhanced oxidation of pyruvate. The formation of radioactive CO_2 from glucose-6-^{14}C by diaphragms from such rats also seems to be increased. From these experiments, a parallel enhancement of the capacity of Krebs-cycle enzymes in adapted rats may be deduced (21).

SOME CHANGES RELATED TO RECOVERY FROM INJURY

During the recovery of animals subjected to the relatively low doses of 400 - 500 revolutions most of the changes return to normal. However, some secondary increases or decreases followed by new normalization were found (2) without any adequate explanation for them (9). We would like to confine our contribution to some changes in the gluconeogenic processes taking place under the influence of adrenal glucocorticoids (32). No differences between non-adapted and adapted rats were found in the post-traumatic decrease in their adrenal ascorbic acid contents (5). In our laboratory, no differences between adapted and non-adapted rats were found before traumatization in the corticosterone levels in adrenal homogenates and plasma. After injury increases were observed in both media. In adrenal homogenates, these increases were less pronounced in adapted groups (17). In plasma immediately after injury, the same increases were found in both groups but the levels in the adapted animals decreased faster and 90 min after injury were significantly lower than in the non-adapted rats (17,22). To explain these differences a more detailed knowledge of the formation, distribution and degradation of steroid hormones seems necessary.

Regarding the peripheral effects of glucocorticoids, a diminished thymolytic and lymphocytolytic response to injury was found in adapted rats (5,11). However, in non-traumatized adapted animals, increased deamination of amino acids was postulated because of the increased blood urea levels in these animals (7) which persisted even after adrenalectomy (8).

Recently a stimulatory effect of glucocorticoid hormones upon the activity of one of the amino acid transaminating enzymes, i.e. tyrosine-α-ketoglutarate transaminase (E.C. 2.6.1.5) (TAT) has been observed (14). We have determined the activity of this enzyme in non-adapted and adapted, non-traumatized and traumatized rats. An increased reactivity of TAT forming systems in response to the increased post-traumatic corticosterone levels seems to exist in the adapted rats. In these animals, a normal or even exaggerated TAT hyperactivity was found 90 min after injury in spite of the lower post-traumatic hypercorticosteronemia (Table 1). However, there was lack of TAT response in adrenalectomized animals even when adapted (results not shown here). In this respect, the presence of the adrenals seems to be indispensable even after adaptation.

Table 1. Plasma corticosterone ("B") level and liver tyrosine-α-ketoglutarate transaminase (TAT) activity of rats; means ± S.E.

		Control	Injured	P
"B" μg/100 ml	Non-adapted	13.3 ± 2.8	84.7 ± 5.7	<0.001
	Adapted	11.6 ± 1.8	44.6 ± 7.5	<0.001
	P	N.S.	<0.001	
TAT activity mμmol. product/1 mg protein/min	Non-adapted	43.1 ± 2.8	69.2 ± 3.5	<0.001
	Adapted	41.5 ± 2.2	75.8 ± 5.6	<0.001
	P	N.S.	N.S.	

SOME ADDITIONAL RESULTS

To elucidate the role of the liver in clearing the blood of the increased amounts of pyruvate and lactate during recovery from injury, experiments were performed on fasted rats with deranged liver function after an oral dose of 0.5 ml/100 g B.W. of Carbon tetrachloride in liquid paraffin, applied 4 hr before drumming. 90 Min after injury when in control animals blood sugar, pyruvate and lactate levels are already in the normal range, the intoxicated animals, if not adapted to trauma, showed the same hypoglycemia and hyperpyruvatemia as immediately after injury and their blood lactate levels were far above the normal values (Table 2) (15). Strangely enough, a normal response, i.e. a return of the blood sugar and lactage to initial values took place in the adapted group. No explanation of this exists as yet. This phenomenon may indicate a sort of non-specific resistance, including one against the toxic effects of Carbon tetrachloride.

Table 2. Blood levels in Carbon tetrachloride treated rats. Means ± S.E.

		Non-adapted	Adapted	P
Blood sugar mg/100 ml	Control	81 ± 3	74 ± 7	N.S.
	0 min after injury	27[++] ± 4	41[++] ± 3	<0.02
	90 min after injury	31[++] ± 5	84 ± 7	<0.001
Pyruvate mg/100 ml	Control	1.3 ± 0.1	1.5 ± 1.1	N.S.
	0 min after injury	3.9[++] ± 0.3	5.7[++] ± 0.2	<0.001
	90 min after injury	4.4[++] ± 0.3	2.4[+] ± 0.4	<0.001
Lactate mg/100 ml	Control	29 ± 1	31 ± 1	N.S.
	0 min after injury	96[++] ± 7	72[++] ± 7	<0.05
	90 min after injury	54[++] ± 3	39 ± 4	<0.01

[+] Statistically significant against control P<0.05
[++] Statistically significant against control P<0.01

Finally, we would like to call attention to some of our results showing the complexity of adaptation. A group of rats was adapted to 1200 revolutions. Each trial however was performed 10 min after the i.p. administration of 0.1 ml/100 g B.W. of 3.3% Pentobarbital (SPOFA, Prague) in saline. These animals, though completely adapted, never had the opportunity to see the drums as they were always tumbled in a state of anesthesia.

Of these animals, 40% died when exposed without anesthesia to a substantially lower trauma dose (900 revolutions) than that survived by them without difficulty in the state of anesthesia. Moreover, the post-traumatic increase in the blood lactate level was more pronounced in such animals than in controls adapted without anesthesia (23). The anesthesia was always light and so the mortality of the animals during adaptation was not increased as in experiments reported in the literature (6), because of the alleged high sensitivity of anesthetized animals to trauma, a slower increase in the dose is recommended. No tumbling without anesthesia of rats adapted under anesthesia was attempted in the experiments reported (6).

In conclusion, adaptation seems to be a complex phenomenon, not limited to a specialized organ or system. Moreover, it seems unjustified to explain the decreased responses of the adapted groups exclusively on the grounds of their reduced vulnerability to the same dose of trauma. In this respect, most important results have been achieved without trauma, such as the decreased sensitivity to exogenous adrenaline, or in which, though under the influence of trauma, changes such as increased oxidation have been observed.

Regarding the opposite question, namely sensitivity to trauma, there is as yet no definite answer but substrate availability and the degree and duration of hypoxia may be very important factors.

SUMMARY

From a review of the literature and the authors' data on the neurohormonal and metabolic changes in rats adapted to Noble-Collip drum trauma, the following points may be emphasized.

A less pronounced hyperlactacidemia and greater hyperpyruvatemia in adapted rats subjected to Noble-Collip drum trauma as compared with non-adapted animals fits well with the smaller hypoxic and acidotic changes found in the blood of adapted rats immediately after injury. Date are presented of a less pronounced catecholamine secretion and increase in the blood with the trauma, and for an increased resistance against these hormones in adapted animals. However, there is not only less hypoxia in these animals after injury, their sensitivity to hypoxia also seems to be decreased, as shown by increased lactate production by diaphragms of adapted, traumatized animals in vitro under conditions of absolute anoxia.

The increase of the plasma corticosterone level immediately after injury is the same in both non-adapted and adapted rats. However, 90 min later increased levels were observed in non-adapted as against nearly normal values in adapted rats. However, the corticosterone induced increase in liver tyrosine-α-ketoglutarate transaminase activity was the same in both groups 90 min after injury. An increased sensitivity of enzyme producing systems of adapted rats against corticosterone is supposed on the basis of these results.

Livers of rats adapted to Noble-Collip drum trauma also seem to be resistant to some of the toxic effects of Carbon tetrachloride.

Rats adapted under Pentobarbital anesthesia may lose a great part of their acquired resistance if exposed to trauma without anesthesia.

REFERENCES

1. Chambers, R., Zweifach, B.W. & Lowenstein, B.E. (1943) Am. J. Physiol. 139:123.

2. Chytil, F. & Hruza, Z. (1958) Can. J. Biochem. Physiol. 36:457.

3. Griswold, R.L. & Decker, W.J. cit. in Greisman, S.E. & Michaelis, M. (1963) Surg. Res. 3:268.

4. Hava, O., Mraz, M. & Triner, L. (1959) Naunyn-Schmiedebergs Arch. 236:81.

5. Hruza, Z. & Chytil, F. (1959) Can. J. Biochem. Physiol. 37:61.

6. Hruza, Z. (1960) Physiol. Bohemoslov 9:251.

7. Hruza, Z. (1961) Physiol. Bohemoslov 10:529.

8. Hruza, Z. (1962) Physiol. Bohemoslov. 11:206.

9. Hruza, Z & Poupa, O. Injured Man In: Adaptation to the Environment. Washington, D.C. American Physiological Society, 1964, p. 939.

10. Hruza, Z., Jelinková, M. & Hlaváčková, V. (1964) Physiol. Bohemoslov. 13:292.

11. Hruza, Z. (1964) Physiol. Bohemoslov. 13:296.

12. Hruza, Z., Albrecht, I, Jelínková, M. & Erdösová, R. (1966) Physiol. Bohemoslov. 15:434.

13. Hruza, Z., Zweifach, B.W. (1969) J. Trauma 9:430.

14. Knox, W.E. (1963) Transactions N.Y. Acad. Sci. 25:503.

15. Magdolenová, A., Németh, Š. & Vigaš, M. (1970) Physiol. Bohemoslov. 19:332.

16. Manning, J.W. & Hampton, J.K. (1957) Am. J. Physiol. 188:99.

17. Mikulaj, L. & Kvetňanský, R. (1966) Physiol. Bohemoslov. 15:439.

18. Morgan, H.E., Randle, P.J. & Regen, D.M. (1959) Biochem. J. 73:573.

19. Mráz, M. (1963) Physiol. Bohemoslov. 12:145.

20. Németh, Š. In: Energy Metabolism in Trauma. A Ciba Foundation Symposium. London: J. and A. Churchill, 1970, p. 7.

21. Németh, Š., Vigaš, M. Macho, L., Babušíková, F. & Štrbák, V. Physiol. Bohemoslov. in press.

22. Németh, Š., Vigaš, M. & Mikulaj, L. (1971) Endocrinologia Experimentalis (Bratislava) 5:179.

23. Németh, Š., Vigaš, M. & Straková, A. Physiol. Bohemoslov. in press.

24. Neufeld, A.H., Toby, C.G. & Noble, R.L. (1943) Proc. Soc. Exp. Biol. Med. 54:249.

25. Noble, R.L. & Collip, J.B. (1942) Quart. J. Exp. Physiol. 31:187.

26. Noble, R.L. (1943) Am. J. Physiol. 138:346.

27. North, W.C. & Levy, E.Z. (1954) Fed. Proc. 13:391.

28. Schumer, W. (1968) Surgery 64:55.

29. Vigaš, M. & Németh, Š. (1968) Endocrinologia Experimentalis (Bratislava) 2:91.

30. Vigaš, M., Németh, Š. & Magdolenová, A. (1970) Physiol. Bohemoslov. 19:354.

31. Vigaš, M. & Németh, Š. Physiol. Bohemoslov. in press.

32. Vigaš, M., Németh, Š. & Straková, A. Physiol. Bohemoslov. in press.

33. Young, J.G. & Gray, I. (1956) Am. J. Physiol. 186:67.

CARBOHYDRATE METABOLISM IN SHOCK RESISTANCE

A. Bardoczi, J. Arokszallasi and S. Karady

Institute of Pathophysiology, University of Szeged

School of Medicine, Szeged, Hungary

At the annual meeting of the Hungarian Society of Physiological Sciences (1970) we reported on the results of our earlier investigation dealing with the carbohydrate metabolism in shock. It has been shown by Karady et al., (1948, 1961) that numerous changes found in shock do not occur or are modified in the so-called "stage of resistance". Therefore it seemed to us interesting and worthwhile to investigate the carbohydrate metabolism in rats rendered resistant to shock, that is in rats 24 hr after having been exposed to sublethal tourniquet shock ("stage of non-specific resistance"). In these experiments we confirmed the changes in carbohydrate metabolism found by us (1970) to be characteristic of shock and established that those changes failed to occur in the stage of non-specific resistance.

The experiments were carried out in male and female rats of R-Amsterdam strain, weighing 170-180 g.

The sublethal tourniquet trauma was elicited by the method of Stoner (1958). The hind limbs of rats were excluded from the circulation for 1.5 hr. The rats were divided into groups of 20. Then we determined the blood level of NAD^+ (Klingenberg, 1962), the sugar concentration (Hultman, 1959), the MDH (malate dehydrogenase; Boehringer test-method). The severity of shock was judged by the drop in rectal temperature.

The values of these parameters, in figures expressed on the ordinates, were determined in normal control rats ("S" in figures) and in rats previously rendered resistant to shock ("R" in figures) at different intervals of time, that is at the moment the sublethal tourniquets were released ("N" in figures), 1 hr 1.5 hr, 6 hr and 24 hr after the release of the tourniquets (shown on the abscissae). The values of the parameters represent the mean values for one group of animals consisting of 20-20 rats. The statistical evaluation was performed according to the Student "t" test.

The results of the experiments are well demonstrated in the figures.

One can see that the blood level of NAD^+ decreases significantly in shock (curve "S"). After 1.5 hr it is so low that it cannot be measured and it does not return to normal even after 24 hr.

In contrast, in the stage of non-specific resistance the blood level of NAD^+ only decreased very slightly after 1 hr and did not change significantly (curve "R").

Figure 2 illustrates that the activity of MDH very markedly increases in shock (curve "S"), whereas in stage of resistance only a moderate and transitory decrease is to be observed (curve "R")

Figure 3 demonstrates the change in blood sugar concentration due to sublethal tourniquet shock of normal and of resistant rats. The blood sugar level in shock shows a large increase (curve "S"), while in the stage of resistance only a very slight increase could be found (curve "R").

Figure 4 shows the changes in rectal temperature. The fall in temperature indicated well the severity of shock. One can observe that the rectal temperature falls steeply in shock (curve "S"), while in the stage of resistance the drop is much less (curve "R").

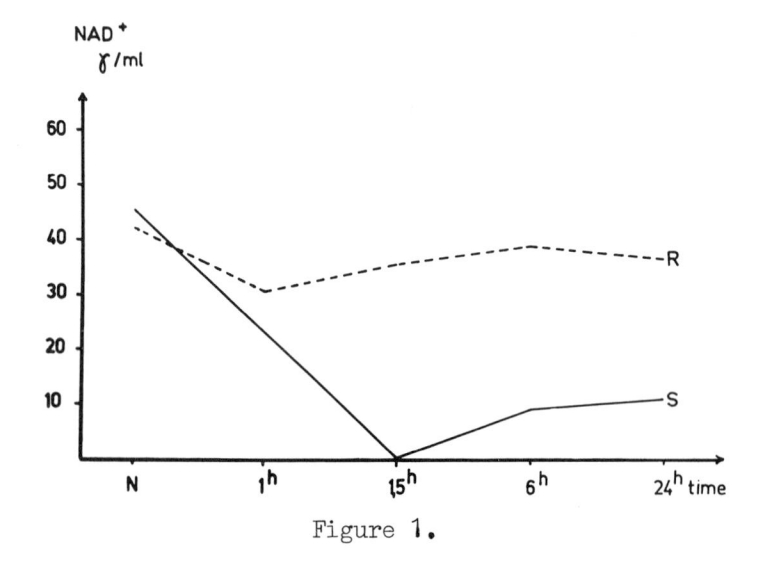

Figure 1.

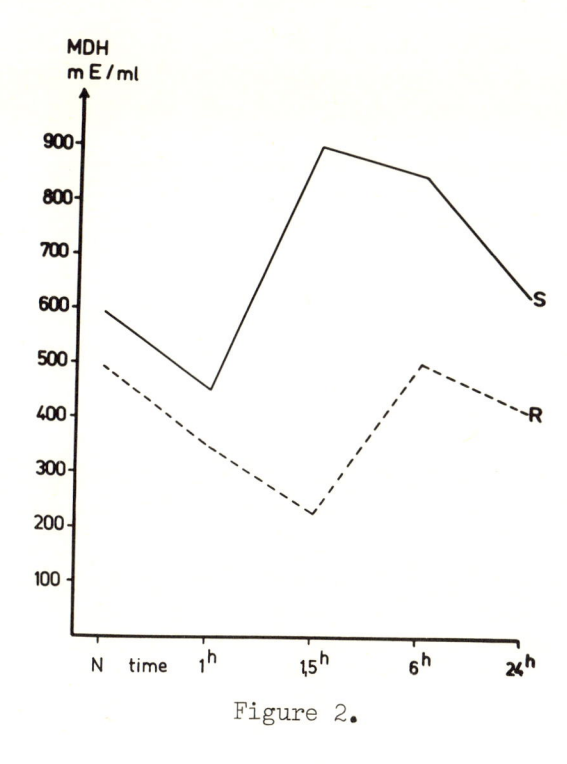

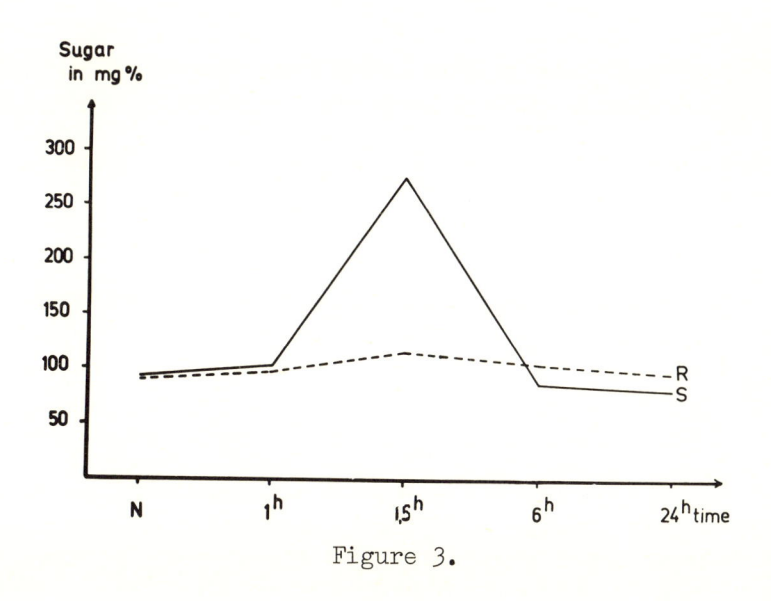

Figure 2.

Figure 3.

Figure 4.

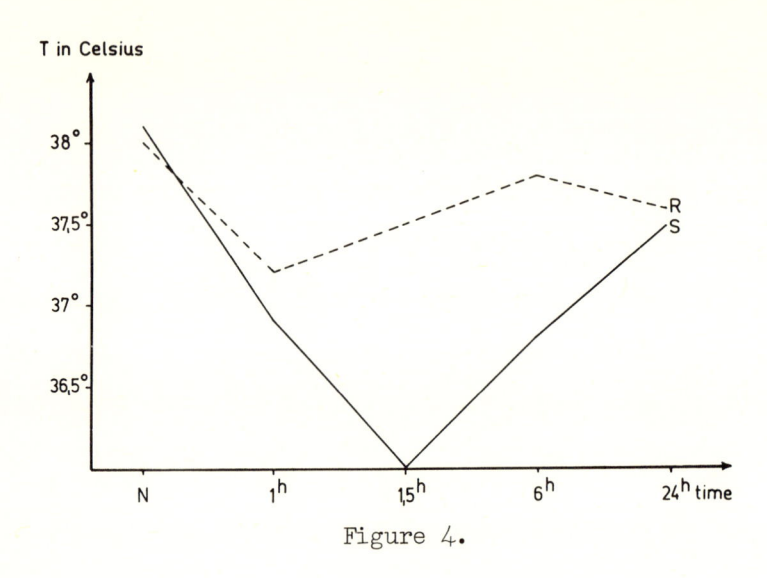

The results of the present experiments concerning the changes in carbohydrate metabolism in shock are in very good agreement with those obtained in our earlier experiments carried out in man, cat, rabbit and rat in shock states of different aetiology (immobilisation, haemorrhage, tourniquet, trauma).

In the present experiments performed on rats rendered resistant by exposure to sublethal shock 24 hr before the shock--inducing stress, the changes in carbohydrate metabolism characteristic of shock could not be demonstrated. The explanation for that seems to us, on the basis of our earlier investigations (1961) to be the increased production of a substance not yet known before, called by us "resistine" possessing antihistamine, antiserotonin and antibradykinin properties and having also the ability to inhibit the release and formation of these endogenous vasoactive substances. Therefore in a resistant organism the shock-inducing stress does not lead to peripheral circulatory disorder and no regulatory epinephrine and norepinephrine release ensues followed by microcirculation changes, the block of acetyl coenzyme A does not take place, metabolic acidosis fails to occur, etc.

Summarizing the results one can state that the changes in carbohydrate metabolism found in shock do not occur or are much less in the stage of resistance. In our opinion this may be attributed to the increased formation of "Resistine" in the stage of resistance. For a deeper understanding of the pathomechanism of this process further investigations are necessary.

REFERENCES

Bardoczi, A., Karady, S. & Kekes-Szabo, A. (1971) Acta phys. Acad.Sci. hung. 39, 189.
Hultman, A. (1959) Nature, London, 183, 108.
Karady, S. & Kovacs, A. (1948) Nature, London, 161, 688.
Karady, S., Prokai, A. & Mustardy, L. (1961) J. Physiol. (Paris) 53, 629.
Klingenberg, M. (1962). Verlag Chemie, 528.
Stoner, H.B. (1958) Brit. J. exp. Path. 39, 251.

CENTRAL NERVOUS INFLUENCE UPON THE ADRENOCORTICAL

REACTION DURING STRESS SITUATIONS

    K. Murgaš and V. Jonec

    Institute of Experimental Endocrinology

    Slovak Academy of Sciences, Bratislava, Czechoslovakia

    The data most frequently met with in literary studies on humoral responses of the organism to stress are those that relate to changes in the function of the adrenal cortex which, to a certain extent, may be conditioned also by Selye's concept of a general adrenal cortex reactivity under stress (33,34). A change in the adrenal weight as such is of course only an indirect indicator of its function (3,31) nor does the ascorbic acid concentration provide a reliable or adequate index of its function (7,21,30). Thus far, a direct follow-up of adrenal cortical steroids in plasma and adrenal tissue has proved to be the most reliable method for studying adrenal cortical activity under stress, and it is generally utilized in practice.

    Direct determination of the concentration of the principal adrenocortical steroid in plasma and adrenal tissue has revealed that with a stress stimulus the adrenal cortex becomes activated in practically all species possessing an adrenal cortex or an analogous structure (13,14).

    A further characteristic of adrenal function in stress situations is that a chronically repeated stress results in decreased activation of the adrenal cortex, despite the fact that the intensity of the stressful stimulus remains unaltered throughout its repeated application. This was found to be the case in rats during chronic intermittent immobilization (17), subcutaneous application of formalin (19), trauma in the Noble-Collip drum (20,29) and during exposure to cold (7,9). A common feature of these findings is an abrupt rise in plasma corticosterone concentration and an increase in corticosterone content of the adrenal tissues during the initial repetitions of acute stress. However, when the same stress stimulus is repeated several times daily over a longer period, these values decrease and gradually tend to regain the starting level, in spite of the unchanged stress stimulus. The term "adaptation" has been chosen to designate the process in which animals are repeatedly (daily) exposed to the same stress stimulus for a definite time over a long period and animals in which the adrenocortical activation decreases with time despite unaltered stress stimuli are said to be "adapted" animals. Even though adaptation of the organism to a stress stimulus cannot be generally equated with activation of the adrenal cortex, a direct correlation has, nonetheless, been found in Noble-Collip drum trauma, where "adapted" animals survived an originally lethal dose of the trauma, and their adrenal response was smaller (20,29), although, of course this need not involve a direct causal relation. An exception here is perhaps the effect of electrical stimuli. Despite a daily short-term electrical stimulation over several weeks, no decrease in adrenocortical activation was noted, rather the contrary, and this applies also to changes in the behaviour of rats subjected to such stress (27).

That there was no question of the secretory or the productive capacity of the adrenal cortex being exhausted, was shown by an exogenous application of ACTH, either during operation in patients and in dogs (15), or in rats (18). An increase in plasma corticosteroid concentration always followed such an ACTH application at whatever stage of the daily stress stimulus.

There is likewise evidence, though only indirect, that no pituitary hypofunction or insufficiency is involved. Thus, for example, it is well known that when animals "adapted" to a certain stress stimulus are exposed to another stress, the adrenal cortex is again strongly activated (35). Another proof of unimpaired pituitary function during long-term stress is provided by data on long-term electrical stimulation, where the adrenal stress response remains unchanged (27).

It would seem that even an adaptation of the organism to a stress stimulus, with simultaneous decline in adrenocortical activation, involves a regulation of the pituitary-adrenocortical axis from higher (hypothalamic) CNS centres. However, a study of the hypothalamic areas and the results observed following lesions, electrical or chemical stimulation of the hypothalamic structures do not adequately explain the mechanism whereby activation of the adrenal cortex decreases during repeated long-term application of the same stress stimulus, even if it is possible to affect the level of this adrenocortical activation by lesions or stimulation of the corresponding areas of the hypothalamus (see 36). This may also be done directly in stress situations, both in the sense of increase and decrease of the adrenal stress response, in particular when the method of telestimulation is employed which eliminates, to some extent, the disturbing elements accompanying other modes of electrical stimulation of brain structures (10). Thus, electrocoagulation of the posterior areas of the hypothalamus brings about a decrease in the adrenocortical response to cold, whereas an electrical stimulation of the posterior areas of the hypothalamus increases the plasma corticosterone concentration (9,11).

However, a study of the hypothalamo-pituitary relationships fail to explain some of our present results. The aim of these experiments had been first to familiarize the animal with the environment in which it would be later exposed to stress stimuli. If animals are daily transferred several times from their cage into a cold chamber which, however, is not cooled initially, and only then are exposed to cold, there is lower adrenocortical activation than in animals exposed without such preparation (the working term is "orienting";(8,12). This applies even to trauma in the Noble-Collip drum. Rats repeatedly placed into a non-revolving drum, showed a lower mortality rate during the subsequent lethal trauma, than "non-oriented" animals, and plasma corticosteroid values in the former are significantly lower than in the controls (22,24).

In another experimental situation thirsty animals were allowed to drink water while exposed to stress (transfer to a foreign environment, cold). Here, either no adrenocortical activation took place at all, or the adrenal stress response was minimal (23). This however, does not apply to the immobilization stress where, even though immobilized animals could drink, no differences in activation of the adrenal cortex were noted, as against the controls.

When following plasma corticosterone concentrations in rats during the course of elaborating a simple defence-avoidance conditioned reflex, we found these values to be lower in those animals that solved the situation in the conditioning box with 80-100% accuracy, than in those with worse scores (24).

It would seem that the pain component of some stress stimuli (trauma, immobilization, subcutaneous formalin application) does not play such an essential role in the activation of the adrenal cortex. In animals (rats) with lesions to both lemnisci or with interrupted ascending sensory pathways in the spinal cord (tested with a needle or a hot plate), activation of the adrenocortical system during acute immobilization did not differ to any considerable extent from that of the controls (Murgaš, Mikulaj & Juráni, unpublished results).

Hence, any attempt to correlate adreno-cortical reactions with the function of higher (suprahypothalamic) structures or CNS functions is unusually welcome. As regards the relation between the function of the adrenals and emotion, no direct correlation has so far been found (1,2,4). These studies are essentially concerned with measurements of emotion with plasma corticosterone concentrations either directly following testing, or more frequently in response to a selected standard stress, e.g. ether (4). No correlation has been found, either because there is none, or because the criteria designed to measure emotion fail to measure it (2). When repeating these procedures, we have not found any correlation between adrenocortical activation and various parameters purporting to measure emotion (5). However, we found repeatedly a direct correlation in rats between the size of the open field box and plasma corticosterone concentration in adrenal tissue homogenate following a 10 min. exposure (peak), or after a 30 min. exposure (28). In the latter case there was a direct correlation also with the frequency of defaecation. The animals had been assigned into individual groups (large or small open field) by random selection without any preceding choice according to emotional behaviour.

Animals with septal lesions exposed to different forms of stress, but particularly to immobilization, showed an increase not only in plasma corticosterone concentration, but also in the activity of the catecholamine-synthesizing enzymes (25). This applies also to the survival rate after trauma in the Noble-Collip drum, which was much higher in animals with septal lesions (26). In animals with lesions to the amygdalae, on the contrary, we found a substantially lower activation level of the adrenal cortex in response to certain types of stress (28). Naturally, following lesions to these structures of the limbic system, we observed changes in aggressiveness or irritability level that ensue from these interventions (6). Hence, we cannot omit to mention here some findings (30,31,37), which admit a relationship between adrenocortical activity and emotion, although, in view of the short post-operative period, we found only an insignificant trend in adrenal weights of individual groups of animals with lesions in the septum or the amygdalae.

Animals that had higher plasma corticosterone values following exposure in a large open field than those in a small one, achieved lower scores when time spent to get through a runway was measured, or their behaviour was followed in the target box (thirsty animals had to run following exposure in the open field along the runway towards drinking water) (28). Our present observations seem to supports Lát's conception of a non-specific excitability of CNS (16). In our view, adrenocortical activation in a given stress situation will be determined precisely by the non-specific excitability level. We presume that this function is an actual factor,and its influence changes from one situation (e.g. open field) to another (e.g. ether stress), particularly if no tightly succeeding or even simultaneous observations are involved.

REFERENCES

1. Ader, R., Friedman, S.B. & Grota, L.J. (1967) Anim. Behav., 15, 37.
2. Ader, R. (1969) Ann. N.Y. Acad. Sci., 159, 791.
3. Anderson, E.E. & Anderson, S.F. (1938) J. Comp. Psychol., 26, 459.
4. Brain, P.F. & Nowell, N.W. (1969) Physiol. Behav., 4, 945.
5. Czako, M., Murgaš, K. & Dobrakovová, M. Čs. fysiol.(in press).
6. Igič, R., Stern, P. & Basagič, E. (1970) Neuropharmacology, 9, 73.
7. Slusher, M.A. (1958) Endocrinology, 63, 412.
8. Jonec, V. (1964) Canad. J. Physiol. Pharmacol., 42, 585.
9. Jonec, V., Murgaš, K. & Kvetňanský, R. (1965) In: Abstracts of the Internat. Symposia on Environmental Physiology, Kyoto.
10. Jonec, V. & Murgaš, K. (1966) Čs. fysiol., 15, 98.
11. Jonec, V., Murgaš, K. & Kvetňanský, R. (1966) Federation Proc., 25, 1200.
12. Jonec, V. & Murgaš, K. (1969) Federation Proc., 28, 987.
13. Juráni, M., Podhradský, V., Mikulaj, L. & Murgaš, K. (1967) Gen. Comp. Endocrinol., 9, 463.
14. Juráni, M. A comparison of adrenocortical activity during stress in some species of vertebrates. Doctoral thesis. Comenius Univ. School of Natural Sciences (Zoology), Bratislava, 1971.
    Kužela, L. & Mikulaj, L. (1966) Klin. Med., 21, 298.
    Lát, J. (1963) Central and Peripheral Mechanisms on Motor
15. Function. Proceedings of the Conference (Ed. E. Gutmann &
16. P. Hník). Czechoslovak Acad. Sci., Prague.
17. Mikulaj, L. & Csiba, J. (1963) Čs. fysiol., 12, 330.
18. Mikulaj, L., Lichardus, B., Mitro, A. & Csiba, J. (1964) J. Physiol. Paris, 56, 612.
19. Mikulaj, L., Javorka, K., Krahulec, P., Kvetňanský, R. & Spišák, S. (1966) Čs. fysiol. 15, 99.
20. Mikulaj, L. & Kvetňanský, R. (1966) Physiol. Bohemoslov., 15, 439.
21. Murgaš, K. & Jonec, V. (1965) Biologia, 20, 862.
22. Murgaš, K. & Jonec, V. (1967) Čs. fysiol., 16, 246.
23. Murgaš, K. & Jonec, V. (1967) Activitas Nerv. Super., 9, 311.
24. Murgaš, K. (1968) Doctoral thesis. Inst. of Experimental Endocrinology, Slovak Acad. Sci., Bratislava.

25. Murgaš, K. & Kvetňanský, R. (1971) In: Proc. of the II. Internat. Meeting of Psycho-Neuro-Endocrinol. Soc., Budapest.
26. Murgaš, K., Dobrakovová, M. & Czako, M. Čs. fysiol., (in press).
27. Murgaš, K., Mikulaj, L. & Juráni, M. Physiol. Bohemoslov., (in press).
28. Murgaš, K., Dobrakovová, M. & Czako, M. Čs. fysiol. (in press).
29. Németh, Š., Vigaš, M., Macho, L., Babušíková, F. & Štrbák, V. Physiol. Bohemoslov., (in press).
30. Paré, W.P. & Cullen, J.W. Jr. (1965) Psychol. Rep., 16, 283.
31. Rogers, P.V. & Richter, C.P. (1948) Endocrinology, 42, 46.
32. Saffran, M. (1966) In: Endocrines and the central nervous system. (ed. R. Levine), Baltimore.
33. Selye, H. (1936) Brit. J. Exp. Path., 17, 234.
34. Selye, H. (1950) Stess. Acta. Montreal.
35. Selye, H. (1961) In: Ergebnisse der allgemeinen Pathologie und Pathologische Anatomie. Band., 41, 208.
36. Schreiber, V. (1963) The hypothalamo-hypophyseal system. (Ed. J. Charvát & V. Jonec) Publ. House of the Czechoslovak Acad. Sci., Prague.
37. Yeakel, E.H. & Rhoades, R.P. (1941) Endocrinology, 28, 337.

INDEX

Adaptation to stress
  adrenal catecholamine production, 603-616
  adrenocortical activity, 619-627, 631-636

Adipose tissue,
  blood flow in hemorrhagic shock, 323-325
  metabolism in endotoxin shock, 337-343
  metabolism in hemorrhagic shock, 335
  norepinephrine stimulated lipolysis by fat cells, 337-343
  regional differences, 326, 335, 337-343
  sympathetic control, 324-327

Adrenocortical activity,
  CNS influence during stress, 655-659
  drum trauma, 489-492
  phylogenetic aspects in adaptation, 619-627

Biogenic amines,
  see Catecholamines

Blood flow,
  adipose tissue, 6, 7, 327-331, 335
  adrenal cortex, 27

Blood flow (cont'd)
  adrenal gland, 9
  atropine, 3, 4
  cerebral, 11, 12, 19-22
  dichloroisoproterenol, 3, 4
  head-forelimb, 4
  hypophysis, 8, 9
  hypothalamus, 5, 12
  ileum, 7, 8
  liver, 5
  local regulation, 46, 47, 53
  measurement by
    ^{14}C-antipyrine, 12
    clearance of ^{133}Xe, 12
    H_2 washout, 3
    thermal conductivity, 3
  myocardium, 5, 6
  pelvis-hindlimb, 4
  phenoxybenzamine, effect of, 3-14
  renal cortex, 6, 7
  salivary gland, 10, 11
  skeletal muscle, 5, 6
  splanchnic, 4, 313-320
  thyroid, 9, 10
  uterus, 61-64

Capillary filtration coefficient (CFC)
  hemorrhagic hypotension, 150-151
  newborn, 590-591

Capillary permeability
See also Microcirculation
increase by histamine,
119-131

Cardiac output,
in hemorrhagic shock,
21, 60
organ fractions and
pregnancy, 58-64

Carotid body,
pO_2 and functional
state, 463-471

Catecholamines,
adrenal production,
517-532, 603-616
norepinephrine (NE)
distribution in
brain, 510
metabolism in the
CNS under stress,
501-507

Central nervous system,
cortical evoked potentials in hemorrhagic
shock, 481-487
influence on adrenocortical activity,
655-659
in resistance to drum
trauma, 489-492
NE metabolism in stress,
501-507
regulating adrenal catecholamine production,
517-532

Cyclic AMP (dibutyryl),
adrenal catecholamine
production, 528-531

Dehydration,
rheological factors,
95-110

2,3-diphosphoglycerate
hypoxia, 420-437

Evoked potentials
cortical and hypothalamic, in hemorrhagic
shock, 481-487

FFA (free fatty acids)
metabolism,
in conscious dogs in
hemorrhagic hypotension, 221-228
in splanchnic area,
293-310
myocardium in endotoxin
shock, 375-385
thyroid function, 175-176

Glucose,
blood level in hemorrhagic shock, 179-185
homeostasis in postoperative state, 209-212
tissue injury in man,
401-406
tourniquet shock, 649-652
transport in hemorrhagic shock, 293-310

Hypo- and hypercapnia,
brain and liver redox
states, 353-360

Hypotension, hemorrhagic,
adrenal venous flow and
corticoid output, 31
arterial glycerol and
lactate in conscious
dogs, 221-228
capillary filtration
coefficient, 150-151
FFA metabolism in conscious dogs, 221-228
O_2 consumption, 225-228

INDEX

Hypotension, Hemorrhagic (cont'd)
  resistance to, in young mammals, 455-461

Hypothalamus,
  catecholamines, and adrenergic innervation, 509-515
  control of adenohypophysial response to stress, 157-162
  corticotropin releasing factor (CRF), 158
  evoked potential in hemorrhagic shock, 481-487
  growth hormone releasing factor (GRF), 160
  prolactin inhibiting factor (PIF), 161
  prolactin releasing factor (PRF), 161
  thermoregulation after injury, 495-499

Hypothermia,
  effect on serum electrolytes and metabolites, 575
  protection against anoxia, 571-585

Hypoxia,
  brain and liver redox states, 353-360
  cold exposure, 574
  and adaptation, 639-645
  convective transport of O_2, 39
  diffusion of O_2 in brain, 35-37
  2,3-diphosphoglycerate, 429-437
  microcirculation, 33

Insulin,
  blood levels in hemorrhagic shock, 179-185

Insulin (cont'd)
  secretion, baboons, 199-207
  secretion, dogs, 187-197

Interstitial fluid,
  in the newborn, 587-591
  regulation of volume and pressure, 111-118

Lactate,
  arterial, in conscious dogs in hemorrhagic hypotension, 221-228
  arterial, in dogs in endotoxin shock, 337-343
  arterial, in dogs, in hemorrhagic shock, 333
  metabolism by heart in endotoxin shock, 375-385

Lipoprotein lipase
  heart, in stress, 387-392

Liver, in hemorrhagic shock
  anaerobic glycolysis, 277-291
  blood flow in conscious dogs, 293-310
  effect of phenoxybenzamine, on 246-247, 249
  gluconeogenesis, 243-250
  glucose, lipid and electrolyte transport, 293-310
  O_2 consumption and redox state, 243-250, 253-261, 277-291
  perfusion, 243-250, 277-291

Lymph,
  effect of histamine on flow and protein components, 121-126
  in development of edema, 117
  pO_2 in normo and hypovolemia, 133-140

Lymph (cont'd)
  regulating interstitial fluid volume and pressure, 114

Lysosomal enzymes
  during hypoxia, 395-399
  effect of RES, 535-542
  in hemorrhagic shock, 537-542
  MDF, 367-373

Microcirculation
  fluid filtration, 143-148
  hypoxia, 33
  reactive hyperemia, 47-51
  rheology, and 65-72, 75-80
  skeletal muscle, 149-155
  vasoconstrictors, and 133-140

Myocardium
  lipoprotein lipase, 387-392
  metabolism in severe hemorrhage, 384
  myocardial depressant factor (MDF), 367-373
  performance in endotoxin shock, 363-365
  substrate utilization in endotoxin shock, 375-385

Newborn,
  CFC, 590-591
  hypothermia in the treatment of asphyxia, 583-585
  resistance to injury, 587-591

Neurotransmitters
  see Catecholamines

Noble-Collip drum trauma,
  adaptation to, 557-563
  adrenocortical reactions 489-492
  metabolic and endocrine functions in adapted rats, 639-645
  metabolic aspects, 423-426
  reticuloendothelial function, 554-563

Postoperative state,
  glucose homeostasis, 209-212

Redox states
  brain and liver, in hypo and hypercapnia, 353-360

  liver, in hemorrhagic shock, 243-250, 253-261, 277-291

Renin-angiotension system,
  in young mammals, 455-461

Reticulo-endothelial function,
  drum trauma, 554-563
  hemorrhagic shock, 546-549, 562-563
  intestinal ischemia, 550-563
  MDF, 373
  release of lysosomal hydrolases, 535-542

Rheology
  chemical stress, 67-72
  dehydration, 95-110
  endotoxin shock, 81-85 88-91
  hemorrhagic shock, 85-91
  microcirculation, 65-72, 75-80

INDEX

Rheology (cont'd)
  tissue injury, 65-72
  587-591

Shock - endotoxin
  adipose tissue, 337-343
  anaphylatoxin, 263, 266, 270
  dexamethasone effect, 263-270
  2,3-diphosphoglycerate (2,3 DPG), 429-437
  energy metabolism and tissue fuel in man, 401-406
  histamine and catecholamine-induced fluid filtration, 145-148
  histamine release, 263-270
  isolated fat cells, 339-343
  lysosomes, 233-235, 240-241
  mitochondrial alterations, 231-241
  myocardial metabolism, 375-385
  peripheral pooling, 363-365
  rheological changes, 81-85, 88-91
  transvascular fluid fluxes, 143

Shock, hemorrhagic
  See also Hypotension, hemorrhagic
  adipose tissue, 323-325
  adrenal venous flow, 31
  baboons, 19
  blood glucose and immunoreactive insulin levels, 179-185
  capillary filtration coefficient, 149-155
  cats, 149
  cell function in brain and lung, 257,258

Shock - hemorrhagic (cont'd)
  cerebral metabolic changes, 22-25
  dogs, 2, 85, 134, 143
  histamine and catecholamine-induced fluid filtration, 146-148
  hypothalamic and cortical evoked potentials, 481-487
  insulin secretion, baboons, 199-207
  insulin secretion, dogs, 187-197
  intestinal necrosis, 313-320
  lymph pO_2, 133-140
  lysosomes, 233-235, 240-241
  metabolism of the non-hepatic splanchnic area, 293-310
  mitochondrial alterations, 231-241, 253-261
  Na^+, $K^\pm$ ATPase, 253-261
  porto-caval shunt, 313-320
  rats, 57-64
  rheological changes, 85-91
  splanchnic blood flow, 313-320
  sympathetic nervous activity, 473-480
  transvascular fluid fluxes, 143
  uterine circulation, 57-64

Stress,
  ACTH, 157
  adrenal catecholamine production, 517-532
  FSH and LH, 161-162
  growth hormone, 159-160
  NE metabolism in the CNS, 501-507
  PGB_x, a new prostaglandin derivative, 215-220

Stress (cont'd)
  phosphatidyl glycerol, 213-215
  prolactin, 160-161

Stress, chemical
  and rheology of human blood, 67-72
  formalin injection, 632

Stress, mechanical,
  See also Noble-Collip drum trauma
  immobilization, 603-616

Stress, thermal
  complete oxyhemoglobin dissociation curves, 441-452
  rheology of human blood, 67-72
  scald, 271-276, 587-591
  thermoregulation, 495-499

Stress, tourniquet
  carbohydrate metabolism, 649-652
  heart lipoprotein lipose, 387-392
  hind-limb ischemia, 271-276
  resistance to, 649-652
  thermoregulation, 495-499

Sympathetic nervous activity
  in hemorrhagic shock, 473-480
  and tolerance to hemorrhage in pigeons, 593-600

Thyroid function,
  FFA, 175-176
  free thyroxine conc., 171-172
  % free T_3, 172
  in trauma, 165-176

Tissue injury
  brain mitochondria, 345-351

Tissue injury (cont'd)
  energy metabolism and tissue fuel in man, 401-407
  environmental temp. 409-415, 417-421
  liver metabolism, 271-276
  plasma proteins, 417-421
  protein, mineral & energy metabolism, 409-415
  rheological factors, 65-72, 587-591

Trauma
  See Stress
  and Tissue injury

Vasoconstrictors
  lymph pO_2 in normo- and hypo-volemia, 133-140

Vasopressin, 158-159